CAMBRIDGE LIBRARY COLLECTION

Books of enduring scholarly value

Mathematical Sciences

From its pre-historic roots in simple counting to the algorithms powering modern desktop computers, from the genius of Archimedes to the genius of Einstein, advances in mathematical understanding and numerical techniques have been directly responsible for creating the modern world as we know it. This series will provide a library of the most influential publications and writers on mathematics in its broadest sense. As such, it will show not only the deep roots from which modern science and technology have grown, but also the astonishing breadth of application of mathematical techniques in the humanities and social sciences, and in everyday life.

Astronomy and Cosmogony

Sir James Jeans (1877–1946) is regarded as one of the founders of British cosmology, and was the first to suggest (in 1928) the steady state theory, which assumes a continuous creation of matter in the universe. He made many major contributions over a wide area of mathematical physics, but was also well known as an accessible writer for the non-specialist. This second edition (1929) of his Astronomy and Cosmogony, originally published in 1928, is an extensive survey which was then at the forefront of the field, with particular reference to the physical state of matter, the structure, composition and life-cycle of stars, and the superstructures of nebulae and galaxies. Intended as a rigourously argued scientific treatise, every effort was made by Jeans to render the results of far-reaching advancements in cosmology intelligible to a broad range of readers.

Cambridge University Press has long been a pioneer in the reissuing of out-of-print titles from its own backlist, producing digital reprints of books that are still sought after by scholars and students but could not be reprinted economically using traditional technology. The Cambridge Library Collection extends this activity to a wider range of books which are still of importance to researchers and professionals, either for the source material they contain, or as landmarks in the history of their academic discipline.

Drawing from the world-renowned collections in the Cambridge University Library, and guided by the advice of experts in each subject area, Cambridge University Press is using state-of-the-art scanning machines in its own Printing House to capture the content of each book selected for inclusion. The files are processed to give a consistently clear, crisp image, and the books finished to the high quality standard for which the Press is recognised around the world. The latest print-on-demand technology ensures that the books will remain available indefinitely, and that orders for single or multiple copies can quickly be supplied.

The Cambridge Library Collection will bring back to life books of enduring scholarly value (including out-of-copyright works originally issued by other publishers) across a wide range of disciplines in the humanities and social sciences and in science and technology.

Astronomy and Cosmogony

JAMES JEANS

CAMBRIDGE
UNIVERSITY PRESS

CAMBRIDGE UNIVERSITY PRESS

Cambridge, New York, Melbourne, Madrid, Cape Town, Singapore,
São Paolo, Delhi, Dubai, Tokyo

Published in the United States of America by Cambridge University Press, New York

www.cambridge.org
Information on this title: www.cambridge.org/9781108005623

© in this compilation Cambridge University Press 2009

This edition first published 1928
This digitally printed version 2009

ISBN 978-1-108-00562-3

ASTRONOMY AND COSMOGONY

ASTRONOMY AND COSMOGONY

BY

Sir JAMES H. JEANS, M.A., D.Sc., LL.D., F.R.S.

SECRETARY OF THE ROYAL SOCIETY, AND RESEARCH
ASSOCIATE OF MOUNT WILSON OBSERVATORY

CAMBRIDGE

AT THE UNIVERSITY PRESS

1929

CAMBRIDGE UNIVERSITY PRESS
Cambridge, New York, Melbourne, Madrid, Cape Town, Singapore, São Paulo, Delhi

Cambridge University Press
The Edinburgh Building, Cambridge CB2 8RU, UK

Published in the United States of America by Cambridge University Press, New York

www.cambridge.org
Information on this title: www.cambridge.org/9780521744706

First edition 1928
Second edition 1929
This digitally printed version 2008

A catalogue record for this publication is available from the British Library

ISBN 978-0-521-74470-6 paperback

CONTENTS

LIST OF ILLUSTRATIONS

[Individual nebulae and clusters are indexed under their N.G.C. identification numbers in the general index (p. 424).]

PREFACE

MY book attempts to describe the present position of Cosmogony and of various closely associated problems of Astronomy, as, for instance, the physical state of astronomical matter, the structure of the stars, the origin of their radiation, their ages and the course of their evolution.

In a subject which is developing so rapidly, few problems can be discussed with any approach to finality, but this did not seem to be a reason against writing the book. Many years have elapsed since the last book on general Cosmogony appeared, and the interval has seen the whole subject transformed by new knowledge imported from observational astronomy and atomic physics. It has also witnessed the growth of an interest in the results of Cosmogony, which now extends far beyond the ranks of professional astronomers, and indeed beyond scientific circles altogether.

With this in my mind, I have tried to depict the present situation in the simplest language consistent with scientific accuracy, avoiding technicalities where possible, and otherwise explaining them. As the book is intended to be, first and foremost, a rigorously argued scientific treatise, the inclusion of a substantial amount of mathematical analysis was inevitable, but every effort has been made to render the results intelligible to readers with no mathematical knowledge, of whom I hope the book may have many.

In a sense the book constitutes a sequel to my *Problems of Cosmogony and Stellar Dynamics* of ten years ago. So much has happened in the intervening decade that a new book seemed to be called for, rather than a new edition of the old. At any rate I allowed myself to be attracted by the idea of a big clean canvas on which I could paint a picture on a more comprehensive scale than had originally been possible in the publication of a Prize Essay. A considerable part of the present book is devoted to examining the consequences of the hypothesis, first put forward in the closing pages of the earlier book, that the energy of stellar radiation arises out of the annihilation of stellar matter. The calculations of stellar ages given in the present book seem to shew that this is the only possible source of stellar energy, since nothing short of the complete annihilation of matter can give an adequate life to the stars. I have, however, tried to explain and discuss all reasonable hypotheses at present in the field, both on this and other subjects, and hope I have been fair and courteous to those whose views I cannot accept. My own personal contributions to the subject represent the outcome of twenty-five years of fairly continuous thought and work, and a considerable number of my results are published for the first time in the present book.

I have to thank many friends for help and courtesies of various kinds, and particularly Dr W. S. Adams, Director of Mount Wilson Observatory, for permission to reproduce a large number of photographs. My thanks are again due to the officials and staff of the Cambridge University Press for extending to the present book the consummate skill and unremitting care by which they transform a mass of muddled manuscript into a masterpiece of typography.

J. H. JEANS

DORKING,
January 25, 1928.

PREFACE TO SECOND EDITION

AS the first edition had the good fortune to be rather speedily exhausted, the preparation of a second edition has been a pleasant and comparatively light task.

I have expanded the book by references to various observational and theoretical results which have appeared since the first edition was written, and have allotted space more liberally to certain problems and investigations which friendly critics thought I had dismissed too briefly in the original book. Finally I have corrected a number of minor errors and misprints, and have to thank many friends and correspondents for bringing these to my notice.

J. H. JEANS

DORKING,
November 14, 1928.

CHAPTER I

THE ASTRONOMICAL SURVEY OF THE UNIVERSE

1. THE moon, our nearest neighbour in the sky, is 240,000 miles away from us; a distance which light, travelling at 186,000 miles a second, traverses in a little over one second. The farthest astronomical objects whose distances are known are so remote that their light takes over one hundred million years to reach us. The ratio of these two periods of time—a hundred million years to a second—is the ratio of the greatest to the least distance with which the astronomer has to deal, and within this range of distances lie all the objects of his study.

As he wanders through this vast range with the aid of his telescope, he finds that the great majority of the objects he encounters fall into well-defined classes; they may almost be said to be "manufactured articles" in the sense in which Clerk Maxwell applied the phrase to atoms. Just as atoms of hydrogen or of oxygen are believed to be of similar structure and properties wherever they occur in nature, so the various astronomical objects—common stars, binary stars, variable stars, star-clusters, spiral nebulae, etc.—are believed to be, to a large extent at least, similar structures no matter where they occur.

The similarity, it is true, is not so definite or precise as that between the atoms of chemistry, and perhaps a better comparison is provided by the different species of vegetation which inhabit a country. Plants and trees, while differing in size, vigour, age and secondary characteristics, nevertheless fall into clearly-defined species. Basing our metaphor on this, we may say that recent extensions in telescopic power have revealed no new species of astronomical objects, but have merely multiplied the numbers of examples of objects which belong to known species. For this reason we may suppose that we are already acquainted with the principal species of astronomical objects in the universe.

The task of the observational astronomer is to survey and explore the universe, and to describe and classify the various types of objects of which it is constituted, discovering what law and order he may in their observed arrangement and behaviour. But only the dullest of human minds can rest content with a mere catalogue of observed facts; the alert mind asks always for the why and the wherefore. How comes it that these various classes of objects exist, but no others? What is the relation between them? Does one of them for instance, produce, or transform into, others? If so, what is the sequence of these changes? How did this universe of objects begin, and what will be its final end? If the heavenly bodies can no longer be regarded as having been created merely to minister to man's pleasure and comfort by illuminating the earth, what purpose, if any, do they serve? And if life, and human life in particular, can no longer be supposed to be the central fact which explains everything, what is its relation to the magnificent, stupendous, almost terrifying, universe revealed by astronomy?

To these obstinate questions observational astronomy provides no answer. Her task is limited to a mere description of the universe, which others may interpret if they can. Another science, cosmogony, provides material that may help to this end, which those who essay to interpret the universe can only disregard at their peril. Cosmogony studies the changes which the play of natural forces must inevitably produce in the objects discovered by the astronomer; it tries to peer back into their past and to foresee their future, guided always by the principle that the laws of nature have moulded the present out of the past, and will in the same way mould the future out of the present. Taking as its starting-point the still picture presented by astronomy, it attempts to create a living cinematograph film which will exhibit the universe growing, developing and decaying before our eyes. The sequence of events depicted in this film will be false unless the relation of each picture to the succeeding one is that of cause to inevitable effect.

Between observational astronomy and cosmogony there intervenes a third science, or branch of science, namely, cosmical physics. Cosmogony proceeds on the supposition that the matter of which the universe is constituted behaves as directed by natural laws, so that a knowledge as to the particular kind of matter with which we are dealing is a prerequisite to knowing what particular laws this matter will obey. As the laws of a liquid are different from those of a gas, a liquid star will behave differently from a gaseous star, and before we can predict the behaviour of a star we must know the state of the matter composing it. Cosmical physics attempts to provide the necessary information by deducing, with such precision as is possible, the physical nature and structure of astronomical bodies, and of cosmical matter in general, from the observations of the astronomer.

The ultimate object of cosmogony, and of the present book in particular, is to construct a sequence of pictures which will provide a contribution towards answering the questions of whence and whither by revealing the past and future of the ever-changing universe. But before attacking the main problem we must study the physical constitution of the bodies with which we are dealing, and as a preliminary to this we shall survey the universe revealed by observational astronomy. Cosmical physics will occupy the first five chapters of our book, but the present chapter will merely describe the picture which observational astronomy provides—for cosmical physics to interpret, and for cosmogony to extend into the past and future.

THE SOLAR SYSTEM.

2. In the foreground of the picture we must place our own solar system. A cursory study of the sky shews that the great majority of the stars retain their positions unchanged relative to one another, at any rate through times far greater than the lives of individual men. None of us will ever see the stars of the Great Bear or of the belt of Orion change their relative positions to any appreciable extent. But against the background provided by this unchanging framework of stars, certain other bodies move with such rapidity that their changes may often be noticed from day to day: these are the planets or wanderers. Like the earth, they describe orbits around the sun. The orbits of these planets and of the earth are approximately circular; all lie approximately in one plane, and all are described in the same direction about the sun. The motion of the planets as they describe their nearly circular orbits, relative to the earth which is itself describing another nearly circular orbit, accounts for the apparently intricate motions of the planets in the sky.

Some sixty-six years after Copernicus had put forward this interpretation of the observed planetary motions, Galileo turned his newly-made telescope on to Jupiter and observed four satellites revolving around it in precisely the way in which Copernicus had maintained that the planets revolved around the sun; all their orbits were nearly circular, all were nearly in one plane, and all were described in the same direction around Jupiter. This provided direct visual proof that the Copernican interpretation of the solar system was tenable and even plausible; many found in it a final and convincing proof of the truth of Copernican theories.

But in verifying Copernicus' solution of one problem Galileo had opened up another deeper and more fundamental problem. For there were now seen to be at least two systems of almost exactly similar structure in the universe, and it was natural to conjecture that similar causes must have been at work to produce two such similar effects. In this way scientific cosmogony had its origin, although nearly two centuries were to elapse before much serious thought was devoted to it.

Modern astronomy has shewn that the similarity between the systems of Jupiter and the sun is far more pronounced than was known to Galileo; actually Jupiter has nine satellites whose general arrangement with respect to Jupiter bears a fairly close resemblance to that of the eight planets in respect to the sun. Moreover, the system of Saturn, again with nine satellites, provides a further instance of precisely the same formation. Not only so, but in addition to the eight great planets, the sun is surrounded by some thousands of minor planets or asteroids*, whose orbits shew the same general regularity as has been already noticed in the motions of the great planets. Some, it is

---

* Up to the end of 1927, 1069 of these were definitely identified with numbers and names, while the discovery of about 1200 others had been reported.

true, have substantially greater inclinations and eccentricities than are found among the great planets. For instance, the planet whose orbit has the greatest inclination to the general plane (invariable plane) of the solar system is Mercury with an inclination of $7° 0'$; against this the orbit of the minor planet Pallas has an inclination of $34° 43'$, and that of Hidalgo an inclination of $43°$. The planet whose orbit shews the greatest eccentricity is Mercury, with an eccentricity of $0·206$, whereas the minor planets Albert and Hidalgo have eccentricities of $0·54$ and $0·65$ respectively.

But the outstanding fact remains that all the orbits are described in the same direction. Adopting an argument which Laplace advanced in his *Système du monde* (1796), we may remark that if the directions of motion of 2000 planets and minor planets were a matter of pure chance, the odds would be $2^{1999} - 1$ to one against the coincidence of the orbits being all described in the same direction. Thus the odds that the directions of the orbits are *not* a matter of pure chance, are far greater than those in favour of well-attested historical events: it is more certain that some definite cause underlies the directions of motion in these orbits than it is, for instance, that the Athenians won the battle of Marathon, or that Queen Anne is dead.

The motion of the satellites of the planets continues, on the whole, the story of ordered arrangement already told by the motions of the planets, although here certain definite exceptions must be noted. These exceptions are limited to the outermost edges of the solar system and the outermost edges of the systems of Jupiter and Saturn. They are as follows:

Neptune, the outermost planet, has only one satellite, and this moves with retrograde motion—i.e. in the direction opposite to that in which Neptune and the other planets move round the sun.

Uranus, the next outermost planet, has four satellites, all moving in the equatorial plane of the planet, which is highly inclined to the general plane of the solar system.

Saturn, which comes next, has nine satellites, the outermost of which (Phoebe) revolving at a mean distance of 217 diameters of Saturn, moves with retrograde motion.

Jupiter has nine satellites, the two outermost of which move with retrograde motion.

THE DISTANCES OF THE STARS.

The nearer Stars.

3. The solar system has occupied the foreground of our picture of the universe, because its members are incomparably nearer to us than other astronomical bodies. As a preliminary to filling in the rest of the picture let us imagine the various objects in the universe arranged in the order of their distances from the earth. Disregarding bodies much smaller than the earth,

such as the moon, other planetary satellites and comets, we must give first place to the planets Venus and Mars, which approach to within 26 and 35 millions of miles of the earth respectively. Next in order comes Mercury with a closest approach of 47 million miles, and then the sun at about 93 million miles. Other planets follow in turn until we reach Neptune at a distance of 2800 million miles.

After this comes a great gap—the gap which divides the solar system from the rest of the universe. The first object on the far side of the gap is the faint star Proxima Centauri, at a distance of no less than 25,000,000 million miles, or more than 8000 times the distance of Neptune. Close upon this come the two components of the binary star α Centauri at 25,300,000 million miles; these, with Proxima Centauri, form a triple system of stars which are not only near together in the sky, but are voyaging through space permanently in one another's company. After these come three faint stars, Munich 15,040, Wolf 359, and Lalande 21,185, at 36, 47 and 49 million million miles respectively, and then Sirius, the brightest star in the sky, at 51 million million miles. Comparing these distances with the distances of the planets, we see that the nearest stars are almost exactly a million times as remote as the nearest planets.

A simple scale model may help us to visualise the vastness of the gulf which divides the planets from the stars. If we represent the earth's orbit by a circle of the size of the full stops of the type used in this book (circles of a hundredth of an inch radius) the sun becomes an entirely invisible speck of dust and the earth an ultra-microscopic particle a millionth of an inch in diameter. On this same scale the distance to the nearest star, Proxima Centauri, is about 75 yards, while that to Sirius is about 150 yards. We see vividly the isolation of the solar system in space and the immensity of the gap which separates the planets from the stars.

Before parting from this model, let us notice that the distance of one hundred million light-years to the farthest object so far discussed by astronomy is represented on the same scale by a distance of about a million miles. In this model, then, the universe is millions of miles in diameter, our sun shrinks to a speck of dust and the earth becomes less than a millionth part of a speck of dust. The inhabitant of the earth may well pause to consider the probable objective importance of this speck of dust to the scheme of the universe as a whole.

4. The ancients were, for the most part, entirely unconscious of the enormous disparity in size between the earth and the rest of the universe. But those few who urged that the earth moved round the sun, saw that this motion must necessarily cause the nearer stars to change their positions against the background provided by the more distant stars, just as a child in a swing observes near objects moving against the distant background of hills and clouds; they also saw that the extent of this motion would make it possible to estimate the distances of the nearer stars. Aristarchus of Samos, who anticipated Copernican doctrines as far back as the third century before Christ, explained clearly that motion of this kind must be observed unless the stars were very remote indeed, and, as no such motion could be detected, laid great stress on the extreme distances of the stars. Four centuries later Ptolemy argued that the impossibility of detecting such motion proved that the earth could not be in motion relative to the stars, and must therefore constitute a fixed centre around which the whole universe revolved. When the Ptolemaic doctrine was finally challenged by Copernicus and Galileo, it became important to detect motion of the kind we have described, both as providing final and conclusive proof that the earth was not the unmoving centre of the universe, and as giving evidence as to the distances of the stars.

The apparent motion caused by the swing of the earth in its orbit is described as parallactic motion; the half of the angle swept out by a star as the earth moves from one extremity of its orbit to the other (or the angle from the mean position to either extreme) is called the "parallax" of the star. A star whose parallax is one second of arc is at a distance at which the mean radius of the earth's orbit subtends an angle of one second of arc. This distance was first introduced as a unit for the measurement of stellar distances by Kobold, and was subsequently named the "parsec" by H. H. Turner. Since there are 206,265 seconds of arc in a radian, the actual length of the parsec is 206,265 times the mean radius of the earth's orbit. The mean radius of the earth's orbit, commonly called the "astronomical unit" being 92,870,000 miles, or 149,450,000 kilometres, the parsec is found to be 19,150,000 million miles or $3 \cdot 083 \times 10^{18}$ centimetres.

Long before the introduction of this unit Herschel had used as unit a quantity which he called "the distance of Sirius," and was supposed to represent the mean distance of "first magnitude" stars (cf. § 5). Seeliger gave precision to this unit, defining it as the distance corresponding to a parallax of 5 parsecs, or 1,031,324 astronomical units. Charlier and various other continental writers call this unit the Siriometer and define it to be 1,000,000 astronomical units or $14 \cdot 94 \times 10^{18}$ cms.

Another unit of astronomical distance, especially used in popular exposition, is the "light-year" or distance which light travels in one year. Since light travels $2 \cdot 998 \times 10^{10}$ cms. in a second, and there are 31,557,600 seconds in a year, the light-year is found to be equal to $9 \cdot 461 \times 10^{17}$ cms. or 5,880,000 million miles.

The relation between the three sets of units is as follows:

One parsec $= 3 \cdot 083 \times 10^{18}$ cms. $= 3 \cdot 259$ light-years.
One light-year $= 9 \cdot 461 \times 10^{17}$ cms. $= 0 \cdot 3069$ parsec.
One Siriometer $= 14 \cdot 94 \ \times 10^{18}$ cms. $= 4 \cdot 848$ parsecs.

5. Long before parallactic motion was detected, it had been clear that such motion must necessarily be of very small amount. Early in the seventeenth century Kepler had maintained that the stars were merely distant suns; if so, the enormous difference between the intensities of sunlight and starlight could only be explained by supposing the stars to be millions of times as distant as the sun. Newton* pointed out that Saturn appears as bright as a first magnitude star, although its size is such that it can only re-emit by reflection about one part in ten thousand million of the total light emitted by the sun, and deduced that "first magnitude" stars, by which he meant the twenty or so brightest stars in the sky, must be about 100,000 times as distant as Saturn. This would assign to them a distance of about 90 million million miles, representing a parallax of 0·21 seconds of arc. The estimate was not a bad one; actually the twenty brightest stars in the sky have a mean parallax of 0·134 seconds. Immediately after this Bradley attempted to measure the parallax of γ Draconis, and although he failed to achieve his primary aim, he proved conclusively that the star's parallax was less than a second of arc.

6. Not until 1838 was the great gulf definitely bridged; in that year Bessel, Struve and Henderson independently found unmistakable positive

Table I. *Stars within five parsecs of the Sun.*

	Star	Parallax	Distance in parsecs
1	Proxima Centauri	0·765	1·31
2	α Centauri	0·758	1·32
3	Munich 15040	0·538	1·86
4	Wolf 359	0·404	2·48
5	Lalande 21185	0·392	2·55
6	Sirius	0·377	2·65
7	B.D. −12°, 4523	0·350	2·86
8	11 h. 12·0 m., −57·2	0·340	2·94
9	Cordoba 5 h. 243	0·317	3·16
10	τ Ceti	0·315	3·17
11	Procyon	0·312	3·21
12	ε Eridani	0·310	3·23
13	61 Cygni	0·300	3·33
14	Lacaille 9352	0·292	3·42
15	Struve 2398	0·287	3·48
16	Groombridge 34	0·282	3·55
17	ε Indi	0·281	3·56
18	Kruger 60	0·256	3·91
19	0 h. 43·9 m., +4·55	0·255	3·92
20	Lacaille 8760	0·253	3·95
21	2 h. 50·3, +52·1	0·239	4·18
22	23 h. 59·5, −37·9	0·220	4·55
23	17 h. 37·0, +68·4	0·213	4·69
24	10 h. 14·2, +20·4	0·207	4·83
25	Altair	0·204	4·90
26	o₂ Eridani	0·203	4·93

* *System of the World* (1727).

parallaxes for the three stars, 61 Cygni, α Lyrae and α Centauri. It had at last been found possible to sound the depths of space and the universe lay open for exploration. As the result of the labours of many astronomers, the parallaxes of over 2000 stars are now known with high accuracy. In the great majority of cases the errors of the determinations lie well within a hundredth of a second of arc, which is the angle that a pin-head subtends at a distance of ten miles.

In Table I (p. 7) will be found a list of all the stars which are at present known to lie within a distance of 5 parsecs of the sun.

Density of Distribution of Stars.

7. There is no reason for expecting any special concentration of stars in the immediate neighbourhood of the sun. If our nearest neighbours in space are distributed approximately at random, the number of stars within a sphere of any radius drawn round the sun should be approximately proportional to the volume of the sphere, and therefore to the cube of its radius.

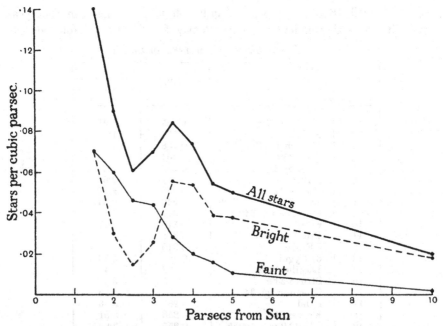

Fig. 1. Density of Distribution of known Stars in the neighbourhood of the Sun.

Fig. 1 shews the distribution of known stars in the neighbourhood of the sun. The abscissa measures distances from the sun in parsecs, while the ordinate gives the number of known stars per cubic parsec within this distance. The uppermost (thick) curve refers to the total of known stars of all kinds, the chain curve refers to stars which emit at least a thousandth part as much light as the sun, while the thin curve refers to stars which emit less than a thousandth part of the light of the sun (see Table IV, p. 33).

8. If all the stars were known we should expect the number of stars per cubic parsec to approach to a definite limit as we receded from the sun. The curves in fig. 1 shew no evidence of such a limit, so that we must conclude that nothing like all the stars in the neighbourhood of the sun are known.

Very faint stars can only be observed if they are quite near to the sun. Disregarding very faint companions of brighter stars, only six stars are known which emit less than a thousandth part of the light of the sun, and of these five are within 3 parsecs of the sun. This explains why the curve of faint stars runs down very rapidly after about 3 parsecs.

The brighter stars can be observed at greater distances. Actually the curve giving the density of bright stars shews no appreciable falling off up to a distance of about 4 parsecs, suggesting that practically all the bright stars within this distance are known. The curve suggests that the density of distribution of such stars is of the order of 0·05 stars per cubic parsec, or one bright star to every 20 cubic parsecs.

It is far more difficult to estimate the true density of distribution of the faint stars. To make a convenient figure for future calculations, we may suppose this to be the same as that for bright stars, so that the density of distribution of stars of all kinds near the sun is one to every 10 cubic parsecs, this requiring 18 stars actually to exist within $3\frac{1}{2}$ parsecs of the sun, of which only 15 are known. These 18 stars are of course additional to the sun itself. In a statistical discussion such as the foregoing we must be careful not to count the sun in our statistics, since its presence is an essential to our being able to make the calculation at all. Our procedure is in effect to draw a small sphere round the sun and discuss the density of distribution of stars in the space bounded by this sphere on the one side and by a larger sphere of variable radius on the other.

Distant Stars.

9. We have seen that the direct method of parallactic measurement has only succeeded in surveying the universe with tolerable accuracy to a distance of about $3\frac{1}{2}$ parsecs, or let us say 10 light-years, from the sun. No doubt this distance will be extended in time, but there is a natural limit to the power of the parallactic method. At best it can only measure the distance of a star whose parallactic motion can be projected on a background of far more distant stars, so that it must inevitably fail for the most distant stars of all. In actual fact it is bound to fail long before this.

The parallactic motion of a star at a distance of 100 parsecs, or 325 light-years, consists in the description of a circle or ellipse whose apparent size in the sky is that of a pin-head held at a distance of 5 miles. The apparent orbit of a star at a distance of 1000 parsecs, or 3250 light-years, is of course only a tenth as great; it is of the size of a pin-head held at a distance of 50 miles. The resources of observational astronomy are strained to the utmost to detect even the former of these parallactic motions, and are totally inadequate to measure it with accuracy, the error of measurement being about equal to the whole quantity to be measured. It is utterly impossible either to detect or measure the smaller parallactic motion of a star a thousand parsecs away, and is likely to remain so for many centuries to come. Yet a thousand parsecs is only a tiny fraction of the whole size of the universe. To survey the remote depths of space something of wider reach than the parallactic method is needed. Quite recently astronomers have discovered other and more far-reaching methods.

10. *Spectroscopic Parallaxes.* One of the most important of these is the method of "spectroscopic parallaxes" discovered by Dr W. S. Adams, now Director of Mount Wilson, and Kohlschütter in 1914. Two stars which are of exactly similar structure in all respects must necessarily emit light of precisely similar quality, so that their spectra must be similar in all respects. If the stars were at different distances, the spectra would naturally differ in brightness, and on measuring the ratio of their two intensities, it would be possible to deduce the ratio of the distances of the stars. Thus if the distance of one star had already been determined by the trigonometrical method already explained, it would be easy to deduce the distance of the other, even though this were so great as to render a direct determination of its parallax utterly impossible. The actual problem is generally far more complicated. When two stars have the same temperatures and the same chemical composition, their spectra are in general almost identical, but they shew minute differences if those parts of their atmospheres which emit their radiation are at different pressures. For reasons which will become clear later, stars of different sizes generally have different pressures in their atmospheres, and so exhibit slightly different spectra. Working backwards from this fact, Dr Adams discovered how to deduce the difference in size of two otherwise similar stars from minute differences in their spectra. As the difference in the intensity of their light arises jointly from differences in size and differences in distance, it is a simple matter to deduce the ratio of the distances of the two stars when once the ratio of their two sizes has been determined. This method is generally called that of spectroscopic parallaxes; it can hardly yet claim the accuracy of the trigonometrical method for near stars, but it has the great advantage of being successful with stars which are too remote for the trigonometrical method to be applicable at all. It is of course only of use for stars which appear moderately bright, but there is no limit to the distances at which it is available.

11. *Cepheid Parallaxes.* An even more far-reaching method of determining stellar distances depends on the peculiar properties of a certain class of stars known as "Cepheid variables" after their prototype δ Cephei. The majority of the stars in the sky shine with a perfectly steady light, but a fair number, known as variable stars, shew fluctuations in brightness. These fluctuations are regular in some stars and irregular in others. Cepheid variables shew perfectly regular fluctuations, flashing out to some two or three times their original brightness at intervals which range from a few hours to several days for different stars, but are always absolutely uniform for the same star. These variables are very common in the mysterious objects known as "globular star-clusters," closely packed groups of stars of approximately globular shape, and also occur in considerable numbers in star clouds such as the greater and lesser Magellanic cloud. Since the various Cepheid variables in any single one of these objects are at approximately the same distance from us, differences in their apparent brightness represent real differences in their output of radiation; no complication arises from the stars being at different distances. In 1912, Miss Leavitt of Harvard, studying the Cepheid variables in the lesser Magellanic cloud, discovered a relation between their time of fluctuation and their brightness. Those which fluctuated most slowly were the brightest, the period being connected with the brightness by a definite law, so that when the brightness of a Cepheid variable had been observed, its period of fluctuation could be predicted with accuracy, and *vice versa*. Dr Shapley, now Director of Harvard Observatory, subsequently proved that this relation was true of Cepheid variables in general. Now a few Cepheids, although only a few, are so near that their distances can be measured by the direct parallactic method, and as the actual output of radiation of these stars is known, it is possible to deduce the output of radiation of any Cepheid variable in the sky whose period of fluctuation is known. For instance, all Cepheids which fluctuate in brightness every 40 hours, emit approximately 250 times as much radiation as the sun, or, to use the technical phrase, their "luminosity" is 250; Cepheids whose period is 10 days have a luminosity of 1600, while those whose period is 30 days have a luminosity of 10,000. The general relation between period and luminosity is known as the "period-luminosity" law; it tells us the luminosity of every Cepheid variable in the sky.

Just as in the method of spectroscopic parallaxes, we can calculate the distance of a star of known luminosity by comparing its apparent brightness with that of a second star whose distance and luminosity are known. For instance, if a star is known to be as luminous as Sirius, but appears only one-hundredth part as bright as Sirius when seen in a telescope, we know that it must be ten times as distant as Sirius, because the apparent brightness falls off inversely as the square of the distance. By this method the distances of all the Cepheid variables can be determined, and so also the distances of the various star clusters and other objects in which they lie. Hertzsprung first used this method in 1913 to determine the distance of the lesser Magellanic Cloud.

Clearly the "period-luminosity" law provides a powerful method for the determination of astronomical distances; like the method of spectroscopic parallaxes, it is especially valuable because it supplements the parallactic method just where the latter fails. However distant Cepheid variables may be, provided only they can be clearly seen in a telescope, the astronomer can pick them out from the main mass of stars by their regular and characteristic light-fluctuations with the same ease and certainty with which the mariner picks out a lighthouse from a confusion of other lights on shore. The subsequent procedure is precisely that of the mariner who, having picked up a lighthouse, looks up its candle-power on a chart and estimates his distance from it by comparing its known candle-power with its apparent brightness; the "period-luminosity" law gives us the candle-powers of the Cepheid variables. The corresponding analogy with the parallactic method would be if the mariner, knowing the speed of his ship, estimated his distance from land by noticing the rate at which an electric standard or other fixed light on the sea-front appeared to move against a distant background of lights. The latter method is independent of the existence of lighthouses of known candle-power, but is obviously useless for ships far out at sea.

12. In 1918 Shapley used the "period-luminosity" law to determine the distances of the globular clusters, and found these to range from 22,000 to 220,000 light-years. At such distances the parallactic method would fail hopelessly: the parallactic orbit of a star at 220,000 light-years distance would have the same apparent size as a pin-head at 3000 miles. This bald statement gives but little real conception of the remoteness of the star-clusters. Their distance is perhaps better conveyed by the reflection that the light by which we now see them left them about the time when primaeval man first appeared on earth. Through the childhood, youth and age of countless generations of men, through the long prehistoric ages, through the slow dawn of civilisation and through the whole span of time which history records, this light has travelled steadily on its course, covering 186,000 miles every second, and is only reaching us now.

Even the distances first mentioned are not the greatest at which Cepheid variables are visible, for in 1924 Dr Hubble of Mount Wilson, detected them in the nearer spiral nebulae, and so was able to shew that the distances of these spirals is of the order of a million light-years, or nearly five times that of the remotest of the star-clusters. Using these nearer spiral nebulae as stepping-stones, he has since found that the remotest of the spiral nebulae which are visible must be over a hundred times as distant as the nearest, bringing us to distances of the order of 140 million light-years (cf. § 17 below).

THE DISTRIBUTION OF STARS IN SPACE.

13. It is natural to wonder how these vast ranges of space are filled. In the neighbourhood of the sun we have found that stars are uniformly scattered at the rate of about one star per 10 cubic parsecs. Known stars establish this uniform distribution up to less than 4 parsecs. After this the density appears to fall off so long as we take account only of stars whose distances are known, but this is merely because most of the stars beyond 4 parsecs have not had their distances measured.

It is nevertheless quite obvious that the uniform distribution of stars which prevails in the near neighbourhood of the sun cannot go on for ever. If it did, the sky would exhibit a uniform blaze of light, since, in whatever direction we looked, we should in time come to a star. Moreover, the infinite mass of stars would produce gravitational forces of infinite intensity, and these would cause the sun and stars to move with infinite speed. Actually the comparative faintness of starlight and the observed finite velocities with which the stars move through space, fix definite calculable limits to the extent of the field of stars surrounding the sun. The limit set by the speed of motion of the stars will be discussed in a later chapter, but we may consider at once the limit set by the observed brightness of the sky.

The naked eye can see some 6000 stars in both hemispheres of the sky. A one-inch telescope has about five times the aperture of the naked eye, and, because it has 25 times the area, admits 25 times as much light. If any star can just be seen with the naked eye, a similar star at five times the distance ought to be just visible in a one-inch telescope. In brief, we may say that the range of vision of the one-inch telescope for a given object is five times as great as that of the naked eye because its aperture is five times as great, and in general the ranges of vision of different telescopes for similar objects are proportional to their apertures.

Thus if the distribution of stars around the sun extended uniformly to infinity, the number of stars visible in different telescopes would be proportional to the cubes of their apertures. A one-inch telescope would shew 5^3, or 125, times as many stars as the naked eye, and, as the naked eye can see 6000 stars, a one-inch telescope should shew 750,000 stars. Actually it only shews about 120,000. Part of the discrepancy no doubt arises from the loss of light caused by passing through the three lenses of the telescope. If this could be eliminated, a one-inch telescope would enable us to see somewhat more than the 120,000 stars which it actually shews, but it would not shew as many as 750,000. A five-inch telescope ought in the same way to shew 125 times as many stars as a one-inch telescope but the ratio actually observed is far less than this. The explanation is that the remaining stars are not there to see; somewhere within the range of a five-inch telescope the star field surrounding the sun must begin to thin out quite perceptibly.

This method of investigation can be extended and refined almost indefinitely. On examining the numbers of stars visible in different directions in space, the star field is found to thin out differently in different directions; a detailed study gives particulars of the way in which the star field thins out in any given direction. The value of such investigations has been recognised since the time of the Herschels. The method of "star-gauges," or counts of stars of different brightnesses in different areas of the sky, instituted by Sir William Herschel* and extended to the southern hemisphere by his son, Sir John†, first established that the system of stars surrounding the sun is of a symmetrical flattened shape like a watch or a ship's biscuit. Investigations on the same lines have been made by Seeliger, Chapman and Melotte, Kapteyn, Seares and van Rhijn and others.

14. *The Galactic System.* We have seen that stars are scattered in the neighbourhood of the sun, at the rate of about one star to every 10 cubic parsecs. We can frame a definite problem by inquiring how far from the sun we should have to go to find the star field thinned out to any definite fraction we please of this density, say, one hundredth part, or one star per 1000 cubic parsecs. In 1922 Kapteyn‡ gave a simple answer to this question; it is far from absolutely accurate, but its simplicity more than compensates for its inexactness.

Looking at the sky on a clear night, we see the band of faint stars known as the Milky Way. This band forms very approximately a great circle in the sky, so that the plane through this circle which is called the galactic plane passes very nearly through the earth, and divides the sky into two approximately equal halves. Now Kapteyn found that if we go for a distance of 8465 parsecs, or about 27,000 light-years, in any direction whatever in the galactic plane, the star-density is reduced to one-hundredth of its value in the regions surrounding the sun. But to reach a corresponding reduction of density in other directions, we need not travel so far; the same reduction of density is reached after travelling only 1660 parsecs, or 5400 light-years, in a direction at right angles to the galactic plane. The various points in space at which the star-density is one-hundredth of that near the sun, lie, according to Kapteyn, on a much flattened spheroid. The cross-section of this spheroid in the plane of the galaxy is a circle of radius equal to the 8465 parsecs already mentioned, but its semi-minor axis, which is at right angles to this plane, is of length only 1660 parsecs.

15. This particular spheroid maps out the regions in space at which the star-density is one-hundredth of that near the sun. If we had selected any other fraction in place of one-hundredth, Kapteyn finds that we should still have obtained a much flattened spheroid whose axes would have been different from those just mentioned, although still in the same proportion of 5·102 to 1. Kapteyn has estimated the axes of these spheroids for ten different densities, with the results shewn in the following table:

---

* "On the Construction of the Heavens," *Phil. Trans.* LXXXV. (1785) or *Sci. Papers*, I. p. 223.

† *Results of Astronomical Observations made during the years 1834–8 at the Cape of Good Hope* (1847), p. 373.

‡ *Astrophys. Journal*, LV. (1922), p. 302 or Mount Wilson Contribution, No. 230.

PLATE I

Yerkes Observatory

The Great Nebula *M* 31 (N.G.C. 224) in Andromeda

The small nebula almost directly over the centre is the companion nebula *M* 32 (N.G.C. 221)

Table II. *Star-densities* (*Kapteyn*).

Ellipsoid No.	Semi-major axis (parsecs)	Semi-minor axis (parsecs)	Relative Star-density	Stars per cubic parsec	Volume of Shell (cubic parsecs)	Number of Stars
Centre	—	—	1·000	0·100		
I	602	118	0·631	0·063	$1·79 \times 10^8$	$0·1 \times 10^8$
II	1010	198	0·398	0·040	6·66	$0·3 \times 10^8$
III	1510	296	0·251	0·025	19·8	$0·6 \times 10^8$
IV	2106	413	0·158	0·016	48·4	$1·0 \times 10^8$
V	2820	553	0·100	0·010	107	$1·3 \times 10^8$
VI	3656	717	0·063	0·006	226	$1·8 \times 10^8$
VII	4600	902	0·040	0·004	388	$2·0 \times 10^8$
VIII	5675	1114	0·025	0·002	702	$2·3 \times 10^8$
IX	6960	1365	0·016	0·002	1270	$2·5 \times 10^8$
X	8465	1660	0·010	0·001	2210	$2·6 \times 10^8$
—	—	—	0·03	0·003	4980	$14·5 \times 10^8$

The first four columns are given by Kapteyn. The next column gives the number of stars per cubic parsec over the spheroid in question, calculated for a central density of one star per 10 cubic parsecs, our estimated density of stars in the neighbourhood of the sun. Kapteyn's own figures were rather less than half of these, as he estimated the central density to be only 0·0451 stars per cubic parsec, or about one star per 22 cubic parsecs.

The general distribution implied in Kapteyn's scheme is shewn diagrammatically in fig. 2. This represents a section through the central axis of the system; if the section is supposed to be $\frac{1}{3000}$ parsec in thickness, each dot represents a single star.

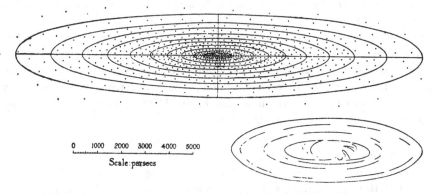

Fig. 2. Density of Distribution of Stars surrounding the Sun (Kapteyn).
(On right, the Andromeda Nebula M. 31 (Plate I) on the same scale.)

The sketch on the right of fig. 2 represents the main outlines of the Andromeda nebula (Plate I) drawn to the same scale. The outermost line of the sketch does not, however, represent the physical boundary of the nebula, the greatest diameter of which is about 15,000 parsecs (cf. § 16 below). Also the distance between the two objects is not drawn to scale; to make the distance conform to the same scale the two objects must be imagined separated by a distance of over six feet.

Although the schemes of Herschel and Kapteyn make no attempt to explain local peculiarities of star distribution, they give an adequate explanation of the general appearance of the night sky. The sun is supposed to be near the centre of the system. The stars which appear brightest in the sky are for the most part relatively near to the sun, occurring at distances within which there is no appreciable thinning out of stars. Sirius, the brightest star of all, is within 3 parsecs, while more than half of the twenty brightest stars are within 20 parsecs. These brightest stars, being well within the first spheroid, appear to be evenly scattered over the sky. On the other hand, stars near the ends of the major axes of the remoter spheroids are so distant that they appear faint, no matter how great their intrinsic luminosity may be. As there is no counterbalancing aggregation of faint stars in other directions, the faint stars appear to be concentrated mainly in a circular band in the sky—the Milky Way.

The plane which passes through the earth and this band forms an obvious plane of reference for the discussion of the distribution of the stars. According to the Harvard determination, the poles of this plane are at

R.A. 12 h. 40 m., Decl. + 28 (in Coma Berenices)

and   R.A.   0 h. 40 m., Decl. − 28 (in Sculptor).

Galactic latitudes are measured from this plane and galactic longitudes from the point in the plane of right ascension 18 h. 40 m. (in Aquila).

Recent investigations have revealed two deficiences in Kapteyn's scheme. It fails to represent the distribution of stars beyond a certain distance. That it does not represent the distribution of all the stars is clear from the scheme itself. Table II shews that the number of stars in each spheroidal shell is greater than that in the shell next inside it; on adding the numbers in the different shells the total shews no tendency to approach a definite limit by the time the tenth shell is reached. Thus the total number of stars accounted for by Kapteyn's scheme, which is just below 1500 million, can be nothing like equal to the number of all the stars.

Even at the limit of visibility of the largest telescopes, the total number of stars is still rapidly increasing. It is estimated that about 1000 million can be noted photographically in the 100-inch telescope at Mount Wilson, but the distribution of luminosity in these is such as to make it clear that a slight increase in the aperture of the telescope would result in an enormous increase in the number of stars. By extrapolation from known observational results, Seares and van Rhijn* have estimated the total number of stars at 30,000 million. As their estimate relies on extrapolations to enumerate totally invisible stars, they do not of course claim great accuracy for it. For instance, they compute that even in the direction of the galactic poles, where the stars thin out most rapidly, there are three stars too faint to be seen in any existing telescope for each one that is telescopically visible; in directions in the galactic plane, they estimate the ratio of invisible to telescopically visible stars to be not less than 70 to 1.

The circular appearance of the Milky Way led Kapteyn to anticipate a general symmetry of arrangement about the galactic pole, and his scheme of star distribution presupposes this symmetry to exist. Recent investigations have shewn that the distribution of stars is by no means the same in the different planes through the galactic pole, being especially disturbed by the local condensations of stars known under the general name of "star clouds." A conspicuous example occurs in Cygnus, at a distance which Pannekoek† estimates as 600 parsecs, and there are others in Sagittarius, Scutum, Monoceros (200 parsecs) and Carina (300 parsecs). Charlier and Shapley have found that the brightest stars of the earliest spectral types form a star cloud which surrounds the sun and appears to have the same general biscuit shape as the main galactic system, but to be inclined to it at an angle of about 12°. The sun appears to be only a small distance, 90 parsecs according to Charlier‡, from its centre. Shapley calls this "the local system," and a large proportion of the stars in the neighbourhood of the sun appear to belong to this system. Seares finds that it is only the nucleus of a much larger "local system," some 6000 parsecs or more in diameter, which is responsible for three-fourths of the stars in the neighbourhood of the sun, and for half of the stars in the galactic plane up to a distance of about 700 parsecs.

As the Milky Way forms almost exactly a great circle in the sky, the sun cannot be far removed from the central plane of the galactic system. Seares§ finds that it lies exactly in this plane to within about eight parsecs, but that it lies some 40 or 50 parsecs north of the central plane of the local cluster.

Kapteyn, treating the sun as lying precisely in this central plane, discussed its distance from the centre of the system, and concluded that the approximate equality of star-density in different galactic latitudes was inconsistent with this distance being more than about 700 parsecs. Seares now finds, however, that the star-density in the main galactic system increases to a maximum at a distance of 1000 parsecs, or even more, from the sun. This main system is far from symmetrical, and its geometrical centre is probably

---

* *Astrophys. Journal*, LXII. (1925), p. 320 or Mount Wilson Contribution, No. 301.

† *Publications of the Astronomical Institute of the University of Amsterdam*, No. 1 (1924).

‡ "The Distances and the Distribution of Stars of Spectral Type B," *Lund Meddelanden*, II. (1916), No. 14.   § *Contributions from Mount Wilson Observatory*, No. 347 (1928).

substantially more than 1000 parsecs from the sun. Seares considers that its diameter may be from 60,000 to 90,000 parsecs, and determines the direction of its centre as galactic latitude 319°.

This is in substantial agreement with Shapley's determination of the system defined by the "globular clusters" (§ 27). He finds that this system lies approximately in the galactic plane, its maximum diameter being about 75,000 parsecs, and its centre lying about 20,000 parsecs away in galactic longitude 325°. Oort* finds that the radial velocities of faint stars suggest orbital motion about a centre some 6000 parsecs distant in galactic longitude 324°, while from a similar study of O and B type stars, J. S. Plaskett† finds evidence of orbital motion about a centre in galactic longitude 324·5°.

THE GREAT NEBULAE.

16. Within the limits of distance set by the Milky Way lie not only the stars we have just discussed, but also an abundance of non-stellar objects. For instance, the irregular nebulae, such as the great nebula in Orion (see Plate V) are found to be comparatively near objects, most if not all of which lie within the confines set by the galaxy.

In addition to these the telescope reveals a great number of nebulae of regular shapes—circular, elliptical, spindle-shaped and spiral. Two of these, the great nebula M 31 (N.G.C. 224)‡ in Andromeda and the nebula M 33 (N.G.C. 598) in Triangulum, are of outstanding brightness and apparent size, both being visible to the naked eye.

The former of these nebulae is shewn in Plate I, enlargements of its central portion and of an outlying region (top left-hand corner) being shewn in Plates VII and VIII below. The nebula in Triangulum is shewn in Plate XI. These nebulae appear to be thin discs in shape; when seen edge-on they present the characteristic appearance exemplified by nebula N.G.C. 981 shewn in Plate II (see also other examples in Plate XIII).

The distances of the two nebulae M 31 and M 33 are revealed by the circumstance that they are found to contain Cepheid variables. Calculating their distances in the manner already explained, Hubble has estimated the distance of M 31§ as 285,000 parsecs, and that of M 33‖ as 263,000 parsecs. By the same method he estimates the distance of the star-cloud N.G.C. 6822¶ to be 214,000 parsecs.

These figures amply shew that these nebulae and star-clouds are quite outside our system of stars; they constitute what Herschel described as

---

* *Bulletin of the Astronomical Institute of the Netherlands*, No. 120 (1926).

† *Monthly Notices of the R.A.S.* LXXXVIII. (1928), p. 395.

‡ The prefix M to a nebula precedes its number in Messier's Catalogue of Nebulae; the prefix N.G.C. similarly refers to Dreyer's New General Catalogue (*Mem. Royal Ast. Soc.* 49 (1888), Part I).

§ *Popular Astronomy*, XXXIII. (1925), No. 4 or *The Observatory*, XLVIII. (1925), p. 139.

‖ *Astrophys. Journal*, LXIII. (1926), p. 236.   ¶ *Ibid.* LXII. (1925), p. 409.

PLATE II

Mt Wilson Observatory

The Spiral Nebula N.G.C. 891 seen edgewise

(This is oriented so as to compare directly with the sideways nebula shewn in Plate I)

"island-universes" distinct from the universe which contains our sun. Knowing their distances, it is easy to calculate their sizes; the diameter of the great Andromeda nebula, which subtends nearly three degrees in the sky, must be about 15,000 parsecs; the diameter of M 33, subtending about one degree, must be about 5000 parsecs. We may call these nebulae "extra-galactic" on account of their distance, or may speak of them as the "great nebulae" on account of their size.

17. As seen in a telescope, the great nebulae differ enormously in apparent size, shape and brightness. But Hubble has recently shown[*] that differences in size and brightness in nebulae of the same shape are almost entirely due to a distance effect. Nebulae of the same shape may thus be thought of as similar manufactured articles, or as astronomical plants belonging to the same species. Hubble further finds that even nebulae of different shape shew only slight differences in intrinsic luminosity, and no great differences in size.

These circumstances make it possible to estimate the distances of all nebulae, down to the very faintest visible, with fair accuracy. The faintest nebulae which can be seen photographically in the Mount Wilson 100-inch telescope give only about a hundred-thousandth part of the light of the brightest. Assuming the difference in light to be due to a distance effect, it is found that the 18th magnitude nebulae must be at a distance of about 140 million light-years. This represents the range of vision of the 100-inch telescope for objects having the luminosity of the great nebulae; it is the greatest distance with which practical astronomy has so far had to deal. Within this distance Hubble estimates that about two million nebulae must lie, these being fairly uniformly spaced at distances of about 570,000 parsecs apart.

We can construct an imaginary model of the system of the great nebulae by taking about 50 tons of biscuits and spreading them so as to fill a sphere of a mile radius, thus spacing them at about 25 yards apart. The sphere represents the range of vision of the 100-inch telescope; each biscuit represents a great nebula of some 4000 parsecs diameter. A few nebulae of exceptional size must be represented by articles rather larger than biscuits, while our system of stars, up to Kapteyn's tenth spheroid, would be represented by a flat cake 13 inches in diameter and $2\frac{1}{2}$ inches in thickness. On this scale the earth is far below the limits either of vision or even of imagination. It is little more than an electron in one of the atoms in our model; and we should have to multiply its dimensions many millions of times to bring it up to the size of even the smallest particles which are visible in the most powerful of microscopes.

18. This completes our brief astronomical survey of the dimensions of the universe. We have dwelt mainly on the lay-out, as regards distance, of the

[*] *Astrophys. Journal*, LXIV. (1926), p. 321. See Chapter XIII below.

main structure of the universe. In rushing in imagination to the depths of space, travelling with millions of millions of times the velocity of light, we have not paused to discuss, or even to mention, except perhaps incidentally, the various minor classes of objects which are found scattered through the universe. Retracing our steps, let us discuss astronomical objects no longer primarily in respect of their distances, but rather in respect of their frequency of occurrence in the sky.

BINARY STARS.

19. The commonest object of all is the simple star, which appears merely as a point of light and shines with a perfectly steady light. If we set a telescope on the sky at random, we shall find that the vast majority of the objects visible in it are simple stars of this kind. We are likely to find, however, that these points of light are not scattered at random in the field. In too large a number of cases to be attributed to mere chance, pairs of stars are found to lie very close together in the sky. Some such pairs undoubtedly appear close as the result of mere chance, the line joining them happening to pass near to the earth at the present moment. If such a pair of stars were watched for several centuries, they would be seen to move steadily apart, and the chance cause of their appearing together would be disclosed; they had merely happened to lie almost one behind the other when we first noticed them. Other pairs of contiguous stars shew no tendency to separate when continuously watched; on the contrary, they are observed perpetually to describe orbits one about the other like partners in a never-ending waltz. Such pairs clearly do not appear close together in the sky as the result of mere chance, but because they actually are close together in space; they are bound together by the force of gravitation and describe regular orbits about one another just as the earth describes an orbit about the sun, and for the same reason. Such systems are spoken of as "binary systems."

The periods in which the orbits of binary stars are described may be too long to be determined by observation at all. Many must be reckoned in thousands of years, and periods of hundreds of years are common. At the other end of the scale, when the period is less than about a year, the stars may be so close as to appear merely as a single point of light in the telescope, so that visual observation cannot detect their binary nature. Methods other than direct visual observation, however, reveal the existence of binary stars of periods extending down to only a few hours.

20. *Spectroscopic Binaries.* The spectrum of a star generally shews a number of sharply-defined dark lines, and if the star is in rapid motion, these lines are observed to be displaced from their normal places by an amount which, by Doppler's principle (cf. p. 50 below), is proportional to the star's velocity relative to the earth. This extremely valuable circumstance makes it possible, within limits set by the power of optical instruments, to determine the speed with which any star is advancing towards, or receding from, the earth. As a result of their motion around one another, the two components of a binary star will generally be advancing towards the earth or receding from it at different rates, and when both components are bright enough for their spectra to be visible the spectrum of a binary star shews two distinct sets of lines, which oscillate about their mean position in a period equal to the period of the star. There are only very few stars whose distance and period are such that they can be observed both visually and spectroscopically as binaries : in these rare cases the period can be determined from either visual or spectroscopic observations. There is a far larger number of stars whose spectra shew the special characteristics of binary stars, although their period is so short, and the two constituents consequently so close together, that their binary nature is not susceptible to direct visual observation. Such stars are called spectroscopic binaries. The periods of spectroscopic binaries range from $2\frac{1}{4}$ hours (γ Ursae Minoris) to 15·3 years (ϵ Hydrae).

21. *The periods of Binary Stars.* The frequency with which the different classes of binary stars figure in star catalogues gives a very misleading idea as to the frequencies with which the stars themselves occur in space. It is a slow and difficult process to determine the orbits of long-period binaries with any accuracy, so that only few of these appear in star catalogues at all. Hertzsprung* considers that out of 15,000 known double stars it is only possible to calculate reliable orbits for 80, while about 1000 more shew traces of orbital motion but with periods so long that it is not yet possible to calculate them with any accuracy. On the other hand the periods of short-period binaries, and especially of fairly bright spectroscopic binaries, are easily and rapidly determined, with the result that these figure to a very disproportionate extent in star catalogues.

Hertzsprung has attempted to obtain information as to the true frequency with which the different types of orbit occur in space by studying the known binaries within a short distance of the sun. Twenty-one binaries are known to be within 10 parsecs of the sun. The actual periods of 13 of these are known, and Hertzsprung estimates periods for the remaining 8 from the observed angular motion or the observed distance between the components. The adopted values for the period are shewn in the following table. The second column contains the period in years where this is known with accuracy, and the third column contains Hertzsprung's estimated value of $\log P$, where P is the period in years. Of the 21 stars only two, χ Draconis and ξ Ursae Maj. are spectroscopic binaries, so that visual binaries outnumber spectroscopic binaries by ten to one. Hertzsprung notes that the distribution of $\log P$ is approximately Gaussian about a median value of $\log P = 2$, with a mean deviation of ± 1. These 21 binaries probably constitute the best sample

* *Bulletin of the Astronomical Institute of the Netherlands*, No. 25, 1922.

available to us of the binaries in space. We accordingly conclude that only a small fraction of binaries are spectroscopic, and that the mean period of visual binaries is of the order of 100 years.

Table III. *Binary Systems within 10 parsecs of the Sun (Hertzsprung).*

Star	Period in years (where known)	log P
χ Draconis	0·770	− 0·1
ξ Ursae Maj.	1·80	0·3
β 395	25·0	1·4
ζ Herculis	34·46	1·5
Procyon	40·23	1·6
β 416	41·47	1·6
μ Herculis	42·23	1·6
Lacertae 353	—	1·7
Sirius	49·32	1·7
Kruger 60	54·9	1·7
α Centauri	78·83	1·9
70 Ophiuchi	87·86	1·9
ξ Bootis	152·8	2·2
o_2 Eridani	180·03	2·3
61 Cygni	—	2·6
Sh. 190	—	2·8
Struve 2398	—	2·9
Struve 1321	—	3·0
κ Tucanae	—	3·0
Groombridge 34	—	3·4
γ Leporis	—	4·1

22. *The number of Binary Stars.* We have just seen that 21 of the 87 stars which are known to be within 10 parsecs of the sun are binary, while of the 26 stars within 5 parsecs of the sun (cf. Table I) 8 are binary. These figures suggest that something like a quarter, or possibly a third, of the stars in the sky are binary. We must be on our guard against putting the proportion too high. It may be true that some of the 87 and 26 stars just mentioned may be binary without their binary nature having yet been discovered; on the other hand there may be a number of stars still undiscovered which must be added to our figures of 87 and 26, and these are far more likely to be simple stars than binary systems. But it seems probable that fully a quarter of the whole number of stars in the sky are binary, while in special classes of stars the proportion is higher. For instance Hertzsprung has found that 9 out of 15 stars in the Ursa Major cluster are binary, while Frost found that half the stars in the Taurus cluster are binary. Next to the simple featureless star, the binary star is the commonest object in the sky.

23. *Eclipsing Variables.* The majority of binary stars shine with a perfectly steady light, the total light emission of the system being the sum of the emissions from the two components separately. In a certain proportion of cases, however, the two components alternately pass in front of one another in their orbits as seen from the earth, and so undergo eclipse at regular intervals. At these moments of eclipse we do not see the total light from both constituents, so that the brightness of the system appears to undergo a diminution. The light received from such a system accordingly varies, and the system is described as an "eclipsing variable," 'or an "eclipsing binary." It will be understood that physically an eclipsing binary is precisely similar to a non-eclipsing binary; the difference between them results solely from the accident of the earth being nearly in the plane of the orbit in the case of the eclipsing binary, and well out of this plane in the case of the non-eclipsing binary.

VARIABLE STARS.

24. While the variation in the light of an eclipsing variable is in a sense accidental, and in no way physically inherent in the system itself, there are other classes of variable stars in which the light variation must be ascribed to actual changes in the light-emission of the stars themselves. The general characteristic of these true variables is that the quality of the light changes as well as the amount received. In extreme cases the light varies through a considerable range of colour as the star varies, while the amount of visual light may vary by a factor as great as 4000 to 1, although the ratio is greatly reduced when the invisible heat-radiation of the star is added to its visible light radiation (cf. § 48 below).

The Cepheid variables already mentioned form the most interesting class of true variables. Both these and other variables are of somewhat rare occurrence in space, but their importance is not to be measured by the frequency of their occurrence, and we shall discuss them further in Chap. XV, below.

GROUPS OF STARS.

Triple and Multiple Systems.

25. We have already noticed that nearly one-third of the nearest stars are binary, and one of these systems, the nearest of all, constitutes in effect a triple system, namely, the binary α Centauri, accompanied in its journey through space by its distant companion Proxima Centauri. The period of the two components of α Centauri is 78·83 years, while its distance shews that the period of Proxima about this binary system is to be reckoned in millions of years.

This provides a rather extreme instance of a triple system, but there are innumerable instances of more normal triple systems in which the periods are shorter, as also of multiple systems in which more than three components journey together through space, describing orbits meanwhile under their mutual gravitational attractions. Russell found that of 800 double stars no fewer than 74, or 9·25 per cent. of the whole, formed part of triple or multiple systems. In Jonckheere's "Catalogue and Measures of Double Stars*" 9·7 per cent. of the total of 3950 systems are either triple or still more complex.

Roughly we may say that of every hundred stars in the sky, about 75 are likely to be single stars, while the remaining 25 will be binary or multiple. And of these 25, some 2 or 3 are likely to be triple or multiple systems.

Moving Clusters.

26. There is a continuous transition from the normal triple star, in which the periods are a few hundred years at most, through systems such as the triple system formed by α and Proxima Centauri, with periods running into millions of years, to systems such as the three well-known stars in Orion's belt which are moving in one another's company through space, although at such great distances from one another that their mutual gravitational attraction may almost be disregarded. We can pass still further and find groups of many more than three stars which are voyaging in company through space. Indeed the stars of Orion's belt are only three members of a great party of such stars, which contains nearly all the bright stars in the constellation of Orion, with the conspicuous exception of the brightest of all, α Orionis (Betelgeux) which appears to be traversing space by itself.

Most of the conspicuous groups of stars in the sky form parties of this kind which are travelling through space in company. The "Great Bear" in Ursa Major is perhaps the most obvious instance, although here again the brightest star of all, α Ursae Majoris is a solitary traveller and not a member of the party. As Hertzsprung and others have shewn, this group, or "moving cluster" to use the usual technical term, contains some twenty stars at least, Sirius almost certainly being a member. H. H. Turner† found that the stars of this cluster form a much-flattened formation lying nearly in one plane. Another striking instance is provided by the Pleiades, the moving cluster containing all the stars which are visible to the naked eye, and many other fainter stars as well. The Hyades again form part of a huge moving cluster, the Taurus cluster. The principal stars in Perseus also belong to a clearly-defined cluster, and there are less clearly-defined clusters in Scorpio-Centaurus and Cygnus. Shapley's "local system" (§ 15) may probably be regarded as forming a single huge moving cluster, and there is a further possibility that the great majority of stars in the neighbourhood of the sun belong either to this or to two inextricably intermingled moving clusters of enormous size and extent. Nearly all known clusters shew the characteristic flattened formation‡. We shall consider the dynamics of moving clusters in a later chapter (Chap. XIV) in which we shall also discuss the general motions of the stars.

* *Memoirs of the Royal Astronomical Society*, LXXI. (1917).

† *The Observatory*, XXXIV. (1911), p. 246.

‡ N. H. Rasmuson, "A Research of Moving Clusters," *Lund Meddelanden*, II. No. 26 (1921).

PLATE III

Mt Wilson Observatory

The Globular Cluster *M* 13 in Hercules

Globular Clusters.

27. In a somewhat different category from the moving clusters are the rather mysterious objects known as "globular" clusters. Whereas the moving clusters just discussed are of a pronouncedly flattened shape, the globular clusters appear at first sight to be strictly of the shape that their name implies. Bailey *, however, noticed a departure from actual spherical symmetry, and Pease and Shapley † have found that five out of six clusters they studied in detail shewed definite departures from sphericity, being apparently of a flattened or spheroidal form. A typical globular cluster, the cluster M 13 in Hercules, is shewn in Plate III.

Whereas the moving clusters are intermingled with the main mass of the stars, the globular clusters are so remote as to be either entirely outside our system of stars or at least in regions where stars are extraordinarily sparse. It has already been mentioned that Shapley has measured their distances by the use of the "period-luminosity" law which is obeyed by the Cepheid variables contained in them, and finds distances varying from 6500 parsecs for ω Centauri to 67,000 parsecs for N.G.C. 7006‡; their diameters are all of the order of 150 parsecs (490 light-years). Their distribution in the sky is peculiar and surprising; Hinks § found that they are practically confined to one hemisphere of the sky, while Melotte‖ further found that more than half lie within 30° of one point in the sky, namely, the point in the galactic plane of galactic longitude 325°. Shapley ¶ has shewn that they lie within an ellipsoidal volume whose centre is at a distance of 20,000 parsecs in this direction and whose major axis is about 75,000 parsecs. The sun is near to one end of the major axis, which explains why the globular clusters all appear to lie in one-half of the sky.

Slipher has measured the velocities of the clusters spectroscopically and finds that their radial velocities vary from a velocity of approach of 410 kms. a second to a velocity of recession of 225 kms. a second. The mean speed of these clusters is 150 kms. a second. The average radial velocities of approach or recession of the stars in the galactic system are only of the order of 10 kms. a second, so that the globular clusters move with far higher velocities than individual stars; on the other hand, as we shall shortly see, their velocities are substantially inferior to those of the spiral nebulae. After allowing for their different distances these globular clusters are found to be remarkably similar in structure and size. The law of density of distribution is found to be approximately the same in all; using counts of stars by Bailey**, Plummer†† has

---

* *Harvard Observatory Annals*, LXXVI. No. 4.   † *Astrophys. Journal* XLV. (1917), p. 225.
‡ *Ibid.* XLVIII. (1918), p. 154, or Mount Wilson Contribution, No. 151.
§ *Monthly Notices of the R.A.S.* LXXI. (1911), p. 693.
‖ *Memoirs of the R.A.S.* LX. (1915), p. 176.   ¶ *l.c.* p. 169.
** *Harvard Observatory Annals*, LXXVI. No. 4.
†† *Monthly Notices of the R.A.S.* LXXVI. (1916), p. 107.

found the star-density at a distance r from the centre to be proportional to $(r^2 + a^2)^{-\frac{5}{2}}$, where the constant a which determines the scale of the cluster is very approximately the same for all clusters. This law gives a density which falls off as r^{-5} at great distances from the centre ; I have, however, found* that the observed distribution in the outer regions of the clusters is better represented by a density falling off as r^{-4}.

The central dense portions of all the clusters is about 5 parsecs in diameter, the star-density remaining appreciable to a distance of 15 or 20 parsecs from the centre. The star-density in the central regions is very high. Within 10 parsecs of the centre of the cluster M 3 there are at least 150,000 stars which are at least four times as bright photographically as the sun. Thus the density of distribution of these bright stars alone is several hundreds of times as great as the density of stars of all kinds in the neighbourhood of the sun. The distances of the globular clusters are so great that only stars of high luminosity are accessible to observation. It seems probable that, as in the galactic system, very bright stars must be accompanied by far greater numbers of fainter stars ; if so, when stars of all degrees of luminosity are taken into account, the star-density in the globular clusters must be very great indeed in comparison with that in the galactic system.

Finally it may be added that only about 100 of these globular clusters are known to exist, practically all of which were known to the Herschels. As they appear to be all of similar dimensions and to lie within a definitely limited region of space it is unlikely that many more, if any, remain to be discovered, at any rate in those parts of space which are near to the sun.

It is not altogether clear to what extent the globular clusters and the moving clusters form distinct formations. In some respects the group of objects known as open clusters seem to form a connecting link between them†. Boss has remarked that if the Taurus cluster were to continue its present motion undisturbed for another 65 million years, it would then appear to us as an ordinary globular cluster 20′ in diameter. On the other hand, it must be remarked that the moving clusters move with ordinary stellar velocities, whereas the globular clusters have far higher than stellar velocities, and that the groups which are clearly established as moving clusters contain few stars, whereas the globular clusters contain hundreds of thousands. We shall return to the problem of the clusters in a later chapter.

NEBULAE.

28. Astronomical objects in general may be divided into two broad classes according as the telescope shews them as points of light or as areas of finite size. The quite small planets are so near as to shew finite discs in the telescope, whereas the far larger stars are so remote as to appear as mere

---

* *Monthly Notices of the R.A.S.* LXXVI. (1916), p. 567.

† Sigfrid Raab, "A Research on Open Clusters," *Lund Meddelanden*, II. No. 28 (1922).

PLATE IV

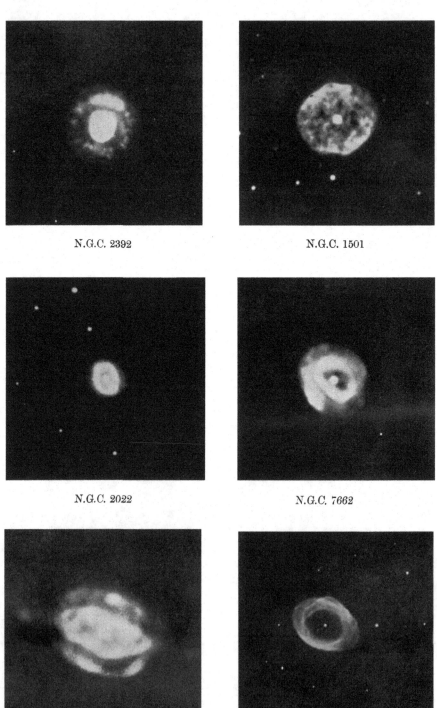

N.G.C. 2392

N.G.C. 1501

N.G.C. 2022

N.G.C. 7662

N.G.C. 7009

N.G.C. 6720

Mt Wilson Observatory

Planetary Nebulae

points of light. The greatest angle which the diameter of any star subtends at the earth is 0·056″, which is the angle subtended by a pin-head at a distance of two miles, and is far too small to shew a finite disc in any telescope which has yet been constructed.

Apart from the planets, all the objects we have so far discussed in detail have consisted of stars and groups of stars, forming the class of objects which shew as points of light in the telescope. There is a second class of objects which appear in the telescope as areas of finite size. These are generally and somewhat unfortunately bundled together under the general name of "nebulae."

Planetary Nebulae.

29. We may mention first the comparatively unimportant class of objects known as "planetary nebulae." The name is doubly unfortunate since the objects are not nebulae in any strict sense, and are not planetary in any sense at all, except that of shewing a disc of planetary size in the telescope.

Planetary nebulae are comparatively rare, only 150 out of 15,000 nebulae investigated by Campbell proving to be "planetary." In general they are of an apparently spheroidal or ellipsoidal shape, many shewing additional features and details of formation. Typical examples are shewn in Plate IV; the ring-shaped nebulae, such as N.G.C. 6720 (the "ring nebula" in Lyra) are probably ellipsoidal shells which are so transparent as only to be visible where the line of sight passes through a considerable thickness of the shell.

Many of the planetary nebulae are near enough for their distances to be estimated by the direct trigonometrical method. That of N.G.C. 7662, for instance, is found by Van Maanen[*] to be 0·023″, from which its diameter may be calculated to be 19 times that of Neptune's orbit. Van Maanen[†] and Newkirk[‡] agree in estimating that the ring nebula N.G.C. 6720 is considerably more distant and substantially larger in linear dimensions. Every planetary nebula shews a star at its geometrical centre, and when the distance of a nebula is known, the luminosity of its central star can at once be calculated. The luminosities of the central stars of the planetary nebulae prove to be well below that of the average star; few, for instance, are as intrinsically bright as our sun.

The planetary nebulae lie within the galactic system of stars, confirmation of this being found in the circumstance that they are found mainly in directions near the galactic plane. They appear to be rather ordinary stars which have in some way become surrounded by an atmosphere or shell of gas of such enormous dimensions as to exhibit finite discs in the telescope. Their velocities of motion are rather above those of average stars: Keeler found the average radial velocity of 13 to be 27·7 kms. a second, or 26·8 kms. a second after

---

[*] *Proc. Nat. Acad. Sci.* III. (1917), p. 133.   [†] *Publ. Astr. Soc. Pac.* XXIX. (1917), p. 209.
[‡] *Lick Obs. Bull.* IX. (1917), p. 100.

correction for the solar motion. Some few have quite exceptionally high velocities; for instance, Campbell has found that N.G.C. 6644 is receding with a velocity of 202 kms. a second, while N.G.C. 47322 is approaching at 141 kms. a second. The majority give spectroscopic evidence of rotation, and in some the rotation is not uniform, the outer layers rotating less rapidly than the inner layers.

Irregular Nebulae.

30. Plate V shews the central portion of the great nebula in Orion, the most striking example of the class of objects described as "irregular nebulae." These lie within the confines of the galactic system of stars, and generally shew stars which appear to be caught within them, or are more probably travelling through them, and illuminating the surrounding nebulosity. Many of these irregular nebulae are fairly near to the sun. The internal motions of the Orion nebula have been studied in great detail by Campbell and Moore* and reveal no high velocities; most parts of the nebulae shew motions of advance or recession of the order of 5 kms. a second, relative to the motion of the nebula as a whole.

The irregular nebulae shew the bright line spectrum which is characteristic of a transparent gas, and so are most probably wisps or clouds of comparatively stagnant gas which are rendered incandescent by the passage of light or other forms of radiation.

The dark streaks or lanes which appear in many of the irregular nebulae, as for instance in the Trifid nebula shewn in the lower half of Plate V, are almost certainly bands of absorbing matter. The alternative interpretation of them as rifts or gaps can hardly be the true one, since it is improbable that there should be so many tunnels through the nebulae all pointing directly towards the earth. The absorbing matter may possibly consist of molecules having absorption bands which cut off all visible light.

31. Closely related to these nebulae from the cosmogonic point of view are probably the "calcium clouds" discovered and investigated by Hartmann, Plaskett†, Otto Struve‡, and others. These appear to surround all stars of above a certain surface temperature; they are approximately stationary in space and are found in all parts of the sky. There is no reason to suppose that calcium especially predominates in the chemical constitution of these clouds. They shew the spectral lines of ionised calcium in their spectra, almost to the exclusion of all others, but this is probably a consequence of the fact that the atom of calcium is ionised with less energy than the atoms of other elements (cf. § 51 below).

---

* *Publications of the Lick Obs.* XIII. (1918), p. 96.

† *Publications of the Dominion Astrophys. Obs.* II. (1924), 335; *M.N.R.A.S.* LXXXIV. (1924), 80.

‡ *Astrophys. Journ.* LXV. (1927), p. 162. This paper contains a bibliography and general discussion of the problem.

PLATE V

Central Portion of Nebula in Orion (N.G.C. 1976)

Mt Wilson Observatory

Trifid Nebula in Sagittarius (*M* 20)

Irregular Nebulae

Extra Galactic Nebulae.

32. The remaining class of nebulae, by far the most important of all, comprises the huge nebulae of regular shape which lie far outside our galactic system of stars. We have already discussed the arrangement of these in space. As it will be convenient to discuss their forms and shapes hand-in-hand with theory, fuller discussion is deferred to a later chapter (Chap. XIII).

33. We have now described, although only in the very barest outline, the universe revealed by the telescope of the astronomer. The contemplation of this universe arouses a whole series of questions to which we are impelled to seek for an answer. What, in ultimate fact, are the stars? What causes them to shine, and for how long can they continue thus to shine? Why are binary and multiple stars such frequent objects in the sky, and how have they come into being? What is the significance of the characteristic flattened shape of the galactic system, and why do some of its stars move in clusters, like shoals of fish, while others pursue independent courses? What is the significance of the extra-galactic nebulae, which appear at a first glance to be other universes outside our own galactic universe comparable in size with it, although different in general quality? And behind all looms the fundamental question: What changes are taking place in this complex system of astronomical bodies, how did they start, and how will they end?

We shall discuss these questions approximately in the order in which they have been stated, devoting our next three chapters to examining the nature of the object which occurs most frequently of all in nature's astronomical museum, the simple star.

CHAPTER II

THE LIGHT FROM THE STARS

STELLAR MAGNITUDES AND LUMINOSITIES.

34. THE ancients thought of the stars as luminous points immovably attached to a spherical shell which covered in the flat earth much as a telescope-dome covers in the telescope, so that when one star differed from another in glory, it was not because the two stars were at different distances from us, but because one was intrinsically more luminous than the other.

Hipparchus introduced the conception of "magnitude" as measuring the brightnesses of the stars, and Ptolemy, in his Almagest, divided the stars into six groups of six different magnitudes. The 20 brightest stars formed the first magnitude stars, while stars which were only just visible to the eye were the sixth magnitude stars. Thus Ptolemy regarded the differences of visible glory as being represented by five steps, each step down being represented as an increase of one magnitude.

35. According to the well-known physiological law of Fechner, the effect which any cause produces on our senses is proportional to the logarithm of the cause. If we can just, and only just, appreciate the difference between 10 and 11, we shall not notice any difference at all between 20 and 21, but shall just be able to detect the difference between 20 and 22, or between 5 and 5½. Our senses do not supply us with a direct estimate of the intensity of the phenomenon which is affecting them, but of its logarithm. It is then not surprising to find that what Ptolemy regarded as equal differences of brightness were actually equal differences in the logarithms of the amounts of light received. Sir John Herschel remarked in 1830 that Ptolemy's "first magnitude" stars were just about 100 times as bright as his sixth magnitude stars, so that his five steps correspond to a difference of 2 in the logarithm of the amount of light received, and actually it is found that each one of his five intermediate steps corresponds very closely to a uniform difference of 0·4 in the logarithms of the light received, and so to a light ratio of $10^{0·4}$ or 2·512.

Accurate measurements of apparent brightness are now expressed on the scale introduced by Pogson* in 1856, on which each step of one magnitude represents a light ratio of exactly 2·512. It is, of course, necessary to admit fractional magnitudes, a tenth of a magnitude representing a light ratio of 1·10† (an easy number to remember), and a hundredth of a magnitude a light ratio of 1·0093.

---

* Catalogue of 53 known variable stars, Radcliffe Observatory, 1856.

† More exactly the number is 1·0965.

36. Thus, if R_1 and R_2 are the amounts of light received from any two stars, their magnitudes m_1, m_2 are connected by

$$m_1 - m_2 = -2\cdot5\,(\log R_1 - \log R_2)\dots\dots\dots\dots\dots(36\cdot1),$$

and as a conventional standard has been selected to fix what is meant by zero magnitude, this relation determines the magnitude of every star in the sky.

The star from which we receive most light is of course the sun, whose magnitude is $-26\cdot72$. Apart from it, the apparently brightest star is Sirius, of magnitude $-1\cdot57$, and then Canopus, far down in the southern sky, with magnitude $-0\cdot86$. The magnitudes of all other stars are expressed by positive numbers, Vega of magnitude $0\cdot14$ coming next, then Capella ($0\cdot21$), Arcturus ($0\cdot24$), and the bright component of α Centauri ($0\cdot33$). At the other extreme we may place the faintest stars which are accessible photographically in the 100-inch Mount Wilson telescope, of which the magnitude is about 21. The difference of about $47\frac{1}{2}$ magnitudes between these and our sun represents a light ratio of 10^{19}.

For comparison with these figures of stellar magnitudes, it may be added that the "magnitude" of the full moon is about $-12\cdot5$, and that of Venus at its brightest is about $-4\cdot0$. The magnitude of a standard candle 100 yards distant is also $-4\cdot0$, and that of a standard candle 6000 miles away is $21\cdot1$.

Absolute Magnitudes.

37. So long as the stars were deemed to be all at the same distance, the relative brightnesses of various stars gave appropriate measures not only of the amount of light we received from them, but also of the amounts of light they emitted; a first magnitude star not only sent two and a half times as much light to the earth as a second magnitude star, but also was supposed to emit two and a half times as much light in all—it might fairly be said to be two and a half times as luminous. We now know that the differences in the amounts of light received from different stars arises largely from their being at different distances, and before we can make any progress with our physical knowledge of the stars we must eliminate this distance effect; in other words, we must consider what would be the amounts of light we should receive from the various stars if they were all placed at the same distance away. The standard distance used for this purpose is generally 10 parsecs, although some continental writers prefer to use the Siriometer (§ 4).

38. Suppose that an amount R_1 of radiation is received from a star whose distance is p parsecs. If the same star were placed at a distance of 10 parsecs, the radiation received, R_2, would be given by

$$R_2 = R_1 \left(\frac{p}{10}\right)^2 \dots\dots\dots\dots\dots\dots\dots(38\cdot1),$$

since the intensity of light falls off inversely as the square of the distance.

If m denotes the actual observed magnitude of the star, and M the magnitude the same star would have if moved to a distance of 10 parsecs from us, then formula (36·1) provides the relation

$$M - m = -2\text{·}5 \log\left(\frac{R_2}{R_1}\right) = -2\text{·}5 \log\left(\frac{p}{10}\right)^2$$

$$= 5 - 5 \log p..$$

If ϖ is the parallax of the star in seconds of arc, $p = \dfrac{1}{\varpi}$ so that

$$M = m + 5 + 5 \log \varpi \quad\dotfill(38\text{·}2).$$

The magnitude M, which would be the observed magnitude of the star if it were moved to a distance of 10 parsecs* from us, is called the absolute magnitude of the star, while m is called its apparent magnitude.

The absolute magnitude M gives a measure of the total light emitted by a star, and again of course a drop of five magnitudes must represent a light ratio of 100. Of stars whose absolute magnitude is known with fair certainty, the most luminous is the bright companion of the binary star B.D. 6° 1309 recently investigated by Plaskett, which has an absolute magnitude of about − 6·4. The star S Doradus, in the greater Magellanic cloud, is almost certainly brighter, its absolute magnitude being estimated at − 9·0.

At the other end of the scale come Proxima Centauri, our nearest neighbour in the sky, with an absolute magnitude of 14·9, the companion to Procyon, with absolute magnitude about 16, and the faint star Wolf 359, whose absolute magnitude is 16·5.

Between the two extremes of absolute magnitude just mentioned, − 6·4 and 16·5, is a range of 22·9 magnitudes representing a light ratio of 1500 millions. The sun with an absolute magnitude of 4·9 comes not far from the middle of this range. Plaskett's star, B.D. 6° 1309, with an absolute magnitude of − 6·4, emits about 30,000 times as much light as the sun, while Wolf 359 emits only 0·00002 times as much light. It is usual to speak of the light emitted in terms of "luminosities," that of the sun being taken to be unity. Thus the luminosities of the two stars just discussed are 30,000 and 0·00002 respectively.

39. We can obtain some idea of the way in which the luminosities of the stars are distributed by considering the luminosities of our nearest neighbours. If we go far afield, our results at once become vitiated, because the fainter stars at great distances are unknown to us. But our previous discussion (§ 8) has made it probable that all fairly bright stars within a distance of 4 parsecs are known, so that if we limit ourselves to the fairly bright stars within a distance of 4 parsecs from the sun, this complication is not likely to enter

* If M_{sir} denotes the absolute magnitude when the Siriometer is taken as standard distance in place of 10 parsecs,
$$M_{sir} = M - 1\text{·}57.$$

to any great extent. The stars within a distance of 4 parsecs at present known are shewn in the following table:

Table IV. *The nearest Stars in order of Distance.*

	Star	Parallax	Visual Mag.	Absolute Mag.	Luminosity	Spectral Type	Mass (Sun as unity)
1	Sun	—	−26·72	4·85	1	G 0	1
2	Proxima Centauri	0·765	10·5	14·9	0·00010	M	—
	a Centauri *A*	0·758	0·33	4·73	1·12	G 0	1·14
	a Centauri *B*	0·758	1·70	6·10	0·32	K 5	0·97
3	Munich 15040	0·538	9·67	13·32	0·00041	M 5	—
4	Wolf 359	0·404	13·5	16·5	0·00002	M 6	—
5	Lalande 21185	0·392	7·60	10·57	0·0051	M 2	—
6	Sirius *A*	0·377	−1·58	1·30	26·3	A 0	2·45
	Sirius *B*	0·377	8·44	11·32	0·0026	A 7	0·85
7	B.D. −12°, 4523	0·350	10	12·7	0·00072	—	—
8	Innes (11 h. 12 m. −57·2)	0·340	12	14·7	0·00011	—	—
9	Cordoba 5 h. 243	0·317	9·2	11·7	0·0018	M 0	—
10	τ Ceti	0·315	3·6	6·1	0·32	K 0	—
11	Procyon *A*	0·312	0·5	3·0	5·5	F 5	1·24
	Procyon *B*	0·312	13·5	16·0	0·00003	—	0·39
12	ε Eridani	0·310	3·8	6·3	0·26	K 0	—
13	61 Cygni *A*	0·300	5·6	8·0	0·055	K 7	—
	61 Cygni *B*	0·300	6·3	8·7	0·029	K 8	—
14	Lacaille 9352	0·292	7·4	9·7	0·011	M 0	—
15	Struve 2398 *A*	0·287	8·7	11·0	0·0035	M 4	—
	Struve 2398 *B*	0·287	9·4	11·7	0·0018	M 4	—
16	Groombridge 34	0·282	8·1	10·4	0·0060	M 2	—
17	ε Indi	0·281	4·7	6·9	0·15	K 5	—
18	Kruger 60 *A*	0·256	9·3	11·3	0·0026	M 3	0·25
	Kruger 60 *B*	0·256	10·8	12·8	0·0007	M	0·20
19	0 h. 43·9 m. +4·55	0·255	12·3	14·3	0·00017	F	—
20	Lacaille 8760	0·253	6·6	8·6	0·032	M 0	—

Counting the components of binary and multiple systems as separate stars, we see that 27 stars are known to lie within a sphere of 4 parsecs radius drawn around the sun as centre. For almost all purposes these 27 stars form the best sample we can obtain of the stars as a whole, for as soon as we go further afield than 4 parsecs, certain types of stars are likely to have remained undiscovered, so that the discovered stars do not form a true sample of the whole.

On dividing the 27 individual stars into groups according to their absolute magnitudes, the numbers in the different groups are found to be as follows:

$$\text{Abs. Mag.} \quad 0 \longleftrightarrow 4 \longleftrightarrow 8 \longleftrightarrow 12 \longleftrightarrow 16.5$$
$$\text{No. of stars} \quad\ \ 2 \qquad 7 \qquad 10 \qquad 8$$

and we notice that stars whose luminosity is less than that corresponding to $M = 4$ are distributed fairly evenly amongst the different absolute magnitudes.

J

There appears to be a slight excess of stars of absolute magnitudes between 8 and 12, but our list is certainly deficient in stars fainter than twelfth magnitude (cf. fig. 1, p. 8), and if all stars were included the excess would probably be in the faintest stars of all. Our sun is well up in the list, being fourth out of twenty-three.

This distribution fails to represent that in the universe as a whole because the number of stars under discussion is so small as to contain no stars of very high luminosity. By counting the stars to far greater distances, Kapteyn and Seares have obtained results summarised in the following table:

Table V. *The Luminosity-Function (Kapteyn-Seares).*

Abs. Mag.	Luminosity	No. of stars per magnitude
−5	10,000	1
−2·5	1000	90
0	100	3,300
2·5	10	42,000
5	1	200,000
7·5	0·1	350,000
10	0·01	500,000
12·5	0·001	600,000

The last column gives the relative number of stars per unit absolute magnitude which have the absolute magnitude shewn in the first column, or the luminosity, in terms of the sun as unity, shewn in the second column. Between absolute magnitudes 4 and 12 the stars are again found to be fairly evenly distributed, although the excess of faint stars which we suspected in the smaller sample is now quite noticeable. But the table further shews the existence, in quite small numbers, of stars of very high luminosity, their numbers falling off steadily and very rapidly for absolute magnitudes less than about 5, i.e. for stars appreciably more luminous than the sun.

The last column of Table IV gives the masses of the stars when these are known from direct observation. The mass of the sun is calculated from the circumstance that its gravitational attraction just suffices to keep the earth in its orbit. In the same way the masses of the two components of a binary system can be calculated from the fact that the gravitational attraction of each just retains the other in its orbit. There is no other means of calculating the masses of stars by direct observation, so that no masses are entered except for the components of binary systems. We shall, however, find later that purely physical considerations make it possible to calculate the masses of single stars with fair accuracy from their luminosities and the temperatures of their surfaces.

STELLAR RADIATION.

40. Of even greater importance than the luminosity, or quantity of light emitted by a star, is the quality of this light, as revealed by analysis in a spectroscope.

Stefan found in 1879 that the amount of radiation emitted by a perfectly radiating surface of any kind is proportional to the fourth power of its absolute temperature T. The radiation per unit area of surface is usually taken to be σT^4 ergs per unit time, where σ is "Stefan's constant" whose value, according to the determinations of Coblentz[*] and Millikan[†], is

$$\sigma = 5 \cdot 72 \times 10^{-5} \text{ erg cm.}^{-2} \text{ degree}^{-4} \quad\ldots\ldots\ldots\ldots(40 \cdot 1).$$

In 1893 Wien brought forward a general thermodynamical argument which shewed that the law of partition by wave-length must be of the form

$$E_\lambda\, d\lambda = f(\lambda T)\, \lambda^{-5}\, d\lambda \quad\ldots\ldots\ldots\ldots\ldots\ldots(40 \cdot 2),$$

and in 1900 Planck discovered the form of the function $f(\lambda T)$, so that the complete law of radiation is now known in the form

$$E_\lambda\, d\lambda = \frac{2\pi h C^2}{e^{\frac{hC}{RT\lambda}} - 1}\, \lambda^{-5}\, d\lambda \quad\ldots\ldots\ldots\ldots(40 \cdot 3),$$

where C is the velocity of light, R is the universal gas-constant and h is another universal constant, "Planck's constant."

The values of these quantities are

$$\left.\begin{aligned} C &= 2 \cdot 998 \times 10^{10} \text{ cms. a second} \\ R &= 1 \cdot 372 \times 10^{-16} \\ h &= 6 \cdot 55 \times 10^{-27} \text{ erg seconds} \end{aligned}\right\} \quad\ldots\ldots\ldots\ldots(40 \cdot 4).$$

The point of primary significance in these formulae is that they are entirely independent of the nature of the matter which emits the radiation, the constants σ, h, R and C all being universal constants of nature. There is a simple physical reason for this. The radiation inside a hot body at a uniform temperature T is absorbed and re-emitted many times before it reaches the surface, so that the radiation which finally emerges is in thermodynamical equilibrium with the matter of the body. Its constitution must thus depend solely on the temperature T. Stefan's law may equally well be stated in the form that radiant energy in thermodynamical equilibrium with matter at temperature T is of amount aT^4 per unit volume, where a is an absolute constant of nature. This radiation, of course, travels equally in all directions. If it all travelled in the same direction, the amount crossing a unit area of surface in unit time would be aCT^4; when allowance is made for the different directions of travel, the amount is found to be $\frac{1}{4}aCT^4$. If this unit area forms part of the surface of a perfect radiator, this amount of energy $\frac{1}{4}aCT^4$ is equal

---

[*] *Phys. Rev.* vii. (1916), p. 694; *Sci. Papers of the Bureau of Standards*, Nos. 357 and 360 (1920).
[†] *Phil. Mag.* xxxiv. (1917), p. 16.

to the emission per unit area of the radiator, which we have taken to be σT^4. Thus the two constants a and σ of Stefan are connected by the relation

$$\sigma = \tfrac{1}{4} aC \quad \dots\dots\dots\dots\dots\dots\dots\dots\dots(40\cdot5).$$

The values of σ and C already given lead to the value

$$a = 7\cdot63 \times 10^{-15} \text{ in C.G.S. centigrade units} \dots\dots\dots(40\cdot6).$$

Effective Temperature.

41. When a star's distance is known, we can calculate its total emission of radiation from measurements of the amount of radiation received on earth. When the diameter of the star is also known, we can deduce its emission of radiation per square centimetre of surface. If this is put equal to σT_e^4, T_e is defined to be the "effective temperature" of the star. With this definition, a star emits as much radiation as a perfect radiator of equal surface raised to the temperature T_e.

According to Abbott and Fowle, the "solar constant" has the mean value 1·938, which means that outside the earth's atmosphere, every square centimetre directly facing the sun receives 1·938 calories of radiation a minute. The mean angular diameter of the sun is 32′0″, so that the earth's mean distance from the sun is equal to

$$\frac{1 \text{ radian}}{32'0''} = 107\cdot4$$

diameters, or 214·8 radii of the sun. Every square centimetre of the sun accordingly discharges $(214\cdot8)^2$ times as much radiation as falls on a square centimetre of the earth's atmosphere. Thus the rate of emission per square centimetre of the sun's surface is $1\cdot938 \times (214\cdot8)^2$ or 89,400 calories a minute. Since a calorie is equal to $4\cdot184 \times 10^7$ ergs, the radiation per square centimetre is $6\cdot24 \times 10^{10}$ ergs a second; each square centimetre of the sun's surface discharges enough energy to work an 8 horse-power engine.

If we put $6\cdot24 \times 10^{10} = \sigma T_e^4$, we find for the effective temperature of the sun's surface

$$T_e = 5750° \text{ absolute} = 5470° \text{ Centigrade.}$$

The foregoing calculation has shewn that to calculate a star's effective temperature, it is not necessary to know a star's radius and distance from the earth separately; it is enough to know the ratio of these quantities as expressed by half of the angular diameter of the star. The ratio of the radiation received by each square centimetre of the earth's surface to the radiation emitted by each square centimetre of the star's surface is the square of the ratio just mentioned, so that a knowledge of this ratio makes it possible to calculate σT_e^4 and hence T_e.

The calculation for an actual star is best performed by using the sun's radiation as a stepping-stone. Let the angular diameter of the star be $1/n$ times the angular diameter of the sun, and let the total radiation received from the star be $1/m$ times that received from the sun. Then each square centimetre of the star's surface must emit n^2/m times as much radiation as a square centimetre of the sun's surface, and hence

effective temperature of star $= n^{\frac{1}{2}} m^{-\frac{1}{4}} \times 5750$ degrees abs.

Of recent years the angular diameters of a number of stars have been measured by means of an interferometer attached to the 100-inch telescope at Mount Wilson, and the measurements so obtained have made it possible to calculate directly the effective temperatures of the stars in question (cf. §§ 54 and 55 below).

The Pressure of Radiation.

42. Maxwell shewed on theoretical grounds that radiation must carry momentum as well as energy with it, and it will subsequently be found (see Chap. x) that this transport of momentum by radiation plays a fundamental part in the dynamics of stellar structures. As a consequence of its carrying momentum, radiation must exert a pressure on any material surface it encounters. Maxwell proved that a beam of radiation, all travelling in the same direction, would exert a pressure in the direction of its motion equal to its energy per unit volume *. For radiation in thermodynamical equilibrium with matter at temperature T this is equal to aT^4. The existence of this pressure was confirmed experimentally by Lebedew† and by Nichols and Hull‡.

In the interior of a body which is at a uniform temperature T, radiation in thermodynamical equilibrium with the matter travels equally in all directions. After allowing for the different directions of travel, the pressure of radiation in any direction whatever is found to be p_R given by

$$p_R = \tfrac{1}{3} aT^4 \quad ....(42{\cdot}1).$$

Great caution is necessary in applying this formula to astronomical problems, since the interiors of astronomical bodies are not at uniform temperatures, with the result that stellar radiation is not generally in perfect thermodynamical equilibrium with matter (cf. § 73 below).

The Partition of Radiant Energy.

43. The partition of energy by wave-length which is expressed by formula (40·3) is exhibited graphically in fig. 3, in which λ is taken as abscissa and E_λ as ordinate. It might at first be thought that there would be a whole series of different curves corresponding to different values of the temperature T. But on drawing the curves it is found that differences of temperature are represented merely by differences in the horizontal and vertical scales on which the curve is drawn; this property follows from the circumstance that the law of partition of energy is of the general type given by formula (40·2).

* *Treatise on Electricity and Magnetism* (1873), § 792.

† *Annalen der Physik*, VI. (1901), p. 433.

‡ *Physical Review*, XIII. (1901), p. 307.

If, however, we agree on a definite scale for λ and E_λ, then different temperatures require different curves, and these lie as shewn in fig. 4. The four curves here shewn are drawn for the temperatures $T = 3000, 4000, 5000$ and 6000 degrees absolute. These curves, which at a first glance look very different from one another, are all derived from the curve of fig. 3 by expansion of its horizontal and vertical scales.

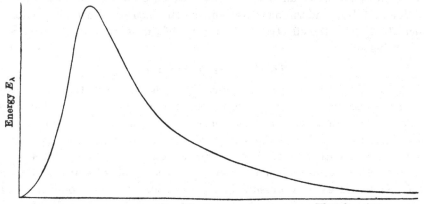

Fig. 3.   Wave-length λ

When the temperature T is specified, formula (40·3), which is represented graphically in fig. 3, gives E_λ as a function of λ. The wave-length for which E_λ is a maximum is of course the value of λ at the peak of the curve shewn in fig. 3; this is generally denoted by $\lambda_{\text{max.}}$. Formula (40·2) shews that, in general, $\lambda_{\text{max.}}$ must vary inversely as T, while by differentiation of formula (40·3) the actual relation is found to be

$$\lambda_{\text{max.}} T = 0\cdot2885 \text{ cm. degrees} \quad\ldots\ldots\ldots\ldots\ldots(43\cdot1).$$

The whole area of the curve shewn in fig. 3 represents the total radiation of energy at a given temperature distributed according to wave-length. The majority of this area is concentrated round the ordinate $\lambda = \lambda_{\text{max.}}$, so that to a rough approximation the radiation from a body at temperature T may be thought of as being all of the same wave-length $\lambda_{\text{max.}}$, determined by equation (43·1), and therefore as being all of the same colour. In fact, we can arrange a scheme in which the colours of radiation are regarded as corresponding to the temperature of the body by which it is emitted as follows:

Colour	Wave-length	Temperature
Reddest visible light	$\lambda = 7500$ Å.	$T = 3850$ degrees abs.
Yellow light.............................	6000	4800
Sunlight	5000	5750
Blue light...............................	4500	6400
Most violet visible light	3750	7700
Limit of atmospheric absorption ...	2950	9800

In this table, as also in fig. 4, wave-lengths are measured in the usual "Ångstrom" unit (Å) of 10^{-8} cms., while temperatures are measured in degrees absolute on the Centigrade scale.

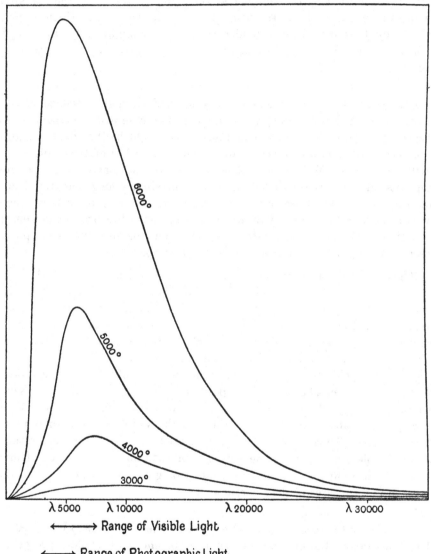

$\lambda\,5000$ $\lambda\,10000$ $\lambda\,20000$ $\lambda\,30000$

⟵⟶ Range of Visible Light

⟵⟶ Range of Photographic Light

⟵ Limit of Atmospheric Absorption

Fig. 4. Distribution of Energy in the Spectra of Perfect Radiators at different Temperatures.

The wave-lengths given for the limits of the visible spectrum have been obtained by slight extrapolation from observations recorded by Nutting[*].

* *Phil. Mag.* xxix. (1915), p. 301.

Nutting finds that light of given energy has maximum visibility when its wave-length is 5550 Å., this being approximately half-way between the two limits of visibility. The wave-length given for sunlight, 5000 Å., has reference to light as emitted by the sun, not to the light received from the sun after partial absorption by the earth's atmosphere. This latter light, which we may call daylight, is considerably redder than the "sunlight" of wave-length 5000 Å., its wave-length being much nearer to the wave-length of maximum visibility, 5550 Å.

44. Some stars appear blue when examined through a telescope; our Table might at first be thought to suggest that these had a temperature of about 6400. Others appear red, and our Table might be thought to suggest that these had a temperature of about 3800. Neither of these inferences would be justified. When the light from a blue star is analysed in a spectroscope, its energy curve does not come to a maximum height in the blue; this height goes on increasing with decreasing wave-length, until it comes to an end somewhat abruptly at about $\lambda = 2950$—not because the star emits no light of shorter wave-length than this, but because our atmosphere absorbs such light before it reaches the spectroscope.

Thus all that the blue colour of a star entitles us to say is that the temperature of such stars is certainly higher than 9800°. There is a similar situation with regard to the red stars, the limit at the red end of the spectrum arising from the circumstance that our eyes cannot perceive radiation of wave-length greater than about $\lambda = 7500$. A better estimate of the temperatures of such stars, and indeed of all stars, is obtained by examining the *shape* of the whole energy curve, or rather of that part of it which is accessible to observation, instead of merely the position of the ordinate which represents maximum energy. The temperatures of the blue stars are in this way found to range up to about 30,000°, while those of the visible red stars may range down to 2500°. Thus there is a range of at least 12 to 1 in the temperatures of the visible stars, and, as the emission of radiation varies as the fourth power of the temperature, this represents a range of over 20,000 to 1 in the emission per unit area of their surfaces.

45. Within the spectral limits already mentioned (about 2950 Å. to 7500 Å.), the curves giving the observed distribution of energy in stellar spectra are found to have shapes which agree tolerably well with the theoretical curves discussed in § 43. This makes it possible to determine the effective temperatures of stars from the shape of their observed energy-curves.

This method has frequently been used to determine the effective temperature of the sun, the temperature so obtained usually being somewhat higher than the temperature of 5750 determined in § 41 from the sun's total emission of radiation. In 1921 Coblentz* determined the sun's effective temperature by this method as 6140° absolute.

The same method has been used to determine the effective temperatures of other stars, by Wilsing, Scheiner, Münch, Rosenberg, Plaskett, and others. In 1923 Sampson† calculated the effective temperatures of 17 stars from the shapes of their energy curves, and in 1925 increased the number to 64. In 1924‡ Abbott analysed the spectra of nine stars by the use of a radiometer attached to the 100-inch telescope at Mount Wilson. Some of the results thus obtained are shewn in Table VI below:

Table VI. *Effective Temperatures of Stars* (*Sampson-Abbott*).

Star	Spectral Type	Effective Temperature	
		Sampson	Abbott
γ Cassiop	*B* 0	30,000	—
γ Pegasi	*B* 2	25,000	—
β Tauri	*B* 8	30,000	—
β Orionis (Rigel)	*B* 8	14,800	16,000
α Lyrae (Vega)	*A* 0	11,600	14,000
α Canis Maj. (Sirius)	*A* 0	12,800	11,000
α Cygni (Deneb)	*A* 2	10,900	—
α Cephei	*A* 5	8900	—
β Cassiop	*F* 4	7800	—
α Canis Min. (Procyon)	*F* 5	8300	8000
α Ursae Min. (Polaris)	*F* 8	6500	—
Sun	*G* 0	6140*	6000
γ Cygni	*G* 0	6200	—
α Aurigae (Capella)	*G* 0	5500	5800
ε Cygni	*G* 7	4600	—
α Bootis (Arcturus)	*K* 0	4200	—
ε Pegasi	*K* 0	3600	—
α Tauri (Aldebaran)	*K* 5	3400	3000
α Orionis (Betelgeux)	*M* 0	3400	2600
β Pegasi	*M* 3	3200	2850
α Herculis	*M* 6	—	2500

* Coblentz's estimate.

MAGNITUDE SCALES.

46. We have already noticed that the eye is only sensitive to light of wave-length lying between the limits 7500 Å. and 3750 Å., this being the range marked "Range of visible light" in Fig. 4. Similarly the photographic plate only records starlight of wave-lengths lying within a range from about 6000 Å. to 2950 Å., the limit of atmospheric absorption, this being the range marked "Range of photographic light" in Fig. 4.

* *Sci. Papers of the Bureau of Standards,* 438 (1922), or *Journ. of the Optical Soc. of America,* v. (1921), p. 272.

† *Monthly Notices of the R.A.S.* LXXXIII. (1923), p. 174, and LXXXV. (1925), p. 212.

‡ *Astrophys. Journ.* LX. (1924), p. 87, or Mount Wilson Contribution, No. 280.

Thus neither visual observation nor photography records the total light coming from a star. Speaking in very general terms, we may say that photographic estimates are concerned only with the amount of blue light coming from a star, while visual estimates are concerned only with the amount of yellow light. Actually the matter is far more complicated than this statement would suggest, for the relative sensitiveness of the eye to light of different colours varies with the intensity of the light; when the light is very faint, blue light produces an abnormally large effect on the eye, whereas when the light is very strong, blue light is comparatively ineffective. This is known as the Purkinje effect; it explains why blue objects are more easily seen than red ones in a faint light, why moonlight makes the whole world look blue, and why artists paint shadows blue regardless of actual colours.

Apart from detail, however, we may say that visual estimates of stellar magnitudes are unduly favourable to stars of reddish or yellow colour, while photographic estimates are unduly favourable to stars of bluish colour.

To avoid this complication, two distinct scales of stellar magnitude are in use, the visual and the photographic. The visual scale deliberately estimates only the light visible to the normal eye, absolutely disregarding all light which is too red or too blue to be seen, while similarly the photographic scale only estimates the light which is recorded on the photographic plate. We have already explained how visual magnitudes are measured, each magnitude representing a light-ratio of 2·512, and the zero-point being fixed by reference to standard stars. On the photographic scale also, each magnitude is made to represent a light-ratio of 2·512, while the zero-point is fixed by the arbitrary convention that the photographic and visual scales shall agree, on the average, for stars of spectral type $A\,0$ (cf. § 52 below), and of visual magnitudes between 5·5 and 6·5.

Colour-Index.

The difference between the visual and photographic magnitudes of a star is called the "colour-index" of the star. Its sign is determined by the equation

Colour-index = (Photographic Magnitude) − (Visual Mag.).

Since magnitudes are measures of the logarithms of the light received, it is clear that the colour-index measures approximately the ratio of the amount of blue light received from a star to the amount of yellow light; it accordingly depends solely on the effective temperature of the star.

Still a third scale of stellar magnitudes must be mentioned, namely, the bolometric scale, which measures the total radiation emitted by a star, as shewn by the total area of the energy curve. Great success has recently attended direct measurements of the total radiant energy received from the stars, the instrument generally used being a thermocouple in a vacuum, a type of instrument introduced by Coblentz. When Coblentz first used such an instrument at Lick in 1914, it was only possible to measure the radiation from stars brighter than the fourth magnitude. With the 100-inch telescope at Mount Wilson, Pettit and Nicholson are now able to measure the total radiation received from stars down to the thirteenth magnitude, this being about equal to the radiation received from a candle 2000 miles away.

Even when it is not possible to obtain direct observational evidence of those parts of the energy curve which lie outside the visual range, our general knowledge of the theory of radiation assures us that they must be there, and those parts of the curve which are accessible to observation enable us to complete the whole. The bolometric scale of magnitude gives a measure of the whole energy emitted by the star, as represented by its completed energy curve.

Bolometric Correction.

Just as the difference between the photographic and visual scales is represented by the colour-index, so the difference between the bolometric and visual scales is represented by a quantity called the "bolometric correction," its sign being determined by the equation

Bolometric correction = (Visual Magnitude) − (Bolometric Mag).

Roughly speaking, the bolometric correction measures the ratio of a star's total radiation to the visual radiation received after absorption by the atmosphere. Thus the bolometric correction, like the colour-index, depends only on the effective temperature. Calculation and observation both shew that the bolometric correction has its minimum value at a temperature of about 6800 degrees. The zero-point of the bolometric scale is fixed by agreeing that the bolometric correction at this temperature shall be zero. The effect of this is to make the bolometric correction always a positive quantity.

Theoretical Evaluation of Colour-Index.

47. Following a method first suggested by Hertzsprung*, both the colour-index and the bolometric correction can be calculated theoretically in terms of the effective temperature T_e.

By Planck's formula (40·3) the partition of energy in the radiation emitted per square centimetre by a surface of effective temperature T_e is

$$E_\lambda d\lambda = \frac{2\pi h C^2}{e^{\frac{hC}{RT_e\lambda}} - 1} \lambda^{-5} d\lambda \quad \ldots\ldots\ldots\ldots\ldots\ldots(47\cdot1).$$

Hence the total intensity of radiation between wave-lengths λ_1 and λ_2 emitted by a star of radius r is E, given by

$$E = 4\pi r^2 \int_{\lambda_1}^{\lambda_2} \frac{2\pi h C^2}{e^{\frac{hC}{RT_e\lambda}} - 1} \lambda^{-5} d\lambda \quad \ldots\ldots\ldots\ldots\ldots(47\cdot2).$$

---

* *Zeitschr. für Wissenschaftliche Photographie,* IV. (1906), p. 43.

If we integrate from about $\lambda_1 = 3750$ Å. to $\lambda_2 = 7500$ Å., we obtain a formula for the emission of radiation which is visible to the eye; let us call this visual radiation. Similarly if we integrate from about $\lambda_1 = 2950$ Å. to about $\lambda_2 = 5000$ Å., we obtain a formula for what we may call the photographic radiation, and finally if we integrate from $\lambda_1 = 0$ to $\lambda_2 = \infty$, we obtain a formula for the bolometric radiation.

In both visual and photographic radiation from bodies at stellar temperatures, a simple calculation shews that $e^{\frac{hC}{RT_e\lambda}}$ is large compared with unity, so that formula (47·2) may be replaced by

$$E = 4\pi r^2 \int_{\lambda_1}^{\lambda_2} 2\pi h C^2 e^{\frac{-hC}{RT_e\lambda}} \lambda^{-5} d\lambda \quad \dots\dots\dots\dots\dots(47\text{·}3).$$

Moreover, since the ranges of integration are comparatively small, we may, to an approximation, suppose the radiation all concentrated in one single wave-length λ_0, selected so as to be near the middle of the range. The formula now becomes

$$E = 4\pi r^2 \left[2\pi h C^2 e^{\frac{-hC}{RT_e\lambda_0}} \lambda_0^{-5} (\lambda_2 - \lambda_1) \right] \dots\dots\dots\dots(47\text{·}4).$$

Since a drop of unity in absolute magnitude represents a light-ratio of $(10)^{0\cdot4}$, the absolute magnitude M on any scale is connected by the emission E on the same scale by a formula of the type

$$0\text{·}4M = -\log E + \text{cons.} \quad \dots\dots\dots\dots\dots(47\text{·}5).$$

Inserting the value of E given by equation (47·4), we obtain, after multiplication by 2·5,

$$M = -5 \log r + 2\text{·}5 \log_{10} e^{\frac{hC}{RT_e\lambda_0}} + \text{cons.},$$

or, introducing the numerical values $\log_{10} e = 0\text{·}4343$ and $hC/R = 1\text{·}432$,

$$M = -5 \log r + \frac{1\text{·}555}{\lambda_0 T_e} + \text{cons.}\dots\dots\dots\dots\dots(47\text{·}6).$$

According to Brill, visual radiation may be supposed to centre round wave-length 5290 Å. On putting $\lambda_0 = 5290$ Å. in equation (47·6) we obtain as the absolute visual magnitude of a star of radius r,

$$M_{\text{vis.}} = -5 \log r + \frac{29{,}500}{T_e} + \text{cons.} \quad \dots\dots\dots\dots(47\text{·}7).$$

Similarly photographic radiation centres round a wave-length 4250 Å., and on putting $\lambda_0 = 4250$ Å. in formula (47·6) we obtain

$$M_{\text{phot.}} = -5 \log r + \frac{36{,}700}{T_e} + \text{cons.} \quad \dots\dots\dots\dots(47\text{·}8).$$

The definition of colour-index now supplies the relation

$$\text{Colour-index} = M_{\text{phot.}} - M_{\text{vis.}} = \frac{7200}{T_e} + \text{cons.}$$

The constant can be determined from the agreed convention that the colour-index shall vanish for stars of spectral type A 0. As their effective temperature is about 11,200°, the constant on the right must be -0.64, and the equation becomes

$$\text{Colour-index} = \frac{7200}{T_e} - 0.64 \quad \dots\dots\dots\dots(47.9).$$

This formula enables us to calculate theoretically the colour-index corresponding to any effective temperature. Thus we have

Effective Temperature	30,000	25,000	20,000	15,000	10,000	6,000	4,000	3,000	2,000
Colour-index	-0.40	-0.35	-0.28	-0.16	$+0.08$	$+0.56$	$+1.16$	1.76	2.96

Either the formula or the table will give the colour-index of a star in terms of its effective temperature. What is of especial importance for physical astronomy is that by using the table backwards, we can deduce a star's effective temperature from its colour-index. The colour-index of a star is of course readily determined observationally; we need only estimate the magnitude of the star twice, once visually and once photographically.

Theoretical Evaluation of Bolometric Correction.

48. Slightly different methods are needed for the discussion of the bolometric magnitude scale, since bolometric radiation cannot be treated as centering round one mean wave-length, but extends throughout the spectrum.

On integrating formula (47.3) from $\lambda = 0$ to $\lambda = \infty$, we merely obtain, as the total bolometric emission of radiation by a star of radius r,

$$E = 4\pi r^2 \sigma\, T_e^4 \quad \dots\dots\dots\dots\dots(48.1),$$

where σ is Stefan's constant introduced in § 40. Using this value for E in equation (47.5) we obtain

$$M_{\text{bol.}} = -5 \log r - 10 \log T_e + \text{cons.} \quad \dots\dots\dots(48.2).$$

This and equation (47.7) now give the relation

$$\text{Bolometric correction} = M_{\text{vis.}} - M_{\text{bol.}} = \frac{29,500}{T_e} + 10 \log T_e + \text{cons.}$$

The quantity on the right attains its minimum value when $T_e = 6790$, and on determining the constant so as to make the bolometric correction vanish at this temperature, we obtain the complete value of the bolometric correction in the form

$$\text{Bolometric correction} = \frac{29,500}{T_e} + 10 \log T_e - 42.53 \quad \dots\dots(48.3).$$

The foregoing are the forms of equation adopted by Russell, Dugan and Stewart[*]. On the other hand, Hertzsprung[†] gave the following formula for $M_{\text{vis.}}$ in 1906:

$$M_{\text{vis.}} = 2 \cdot 3 \left(\frac{14,300}{T_e}\right)^{0 \cdot 93} - 5 \log r + \text{cons.} \quad \ldots\ldots\ldots(48 \cdot 4),$$

and from this and equation (48·2), we obtain

$$\text{Bolometric correction} = 2 \cdot 3 \left(\frac{14,300}{T_e}\right)^{0 \cdot 93} + 10 \log T_e + \text{cons.} \ldots(48 \cdot 5).$$

The bolometric correction, as given by this formula, is found to be a minimum when $T_e = 6800$. Adjusting the constant on the right of formula (48·5) to make the bolometric correction vanish when it is a minimum, the formula becomes

$$\text{Bolometric correction} = 2 \cdot 3 \left(\frac{14,300}{T_e}\right)^{0 \cdot 93} + 10 \log T_e - 42 \cdot 9.$$

Eleven years later Eddington[‡] calculated the values of the bolometric correction which were required by the purely physical observations of Nutting[§] on the visibility of light of various wave-lengths. As was subsequently remarked by Seares[||], the empirical table constructed by Eddington in this way agreed almost precisely with a table calculated from Hertzsprung's formula (48·5). The quality of the agreement is shewn in the first two lines of the following table, the values given by formula (48·3) being shewn in the third line.

T_e	2540	3000	3600	4500	6000	7500	9000	10500	12000
Δm (Hertzsprung)	2·62	1·67	0·94	0·35	0·02	0·02	0·16	0·36	0·58
Δm (Nutting-Eddington)	2·59	1·71	0·95	0·35	0·00	0·02	0·12	0·31	0·53
Δm (Formula (48·3))	3·00	1·94	1·28	0·42	0·04	0·02	0·16	0·36	0·59

Each set of values indicates that Δm is a minimum at about 6800 degrees, and the zero-point in each has been arranged so that the bolometric correction is zero at this temperature. As a consequence a star's bolometric magnitude is always less than its visual magnitude, or again a star is always more luminous bolometrically than visually. At temperatures below 6800° the bolometric excess is of course mainly due to the heat and infra-red radiation which does not affect the eye at all, while at the other extreme of high temperatures the excess is mainly due to the ultra-violet radiation cut off by atmospheric absorption.

The bolometric correction assumes its greatest importance in the case of stars whose effective temperature is so low that they are almost invisible.

---

[*] *Astronomy*, p. 736.   [†] *Zeitschr. für Wissenschaft. Photog.* IV. (1906), p. 43.
[‡] *Monthly Notices of the R.A.S.* LXXVII. (1917), p. 605.   [§] *Phil. Mag.* 29 (1915), p. 301.
[||] *Astrophys. Journ.* LV. (1922), p. 197.

An extreme instance is provided by the variable star χ Cygni. Visually this varies from the 5th to the 14th magnitude, so that its visual luminosity is 4000 times greater at maximum than at minimum. Yet the bolometric observations of Pettit and Nicholson * at Mount Wilson shew that the total radiation varies by a factor of only 1·7. Thus of the apparent visual change of 9 magnitudes, no less than 8·4 mags. must be ascribed to the bolometric correction. The radiation of this star at minimum consists almost entirely of invisible heat—in fact, it sends us some 50,000 times as much heat as Sirius would if the two stars were placed at such distances as to appear to be of equal brightness.

As with the formula for colour-index, one of the main points of interest about the formula for the bolometric correction is that it can be used to determine a star's effective temperature when its bolometric magnitude has been determined observationally by radiometric measurements. In this way Nicholson and Pettit have determined the temperatures of a number of stars, especially long period variables such as χ Cygni just mentioned. The bolometric corrections for these stars are generally found to be exceedingly great, averaging 4·3 magnitudes at maximum, and 7·8 at minimum, from which Nicholson and Pettit deduce temperatures of 2300° and 1650° respectively.

The Calculation of Stellar Radii.

49. Before leaving the group of formulae we have just been discussing, we may notice that the group of formulae (47·7), (47·8) and (48·2) provide a means of determining the radius of a star when either its visual, photographic or bolometric absolute magnitudes are known.

In each formula the final constant remains to be determined, and this is conveniently done by taking the sun as a standard case. The sun's apparent visual magnitude is $- 26\cdot72$. The standard distance with reference to which absolute magnitudes are reckoned is 10 parsecs, or 2,062,648 mean radii of the earth's orbit, this being of course the number of seconds of arc in 10 radians. It follows that the sun's absolute visual magnitude is

$$- 26\cdot72 + 5 \log (2{,}062{,}648) = 4\cdot85.$$

The sun's effective temperature being close to 6000, the bolometric correction is negligible, and we may take the sun's bolometric absolute magnitude also to be 4·85. The colour-index at 6000° is about 0·56, so that the sun's absolute photographic magnitude is about 5·41.

Using these values, taking the sun's effective temperature to be 6000° and the sun's radius to be unity, we find that the radius R of any star in terms of the sun's radius as unity is given by any one of the formulae

$$\log R = - 0\cdot2 \, M_{\text{bol.}} - 2 \log T_e + 8\cdot53 \dots\dots\dots\dots(49\cdot1),$$

* *Report of the Director of Mount Wilson Observatory* (1924), p. 101.

$$\log R = - \, 0\!\cdot\!2 \, M_{\text{vis.}} + \frac{5880}{T_e} - 0\!\cdot\!01 \dots\dots\dots\dots(49\!\cdot\!2),$$

$$\log R = - \, 0\!\cdot\!2 \, M_{\text{phot.}} + \frac{7320}{T_e} - 0\!\cdot\!14 \dots\dots\dots(49\!\cdot\!3).$$

We may also determine the constant in equation (47·5), namely,

$$\log E = - \, 0\!\cdot\!4 \, M + \text{cons.},$$

by reference to the sun as standard. As regards bolometric magnitude, we put $M = 4\!\cdot\!85$, the value just found, and $E = 3\!\cdot\!80 \times 10^{33}$, the total bolometric radiation of the sun in ergs per second. The constant is then found to be 35·52, so that the total radiation of energy $E_{\text{bol.}}$ of any star is given by

$$\log E_{\text{bol.}} = - \, 0\!\cdot\!4 \, M_{\text{bol.}} + 35\!\cdot\!52 \dots\dots\dots\dots(49\!\cdot\!4).$$

STELLAR SPECTRA.

50. From the foregoing theory it might reasonably be expected that every star should exhibit a continuous spectrum with the distribution of energy shewn in fig. 3. To an exceedingly rough approximation this is indeed found to be the case, but the spectrum is invariably complicated by the presence of dark lines, and sometimes of bright lines as well (see the typical spectra shewn in Plate VI*). Modern physical theory explains very clearly how the former at least of these complications originates.

Bohr's theory of atomic structure, which is now generally accepted by physicists as providing at least a useful working model of atomic processes, supposes the atom to be capable of existing only in certain definite states, each of which has a specific amount of energy associated with it. Whenever an atom passes from one to another of these states, radiation is emitted or absorbed. The frequency ν of this radiation is related to the energies E_1, E_2 of the two states by Einstein's equation

$$E_1 - E_2 = h\nu \dots\dots\dots\dots\dots\dots\dots(50\!\cdot\!1),$$

where h is Planck's constant, and E_1 refers to the state of higher energy; if this is the original state of the atom, the change of state is accompanied by an emission of radiation—in the reverse case by absorption of radiation.

In a mass of matter which is at uniform temperature throughout, the atoms distribute themselves amongst the various possible states in such a way as to give maximum entropy. If 1, 2 are states of higher and lower energy respectively, a certain number of atoms in state 2 are at any instant absorbing energy and jumping up to state 1. To maintain the entropy at its constant maximum value, an exactly equal number must be falling from state 1 to state 2 and emitting energy in the process, so that statistically there is no change on balance.

* These are reproduced from photographs by Curtiss and Rufus at the University of Michigan.

PLATE VI

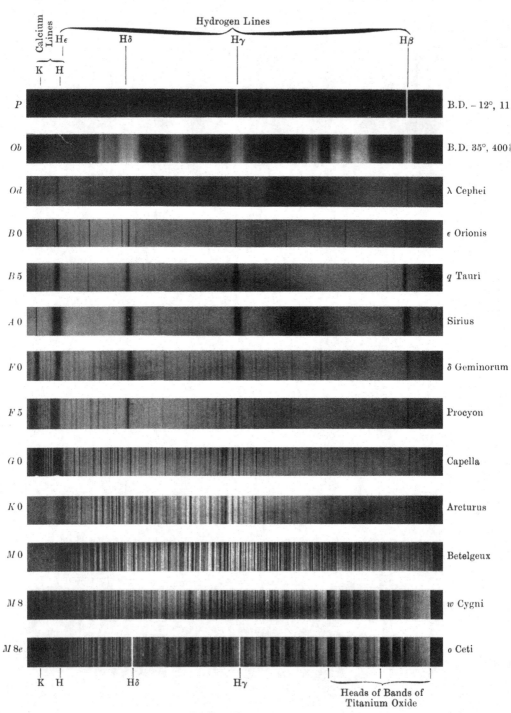

Stellar Spectra

A star, however, is not a mass of matter at uniform temperature. Heat continually flows from its interior to its surface. Since heat always flows in the direction in which the temperature is falling, just as water flows in a direction in which the height is falling, a star's temperature falls as the surface is approached. Energy which has been emitted in some internal layer of high temperature, and has the constitution appropriate to that high temperature, passes to outer space through surface-layers of lower temperature. These layers absorb radiation which is of a wave-length suitable for absorption— i.e. radiation of those particular frequencies which represent jumps of the atoms from one possible state to another—but, on account of their lower temperatures, they do not emit an equal amount of radiation of the same wave-lengths. There is now no longer a statistical balance between the absorption and emission of the radiation of these particular wave-lengths, but definite deficiencies in the radiation of the wave-lengths in question, and, when the radiation reaches outer space, these deficiencies reveal themselves by dark lines crossing the spectrum at the places corresponding to these special wave-lengths. These dark lines are called "absorption-lines." The outer layer we have just described, the seat of the absorption by which these lines are produced, is called the "reversing-layer" of the star; the inner layer which produces the light which undergoes this absorption is called the "photosphere."

We should accordingly expect the spectrum of a star to consist of the continuous bright spectrum already discussed, modified by the presence of absorption-lines. Such absorption-lines were first observed in the spectrum of the sun by Fraunhofer in 1814.

The "flash spectrum" which appears at a total eclipse of the sun provides proof that the above explanation of the dark lines is the true one. After the photosphere has been covered by the advancing moon, the only light which reaches the earth is that emitted by the reversing layer, and the spectrum of this light is found to consist of bright lines on a dark background, these lines having precisely the same wave-lengths as the Fraunhofer dark lines. This proves that the atoms of the reversing layer are themselves emitting radiation of the wave-lengths in question; the Fraunhofer lines appear relatively dark merely because the reversing layer emits less radiation of these wave-lengths than it extracts from the light of the photosphere.

Absorption-Lines.

51. Since Fraunhofer first observed absorption-lines in the spectrum of the sun, it has been found that not only the spectrum of the sun, but practically all stellar spectra are of the general type suggested by theory, consisting, that is to say, of a continuous bright background with absorption-lines superposed.

These absorption-lines are among the most informative objects in the whole of astronomy. In accordance with Doppler's principle, a difference of velocity v between a source emitting light and an instrument receiving the light, causes the wave-length λ' of the light received, to be different from the wave-length λ of the emitted light, the two being connected by the relation $\lambda' = \lambda \left(1 + \dfrac{v}{C}\right)$, where C is the velocity of light. Consequently the motion of a star causes the lines in its spectrum to shew a slight displacement from their standard positions, and by measuring the amount of this displacement it is possible to determine the star's velocity of recession or approach, relative to the moving earth. We have already noticed how W. S. Adams and Kohlschütter found that certain peculiarities in these lines made it possible to determine the absolute magnitude of the star by which they are emitted, and hence its distance. Further the Einstein theory of relativity requires that each spectral line should shew a displacement towards the red end of amount proportional to M/r, the gravitational potential at the surface of the emitting star, so that if the displacement can be measured in the light emitted by a particular star, the value of M/r is at once known. This method has recently been used to determine the value of M/r for the companion of Sirius. The mass M of this star was already known, from its gravitational pull on Sirius, to be about 0·85 times the mass of the sun, so that it became possible to calculate the star's radius r. This proved to be only about 20,000 kms., shewing that the mean density of the star must be about 50,000 times that of water.

Finally, a comparison of the positions of the absorption bands of a stellar spectrum with those obtained from known chemical substances in the laboratory, makes it possible to identify the atoms or molecules which absorb the light in the star's surface. Practically all of the lines in the solar spectrum have been identified in this way, and are found to originate from the atoms of elements known on earth; a large number of the lines originating from the upper layers of the sun's atmosphere have their origin in ionised atoms, particularly those of calcium, strontium and barium.

Table VII (opposite)* gives the 90 chemical elements which are known on earth, arranged in order of their atomic number, with an indication of whether they are represented (P) or absent (A) in the spectrum of the sun.

Saha† and, more recently, R. H. Fowler and Milne‡, have shewn how this identification of absorption lines in stellar spectra makes it possible to estimate the temperature of the absorbing atoms, and hence the effective

* Compiled from Russell, Dugan and Stewart, *Astronomy*, p. 503; Miss Payne, *Stellar Atmospheres*, pp. 5 and 184; *Publicat. Ast. Soc. Pac.* 39 (1927), p. 238; *Astrophys. Journ.* LXVIII. (1928), p. 327, etc.

† *Phil. Mag.* XL. (1920), pp. 472, 809. *Proc. R.S.* 99A (1921), p. 135. *Zeit. f. Phys.* VI. (1921), p. 40.

‡ *M.N., R.A.S.* LXXXIII. (1923), p. 403.

temperatures of the stars. For instance, the solar spectrum and most stellar spectra (see Plate VI) shew two very conspicuous lines, the H and K lines of

Table VII. *The Chemical Elements in the Solar Spectrum*

Atomic Number	Element	Present or Absent	Atomic Number	Element	Present or Absent
1	Hydrogen	P	47	Silver	P
2	Helium	P	48	Cadmium	P
3	Lithium	P	49	Indium	P
4	Beryllium	P	50	Tin	P
5	Boron	P	51	Antimony	P
6	Carbon	P	52	Tellurium	A
7	Nitrogen	P	53	Iodine	A
8	Oxygen	P	54	Xenon	A
9	Fluorine	A	55	Caesium	A
10	Neon	A	56	Barium	P
11	Sodium	P	57	Lanthanum	P
12	Magnesium	P	58	Cerium	P
13	Aluminium	P	59	Praesodymium	P
14	Silicon	P	60	Neodymium	P
15	Phosphorus	A	61	Illinium	A
16	Sulphur	A	62	Samarium	P
17	Chlorine	A	63	Europium	P
18	Argon	A	64	Gadolinium	A
19	Potassium	P	65	Terbium	A
20	Calcium	P	66	Dysprosium	A
21	Scandium	P	67	Holmium	A
22	Titanium	P	68	Erbium	A
23	Vanadium	P	69	Thulium	A
24	Chromium	P	70	Ytterbium	A
25	Manganese	P	71	Lutecium	A
26	Iron	P	72	Hafnium	A
27	Cobalt	P	73	Tantalum	A
28	Nickel	P	74	Tungsten	A
29	Copper	P	75	Rhenium	A
30	Zinc	P	76	Osmium	A
31	Gallium	P	77	Iridium	A
32	Germanium	P	78	Platinum	A
33	Arsenic	A	79	Gold	A
34	Selenium	A	80	Mercury	A
35	Bromine	A	81	Thallium	P
36	Krypton	A	82	Lead	P
37	Rubidium	P	83	Bismuth	A
38	Strontium	P	84	Polonium	A
39	Yttrium	P	85	—	—
40	Zirconium	P	86	Radon	A
41	Niobium	P	87	—	—
42	Molybdenum	P	88	Radium	A
43	Masurium	A	89	Actinium	A
44	Ruthenium	P	90	Thorium	A
45	Rhodium	P	91	Protoactinium	A
46	Palladium	P	92	Uranium	A

Fraunhofer, which a comparison with laboratory spectra assigns to ionised calcium. Calculation shews that ionised calcium is only active in absorbing radiation within a limited range of temperature which extends approximately from 3000° to 15,000°; according to Fowler and Milne the maximum activity occurs at 6290°. Other lines in stellar spectra provide the means of fixing other ranges of temperature. From a combination of all the evidence, it is possible to fix the temperature of the sun's "reversing layer" at about 6000°, and to estimate the temperatures of other stars.

Spectral Type.

52. The circumstance that the H and K lines of calcium were specially prominent in the sun was at one time interpreted to mean that the sun was in some way especially rich in calcium. That three different stars shewed prominently the lines of, say, hydrogen, calcium, and titanium oxide, was supposed to indicate that the composition of a star changed as it aged, so that the atoms of which it was made changed, by transmutation of the chemical elements, from hydrogen into calcium, from calcium into titanium oxide, and so on. But it is now clear that the star's spectrum reveals its surface-temperature rather than its chemical composition. If the sun's temperature were suddenly doubled, without any change in its chemical composition taking place, the calcium lines would almost disappear from its spectrum and hydrogen lines would take their place; if the sun's temperature were halved, the calcium lines would again disappear and the solar spectrum would be dominated by the bands of titanium oxide.

Two stars whose outer layers were at the same temperature and in the same physical state would shew different spectra if they contained different chemical elements, or even if they contained the same chemical elements in different proportions. A few cases are known in which the lines of certain elements appear in abnormal strength in the spectra of particular stars but apart from these, it appears to be a general rule that the elements appear in the same relative proportions in all the stars [*]. As a consequence stellar spectra fall approximately into a single continuous sequence, different positions in the sequence merely representing different temperatures of the star's surface.

It is not yet convenient to express a star's spectrum as a temperature, although possibly this means of identification may come into use in time. At present a star's spectrum is described by reference to arbitrarily selected standard points in the continuous series of spectra. As regards the spectra of normal stars, six selected points are designated by the letters B, A, F, G, K, M, in this order; the statement that a star's spectrum is $B\,3$ means that it is three-tenths of the way on from B to A. A star whose spectrum is exactly at B is generally said to be of type $B\,0$; a star whose spectrum is described merely as being of type B may be anywhere from $B\,0$ to $B\,9$. Type O comes in front of type B, its various sub-classes being temporarily designated as Oa, Ob, Oc, Od and Oe, Oe itself being further divided with ten sub-divisions. Typical spectra of the various classes are shewn in Plate VI facing p. 49.

---

[*] Miss Payne, *Stellar Atmospheres*, chap. XIII.

The characteristics of the various types of spectra are briefly indicated in the following table:

Type	Elements shewn	Typical stars		Probable effective Temp.
O	Ionised helium. Doubly and trebly ionised oxygen and nitrogen	*Oa* *Od* *Oe* 5 *Oe* 8	$BD+35°$ 4013 λ Cephei λ Orionis Plaskett's star	About 30,000 (?) 28,000
B	Hydrogen and Helium (strong) ionised silicon, oxygen, nitrogen, magnesium and calcium	*B* 0 *B* 1 *B* 8	ε Orionis *V* Puppis Rigel	23,000 22,000 15,000
A	Hydrogen (strong), ionised and neutral metals (weak)	*A* 0 *A* 2 *A* 5	Sirius Deneb Altair	11,200 10,900 8,600
F	Ionised calcium (very strong), hydrogen and metallic lines	*F* 0 *F* 5	Canopus Procyon	7,500 8,000
G	Neutral metals and ionised calcium, hydrogen (weak)	*G* 0	{ Capella *A* { Sun	5,650 6,000
K	Neutral metals and ionised calcium (very strong), hydrogen (very feeble). At *K* 5 titanium oxide bands begin to appear	*K* 0 *K* 5 *K* 7	Arcturus Aldebaran 61 Cygni	4,200 3,300 4,000
M	Titanium oxide (strong). Continuous spectrum very weak at violet end	*M* 0 *M* 3 *M* 6	Betelgeux Kruger 60 α Herculis	3,000 3,200 2,500

The table shews that the sequence of spectral types can be regarded not only as one of decreasing temperature, but also as one of decreasing ionisation or increasing aggregation. In class *O* trebly and doubly ionised atoms are prominent; later we come to singly ionised atoms, then to complete atoms and finally to complete molecules exhibiting band-spectra. The complete molecular structures are built up as the temperature diminishes.

53. One reason why a star's spectrum cannot conveniently be specified simply as a temperature, is that the spectrum of a mass of hot gas does not depend solely on its temperature; it depends to an appreciable degree on its density as well. The theoretical reason for this is provided at once by the work of Saha, R. H. Fowler and Milne, to which reference has already been made. We shall discuss the theory of ionisation more fully below (§ 137) when it will be seen that the ionisation temperature for the atoms of any particular element depends very largely on the density of the matter of which the atoms form part. Thus stars of the same spectral class have different effective temperatures when their atmospheres have different densities, the differentiation being most marked when the effective temperature is low. As a consequence, the colour-index and bolometric correction do not depend solely on a star's spectral class. As regards colour-index, the deviation is shewn in the following table given by Seares[*].

| Type | Colour-index | | T(dense) $-$ T(diffuse) |
	dense	diffuse	
B0	-0.32		
B5	-0.17		
A0	0.00		
F0	0.38		
G0	0.72	0.86	470
G5	0.83	1.15	890
K0	0.99	1.48	1020
K5	1.26	1.84	850
M0	1.76	1.88	250

The temperature difference in the last column is somewhat provisional, being based on somewhat uncertain data, but it suffices to indicate the extent to which the spectra of stars of types G and K depend on the densities as well as on the effective temperatures of the stars.

ESTIMATES OF EFFECTIVE TEMPERATURES.

54. The effective temperature T_e of a star has been defined to be such that the radiation emitted per second from each square centimetre of the star's surface is σT_e^4, this being the amount of radiation that would be emitted from a perfect radiator at temperature T_e.

We have found four ways of estimating the effective temperature of a star:

(1) From a comparison of a star's observed angular diameter with the total radiation received from it (§ 41).

(2) From the shapes of the energy curves obtained by analysing the distribution of energy in the star's spectrum (§ 45).

(3) From the colour-index, which gives, in a general way, the ratio of photographic to visual light emitted by a star (§§ 47, 53).

(4) From the bolometric correction, which gives a measure of the ratio of a star's total radiation to its emission of visual light (§ 48).

That these four methods give fairly accordant values for the effective temperatures of stars of different spectral types, is shewn in the following table which is given by Russell, Dugan and Stewart[†].

[*] *Astrophys. Journ.* LV. (1922), p. 198.
[†] *Astronomy*, p. 753.

Table VIII. *Effective Temperatures (Russell, Dugan and Stewart).*

Spectral Type	Effective Temperature deduced from				Adopted effective Temperature
	Colour-index	Bolometric Radiation	Energy Curves	Angular Diameters	
B 0	23,000	—	—	—	23,000
B 5	15,000	—	—	—	15,000
A 0	11,200	—	12,500	10,000	11,000
A 5	8,600	—	—	8,000*	8,600
F 0	7,400	—	—	—	7,400
F 5	6,500	—	8,000	—	6,500
Giant *G* 0	5,500	5,500	5,800	—	5,600
G 5	4,700	—	—	—	4,700
K 0	4,100	4,200	—	4,300	4,200
K 5	3,300	—	3,000	3,800	3,400
M 2	3,050	3,100	2,700	3,100	3,100
M 7	—	2,750	2,500	2,650	2,700
Long period variables:					
max.	—	2,300	—	2,400	2,300
min.	—	1,650	—	1,700	1,650
Dwarf *G* 0	6,000	5,750†	6,000†	—	6,000
G 5	5,600	—	—	—	5,600
K 0	5,100	—	—	—	5,100
K 5	4,400	—	—	—	4,400
M	3,400	—	—	—	3,400

* Only one star is used, the companion of Sirius, and its spectral type is uncertain.
† Values for sun (§§ 41, 45).

RADIATION AND PHYSICAL CONSTANTS OF VARIOUS STARS.

55. We shall now consider the radiation and physical constants of various stars as given by the foregoing formulae.

The Sun. For convenience the various data already given for the sun may be collected together.

Sun's diameter = 864,000 miles = $1 \cdot 391 \times 10^{11}$ cms.

„ surface = $6 \cdot 08 \times 10^{22}$ sq. cms.

„ volume = $1 \cdot 412 \times 10^{33}$ cu. cms.

„ mass = $2 \cdot 00 \times 10^{33}$ grammes.

„ mean density = $1 \cdot 416$ grammes per cu. cm.

„ emission of energy = $3 \cdot 80 \times 10^{33}$ ergs per second.

„ emission of energy per gramme = $1 \cdot 90$ ergs per second.

„ emission of energy per sq. cm. of surface = $6 \cdot 24 \times 10^{10}$ ergs per second.

„ apparent visual mag. = $- 26 \cdot 72$.

„ absolute visual mag. = $4 \cdot 85$.

„ absolute bolometric mag. = $4 \cdot 85$.

„ absolute photographic mag. = $5 \cdot 4$.

Sun's effective temperature

(total radiation) = 5750 absolute

(energy curves) = 6000 absolute.

Betelgeux (α Orionis). This is the twelfth brightest star in the sky, its apparent visual magnitude being 0·92. Its spectral type is M 0. From observations on the energy curve Sampson and Abbott estimate effective temperatures of 3400 and 2600 respectively; Coblentz, from direct bolometric measurements, estimates an effective temperature of from 2800 to 3300. The angular diameter of the star as determined by interferometer measurements at Mount Wilson, is 0·047″, which would represent an effective temperature of exactly 3000. Thus all methods agree in assigning to the star an effective temperature of about 3000°.

The probable parallax of the star is 0·017″. Combining this with its angular diameter of 0·047″, the diameter of the star must be $\frac{47}{17}$ times the radius of the earth's orbit, or about 250 million miles. With this parallax, the star's luminosity must be 1320 times that of the sun, its bolometric absolute magnitude being − 4·6 and its total emission of radiation about 6000 times that of the sun. The star is a variable, but shews only slight changes of luminosity.

Antares (α Scorpii), the sixteenth brightest star in the sky, is in many respects similar to Betelgeux. Its spectral type is M 0, its apparent visual magnitude being 1·2. The angular diameter, as measured with the interferometer, is 0·040″, which corresponds to an effective temperature of about 3020 degrees. Its trigonometrically measured parallax* of 0·026″ gives the star a diameter of 140 million miles, absolute magnitude − 1·7 and luminosity about 440; the corresponding bolometric magnitude is − 3·4 and total emission of radiation about 2000 times that of the sun.

Seares, Russell and others disregard its measured parallax, and, treating it as a member of the Scorpius cluster, assign it the parallax 0·0085″ of the cluster. This gives it an absolute magnitude of − 4·2, and an absolute bolometric magnitude of − 5·9, with an emission of radiation of about 20,000 times that of the sun.

Arcturus (α Bootis) is the third brightest star in the northern sky, and the fifth brightest in the whole sky. Its spectral type is K 0 and its apparent visual magnitude 0·2. Coblentz has determined its effective temperature by thermocouple measurements as from 3500 to 4500; Sampson from energy curves gives 4200. Its angular diameter, as measured with the interferometer, is 0·022″, again indicating an effective temperature of 4200.

Assuming an effective temperature of 4200, its parallax of 0·080″ gives it a diameter of 26 million miles, an absolute bolometric magnitude of − 0·8, and a total emission of radiation equal to 180 times that of the sun.

---

* *General Catalogue of Parallaxes*, Schlesinger (1924).

Capella (α Aurigae), the second brightest star in the northern sky, is a binary system which has the rare distinction of being both a visual and a spectroscopic binary. Its parallax is 0·063″ and its period 104 days.

The brighter component is of visual magnitude 0·8, its spectral type being *G* 0, the same as that of the sun. From the shape of its energy curves Sampson and Abbott estimate its effective temperature as 5500 and 5800 respectively. Its absolute magnitude, both visual and bolometric, is − 0·2, its luminosity and emission of radiation being about 105 times that of the sun. With an effective temperature of 5650, its surface must be 120 times that of the sun, and its diameter 11 times that of the sun, or say 9½ million miles. Its mass is 4·18 times the mass of the sun, whence its mean density must be 0·004.

The fainter component is of spectral type *F* 0, representing an effective temperature of about 7400. Its visual magnitude is 1·1, so that its absolute magnitude, both visual and bolometric, is 0·1. Its radius is about 5½ times that of the sun. Its mass being 3·32 times that of the sun, its mean density is 0·028.

B.D. 6° 1309 (Plaskett's star)*. This is a spectroscopic binary with a period of 14·414 days, which is the most massive and the absolutely brightest star whose elements are known with fair certainty†. The minimum values possible for the masses are 75·6 and 63·2 times that of the sun. These are not computed masses, but values of $M \sin^3 i$, where i is the inclination of the orbit, so that the actual masses are probably considerably greater. The spectral type is *O* 8, which on the theoretical scale of Fowler and Milne, would represent a temperature of about 28,000 degrees.

H.D. 1337 (Pearce's star)‡. This is a spectroscopic and eclipsing variable of period 3·5234 days, which has the distinction of being the most massive and the absolutely brightest star whose mass and luminosity are known with precision. The masses of the two components are found to be 36·3 and 33·8 times the mass of the sun, with a probable error of about 5 per cent. The visual absolute magnitudes are − 5·95 and − 4·84, the bolometric absolute magnitudes being estimated at − 8·82 and − 7·71, so that the total emission of radiation by the two components are respectively 294,000 and 73,000 times that of the sun. The spectroscopic type is *O* 8½, which may again be supposed to indicate an effective temperature of 28,000 degrees.

An analysis of the light curve shews that the star consists of two ellipsoidal components which are almost in contact, their semi-major axes being 23·8

---

* *Dominion Astrophys. Observ. Publications*, ii. (1922), No. 4.

† Otto Struve (*Astrophys. Journ.* lxv. (1927), p. 273, and lxviii. (1928), p. 109) finds that the quadruple system 27 Canis Majoris has probably at least 950 times the mass of the sun. He discusses other possible interpretations of the observations, but can find no satisfactory explanation which would give a total mass substantially less than that just mentioned.

‡ *Dominion Astrophys. Observ. Publications*, iii. (1926), No. 13.

and 15·5 radii of the sun, and the distance between their centres being 40·1 radii of the sun, so that their closest points are only 0·8 radii of the sun apart.

V Puppis. This is a spectroscopic and eclipsing binary whose elements are well determined. Both components have spectral type B 1, and Plaskett assigns an effective temperature 22,000 degrees to both. The masses are 19·2 and 17·9 times the mass of the sun, and the absolute bolometric magnitudes are − 5·26 and − 5·05, so that the emission of radiation of the two components are 11,000 and 9100 times that of the sun respectively.

Sirius (α Canis Majoris), the brightest star in the sky, is a visual binary with a period of 49·3 years.

Its brighter component is of spectral type A 0; Sampson and Abbott have determined effective temperatures of 12,800 and 11,000 respectively, while its colour-index indicates an effective temperature of 11,200. Assuming an effective temperature of 11,200, its absolute bolometric magnitude is 0·9. Its diameter is then 1·58 times that of the sun, its mass being 2·45 times that of the sun, so that its mean density is 0·9.

The faint companion is of absolute visual magnitude 11·3. Moore finds its spectral type to be A 5, or possibly A 3 or A 4. Its mass is about 0·85 times the mass of the sun. The ratio of mass to radius, as determined by the Einstein shift of spectral lines, fixes its radius as 0·030 times that of the sun; hence its mean density must be about 50,000. Its bolometric absolute magnitude must be approximately the same as its absolute visual magnitude, and its total emission of radiation must be 0·0028 times that of the sun. Combining this with its known size, we can deduce an effective temperature of 8000°. This corresponds to a spectral type of about A 7, in agreement with Adams' determination.

Kruger 60*. This is a visual binary of period 54·9 years. Its two components are the least massive of all stars whose masses are known with fair accuracy.

The brighter component has a mass equal to 0·25 times that of the sun. Its spectral type M 3 corresponds to an effective temperature of about 3200. This gives it an absolute bolometric magnitude of 10·0, so that its emission of radiation is 0·009 times that of the sun. Its radius must accordingly be one-third of that of the sun, and its mean density about 10.

The fainter component has a mass equal to 0·20 times that of the sun, its spectral type and effective temperature being approximately the same as those of its companion. Its absolute bolometric magnitude is 11·5, so that it has an emission of radiation equal to 0·002 times that of the sun, and a radius equal to one-sixth that of the sun. Its mean density is accordingly about 60.

The foregoing data, together with some others, estimated and calculated by similar methods, are collected in the following table:

* Aitken, *Lick Observ. Bull.* 365 (1925).

Physical Data

Table IX. Physical Data for Selected Stars*.

Star	Spectral Type	Visual Mag.	Parallax	Absolute vis. mag.	Effective Temp.	Absolute bol. mag.	Radius (Sun=1)	Mass (Sun=1)	Mean density	Energy radiated (ergs per gm.)
Betelgeux	M0	0·92	0·017	[−2·9]	3,000	−4·6	290	[40]	[0·000002]	[300]
Antares	M0	1·2	[0·026]	[−1·7]	3,000	[−3·4]	[160]	—	—	—
Arcturus	K0	0·2	0·080	−0·3	4,200	−0·8	30	4·18	0·004	48
Capella A	G0	0·8	0·063	−0·2	5,650	−0·2	11	3·32	0·028	41
,, B	F0	1·1	0·063	0·1	7,400	0·1	5·5	—	—	—
B.D. 6° 1309 A	O8	6·71	0·00035	−6·43	28,000	−9·31	—	75·6+	—	11,000 –
,, B	O8	—	0·00035	—	28,000	—	—	63·3+	—	—
H.D. 1337 A	O8½	6·56	0·00032	−5·94	28,000	−8·82	23·8	36·3	0·004	15,000
,, B	O8½	7·67	0·00032	−4·83	28,000	−7·71	15·5	33·8	0·013	6,000
V Puppis A	B1	—	—	—	22,000	−5·26	7·6	19·2	0·06	1,100
,, B	B1	—	—	—	22,000	−5·05	6·8	17·9	0·07	1,000
β Aurigae A	A0	2·8	0·034	0·6	11,200	0·2	2·80	2·38	0·13	57
Sirius A	A0	−1·6	0·377	1·3	11,200	0·9	1·58	2·45	0·93	29
,, B	A5	8·4	0·377	11·3	8,000	11·2	0·03	0·85	50,000	0·007
Procyon A	F5	0·5	0·312	3·0	7,000	3·0	1·80	1·24	0·28	10
,, B	—	13·5	0·312	16·0	—	[15]	—	0·39	—	[0·0005]
Sun	G0	−26·72	206265	4·85	6,000	4·85	1·00	1·00	1·42	1·90
α Centauri A	G0	0·3	0·758	4·7	6,000	4·7	1·07	1·14	1·34	1·91
,, B	K5	1·7	0·758	6·1	4,400	5·7	1·22	0·97	0·76	0·90
o₂ Eridani A	G5	4·5	0·203	6·0	5,600	5·9	0·70	0·9	3·7	0·8
,, B	A0	9·7	0·203	11·2	11,200	10·8	0·018	0·44	98,000	0·002
,, C	M6	10·8	0·203	12·3	2,800	10·3	—	0·20	—	0·07
Kruger 60 A	M3	9·3	0·256	11·3	3,200	10·0	0·33	0·25	9·6	0·068
,, 60 B	M	10·8	0·256	12·8	3,200	11·5	0·17	0·20	60	0·021
Proxima Centauri	M	10·5	0·765	14·9	3,000	13·2	0·07	—	—	—
Van Maanen's Star	F	12·3	0·255	14·3	7,000	14·3	0·009	—	—	—

* Entries in square brackets are uncertain or partly conjectural.

THE TEMPERATURE-LUMINOSITY DIAGRAM.

56. The foregoing table has given the absolute bolometric magnitudes and the effective temperatures of certain stars. The former of these two quantities serves to specify the quantity of radiation emitted by a star, while the latter specifies its quality, at any rate in respect of its more essential features.

We can conveniently exhibit the quantity and quality of light emitted by a number of stars by representing each star as a point in a plane diagram. We shall find it convenient to take the stars's absolute bolometric magnitude M as ordinate, and the logarithm of its effective temperature, $\log T_e$, as abscissa. Further, we shall place the stars which radiate most energetically at the top of the diagram, and those of lowest effective temperature to the right. This is a modification of a procedure introduced by Russell in 1913. Russell took absolute visual magnitude as ordinate and spectral type as abscissa, but the quantities we have selected are far more fundamental, and consequently far more intimately connected with the physical condition of the star.

We have seen (§ 49) that a star's emission of radiation E in ergs per second is given by

$$\log E = 35\cdot52 - 0\cdot4M.$$

It is also given by

$$E = 4\pi r^2 \sigma T_e{}^4,$$

where r is the radius of the star, σ is Stefan's constant, and T_e is the star's effective temperature. From these two equations we obtain

$$\log r = 19\cdot33 - 2\log T_e - 0\cdot2M \quad\ldots\ldots\ldots\ldots\ldots(56\cdot1),$$

which gives a star's radius in terms of T_e and M. If R is the radius measured in terms of the radius of the sun ($6\cdot95 \times 10^{10}$ cms.), the equation takes the form

$$\log R = 8\cdot49 - 2\log T_e - 0\cdot2M \quad\ldots\ldots\ldots\ldots\ldots(56\cdot2).$$

We see that stars of a specified radius lie on a straight slant line in a diagram in which M and $\log T_e$ are taken as ordinate and abscissa.

Such a diagram is shewn in fig. 5. The slant lines represent the lines of constant radii equal to $0\cdot01$, $0\cdot1$, 1, 10, 100 and 1000 times the radius of the sun.

The points surrounded by small circles represent the stars within 4 parsecs of the sun, as given in Table IV. Three of these are omitted owing to their spectral type and effective temperature being insufficiently known. The remaining points, marked by small crosses, represent stars which appear in Table IX, but not in Table IV. These stars are of exceptional interest but are in no sense typical of the whole mass of stars. The best sample of the stars as a whole is formed by the stars within 4 parsecs of the sun, and so by the stars enclosed in circles in fig. 5.

Giants and Dwarf Stars.

57. In 1905 Hertzsprung* discovered that the red stars fell into two distinct classes, one of which he called Giants on account of their great size,

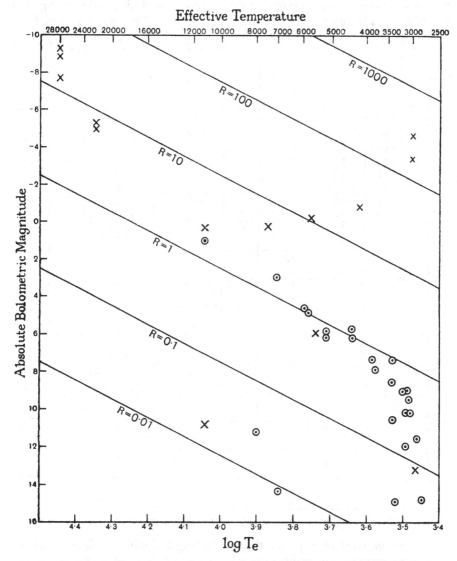

Fig. 5. Effective Temperatures, Absolute Bolometric Magnitudes and Radii of Stars.

⊙—Stars within four parsecs of Sun.
×—More distant stars.

and the other Dwarfs on account of their small size. This division is clearly shewn in our diagram, there being no *M*-type stars of magnitudes between

---

* *Zeitsch. für Wissenschaftliche Photog.* III. p. 442. See also v. p. 86, and *Ast. Nach.*, No. 4296.

$-3\cdot4$ (Antares) and $7\cdot4$ (Lacaille 8760). Similarly there are no K-type stars between $-0\cdot8$ (Arcturus) and $+5\cdot7$ (τ Ceti and α Centauri B).

In 1914 H. N. Russell* confirmed Hertzsprung's conclusion, using more extensive material, and again, in 1917, Adams and Joy†, discussing the spectroscopically determined absolute magnitude of 500 stars, reaffirmed the

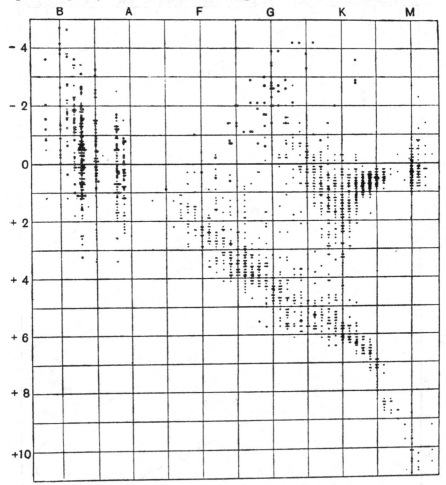

Fig. 6. Absolute Magnitudes and Spectral Type of 2100 stars (Mount Wilson Observatory).

conclusions of Hertzsprung and Russell, finding a distinct division to occur in stars of spectral type redder than K 4. Fig. 6‡ shews the absolute visual magnitudes of 2100 stars plotted against their spectral types, and the division again shews with remarkable clearness.

* *Nature*, April 30, 1914.

† *Astrophys. Journ.* XLVI. (1917), p. 313.

‡ This diagram was prepared by Mr Hubble and appears in the *Carnegie Institution Year-book*, No. 20 (1921), p. 270.

A general discussion by Adams, Joy and Humason[*], in 1925 shews that the absolute magnitude of the faintest of the giant M-type stars is $+0.7$, while that of the brightest of the dwarfs is 6.9, leaving a clear gap of 6.2 magnitudes. For stars of type $M5$ or later this interval increases to about 11.0 magnitudes.

Thus the red stars fall into two distinct groups—giants, such as Betelgeux and Antares, which emit a large amount of radiation and so require a very large surface to discharge it, and dwarfs such as the two components of Kruger 60, which emit but little radiation and so need only small surfaces.

The Main Sequence.

We notice that the majority of stars in figures 5 and 6 lie on a narrow slant-band which runs across the diagram. This band is commonly called the "Main Sequence." Various attempts have been made to fix its exact position. By comparing the magnitude-differences of the two components of binary systems with their differences of spectral type, H. O. Redman[†] has recently obtained the following determination of the central line of the main sequence:

Spectral type M_{vis}	$B0$	$B5$	$A0$	$A5$	$F0$	$F5$	$G0$	$G5$	$K0$	$K5$	$M0$	$M5$
	-0.62	0.20	1.30	2.52	2.80	3.32	4.21	5.09	5.17	6.79	7.64	11.80

This method has the great advantage of being practically unaffected by errors in the determination of stellar distance. Redman further finds that in the region $F0$ to $K0$ the spread of the main sequence is about two-thirds of a magnitude on either side of the central line.

The White Dwarfs.

When a number of stars are entered in a temperature-luminosity diagram of the type shewn in fig. 5, practically all are found to lie either on the main sequence or to the right of it. Fig. 5 shews three stars occupying isolated positions to the left of the main sequence.

Stars of this type are described as "white dwarfs." Only four are known with certainty, namely, the three shewn in the diagram, and the companion to o Ceti, to which very possibly Procyon B ought to be added. These stars are, however, of very low luminosity, so that white dwarfs are likely to escape detection except when they are near to the sun. Table IV (p. 33) shews that there are no fewer than three white dwarfs (counting Procyon B) within 4 parsecs of the sun, and this suggests that the white dwarfs may be common objects in space, although not in star-tables.

[*] *Report of the Director of Mount Wilson Observatory*, 1925-6, p. 128.
[†] *M.N., R.A.S.* LXXXVIII. (1928), p. 722.

CHAPTER III

GASEOUS STARS

General Configurations of Equilibrium.

58. THE preceding chapter has exhibited the stars as a series of bodies shewing very wide ranges of size and density and considerable ranges of mass and surface-temperature. Much labour has been expended in building up a physical theory of the structure of the stars. The first attempts treated the stars as spheres of gas, in which the pressure of the gas resisted the tendency of the star to collapse under its own gravitational attraction. And as the study of the internal constitution of the stars is still best approached through this conception, we shall examine it in some detail before turning to more realistic, but also more complex, theories of stellar structure.

We start, then, with the consideration of a sphere of gas which is at rest throughout, the pressure of the gas exactly balancing gravitational forces at every point. If p is the pressure, ρ the density and g the force of gravity at a point distant r from the centre of the star, the condition that gas-pressure and gravitational force shall exactly balance is expressed by the usual equation of hydrostatic equilibrium

$$\frac{dp}{dr} = -g\rho \quad\dotfill(58\cdot1).$$

The value of g, the force of gravity, is

$$g = -\frac{\gamma}{r^2}\int_0^r 4\pi\rho r^2\, dr \quad\dotfill(58\cdot2),$$

where γ is the gravitation constant ($\gamma = 6\cdot66 \times 10^{-8}$).

Substituting this value for g into (58·1), we readily obtain

$$\frac{1}{r^2}\frac{d}{dr}\left(\frac{r^2}{\rho}\frac{dp}{dr}\right) + 4\pi\gamma\rho = 0 \quad\dotfill(58\cdot3).$$

This differential equation enables us to find p if ρ is given, or ρ if p is given, but does not determine the arrangement of both p and ρ of itself; for this some further relation or datum is necessary.

Lane's Law.

59. If we replace r by αr, the new co-ordinate αr is suited for the study of the original mass of gas after its linear dimensions have been uniformly stretched α-fold. This stretching changes the original density ρ into $\alpha^{-3}\rho$. Now if we replace r and ρ by αr and $\alpha^{-3}\rho$, and at the same time replace p by $\alpha^{-4}p$, equation (58·3) becomes

$$\frac{1}{(\alpha r)^2}\frac{d}{d(\alpha r)}\left[\frac{(\alpha r)^2}{\alpha^{-3}\rho}\frac{d(\alpha^{-4}p)}{d(\alpha r)}\right] + 4\pi\gamma(\alpha^{-3}\rho) = 0 \quad\dots\dots(59\cdot1),$$

which is exactly identical with equation (58·3), the factors in α cancelling throughout. Thus if equation (58·3) is satisfied, so also is equation (59·1). It follows that if a mass of gas, initially in equilibrium, is stretched uniformly to α times its original linear dimensions, and the pressure is at the same time changed to $p\alpha^{-4}$, the new mass of gas so obtained will be in equilibrium.

In a gas in which Boyle's law is obeyed, the pressure is given by

$$p = \frac{R}{m\mu}\rho T \ \dots\dots\dots\dots\dots\dots\dots\dots (59·2),$$

where R/m is the gas-constant ($8\cdot 26 \times 10^7$) for a gas of molecular weight unity and μ is the molecular weight of the gas. If p is to be changed to $\alpha^{-4}p$, while ρ is changed to $\alpha^{-3}\rho$, then T must be changed to $\alpha^{-1}T'$ in order that equation (59·2) may still be satisfied.

Thus if p, ρ, T, r are the values of the pressure, density, temperature and distance from the centre in a mass of gas in equilibrium, these values can be changed to p', ρ', T', r' given by

$$p' = \alpha^{-4}p, \ \ \rho' = \alpha^{-3}\rho, \ \ T' = \alpha^{-1}T, \ \ r' = \alpha r \ \ \dots\dots\dots (59·3),$$

and the mass of gas will still be in equilibrium. This was pointed out by Homer Lane in 1870*, and is commonly known as *Lane's law*.

Homologous Configurations.

60. Two configurations (p, ρ, T, r) and (p', ρ', T', r') of a spherical mass of gas, which are related by equations (59·3) are called *homologous* configurations.

As we give all possible values to α ranging from $\alpha = 0$ to $\alpha = \infty$ in these equations, we pass through an infinite number of homologous configurations. This series of configurations may be spoken of as a *homologous series*. The equations shew that throughout the length of a homologous series

$$rT \text{ is constant,}$$
$$T/\rho^3 \text{ is constant,}$$
$$p/T^4 \text{ is constant,}$$

for each element of matter forming the star.

At one end ($\alpha = 0$) of a homologous series, the radius of the mass of gas is zero and its temperature infinite. At the other end ($\alpha = \infty$), the radius is infinite and the temperature zero. In general there will be an infinite range of energy between the two ends, but it is not altogether a simple matter to determine which end corresponds to high energy and which to low.

If we imagine a star to move a slight distance along a series of homologous configurations, so that α changes to $\alpha + d\alpha$, the radius increases, so that work must be put into the star to expand it against its own gravitational attraction, each element of gas being moved against the pressure of the neighbouring elements. On the other hand, the gas cools so that heat is yielded up by the

* *Amer. Journ. of Science*, LIII. (1870), p. 57.

mass of the star. The amount of heat dQ which must be added to the gaseous star to cause it to move from the configuration α to $\alpha + d\alpha$ is readily found to be given by

$$dQ = \frac{d\alpha}{\alpha} \iiint [3p' - \rho' C_v' T'] \, dv' \quad\ldots\ldots\ldots\ldots\ldots(60\cdot1),$$

where C_v' is the specific heat at constant volume and the integration is throughout the whole mass of the star. The question of whether the internal energy of the star increases or decreases as we proceed along a homologous series, turns on the sign of the integral in the above equation.

The average energy of a molecule of the gas is

$$\tfrac{3}{2}(1+\beta)\, RT \ldots\ldots\ldots\ldots\ldots\ldots\ldots\ldots(60\cdot2),$$

where $\tfrac{3}{2}RT$ is the energy of translation and $\tfrac{3}{2}\beta RT$ is the remaining energy of the internal vibrations, etc. of the molecule, so that β is the ratio of the internal to translational energy. From the usual formula for C_v we obtain

$$\rho C_v T = \tfrac{3}{2}(1+\beta)\frac{R}{m\mu}\,\rho T = \tfrac{3}{2}(1+\beta)p,$$

and the value of dQ is seen to be

$$dQ = \frac{d\alpha}{\alpha}\iiint \tfrac{3}{2}(1-\beta)\,p'dv'.$$

By a known formula*, κ, the ratio of the two specific heats of a gas, is given by

$$\kappa = 1 + \frac{2}{3\,(1+\beta)} \quad\ldots\ldots\ldots\ldots\ldots(60\cdot3),$$

so that the value of dQ can be written in the alternative form

$$dQ = \frac{d\alpha}{\alpha}\iiint \frac{3\kappa - 4}{\kappa - 1}\, p'\,dv' \quad\ldots\ldots\ldots\ldots(60\cdot4).$$

If $3\kappa - 4$ is positive throughout the star, dQ is positive, so that the expanded state of the star is a state of higher energy than the contracted state. Thus if we assume a star to be limited to configurations on a single homologous series, it appears that when κ is greater than $\tfrac{4}{3}$, a loss of energy, as for instance by radiation from its surface, will cause the star to contract. By Lane's law, the temperature simultaneously increases at every point of the star, and we have the seemingly paradoxical result that the more the star radiates heat from its surface, the hotter it becomes. The explanation is, of course, the simple one that the contraction of the star sets free enough gravitational energy to account for both the radiated energy and the heating of the star.

If $3\kappa - 4$ is negative throughout the star, the reverse conditions prevail; the expanded state of the star is a state of lower energy than the contracted state, and if the star loses heat by radiation, it expands and cools. We shall return to a fuller discussion of these questions later.

---

* Jeans, *Dynamical Theory of Gases*, p. 185.

Poincaré's Theorem.

61. The foregoing results as well as others of a more general kind may also be obtained from a theorem first given by Poincaré*.

Consider a collection of detached masses moving under no forces except their own mutual gravitational attraction. The masses may be stars, molecules, dust particles, atoms or electrons; for convenience we shall speak of them as molecules.

The equations of motion of a single molecule of mass m are

$$m\frac{d^2x}{dt^2} = X \text{ etc. } \quad\text{...........................(61·1)},$$

where X, Y, Z are the components of the force acting on it. Using these equations we readily find that

$$\frac{1}{2}\frac{d^2}{dt^2}[\Sigma m\,(x^2 + y^2 + z^2)] = 2T + \Sigma\,(xX + yY + zZ)\text{.........(61·2)},$$

where the summation extends over all the molecules, and

$$2T = \Sigma m\left[\left(\frac{dx}{dt}\right)^2 + \left(\frac{dy}{dt}\right)^2 + \left(\frac{dz}{dt}\right)^2\right] \quad\text{...............(61·3)},$$

so that T is the total kinetic energy of translation of the system.

If the only forces which act on the molecules are those arising from their mutual gravitation, we have

$$X = -\frac{\partial W}{\partial x} \text{ etc.,}$$

where W is the total gravitational potential energy of the system. This is equal to $-\gamma\Sigma\Sigma m_1 m_2/r_{12}$ where m_1, m_2 are any pair of molecules, r_{12} is their distance apart, and the summation extends over all pairs of molecules. Since W is homogeneous in x, y, z and of dimensions -1, it follows from a well-known theorem that

$$\Sigma\left(x\frac{\partial W}{\partial x} + y\frac{\partial W}{\partial y} + z\frac{\partial W}{\partial z}\right) = -W \quad\text{...............(61·4).}$$

Equation (61·2) now assumes the form

$$\frac{1}{2}\frac{d^2}{dt^2}[\Sigma m\,(x^2 + y^2 + z^2)] = 2T + W \quad\text{...............(61·5).}$$

If the system has attained to a steady state, the left-hand member vanishes, and the equation assumes the form

$$2T + W = 0\text{.............................(61·6).}$$

This is known as Poincaré's theorem. Eddington has remarked† that, from equation (61·5), it can be extended in the form

$$\frac{1}{2}\frac{d^2I}{dt^2} = 2T + W \quad\text{...........................(61·7),}$$

---

* *Leçons sur les hypothèses Cosmogoniques*, p. 94.   † *M.N.* LXXVI. (1916), p. 525.

where I stands for $\Sigma m (x^2 + y^2 + z^2)$, in which form it is not restricted to states of steady motion.

If, as before, β denotes the ratio of the total internal energy of the molecules to their energy of translation, the total heat-energy of the gas is $(1 + \beta) T$, and the total energy E is given by

$$E = (1 + \beta) T + W \qquad \text{(61·8)}.$$

In the steady state in which $2T + W = 0$ it follows that

$$T (\beta - 1) = E \qquad \text{(61·9)},$$

so that E increases with T if β is greater than unity, but decreases with T increasing, if β is less than unity. This brings us directly to the results already given in the last section.

62. We can write the kinetic energy T in the form $\frac{1}{2}\Sigma mv^2$, where v is the velocity of translation of a molecule of mass m. The potential energy W may similarly be written in the form $-\frac{1}{2}\Sigma mV$, where V is the gravitational potential at the point occupied by the mass m. Thus Poincaré's theorem takes the form that

$$\Sigma m (v^2 - \tfrac{1}{2} V) = 0 \qquad \text{(62·1)},$$

so that, in the steady state, the average value of v^2, averaged over all the separate masses, is equal to the average value of $\frac{1}{2} V$.

If the system is of total mass M and has a mean radius r, the average value of $\frac{1}{2} V$ is of the order of magnitude of $\gamma M/r$, so that the average value of v^2 is of this order of magnitude. This provides a convenient rough measure of the average velocity of agitation of a system of gravitating masses in a steady state: it is equally applicable to systems of stars, star-clusters, nebulae, and masses of gravitating gas.

If the particles which constitute the system are taken to be the molecules of a gas, or other independently moving units such as atoms, free electrons, etc., v^2 is equal to $3R/m\mu$ times the temperature of the gas, where μ is its mean molecular weight. Thus the mean temperature of the gas is of the order of magnitude of

$$\frac{\gamma M}{r} \left(\frac{m\mu}{3R} \right) \qquad \text{(62·2)},$$

so that the mean internal temperatures of different stars are approximately proportional to the values of $\mu M/r$ for these stars. Lane's law is included as a special case.

As regards absolute values, we find from this formula that if the sun (for which $M = 2 \times 10^{33}$, $r = 6\cdot95 \times 10^{10}$) is supposed to be formed of hydrogen molecules for which $\mu = 2$, its mean temperature must be of the order of 15,000,000 degrees; if it is formed of molecules of air, the mean temperature will be about fourteen times this, or 210,000,000 degrees.

More accurate figures will be given later as the result of a detailed study of stellar models. But at the outset of this study we may notice that, apart from all particular models, the gravitational attraction of the molecules of a star endows the stellar matter with a temperature of the order of 10^7 degrees centigrade.

ADIABATIC EQUILIBRIUM.

63. Let us now return to the general equation of equilibrium (58·3), namely

$$\frac{1}{r^2}\frac{d}{dr}\left(\frac{r^2}{\rho}\frac{dp}{dr}\right) + 4\pi\gamma\rho = 0 \quad \dots\dots\dots\dots\dots\dots(63\cdot1),$$

As we have already noticed, this does not of itself enable us to determine the distribution of density inside a sphere of gravitating gas; we can only do this by introducing some further assumption as to the conditions in the star's interior.

In the early discussions of the problem the supposition usually made was that the star's interior was in a state of "adiabatic" or "convective" equilibrium such as prevails in the lower regions of the earth's atmosphere. As ordinary gaseous conduction is easily shewn to be too slow a process to account for the violent flow of heat to the star's surface (cf. § 71 below), it was supposed that the greater part of the heat was transferred from the star's centre to its surface by convection currents. If the whole of a star's interior is kept mixed and stirred up by such convection currents, the state of equilibrium in its interior is readily determined. We can imagine any two elements of the star of equal mass to become interchanged in the process of mixing; after they have expanded or contracted until their pressures are suitable to their new positions the stars must again be in equilibrium as before. The process of expansion is so rapid in comparison with the process of transfer of heat by conduction, that any transfer of heat by conduction may be neglected, and the condition for equilibrium is simply that the energy-contents of the two elements of gas must originally have been identical.

When this is the case throughout the star, the pressure p and density ρ are connected by a relation of the form

$$p = K\rho^\kappa \quad \dots\dots\dots\dots\dots\dots\dots\dots\dots(63\cdot2),$$

where K is a constant and κ is the ratio of the specific heats of the gas. Here, then, is a possible further relation between p and ρ, and from equations (63·1) and (63·2) both p and ρ can be determined.

Replacing p by its value $K\rho^\kappa$, equation (63·1) assumes the form

$$\frac{K\kappa}{r^2}\frac{d}{dr}\left(r^2\rho^{\kappa-2}\frac{d\rho}{dr}\right) + 4\pi\gamma\rho = 0 \quad \dots\dots\dots\dots\dots(63\cdot3),$$

a differential equation which determines the way in which ρ depends on r.

64. It is generally only possible to solve this equation by quadratures. There are, however, two exceptions. When $\kappa = 2$, the equation assumes the simple linear form

$$\frac{K}{r^2}\frac{d}{dr}\left(r^2\frac{d\rho}{dr}\right) + 2\pi\gamma\rho \quad\quad\quad\text{......................(64·1)}$$

which was discussed by Laplace. This equation is linear in ρ, and its general solution is easily found to be

$$\rho = A\,\frac{\sin\,(cr - \epsilon)}{r} \quad\quad\quad\text{......................(64·2)},$$

where $c^2 = 2\pi\gamma/K$, and A and ϵ are constants of integration. We are not at present interested in the most general solution either of the general equation (63·3) or of the simpler equation (64·1). For if ϵ has any value other than zero in the solution (64·2), ρ runs up to an infinite value at the centre of the star. At the centre of an actual star, ρ must remain finite, reaching a maximum of the usual type at which $d\rho/dr = 0$. The solution expressed by equation (64·2) only satisfies this condition when $\epsilon = 0$, in which case it reduces to

$$\rho = A\,\frac{\sin\,cr}{r} \quad\quad\quad\text{..........................(64·3)}.$$

Again when $\kappa = 1\cdot2$, equation (63·3) has a solution in finite terms, first given by Schuster[*], for which $d\rho/dr$ is zero at the centre, namely,

$$\rho = \rho_0\left(1 + \frac{r^2}{x^2}\right)^{-\frac{5}{2}} \quad\quad\quad\text{..........................(64·4)}.$$

For other values of κ the equation can only be solved by quadratures. Starting at the centre, taking an arbitrary value of ρ and assuming that $d\rho/dr = 0$, we find that ρ must steadily decrease as we pass outwards; as soon as ρ reaches a zero value we know that the star's surface has been reached and at this point the quadrature stops.

65. The solution of the general equation (63·3) by quadratures has been very fully investigated by Emden in his book *Gas Kugeln*[†]. Put $\dfrac{R}{m\mu}\,\Theta$ for the quantity we have denoted by K, R being the universal gas-constant, μ the molecular weight of the substance of which the star is supposed to be formed, and m the mass of a molecule of molecular weight unity. The adiabatic relation (63·2) between p and ρ now becomes

$$p = \frac{R}{m\mu}\,\Theta\rho^{\kappa} \quad\quad\quad\text{..........................(65·1)}.$$

The usual Boyle-Charles law is expressed by

$$p = \frac{R}{m\mu}\,T\rho \quad\quad\quad\text{..........................(65·2)},$$

---

[*] *British Association Report*, 1883, p. 428.

[†] Teubner (Leipzig, 1907). See also *Encyc. der Math. Wissen.*, vol. VI 2 B, part 2, reprinted in book form "Thermodynamik der Himmels Körper" (Leipzig and Berlin, 1926).

where T is the temperature corresponding to the given values of p and ρ. Comparing equations (65·1) and (65·2), we see that

$$T = \Theta \rho^{\kappa-1} \quad \dots\dots\dots\dots\dots\dots(65\cdot3),$$

so that Θ is the value of the temperature of any element of the star when it is compressed to the density $\rho = 1$, and so is the actual temperature inside the star at the point at which the density is unity.

If we now write u for $\rho^{\kappa-1}$ and introduce $\mathfrak{r}$ defined by

$$\mathfrak{r}^2 = r^2 \frac{4\pi\gamma(\kappa-1)m\mu}{R\Theta\kappa} \quad \dots\dots\dots\dots(65\cdot4),$$

equation (63·3) is found to assume the form

$$\frac{d^2u}{d\mathfrak{r}^2} + \frac{2}{\mathfrak{r}}\frac{du}{d\mathfrak{r}} + u^n = 0 \quad \dots\dots\dots\dots\dots(65\cdot5),$$

where n stands for $\dfrac{1}{\kappa-1}$, so that $\rho = \left(\dfrac{T}{\Theta}\right)^n$.

For any given value of n, the most direct procedure would be to assume an initial value u_c at the centre, together with the condition $du/d\mathfrak{r} = 0$ at the centre and calculate values of u and $\mathfrak{r}$ by successive stages outward. But it is readily seen that the solutions so obtained would fall into homologous series, the central density ρ_c, and so also u_c, the value of u at the centre, having all values from zero to infinity as we pass along any one series. It is accordingly sufficient to calculate the solution for any one standard value of u_c, when the solutions for all other values can be derived immediately by homologous contraction or expansion. Emden obtains his standard solution by taking $u_c = 1$, so that the central density ρ_c is also unity. If u_1, ρ_1, $\mathfrak{r}_1$ are the values of u, ρ and $\mathfrak{r}$ for this particular solution, the values corresponding to any other central density ρ_c are given by

$$u = u_c u_1 = \rho_c{}^{\overline{n}}u_1; \quad \rho = \rho_c\rho_1; \quad \mathfrak{r} = u_c{}^{-\frac{1}{2}(n-1)}\mathfrak{r}_1 = \rho_c{}^{-\frac{n-1}{2n}}\mathfrak{r}_1 \quad (65\cdot6).$$

66. As a first illustration of the use of this solution, Emden studies the internal arrangement of the sun on the improbable supposition that it is made of atmospheric air. For this, κ, the ratio of the specific heats, is equal to 1·4, so that $n = 2\cdot5$.

His numerical solution for the value $n = 2\cdot5$ shews that the density at the centre is 24·07 times the mean density. As the mean density of the sun is 1·416, the density at the centre on this model must be $\rho_c = 34\cdot02$. The numerical solution gives the radius of the mass of gas to be $\mathfrak{r}_1 = 5\cdot417$, so that the last of equations (65·6) gives

$$\mathfrak{r} = 5\cdot417\rho_c{}^{-0\cdot3}$$

The actual radius of the sun is $r = 6\cdot95 \times 10^{10}$ cms., and on inserting these values for r and $\mathfrak{r}$ in equation (65·4) and putting $\kappa = 1\cdot4$, we obtain

$$\frac{R}{m\mu}\Theta = 3\cdot22 \times 10^{14} \quad \dots\dots\dots\dots\dots(66\cdot1).$$

The value of $R/m\mu$ for atmospheric air is $2\cdot87 \times 10^6$, so that

$$\Theta = 1\cdot1226 \times 10^8 \text{ degrees} \quad \dots\dots\dots\dots\dots(66\cdot2).$$

This, as we have seen, would be the temperature at a point inside the model sun at which the density is unity. From this it follows at once that the temperature, density and pressure at the sun's centre are:

Temperature at centre of sun = 455 million degrees.
Density   „   „   „   „   = 34·02.
Pressure   „   „   „   „   = 4·3 × 10¹⁶ dynes.
                             = 43,000 million atmospheres.

IONISATION IN STELLAR INTERIORS.

67. These last figures refer only to a very special model of a single star, but calculations for other stars and other models give very similar temperatures for stellar centres. A large number of detailed calculations will be found in Emden's book. As was first pointed out by the present writer in 1917 [*], neither molecules nor atoms could retain their existence as such at temperatures as high as this. Whatever model we take, a simple calculation shews that the temperature throughout the greater part of a star's interior must produce a very high degree of electronic dissociation, the molecules and atoms being almost completely broken up into their constituent electrons and nuclei, which will now all move about independently like the molecules of a gas. In more peaceful surroundings their electrostatic attractions would rapidly unite the wandering nuclei and electrons into complete atoms and molecules, but these are powerless in the general whirl of rapidly moving projectiles and in face of the shattering blows of the quanta of high-frequency radiation which the high temperatures of the stellar interiors generate. It is no more possible to build up an atom in the interior of a star than to build a house of cards in a hurricane.

When I first put forward this view I believed it to be entirely novel, but I have since found that in 1644 Descartes had conjectured [†] that the sun and fixed stars were made of matter "which possesses such violence of agitation that, impinging upon other bodies, it gets divided into indefinitely minute particles."

68. This view of the constitution of stellar interiors reduces the central temperature to a value far below that just calculated.

In breaking up a molecule of, say, nitrogen into its ultimate constituents, we replace a single moving unit by sixteen separately moving parts, two positive nuclei and fourteen free electrons. If the temperature remains unaltered each of these sixteen parts exerts, statistically, the same pressure as the original molecule, so that the pressure corresponding to a given

---

[*] *Observatory*, XL. (1917), p. 43 and *Bakerian Lecture* (1917), *Phil. Trans.* 218 A, p. 209.
[†] *Principiorum Philosophiae*, Part III, Chap. 52.

temperature is increased sixteen-fold. Or, to put the same thing in the form immediately suited to our problem, the temperature corresponding to a given pressure is decreased to one-sixteenth.

The total mass of the sixteen moving parts is equal to the original mass of the nitrogen molecule of molecular weight 28. Thus we can regard the positive nuclei and negative electrons as forming a gas of mean molecular weight $\frac{28}{16}$ or 1·75. If the original gas had been hydrogen, the mean molecular weight of the broken pieces would have been 0·5; similarly the mean molecular weight of broken up helium is 1·33, of broken up calcium is 1·90, of broken up iron 2·07, and of broken up lead 2·50. There is of course very much greater equality between the mean molecular weights of the broken up pieces of molecules than between those of the original molecules themselves, since the more massive molecules break into a greater number of separate fragments than the less massive ones; indeed, the atomic number of an atom of atomic weight n in general approximates to $\frac{1}{2}n$ except for hydrogen and the very massive atoms, and in consequence the mean molecular weight of the broken mixture can never differ very greatly from 2.

For a general mixture of elements such as are known to us on earth, we should not be far wrong in assuming a mean molecular weight of 2 for the broken up fragments. The corresponding value of $R/m\mu$ is $4·13 \times 10^7$ in place of the much lower value assumed by Emden and used above, and on inserting this value in equation (66·1) we obtain $\Theta = 7·76 \times 10^6$ degrees. The central temperature T_c is now found to be

$$T_c = 31·5 \text{ million degrees,}$$

the values of p_c and ρ_c remaining unaltered.

We shall find reasons for supposing that the stars in general do not consist of a general mixture of elements of this kind, but that the great majority of their atoms have atomic weights comparable with that of uranium and the radioactive elements. We shall further find that in all probability the atoms are not completely broken up, although very nearly so. When we allow for these new factors the mean molecular weights of the broken up fragments is about 2·6 rather than 2, and this increases the value just calculated for T_c to about 41 million degrees. We shall find that further adjustments have to be made in the value of T_c, but these nearly neutralise one another leaving the final value of T_c in the neighbourhood of 40 million degrees.

69. There would seem to be little room for doubt that the foregoing view of the interior structure of a star is the correct one. Convincing evidence in its favour is provided by the high density of many stars, and in particular, as Eddington first pointed out, by that of the companion of Sirius, which direct observation shews to be something like 50,000. It is inconceivable that such high densities could occur in a gas formed of complete atoms or molecules, although it is quite natural that they should be found when the molecules of diameter 10^{-8} cms. are replaced by crowds of free electrons of diameters about 4×10^{-13}, of nuclei whose diameters are less even than this and of atoms whose rings of electrons have nearly all been stripped away.

Even at a density of 50,000 times that of water, the electrons and nuclei are at mean distances apart which are of the order of 10^{-10} cms., and this is some hundreds of times the probable diameter of both free electrons and nuclei. If then all atoms were stripped absolutely bare of electrons, so that stellar matter consisted solely of electrons and bare nuclei, it would seem reasonable to assume the ordinary gas-laws to hold, even at these monstrously high densities, at any rate as a first approximation. Anderson* has pointed out that the electron pressure in stellar interiors is that of a "degenerate" gas†, so that the pressure can in no case be given exactly by Boyle's law, the divergence increasing from one of a few per cent. in ordinary stars to a factor of about 2 for Kruger B and Sirius B.

We shall, however, find that, although the mean density of Sirius B may be 50,000, the density in its central regions is probably to be measured in millions, so that here Boyle's law can hardly hold, even as an approximation. Furthermore, the temperatures at the centres of most stars are not adequate to break the atoms down to their ultimate constituents. Stellar matter consists in part of free electrons and perhaps some bare nuclei, but these are mixed with miniature atoms in which some rings of electrons remain to clear a space about the nucleus, and when such atoms are present it is not obvious without further investigation whether Boyle's law will give a valid approximation to the pressure-density relation or not.

For the moment we shall proceed on the supposition that the pressures in stellar interiors are those given by Boyle's law, but we shall soon find it necessary to discard this supposition. For the two reservations just mentioned prove to be of the utmost importance in stellar physics, and probably dominate the whole dynamics of stellar interiors.

THE PRESSURE OF RADIATION.

70. We have been led to picture the far interior of a star as consisting of a crowd of dissociated electrons, of bare nuclei and of atoms stripped of electrons almost down to their nuclei. The electrons move about from one broken atom to another, seldom staying with any one atom for long, since the majority of the atoms are either completely or almost completely bare of electrons. As we approach the surface, we come to regions of lower temperatures where the disintegration is far less complete, and here we find semi-formed atoms. Finally, close to the surface we come upon ionised atoms from which only a few of the outer electrons are missing, and possibly fully formed atoms. In the M-type stars, the surface layers are so cool that even fully formed molecules are found, such as those of titanium oxide and magnesium hydride.

---

* *Zeitsch. f. Physik*, L. (1928), p. 874.   † E. Fermi, *Zeitsch. f. Physik*, XXXVI. (1926), p. 911.

Before accepting this view of a star's constitution as complete, a further complication must be taken into account. We have already noticed that the presence of radiation results in a pressure $\frac{1}{3}aT^4$, where $a = 7.63 \times 10^{-15}$. If T has the value just calculated for the centre of the sun, namely, 4.1×10^7 degrees, the corresponding pressure of radiation is found to be 7.2×10^{15} dynes, which is about one-sixth of the total pressure given by the rough calculation of § 66. Thus the mechanical effects of the pressure of radiation, while not great, are just too large to be disregarded entirely. The present writer drew attention to Emden's neglect of the pressure of radiation when reviewing his *Gas Kugeln* in 1909[*], and gave the first reasonably accurate estimate of its importance in 1917[†]. Eddington had attempted a calculation some months earlier, but had obtained values which were thousands of times too large through supposing the stellar matter to consist of unbroken atoms[‡].

When the pressure of radiation is taken into account, the total pressure p inside a star whose material is of mean molecular weight μ is given by

$$p = \frac{R}{m\mu} \rho T + \tfrac{1}{3} a T^4 \quad\dots\dots\dots\dots\dots\dots(70.1),$$

where the first term on the right represents the usual gas-pressure, which we shall henceforth denote by p_G and shall, for the present, suppose to be given by Boyle's law, and the second term denotes the pressure of radiation, which we shall call p_R. If the ratio of gas-pressure to pressure of radiation is denoted by λ,

$$p = p_G + p_R = \left(1 + \frac{1}{\lambda}\right) p_G = \frac{R(1+\lambda)}{m\mu\lambda} \rho T \quad\dots\dots\dots(70.2),$$

so that the effect of taking radiation-pressure into account is the same as that of reducing the molecular weight μ by a factor $\lambda/(1+\lambda)$. We may, if we please, treat the gas as though it had a fictitious molecular weight μ' given by

$$\mu' = \frac{\mu\lambda}{1+\lambda} \quad\dots\dots\dots\dots\dots\dots\dots(70.3),$$

and neglect the pressure of radiation entirely. By so doing we should be treating the pressure of radiation as though it arose from molecules of molecular weight zero, and on averaging over these fictitious molecules and the real material molecules of molecular weight μ we obtain the average molecular weight μ' given by equation (70.3).

The very rough calculation given above has suggested that λ is about 5 at the sun's centre, so that we could allow for the pressure of radiation by reducing the effective molecular weight at the sun's centre by about 17 per cent.

In more massive stars the pressure of radiation assumes greater importance. To see this let us start from a standard star such as the sun and increase its

[*] *Astrophys. Journ.* xxx. (1909), p. 72.
[†] *Bakerian Lecture* (1917), *Phil. Trans.* 218 A, p. 209.
[‡] *M.N.* lxxvii. (1917), p. 16.

density θ-fold at every point, from ρ to $\rho\theta$, keeping the dimensions of the star fixed. At the same time let us increase its mean molecular weight ψ times, from μ to $\mu\psi$.

The new star so obtained has θ times the mass of the standard star, so that the value of gravity has been increased θ-fold at every point. To satisfy the dynamical equation

$$\frac{dp}{dr} = -g\rho \quad \dots\dots\dots\dots\dots\dots\dots\dots\dots(70\cdot4)$$

the pressure must be increased θ^2 times at each point. If this is assumed to be given by Boyle's law,

$$p = \frac{R}{m\mu}\rho T \quad \dots\dots\dots\dots\dots\dots\dots(70\cdot5),$$

we see that T/μ must be increased θ times, so that T must be increased $\theta\psi$ times. The ratio of pressure of radiation to gas-pressure, which is proportional to $\frac{\frac{1}{3}aT^4}{p}$ or to $\frac{T^3\mu}{\rho}$ has now been increased $\frac{(\theta\psi)^3\psi}{\theta}$ or $\theta^2\psi^4$ times.

As explained in § 59, we can let this star undergo any homologous contraction we please and so assume any radius we please. Equations (59·3) shew that when the radius of the star is increased α-fold, the pressure of radiation, which is proportional to T^4, becomes multiplied by a factor α^{-4}, while gas-pressure is also multiplied by a factor α^{-4}. Thus the ratio of gas-pressure to pressure of radiation remains unchanged by homologous contraction.

Hence changing the density by a factor θ, the molecular weight by a factor ψ and the radius of a star by any factor we please, changes the ratio of radiation-pressure to gas-pressure by the factor $\theta^2\psi^4$ already calculated. In other words, in stars of different masses M and different molecular weights μ, the ratio of pressure of radiation to gas-pressure is proportional to $M^2\mu^4$. Although this ratio is only of the order of $\frac{1}{8}$ in the sun, it becomes considerable in stars whose mass is several times that of the sun. We must, however, notice that formula (70·5) is only accurate so long as radiation-pressure is negligible in comparison with gas-pressure; as soon as the ratio becomes appreciable our calculations fail to give its exact value. Definite exact calculations will be given later.

Whatever the relative importance or amount of the radiation-pressure may be, we get a true picture of stellar structure by thinking of the layers of stellar matter as held up against gravitation by the incessant impact of a certain number of atomic nuclei or partially stripped atoms, the "molecular weight" of which is practically the same as that of the corresponding complete atoms, together with a far greater number of free electrons of standard "molecular weight" $\frac{1}{1840}$ or 0·00055, and a rather small number of "molecules" of radiation, the molecular weight of which is negligibly small. The combined impacts of these three types of projectiles prevent the star from falling in under its own gravitational attraction.

This gives us a good snapshot picture of a star's structure. We obtain the corresponding picture of its mechanism by thinking of the nuclei as α-ray particles, of the free electrons as β-ray particles, and of the radiation as γ-rays (although in most stars the main bulk of the radiation has the wave-length of X-rays); and, precisely as in laboratory work, the β-rays are more penetrating than the α-rays, and the γ-rays are more penetrating than either.

RADIATIVE TRANSFER OF ENERGY.

71. Apart from precise figures, the temperature inside a star must range from some million degrees at its centre to a few thousand degrees at its surface. The various solutions given by Emden agree in indicating* that there is a fairly uniform fall of temperature along a star's radius except close to the centre, where the rate of fall is necessarily zero. If the sun has the central temperature of about 41 million degrees, calculated in § 68, the average fall of temperature along its radius of 6.95×10^{10} cms. would be about 0.0006 degrees per centimetre.

The flow of heat consequent on a temperature gradient $\partial T/\partial r$ is

$$-\vartheta \frac{\partial T}{\partial r}$$

per unit time per unit area, where ϑ is the coefficient of conduction of heat, and for a temperature gradient of the order of 0.0006 degrees per centimetre to produce the observed flow of heat at the surface of the sun (6.25×10^{10} ergs per sq. cm. per second) the value of ϑ in ergs must be of the order of 10^{14}.

The flow of heat in a star will be the aggregate of the amounts transported by atoms and nuclei, by free electrons and by radiation, so that the total value of ϑ will be the sum of contributions made by these three types of carriers respectively.

The contribution arising from the transport of heat by material carriers, atoms, nuclei and free electrons, is given by the usual kinetic theory formula†

$$\vartheta = \tfrac{1}{3}\nu \bar{c} l \frac{d\bar{E}}{dt} = \tfrac{1}{2} R \nu \bar{c} l \quad\dots\dots\dots\dots\dots\dots(71 \cdot 1),$$

where $\bar{c}$ is the mean velocity of the type of carrier in question and l is its mean free path. A simple calculation shews that the transport of energy by free electrons far exceeds that by atoms and nuclei, so that as regards material transport the coefficient of conduction is given by formula (71·1) where $\bar{c}$ and l refer to the free electrons. On inserting any reasonable values for $\bar{c}$ and l into this formula, the resulting value of ϑ is found to be far below the value of 10^{14} which is necessary to account for the observed flow of heat to the sun's surface.

---

* See in particular a diagram given by Emden, *Gas Kugeln*, p. 86.
† Jeans, *Dynamical Theory of Gases* (4th Ed.), p. 291.

72. To discuss the transfer of energy by radiation, it is convenient to replace equation (71·1) by the equivalent equation

$$\vartheta = \tfrac{1}{3}\rho' C_v \bar{c} l \quad\text{..............................(72·1)},$$

where ρ' is the density, and $\rho' C_v$ is the specific heat of the carriers per unit volume, $\bar{c}$ as before denotes their mean velocity, and l is their mean free path.

The nuclei and the free electrons have, of course, quite definite free-paths. The same is true of the radiation if this is regarded as consisting of discrete quanta; when a quantum is emitted a free-path begins, and when it is re-absorbed the free-path ends. Whether we think in terms of undulatory theory or quanta, we may suppose that a beam of radiation is reduced in intensity by a factor $e^{-k\rho x}$ on passing through a thickness x of matter of density ρ, where k is the "coefficient of opacity" of the matter. In ordinary kinetic theory a stream of molecules is reduced to $e^{-x/l}$ of its original strength after traversing a distance x, where l is the free-path. By comparison the free-path of our molecules of radiation must be supposed to be $1/k\rho$. The energy of these fictitious molecules per unit volume is aT^4, so that the specific heat per unit volume is $\dfrac{d}{dT}(aT^4)$ or $4aT^3$, which may be compared with the specific heat per unit volume ρC_v of ordinary material molecules. The velocity of these radiation-molecules is of course uniformly equal to C, the velocity of light.

If now we make the appropriate substitutions in formula (72·1), replacing $\bar{c}$ by C, l by $1/k\rho$ and $\rho' C_v$ by $4aT^3$, the formula becomes

$$\vartheta = \frac{4}{3}\frac{aT^3 C}{k\rho} \quad\text{..............................(72·2)},$$

so that the transfer of energy by our fictitious radiation molecules is the same as if there were a coefficient of conduction having this value. On inserting numerical values suitable for the sun's interior (say $T = 3 \times 10^7$, $k = 1000$, $\rho = 10$) and putting $aC = 4\sigma = 2\cdot3 \times 10^{-4}$, we find that this coefficient of conduction is of the order of 10^{14}. This so entirely outweighs the coefficients of conduction of heat by material conduction that the latter may be neglected by comparison. Further, the flow of heat produced by radiative conduction is at least of the same order of magnitude as that actually observed. Whether the two quantities are in complete agreement will be the subject of a careful enquiry below.

With formula (72·2) for the coefficient of radiative conductivity, the radiative flow of heat per unit area becomes

$$-\vartheta\frac{\partial T}{\partial r} = -\frac{4aT^3 C}{3k\rho}\frac{\partial T}{\partial r} \quad\text{..............................(72·3)}.$$

Both these formulae are only approximate; indeed the fundamental formula (72·1) from which they are derived was only approximate.

73. As far back as 1894[*] Sampson had pointed out that the transfer of heat by radiation inside a star must far exceed the transfer by ordinary material conduction, but his detailed discussion of the problem was vitiated by his assuming an erroneous law for temperature radiation. In 1906[†] Schwarzschild independently advanced the same idea, describing the vehicle of transfer of energy in the sun as "a mighty stream of radiant energy, springing from unknown sources in the sun's interior and pressing through his atmosphere into the space beyond."

Our equation (72·2) is the mathematical expression of this physical concept. This, or rather its equivalent, formula (72·3), was first given in 1917 by Eddington[‡], who derived it by a method which followed Schwarzschild's original analysis much more closely than does that just given. Eddington believed this equation to be true for all stellar matter, including the case, not considered by Schwarzschild, in which the stellar matter is continually§ generating energy. In 1917 the present writer shewed§ that equation (72·3) is accurate only in stellar matter in which there is no internal generation of energy. This was precisely the type of matter to which Schwarzschild's original analysis referred, since he supposed his equations to apply only to the sun's atmosphere, while the generation of solar energy was assumed to occur in layers of greater depth. When there is an internal generation of energy at a rate G per unit mass, the formulae are in error by terms of the order of magnitude of G‖.

As a result of the temperature gradient which exists in a star's interior, matter and radiation are not in precise thermodynamical equilibrium, and there is room for some arbitrariness in the definition of temperature. As a matter of convenience it is best defined in terms of the emission of radiation by matter, the emission per unit mass at any point being taken to be $kaCT^4$. When thermodynamical equilibrium exists, the quantity T defined in this way of course becomes identical with the ordinary temperature.

With this definition, I have found¶ that the outward flow of energy per unit area H, the radiant energy per unit volume R, and the pressure of radiation normal to the star's radius, p_R, are given by

$$H = \frac{1}{3}\frac{\partial}{\partial \nu}(aCT^4) + \left(\frac{4\pi}{5}\frac{\partial}{\partial \nu} + \frac{4\pi}{7}\frac{\partial^3}{\partial \nu^3} + \ldots\right)\left(-3 + \frac{9}{5}\frac{\partial^2}{\partial \nu^2} + \ldots\right)\left(\frac{G}{k}\right) \quad (73\cdot1),$$

$$R = aT^4 + \frac{1}{C}\left(\frac{1}{3} + \frac{1}{5}\frac{\partial^2}{\partial \nu^2} + \frac{1}{7}\frac{\partial^4}{\partial \nu^4} + \ldots\right)\left(-3 + \frac{9}{5}\frac{\partial^2}{\partial \nu^2} + \ldots\right)\left(\frac{G}{k}\right) \quad (73\cdot2),$$

[*] *Memoirs R.A.S.* LI. (1894), p. 123.

[†] *Göttingen Nach.* (1906), p. 41.

[‡] *M.N.* LXXVII. (1917), p. 16.

[§] *M.N.* LXXVII. (1917), p. 28; *M.N.* LXXXVI. (1926), p. 574.

[‖] This has also been confirmed by Ambarzumian and Kosirev, *M.N.* LXXXVII. (1927), p. 651.

[¶] *M.N.* LXXXVI. (1926), p. 576.

$$p_R = \frac{1}{3}aT^4 + \frac{1}{C}\left(\frac{1}{5} + \frac{1}{7}\frac{\partial^2}{\partial\nu^2} + \frac{1}{9}\frac{\partial^4}{\partial\nu^4} + \dots\right)\left(-3 + \frac{9}{5}\frac{\partial^2}{\partial\nu^2} + \dots\right)\left(\frac{G}{k}\right) \dots(73\cdot3),$$

where $\dfrac{\partial}{\partial\nu} = -\dfrac{1}{k\rho}\dfrac{\partial}{\partial r}$.

Besides failing owing to the neglect of terms in G, equations (72·2) and (72·3) fall further into error near the surface of the star, owing to there being no true approximation to thermodynamical equilibrium between matter and radiation in these regions. The extent of this error has been discussed by the present writer[*], Milne[†], Freundlich, Hopf and Wegner[‡], and others.

THE CONFIGURATIONS OF A STAR IN RADIATIVE EQUILIBRIUM.

The General Equations.

74. We have seen how radiation completely outstrips the material carriers in the transport of energy to the star's surface. As a consequence the ordinary coefficient of conduction of heat is of no importance, and the build of a star is entirely determined by the values of k, the coefficient of opacity in its interior. If this coefficient is everywhere zero, the star is entirely transparent, and so cannot retain any heat; we have a star of zero temperature and therefore of infinite extent. If, on the other hand, the coefficient of opacity is everywhere infinite, the star is completely opaque, so that all radiation accumulates where it is generated until the star's temperature becomes infinite, and we have a star of infinite temperature and consequently of infinitesimal radius. Naturally, only the intermediate values are of any practical interest, but the two extreme cases just mentioned shew how the whole build of a star depends on the value of the opacity coefficient k. So much is this the case that attempts to investigate the build of stars before the value of this coefficient was known can only be regarded as speculation; much of it was, moreover, unfortunate speculation, since the results obtained were mostly at variance with the results subsequently obtained by using the true value of the coefficient of opacity.

75. We have already noticed that the problem of determining the equilibrium configurations of a sphere of gas only becomes definite when some relation, outside the dynamical equations, is introduced to fix the temperature of the gas at every point. Emden's solutions assumed the adiabatic relation; each element of a star was supposed to have the temperature which it would assume if the stellar material was being continually stirred up. It is now clear that the proper relation to take is one expressing that the temperature of each element of the star is that determined by the flow of radiation. This flow of radiation is in turn determined by the rate at which energy is being generated in the star's interior. The outward flow of energy across a sphere

---

[*] *M.N.* LXXVII. (1917), p. 32.   [†] *Ibid.* LXXXI. (1921), p. 361.
[‡] *Ibid.* LXXXVIII. (1927), p. 139.

of radius r must be equal to the rate at which energy is being generated inside the sphere of radius r; this leads to the equation

$$- 4\pi r^2 \frac{4aT^3 C}{3k\rho} \frac{dT}{dr} = \int_0^r 4\pi \rho r^2 G \, dr \dots\dots\dots\dots(75\cdot1),$$

where G is the rate of generation of energy per unit mass at a distance r from the centre of the star. As we have seen in § 73, terms in G ought to be added to the left-hand member of this equation, but a numerical discussion shews that in the present problem they are inappreciable in comparison with the term on the right, and so may legitimately be disregarded.

The approximate equation (75·1) may accordingly be taken to be the equation of radiative equilibrium. The physical meaning of this equation is that the temperature gradient dT/dr at every point must be just that required to discharge the stream of radiation generated at all interior points.

To examine the build of a star in radiative equilibrium we combine this equation with the dynamical equation which expresses the condition that the star shall be in dynamical equilibrium, namely (§ 58),

$$\frac{dp}{dr} = - \frac{\gamma\rho}{r^2} \int_0^r 4\pi\rho r^2 dr \dots\dots\dots\dots\dots(75\cdot2),$$

in which p must now be taken to represent the total pressure, comprising both gas-pressure and pressure of radiation. The physical meaning of this equation is of course that the total pressure p is just adequate to support the weight of the whole column of gas standing above it.

It can be shewn that the pressure of radiation is, with sufficient accuracy for the present problem, equal to its value when there is a state of thermodynamical equilibrium, namely $\frac{1}{3}aT^4$. If we still assume the gas-pressure to be given by the ordinary laws of Boyle and Charles for an ideal gas, the total pressure p is

$$p = \frac{R}{m\mu}\rho T + \tfrac{1}{3}aT^4 \dots\dots\dots\dots\dots(75\cdot3).$$

Writing $\bar{G}$ for the mean value of G throughout a sphere of radius r, the equation of radiative equilibrium (75·1) assumes the form

$$\frac{4\pi r^2 C}{k\rho}\frac{d}{dr}(\tfrac{1}{3}aT^4) = - \bar{G}\int_0^r 4\pi\rho r^2 dr \dots\dots\dots\dots(75\cdot4),$$

or, transformed by the use of equation (75·2),

$$\frac{4\pi C\gamma}{k\bar{G}}\frac{d}{dr}(\tfrac{1}{3}aT^4) = \frac{dp}{dr} \dots\dots\dots\dots\dots(75\cdot5).$$

If we write p_G, p_R, for the gas-pressure and the pressure of radiation $(\tfrac{1}{3}aT^4)$ respectively, this may be put in the form

$$\frac{4\pi C\gamma}{k\bar{G}}\frac{dp_R}{dr} = \frac{d}{dr}(p_G + p_R)\dots\dots\dots\dots(75\cdot6).$$

This and the dynamical equation (75·2) determine the configurations of equilibrium of the gaseous star.

Eddington's Model.

76. The first attack on these equations was made by Eddington[*], who has discussed their solution subject to the special assumption that $k\bar{G}$ has a constant value throughout the star. This special assumption facilitates the mathematical treatment of the problem enormously, for when it is made, equation (75·6) admits of immediate integration in the form

$$\frac{4\pi C\gamma}{k\bar{G}} p_R = p_G + p_R \quad \dots\dots\dots\dots\dots\dots(76\cdot1).$$

This shews that when $k\bar{G}$ is assumed to be constant throughout a star, the ratio of p_G to p_R is also constant. Inserting their values for p_G and p_R the equation becomes

$$\rho = \frac{am\mu}{3R}\left(\frac{4\pi C\gamma}{k\bar{G}} - 1\right) T^3 \quad \dots\dots\dots\dots\dots(76\cdot2).$$

Thus ρ varies as T^3, so that p varies as $\rho^{\frac{4}{3}}$ throughout the star, and this is the relation of pressure to density which prevails in a sphere of gas arranged in adiabatic equilibrium with $\kappa = \frac{4}{3}$. Thus Eddington's assumption restricts us to stars arranged in the same way as a sphere of gas in adiabatic equilibrium with $\kappa = \frac{4}{3}$ or $n = 3$. But, as we shall now see, such a sphere of gas has very weird and wonderful properties, which make it unsuitable to be used as a model of actual stars.

In general, the total mass of a sphere of gas is

$$M = 4\pi \int \rho r^2 dr,$$

and, on inserting the appropriate values for ρ and r from equations (65·4) and (65·6), this becomes

$$M = 4\pi \rho_c^{\frac{(3-n)}{2n}} \left(\frac{R\Theta\kappa}{4\pi r\gamma (\kappa - 1)\, m\mu}\right)^3 \int \rho_1 r_1^2 dr_1 \quad \dots\dots(76\cdot3).$$

This shews that in general M depends on ρ_c and on Θ. For a mass of matter of given heat-energy Θ is given. As a star formed of such matter expands or contracts adiabatically, Θ retains its value unchanged through the expansion or contraction, but ρ_c, the central density, changes. Equation (76·3) shews that a given mass of matter for which the value of Θ is assigned can always find a position of equilibrium by changing its value of ρ_c—i.e. by expanding or contracting. If n is greater than 3, $(3-n)/2n$ is negative, so that large masses are diffuse and small ones are compact. When n is less than 3, the reverse is the case; large masses are compact and small ones are diffuse. The closer the value of n is to 3, the more rapidly the density varies with the mass. Finally as n touches the value 3, the addition or subtraction of the slightest amount of mass causes the star to rush through the whole range of values from $\rho_c = 0$ to $\rho_c = \infty$.

---

[*] *M.N.* LXXVII. (1917), pp. 16 and 596, and LXXXIV. (1927), pp. 104 and 308; see also *The Internal Constitution of the Stars*, Chap. VI.

In the special case of $n = 3$, M depends on Θ only. Combining the general equation (65·3) with Eddington's integral (76·2), we readily find that

$$\left(\frac{4\pi C\gamma}{k\bar{G}} - 1\right)(\tfrac{1}{3}a\Theta^3) = \frac{R}{m\mu} \quad\dots\dots\dots\dots\dots(76\text{·}4).$$

By equation (76·3) all stars having the same mass M must have the same value of Θ^3, so that if equation (76·4) were true, they would have also the same value of $k\bar{G}$. Thus Eddington's assumption that $k\bar{G}$ is constant throughout each single star involves that $k\bar{G}$ is the same in all stars of the same mass.

Eddington's earlier investigations were based on the supposition that k, the coefficient of opacity, was also constant. The constancy of $k\bar{G}$ for stars of a given mass then involved the constancy of $\bar{G}$, so that the average rate of generation of energy of stellar matter could depend only on the mass of the star of which it formed part and stars of equal mass were necessarily of equal luminosity. Indeed, Eddington actually announced this conclusion under the name of the "mass-luminosity" law. But it is clear that it must be a consequence merely of the special assumption that $k\bar{G}$ is constant, and cannot have reference to actual stellar conditions; a star's average generation of energy will be the average of the generations of its separate elements, and these will not be determined by the mass of the star, but by the substance of which it is made. A star made of pure uranium will necessarily generate 100 times as much energy as one of equal mass which is only one per cent. uranium and the rest lead. The former star will consequently emit 100 times as much radiation as its fellow of equal mass, whereas on Eddington's model the two stars would necessarily emit equal amounts of radiation.

This brings out the main defect of this special model. A model which cannot take into account that different substances, such as lead and uranium, have different capacities for generating radiation, does not seem a suitable model to assume for stellar structure.

After the appearance of Kramers' theory, which we shall now discuss, Eddington abandoned the supposition that k was constant, although retaining the constancy of $k\bar{G}$. As we shall see (§ 90), this causes the star's luminosity to depend on other things than its mass, so that the supposed "mass-luminosity" law disappears, and a star of given mass M can have any luminosity from 0 to ∞.

The Absorption Coefficient.

77. In 1923 Kramers [*] investigated the process of absorption of X-radiation in matter. As the ordinary radiation inside a star has about the same wavelength as the X-radiation of the laboratory, the investigation immediately

---

[*] *Phil. Mag.* XLVI. (1923), p. 836. A valuable account and critical discussion of Kramers' theory will also be found in Eddington's *Internal Constitution of the Stars*, Chap IX, p. 229 et seq.

provided a value for the coefficient of opacity k for ordinary stellar material. His theoretical investigation shews that k must be of the form

$$k = \frac{cF\rho}{\mu T^{3\cdot5}} \qquad \dots\dots\dots\dots\dots\dots(77\cdot1),$$

where c is a constant, F a numerical factor, and μ, ρ, T have the same meanings as before. The value of the constant c is found to be

$$c = \frac{16\pi^2}{3\sqrt{3}} \frac{N^2}{A} \frac{e^6 u_0}{m^2 aC^4 m_e h(1+f)} \qquad \dots\dots\dots\dots(77\cdot2),$$

where N is the atomic number and A the atomic weight, e is the charge and m_e the mass of an electron, u_0 is the mean velocity of an electron at $1°$ absolute. The number F represents the ratio of the total absorption to that caused by free electrons, and $1/f$ is the number of free electrons per atomic nucleus.

The formula has been tested under laboratory conditions and is found to agree well with observation. But under laboratory conditions there are practically no free electrons so that F and f are both very large and cancel out in the value of k. Under stellar conditions, where most of the atoms are highly ionised, f is quite small, so that the factor $1+f$ may be put equal to unity. The value of F can now only be calculated from pure theory, laboratory experiments providing no means of checking this value.

A preliminary investigation by Eddington[*] suggested that F had a fairly uniform value of from 8 to 10, but Milne[†] subsequently shewed that this value was too small by a factor of the order of 2 or 3. If we absorb in the factor F a further small correction in the opposite sense which Rosseland[‡] shewed to be necessitated by the distribution of radiant energy over different wave-lengths, it is found that the value of F must be about 20[§].

In most investigations the numerical value of F is unimportant, the results depending only on the general form of the law for k, which has been tested by experiment. And when the value for F enters, it is generally through so low a power that small errors in the adopted value for F are unimportant. For instance, we shall find ultimately that the central temperature of a star is proportional to $F^{\frac{2}{15}}$, so that an error of 100 per cent. in the assumed value for F results only in an error of $9\cdot7$ per cent. in the central temperatures of the stars.

Adopting the value $F = 20$, and inserting numerical values into c for all quantities except N, A and μ, equation $(77\cdot1)$ assumes the form

$$k = 4\cdot46 \times 10^{25} \left(\frac{N^2}{A}\right) \frac{\rho}{\mu T^{3\cdot5}} \qquad \dots\dots\dots\dots(77\cdot3).$$

[*] *M.N.* LXXXIV. (1924), p. 104. [†] *Ibid.* LXXXV. (1925), p. 750.
[‡] *Ibid.* LXXXIV. (1924), p. 525. [§] *Ibid.* LXXXVI. (1926), p. 561.

GENERAL STELLAR EQUILIBRIUM.

78. On introducing the value of the opacity coefficient from equation (77·1), the general equation of radiative equilibrium (75·6) assumes the form

$$\frac{d}{dr}(p_G + p_R) = \frac{4\pi C\gamma\mu T^{3\cdot5}}{cF\bar{G}\rho}\frac{dp_R}{dr} \quad\text{................(78·1)}.$$

Let us again replace p_G by λp_R, so that λ is given by

$$\lambda = \frac{R\rho}{\tfrac{1}{3}am\mu T^3}.$$

On inserting the value of ρ given by this equation, and dividing throughout by dp_R/dr, equation (78·1) becomes

$$\frac{d}{dp_R}[p_R(\lambda + 1)] = \frac{12\pi C\gamma R T^{\frac{1}{2}}}{cF\bar{G}am\lambda} \quad\text{................(78·2)}.$$

79. The case which is physically simplest and most natural occurs when c, F, $\bar{G}$ and μ are all constant throughout the star. We shall not limit our discussion to this case, but shall allow for some possibility of variation in these quantities by assuming that

$$\mu = \mu_0 T^j \quad\text{............................(79·1)},$$
$$cF\bar{G} = \beta T^l \quad\text{............................(79·2)}.$$

The effective molecular weight μ does not appear at all in the equation of radiative equilibrium just considered, but reappears when we return to the dynamical equations.

If we insert the value just assumed for $cF\bar{G}$, and replace T by its value as given by $p_R = \tfrac{1}{3}aT^4$, equation (78·2) becomes

$$p_R\frac{d\lambda}{dp_R} = \frac{12\pi C\gamma R}{\beta am\lambda}\left(\frac{3p_R}{a}\right)^{\frac{1}{4}(\frac{1}{2}-l)} - (\lambda + 1) \quad\text{...........(79·3)}.$$

If for brevity we put

$$\frac{12\pi C\gamma R}{\beta am}\left(\frac{3p_R}{a}\right)^{\frac{1}{4}(\frac{1}{2}-l)} = x^2 \quad\text{....................(79·4)},$$

so that x is a function of p_R only, and so only of T, the equation assumes the form

$$\tfrac{1}{8}(\tfrac{1}{2}-l)x\lambda\frac{d\lambda}{dx} = x^2 - \lambda(\lambda + 1) \quad\text{................(79·5)}.$$

80. The general nature of the solution is best seen by graphical methods. For this it will be sufficient to consider the case in which $\tfrac{1}{2}-l$ is positive, so that x, as defined by equation (79·4), is small near the surface of the star, and increases steadily as we pass into the star's interior.

Taking x and λ as abscissa and ordinate respectively, I have found* that the solutions lie as shewn in Fig. 7. There is found to be an asymptotic

* *M.N.* LXXXV. (1925), p. 201.

solution, passing through O in the figure, to which all other solutions approximate very rapidly as x increases.

The physical interpretation of this is very simple. The surface of the star is determined by $p_G = 0$, and so by $\lambda = 0$, so that the axis of x represents the star's surface. The value of x, although small at the star's surface, is not actually zero, and the various solutions correspond to different values of x at the surface, and so to different effective temperatures of the star. But we see that very shortly they all run together; the influence of the special conditions which prevail at the surface soon disappears as we pass inwards into the star, and at a short distance from the surface we are on the asymptotic solution, so that λ depends only on x, and no longer on the effective temperature of the star's surface.

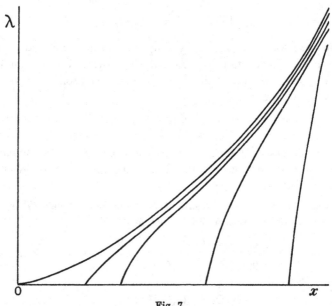

Fig. 7.

The star's surface temperature and effective temperature T_e are determined by the star's emission of radiation, and so by the rate at which energy is being generated in the star's interior. The bunch of solutions shewn in Fig. 7 branch out from the asymptotic solution, and it is this branching out which enables us to find a solution suitable for every given rate of internal generation of energy by the star.

81. Equation (79·5) may also be solved by analytical methods. To do this it is convenient to regard $\frac{1}{8}(\frac{1}{2} - l)$ as a small quantity and search for an expansion in powers of this small quantity.

If, to a first approximation, we neglect $\frac{1}{8}(\frac{1}{2} - l)$ altogether, we find that λ is a root of

$$x^2 = \lambda(\lambda + 1) \qquad \ldots\ldots\ldots\ldots\ldots\ldots\ldots(81\cdot1).$$

This approximate solution fails near the star's surface, because $d\lambda/dx$ may become infinite here. This circumstance, however, only represents the branching out which has already been noticed, and equation (81·1) will give one solution, which we may regard as the asymptotic solution, right up to the boundary of the star.

As far as first powers of the small quantity $\frac{1}{8}(\frac{1}{2} - l)$, it is readily shewn[*] that the asymptotic solution is

$$\lambda = \left[1 - \frac{\frac{1}{4}(\frac{1}{2} - l)\,x^2}{4x^2 + 1}\right][(x^2 + \tfrac{1}{4})^{\frac{1}{2}} - \tfrac{1}{2}]\dots\dots\dots\dots (81\cdot2).$$

The second factor alone would of course give the solution already obtained, the root of equation (81·1). This further approximation shews that the amount of error involved in the earlier approximation is quite small. The maximum error occurs when $x = \infty$, in which case it is of amount $\Delta\lambda$ such that

$$\frac{\Delta\lambda}{\lambda} = 1 - \tfrac{1}{16}(\tfrac{1}{2} - l).$$

When $l = 0$, this is of the order of 3 per cent., and for all small values of l the error is so small that we may properly neglect it.

To a similar degree of error we might have determined λ directly from equation (78·2) and should have obtained λ as a root of the equation

$$\lambda(\lambda + 1) = \frac{12\pi C R \gamma T^{\frac{1}{2}}}{c F \bar{G} a m} \dots\dots\dots\dots\dots(81\cdot3).$$

To within the degree of error just mentioned, this gives the value of λ throughout a star's interior, no matter in what way c, F and $\bar{G}$ vary, provided only that their changes are not too abrupt.

82. We now turn to the dynamical equations.

The total pressure p is given by

$$p = \left(1 + \frac{1}{\lambda}\right)\frac{R}{m\mu}\rho T \dots\dots\dots\dots\dots(82\cdot1).$$

If T' is defined by

$$T' = \left(1 + \frac{1}{\lambda}\right)\frac{\mu}{\mu_0}T \dots\dots\dots\dots\dots(82\cdot2),$$

where μ_0 is any constant,

$$p = \frac{R}{m\mu_0}\rho T' \dots\dots\dots\dots\dots(82\cdot3).$$

so that the pressure is identical with that in a fictitious mass of gas in which radiation pressure is non-existent, the density ρ is the same as in the actual star, the molecular weight has a uniform value μ_0 throughout, and the temperature is T'. Emden's numerical solutions give the equilibrium of just such masses of gas.

* *M.N.* LXXXV. (1925), p. 203.

If we suppose the variations in μ to be given by our previously assumed relation

$$\mu = \mu_0 T^j \quad \dots \dots \dots \dots (82\cdot4),$$

the value of T' becomes

$$T' = \frac{\lambda + 1}{\lambda} T^{1-j} \quad \dots \dots \dots \dots (82\cdot5),$$

and the relation $p_G = \lambda p_R$ assumes the form

$$\frac{R}{m\mu_0} \rho = \tfrac{1}{3} a\lambda T^{3+j} = \tfrac{1}{3} a\lambda \left(\frac{\lambda T'}{\lambda + 1}\right)^{\frac{3+j}{1-j}} \quad \dots \dots \dots (82\cdot6).$$

If, as in § 79, we put $cF\bar{G}$ equal to βT^l, equation (81·3) becomes

$$\lambda(\lambda + 1) = \frac{12\pi CR\gamma T^{\frac{1}{2}-l}}{ma\beta} \quad \dots \dots \dots \dots (82\cdot7),$$

or, replacing T by its value in terms of T' and λ as given by equation (82·5),

$$\lambda^{\frac{1}{2}-j+l}(\lambda + 1)^{\frac{3}{2}-j-l} = \left[\frac{12\pi CR\gamma}{ma\beta}\right]^{1-j} T'^{(\frac{1}{2}-l)} \quad \dots \dots (82\cdot8).$$

This equation gives the value of λ in terms of T', and with this value of λ equation (82·6) gives the value of ρ as a function of T'.

Stars of Small Mass.

83. We have seen (§ 70) that λ, the ratio of gas-pressure to radiation-pressure, is fairly large at the centre of the sun, and for stars of different masses is approximately proportional to $1/M^2$. For stars of mass considerably less than that of the sun, λ will be quite large, so that $1/\lambda$ may be neglected.

In this case the left-hand member of equation (82·8) takes the simple form λ^{2-2j}, so that this is proportional to $T'^{(\frac{1}{2}-l)}$ throughout the star. From equation (82·6) we now find that

$$\rho^{1-j} \propto \lambda^{1-j} T'^{(3+j)} \propto T'^{3\cdot25+j-\frac{1}{2}l}.$$

Thus so far as its distribution of density is concerned a star of small mass arranges itself like a sphere of gas, without radiation pressure, in adiabatic equilibrium with

$$n = \frac{3\cdot25 + j - \frac{1}{2}l}{1-j} \quad \dots \dots \dots \dots (83\cdot1).$$

When μ and $cF\bar{G}$ are uniform throughout the star, j and l are both zero, so that $n = 3\cdot25$.

Stars of Very Great Mass.

84. Inside an ideal star of enormously great mass, λ is very small, so that, from equation (82·8)

$$\lambda^{\frac{3}{2}-j+l} \propto T'^{(\frac{1}{2}-l)}.$$

Configurations of Equilibrium

Equation (82·6) now shews that $\rho \propto T''^n$, where

$$n = \frac{3\cdot 5 - 3j - l}{(1-j)(\frac{1}{2}-j+l)} + \frac{j}{1-j} \quad \dots\dots\dots\dots(84\cdot1).$$

When μ and $cF\bar{G}$ are assumed uniform throughout the star, $j = l = 0$ so that $n = 7\cdot0$. Our analysis shews that in this case the star would arrange itself like a sphere of gas in adiabatic equilibrium with $n = 7$. The radius of such a sphere is, however, infinite, so that the star must find some means of evading the constitution which our analysis would try to impose on it.

Stars of Moderate Mass.

85. The case of a star of intermediate mass, in which λ can neither be treated as very small nor very great, presents more serious difficulties. We can obtain the relation between ρ and T' by eliminating λ between equation (82·6) and (82·8), but the result will not in general be of the desired form

$$\rho \propto T''^n \dots\dots\dots\dots\dots\dots\dots(85\cdot1),$$

which corresponds to a sphere of gas in adiabatic equilibrium and so, thanks to Emden's tables, is suited to easy numerical computation.

The ρ, T' curve which results from the elimination of λ between the two equations just mentioned can always be made to agree with equation (85·1) at the centre of the star by a suitable choice of the constant multiplying T''^n in this equation. We can further make the tangents to the two curves coincide at the centre of the star by a suitable choice of n. Moreover, the two curves agree in any case at the surface of the star ($\rho = 0$, $T' = 0$).

Thus by a suitable choice of n we can make the curves agree at three points, two of which are close to the centre of the star, round which the main part of the mass is congregated. In the special cases of λ small and λ great, the curves not only agree at these three points, but coincide throughout their lengths.

For stars of intermediate mass we shall determine n in the way just described, and then assume the ρ, T' relation to be that given by equation (85·1). Analysis of definite cases suggests that this procedure gives quite a good approximation for stars of small and medium mass.

The value of n will of course be the value of

$$\frac{T'}{\rho}\frac{\partial\rho}{\partial T'},$$

obtained by differentiating equations (82·6) and (82·8) and eliminating $d\lambda$. After simple algebra its value is found to be

$$n = \frac{3+j}{1-j} + \frac{1-2l}{1-j}\left[\frac{4+\lambda_c(1-j)}{1+4\lambda_c(1-j)+2l-2j}\right] \quad \dots\dots(85\cdot2).$$

In passing we may notice that for a star of small mass (λ_c large) this gives the value of n already obtained in equation (83·1); for a star of large mass (λ_c small) it gives the same value for n as equation (84·1).

In the special case in which μ and $cF\bar{G}$ have uniform values throughout the star, so that $j = l = 0$, this reduces to

$$n = 3 + \frac{4 + \lambda_c}{1 + 4\lambda_c} \quad \ldots\ldots\ldots\ldots\ldots\ldots(85\cdot3).$$

Thus as λ_c varies from ∞ to 0—i.e. as we pass from stars of very small mass to stars of very great mass, n varies continuously between $n = 3\cdot25$ and $n = 7$, the two limiting values already obtained.

We can obtain somewhat higher accuracy as follows. The contributions $4\pi\rho r^2 dr$ made to the total mass by small equal steps dr in r, increase from 0 at the centre ($r = 0$) up to a maximum, and afterwards decrease until they again become zero at the surface. A study of Emden's solutions suggests that the maximum occurs near to a value r' of r such that half the total mass of the star is contained within the sphere of radius r'. Thus we may say that the main part of the star's mass is concentrated about the value r' of r.

A further study of Emden's solutions shews that the temperature at a distance r' from the centre ranges from $0\cdot59\,T_c$ for stars of very small mass up to $0\cdot61T_c$ for stars of very great mass. We shall obtain sufficient accuracy for our present purpose by supposing this temperature to be uniformly equal to $0\cdot60T_c$. To the accuracy of this approximation, half of the star's mass is at a temperature greater than $0\cdot60T_c$, whilst half is at a lower temperature, the temperatures of different elements of the star being ranged fairly closely around the value $0\cdot60T_c$. If λ' is the value assumed by λ at a distance r' from the centre, equation (82·7) gives the relation

$$\lambda'(\lambda'+1) = (0\cdot60)^{\frac{1}{2}-l}\lambda_c(\lambda_c+1)\ldots\ldots\ldots\ldots(85\cdot4).$$

When μ and $cF\bar{G}$ have uniform values throughout the star, $l = 0$, and this equation becomes

$$\lambda'(\lambda'+1) = 0\cdot775\lambda_c(\lambda_c+1)\ldots\ldots\ldots\ldots\ldots(85\cdot5),$$

which reduces further in the special cases of λ large and λ small, giving

for a star of very small mass, $\lambda' = 0\cdot880\lambda_c$,

    ,,   ,,   ,,   large   ,, $\lambda' = 0\cdot775\lambda_c$.

Clearly on replacing λ_c by λ' in formulae (85·2) and (85·3) we shall obtain a better approximation to the arrangement of stars of moderate mass.

When $n = 5$, the standard differential equation (65·5) can be solved in finite terms, the solution being Schuster's solution

$$\rho = \rho_c\left(1 + \frac{r^2}{a^2}\right)^{-\frac{5}{2}}.$$

For all values of n less than 5, the differential equation (65·5) gives the star an infinite radius; for values of n less than 5, the radius is finite. From equation (85·3), the critical value $n = 5$ corresponds to $\lambda' = 0·286$ and, by equation (85·5), the corresponding value of λ_e is 0·35. Even this improved approximation, however, fails for the limiting case of a star of infinite radius, for I have found, as the result of numerical calculations by quadratures, that the actual value of λ_e for which the radius first becomes infinite is quite close to $\lambda = 0·5$. For this reason we revert to the simpler approximation again from § 88 onwards.

Distribution of Density.

86. Emden's numerical solutions give the following values for the radius r_1 of masses of gas arranged in adiabatic equilibrium with different values of n, and for $\rho_c/\bar{\rho}$, the ratio of the central density to the mean density:

$n =$	3·25	4·0	4·5	4·9	$\geqslant 5·0$
$r_1 =$	8·00	15·00	32·14	169·47	∞
$\rho_c/\bar{\rho} =$	88	623	6378	934800	∞

We see that gaseous stars of large mass have very high central condensation of mass; the greater part of their mass is condensed in a very small central region, while their outer layers are mere tenuous unsubstantial veils drawn round small and massive central cores.

For instance, from Emden's solution[*] for a star arranged according to the law $n = 4·9$, I have calculated the following values for M_r/M, the ratio of the mass enclosed within a sphere of radius r to the total mass:

$r_1 = 0$	1	2	3	4	5	10	50	169·5
$M_r/M = 0$	0·125	0·436	0·657	0·788	0·850	0·958	0·997	1·000

The main part of the mass is concentrated very near to the centre of the star. Ninety-six per cent. of the whole is enclosed within a sphere whose radius is one-seventeenth of that of the star, the remaining space, which is 99·98 per cent. of the total volume of the star, containing only 4 per cent. of the star's mass.

To make the example more concrete, let us imagine that Betelgeux, with a radius 300 times that of the sun, is arranged internally according to the law $n = 4·9$. The above figures shew that a sphere of radius equal to that of the sun ($r_1 = 0·56$) drawn round the centre of Betelgeux will contain over a twentieth part of the star's mass, or a mass which is almost certainly greater than that of the sun itself. Thus to within a sun's radius from its centre, Betelgeux would, if it had the structure we have imagined, be denser than the sun. The remaining 299 radii are comparatively devoid of matter and it is their inclusion in the star's volume which reduces its mean density to the low value of 0·000002.

---

[*] *Gas Kugeln*, p. 81.

The masses of stars such as Betelgeux, Plaskett's star, Pearce's star and Otto Struve's star (27 Canis Majoris) are so great that if the gas-laws were obeyed, their internal arrangements might well approximate to that represented by the law $n = 4.9$. The central densities of such stars being a million times their mean density, the last three of these stars would have a central density of the order of 100,000, which is greater than the mean density of the companion to Sirius. These densities do not actually occur, because as we shall see later, the gas-laws break down long before they are reached.

ABSENCE OF CONVECTION CURRENTS.

87. To the accuracy of the approximation considered in §85, the distribution of pressure p and density ρ in every star is of the type determined by the relation

$$p = K\rho^\kappa \dots\dots\dots\dots\dots\dots\dots(87\cdot1),$$

where $\kappa - 1 = 1/n$.

The ratio of the specific heats of the stellar material, γ, must be very nearly equal to $1\frac{2}{3}$, since the atoms are very nearly broken up into their ultimate constituent nuclei and free electrons, which possess no capacity for internal energy.

Equation $(87\cdot1)$ can be written in the form

$$p = L\rho^\gamma \dots\dots\dots\dots\dots\dots\dots(87\cdot2),$$

where

$$L = K\rho^{-(\gamma-\kappa)} \dots\dots\dots\dots\dots\dots(87\cdot3).$$

We have seen that in general n has values greater than $3\cdot25$, so that κ is less than $1\cdot308$ for all stars, while γ, being nearly equal to $1\cdot667$, is certainly larger than $1\cdot308$. Thus $\gamma - \kappa$ is positive, so that L increases as ρ decreases and *vice versa*.

Now a spherical mass of gas in which p and ρ are arranged according to the relation $(87\cdot2)$ will be stable if L increases everywhere as we pass from centre to surface, but if L decreases in any region, convection currents will be set up until the stellar matter has become thoroughly stirred up and the matter rests in equilibrium with L constant over the region in question*.

Our analysis has shewn that L increases everywhere on passing from centre to surface, so that there will be no convection currents. We can see in a general way why this must be. Convection occurs in a kettle of water which is being heated, when the hot water at the bottom is of lower density than the cool water at the top; it is absent in a star (except perhaps quite close to the star's surface) because the hot matter near the centre, notwithstanding its intense heat, is still enormously more dense than the comparatively cool matter near its surface.

Thus we see that the mixture of matter in a star's interior is not analogous to that in the earth's lower atmosphere in which the constituent gases are

---

* *Problems of Cosmogony*, p. 193.

kept thoroughly mixed by winds and convection currents, but rather to the serene upper atmosphere in which the lightest elements float to the top while the heaviest sink downwards under gravity.

Our discussion has applied only to the interior of a star. On approaching the star's surface, our equations of radiative equilibrium begin to fail, so that the discussion gives no information as to the occurrence of convection-currents near a star's surface. Solar physics suggests that there may be quite appreciable convection currents near a star's surface, but this neither confirms nor disproves our theoretical result, which has reference only to stellar interiors.

We cannot overlook the possibility that other factors, which our idealised discussion has ignored, may produce a tendency to establish convection currents in stellar interiors.

Electric Field.

We have treated atoms and electrons merely as gravitating particles, thus ignoring the electric field in a star's interior. The electric forces between electrons and positively-charged atoms completely outweigh the gravitational forces, and this causes the charges to arrange themselves so that the positive and negative charges nearly neutralise one another, leaving only a small residual field. This, however, provides no justification for neglecting this residual field altogether.

Just as the molecules of hydrogen and helium have diffused upwards in the earth's atmosphere on account of the smallness of their masses, so the free electrons, having far smaller masses than nuclei and atoms, must diffuse outwards in a star. As a result, the inner regions of a star must become positively charged, and the outer regions negatively charged. The process is held in check by the electric fields which electronic diffusion creates in the star's interior, and as soon as the central regions of a star acquire an appreciable positive charge, the tendency for the free electrons to wander away to the surface is counteracted, and a state of equilibrium is attained. The problem is further complicated by the tendency of the free electrons near a star's surface to escape from the star altogether, precisely as the hydrogen molecules have escaped altogether from the earth's atmosphere. This results in the star acquiring a positive charge*.

Pannekoek† and Rosseland‡ have studied the equilibrium of a star, taking the electric forces into account. They find that a quite simple solution exists, such that the inward force on an electron at any point is precisely equal to the outward force on a positively-charged atom at the same point. Rosseland has investigated stellar equilibrium on the supposition that this condition

---

* Jeans, *Dynamical Theory of Gases* (4th Edn.), p. 348.
† *Bull. Ast. Netherlands*, XIX. (1922), p. 107.
‡ *M.N., R.A.S.* LXXXIV. (1924), p. 720.

determines the state of actual stars, and finds that atoms of different weights would be nearly evenly distributed throughout the star, even if there are no convection currents. I have, however, calculated* that this special solution assigns a positive charge of about 9×10^{11} coulombs to the sun, whereas the limit of charge, fixed by the condition that it shall just restrain negative electrons from leaving the sun, is 4×10^9 coulombs, or only 0·0045 times that given by Rosseland's solution; the factor of 0·0045 is, moreover, independent of the special constitution of the sun, and is the same for the stars†. Thus in an actual star the effect discussed by Rosseland only goes about a two-hundredth part of the way towards producing a general mixing of atoms, and so may be neglected for all practical purposes.

v. Zeipel's Theorem.

In 1924, H. v. Zeipel‡ published a remarkable series of investigations which claimed to shew that the generation of energy per unit mass G in a uniformly rotating star is given by

$$G = c \left(1 - \frac{\omega^2}{2\pi\gamma\rho}\right) \dots\dots\dots\dots\dots\dots\dots(87\cdot4),$$

where ω is the angular velocity and c is constant throughout the star. The theorem is obviously impossible when ω has any appreciable value different from zero, since it makes G assume violently negative values near the surface of the star ($\rho = 0$). In the special case of a very slow rotation, v. Zeipel's result takes the form that G must be constant throughout the star. If, for instance, a star consisted only of atoms of lead and of uranium, there could be no equilibrium until complete mixing of these atoms was attained.

Milne and Eddington have interpreted v. Zeipel's condition as one which the star would evade rather than conform to§, the latter suggesting that an actual star would evade it by setting up a system of rotatory currents in meridional planes.

The simple explanation of the puzzle appears, however, to be that v. Zeipel did not derive his theorem from the exact equations of equilibrium (73·1) but from Eddington's inexact equations (72·3). The latter, as we have seen, are only true in the special case of $G = 0$. Thus to reconcile v. Zeipel's theorem with the assumptions from which it is derived, we must assign to the undetermined constant c the special value $c = 0$. The theorem now takes the harmless form that when $G = 0$, then $G = 0$. If the exact equations (73·1) are introduced, the theorem is found to fail altogether.

Thus stellar matter appears to be free to arrange itself under the influence

---

* *M.N., R.A.S.* lxxxvi. (1926), p. 561.
† *l.c.* p. 562.
‡ *M.N., R.A.S.* lxxxiv. (1924), p. 665, and subsequent papers.
§ *Observatory*, xlviii. (1925), p. 73.

of the gravitational and electric forces acting upon its constituent particles, and we have seen that the gravitational forces predominate, so that the heavier elements tend to sink to the star's interior.

This consideration at once suggests that the elements which proclaim their existence spectroscopically in the outermost layers of the sun and stars are likely to be only the lightest of the elements contained in the star; beyond these there ought to be other heavier elements, too heavy to rise to the surface, which have found their natural place near the centre of the star. We should not expect to find these elements on the earth since the earth, to the best of our knowledge, has been formed quite recently from the outermost layers of the sun.

INTERNAL TEMPERATURE, DENSITY, ETC.

88. The foregoing analysis has determined the model on which a gaseous star is built; we now turn to the consideration of the absolute values of temperature, density, etc.

When the relation between T' and ρ is $\rho \propto T'^n$, equation (82·3) shews that the configuration of the star is that of a sphere of gas in adiabatic equilibrium with a mean molecular weight equal to μ_0 throughout. Thus the general relations already developed in § 65 are directly applicable with T' replacing T throughout.

In particular we have (cf. equation (65·3))

$$T_c' = \Theta \rho_c^{\kappa-1} \quad\dots\dots\dots\dots\dots\dots\dots\dots(88\cdot1),$$

where Θ is given by equation (65·4) with μ put equal to μ_0, and $\kappa - 1 = \dfrac{1}{n}$.

Using this value of Θ and equations (65·6), the equation becomes

$$T_c' = \frac{4\pi\gamma m\mu_0}{R(n+1)} \frac{r^2}{\mathfrak{r}_1{}^2} \rho_c \quad\dots\dots\dots\dots\dots\dots(88\cdot2),$$

or again, introducing the mean density $\bar{\rho}$, defined by $M = \tfrac{4}{3}\pi\bar{\rho}r^3$,

$$T_c' = \frac{3\gamma m\mu_0}{R(n+1)\mathfrak{r}_1{}^2} \left(\frac{\rho_c}{\bar{\rho}}\right) \frac{M}{r} \quad\dots\dots\dots\dots\dots(88\cdot3).$$

Passing from T' to T by the use of relation (82·2), we find that T_c is given by

$$T_c^{1-j} = \frac{3\gamma m}{R(n+1)\mathfrak{r}_1{}^2} \left(\frac{\lambda_c}{1+\lambda_c}\right) \left(\frac{\rho_c}{\bar{\rho}}\right) \frac{M\mu_0}{r} \quad\dots\dots\dots(88\cdot4),$$

or, multiplying throughout by $T_c{}^j$

$$T_c = \frac{3\gamma m}{R(n+1)\mathfrak{r}_1{}^2} \left(\frac{\lambda_c}{1+\lambda_c}\right) \left(\frac{\rho_c}{\bar{\rho}}\right) \frac{M\mu_c}{r} \quad\dots\dots\dots(88\cdot5),$$

where μ_c is the value of the effective molecular weight at the centre of the star.

Inserting the numerical values $R/m = 8\cdot26 \times 10^7$ and $\gamma = 6\cdot66 \times 10^{-8}$, and further measuring M and r in terms of the mass ($2\cdot00 \times 10^{33}$) and radius ($6\cdot95 \times 10^{10}$) of the sun, this assumes the form

$$T_c = 69\cdot6 \times 10^6 \frac{\lambda_c}{1+\lambda_c} \left[\frac{1}{(n+1)\,\mathfrak{r}_1{}^2} \left(\frac{\rho_c}{\bar\rho}\right) \frac{M\mu_0}{r} \right] \quad \dots\dots\dots(88\cdot6),$$

in which M and r are measured in terms of the mass and radius of the sun.

From equation (82·6) we obtain at once

$$\lambda_c{}^4 \left(\frac{T_c'}{1+\lambda_c}\right)^{3+j} = \left(\frac{R\rho_c}{\tfrac{1}{3}am\mu_0}\right)^{1-j},$$

so that, on substituting for T_c' from (88·2)

$$\lambda_c{}^4 \mu_0{}^4 M^{2+\frac{2}{3}j} (\bar\rho)^{\frac{2}{3}j} = B (1+\lambda_c)^{3+j} \quad \dots\dots\dots\dots\dots(88\cdot7),$$

where

$$B = \left(\frac{R}{m}\right)^4 \left[\frac{(n+1)\,\mathfrak{r}_1{}^2}{3\gamma}\right]^{3+j} (\tfrac{1}{3}a)^{-1+j} \left(\frac{\bar\rho}{\rho_c}\right)^{2+2j} \quad \dots\dots\dots(88\cdot8).$$

This completes the analysis necessary for the evaluation of internal temperature, density, etc. Starting from any values of n, we calculate the value of λ_c from relation (85·2); equation (88·7) next gives the mass of the corresponding star, and equation (88·6) then gives the central temperature for a star of this mass.

Stars with uniform effective molecular weight.

89. The problem assumes its simplest form when the effective molecular weight is constant throughout the star. In this case $j = 0$, so that from equations (88·7) and (88·8),

$$\lambda_c{}^4 \mu^4 M^2 = B (1+\lambda_c)^3 \quad \dots\dots\dots\dots\dots\dots(89\cdot1),$$

where

$$B = \left(\frac{R}{m}\right)^4 \left[\frac{(n+1)\,\mathfrak{r}_1{}^2}{3\gamma}\right]^3 \frac{3}{a} \left(\frac{\bar\rho}{\rho_c}\right)^2 \quad \dots\dots\dots\dots(89\cdot2).$$

The value of B now depends only on n, and is readily determined from Emden's numerical solutions.

Emden gives solutions only for the values $n = 2\cdot5,\ 3,\ 4,\ 4\cdot5,\ 4\cdot9$ and 5, so that solutions for intermediate values of n must be obtained by interpolation. The interpolation becomes easy on noticing that the values of $1/\mathfrak{r}_1$, and of $\mathfrak{r}_1{}^3 \left(\frac{\bar\rho}{\rho_0}\right)$ run quite smoothly with n. Table X (p. 97) gives values calculated from Emden's solutions, those corresponding to the values $n = 3\cdot25$ and $3\cdot5$ being obtained by interpolation.

The table refers to the case in which μ is constant throughout the star so that $j = 0$, and the relation (85·2) between n and λ_c reduces to

$$n = 3 + (1 - 2l) \frac{4 + \lambda_c}{1 + 4\lambda_c + 2l} \quad \dots\dots\dots\dots(89\cdot3).$$

Table X. *Solutions for various internal arrangements of stars.*

κ	n	τ_1	$\dfrac{\rho_c}{\bar{\rho}}$	$\tau_1{}^3\left(\dfrac{\bar{p}}{\rho_c}\right)$	$\dfrac{1}{(n+1)\tau_1{}^2}\left(\dfrac{\rho_c}{\bar{\rho}}\right)$	B	$B\,(\odot=1)$
1·4	2·5	5·42	24·1	6·60	0·235	$1\cdot02\times10^{69}$	255
1·33	3·0	6·90	54·4	6·04	0·285	$1\cdot27\times10^{69}$	319
1·31	3·25	8·00	88	5·84	0·322	$1\cdot43\times10^{69}$	358
1·29	3·5	9·52	152	5·67	0·374	$1\cdot60\times10^{69}$	400
1·25	4·0	15·00	623	5·42	0·554	$2\cdot00\times10^{69}$	500
1·222	4·5	32·14	6378	5·20	1·124	$2\cdot46\times10^{69}$	614
1·204	4·9	169·47	934,800	5·20	5·524	$3\cdot03\times10^{69}$	758
1·200	5·0	∞	∞	5·20	∞	$3\cdot19\times10^{69}$	799
1·14	7·0	∞	∞	5·20	∞	—	—

Further progress can only be made by assigning a definite value to l. As the labour of continuing the calculations for all values of l would be excessive, we may confine our attention to two values for l, namely $l=\frac{1}{2}$ and $l=0$, trusting to interpolation or extrapolation to give an adequate idea of the solutions for other values of l.

Solutions when $j=0$ and $l=\frac{1}{2}$ (Eddington's model).

90. When $l=\frac{1}{2}$, equation (89·3) gives $n=3$, independently of the value of λ_c; this, as we have already seen (§ 76), is the model discussed by Eddington. For this value of n, Table X gives $\rho_c=54\cdot4\,\bar{\rho}$, so that the central density is always equal to 54·4 times the mean density, and $B=319$.

Equation (89·1) now becomes

$$\lambda_c{}^4\mu^4 M^2 = 319\,(1+\lambda_c)^3 \quad\ldots\ldots\ldots\ldots\ldots\ldots\ldots(90\cdot1),$$

which gives λ_c at once in terms of M and μ, and we have already noticed that λ is constant throughout the star, as is also obvious from equation (82·7). Equation (88·6) gives the central temperature in the form

$$T_c = 19\cdot8\times10^6\,\frac{\lambda}{1+\lambda}\left(\frac{M\mu}{r}\right)\ldots\ldots\ldots\ldots\ldots\ldots(90\cdot2)$$

The total emission of radiation E is in general equal to $M\bar{G}$, where $\bar{G}$ is measured at the surface of the star, and when all stars are supposed built on the same special model ($n=3$), this is proportional to $M\bar{G}$ where $\bar{G}$ is measured at the star's centre. Inserting the value of $\bar{G}$ from equation (81·3), or from Eddington's integral (76·1), we find that

$$E \propto \frac{MT_c{}^{\frac{1}{2}}}{\lambda\,(\lambda+1)} \quad\ldots\ldots\ldots\ldots\ldots\ldots\ldots(90\cdot3).$$

Using equations (90·1) and (90·2), and also the relation $E=4\pi\sigma r^2 T_e{}^4$, where T_e is the effective temperature of the star's surface, we find that

$$E \propto M^{\frac{7}{5}}(\lambda+1)^{-\frac{3}{2}}\mu^{\frac{4}{5}}T_e{}^{\frac{4}{5}} \quad\ldots\ldots\ldots\ldots\ldots(90\cdot4)$$

Transforming to absolute bolometric magnitude m, by using the relation $m = -2.5 \log E + $ a constant, this becomes

$$m = \text{cons} - \tfrac{7}{2} \log M + \tfrac{15}{4} \log (\lambda + 1) - 2 \log \mu - 2 \log T_e \quad \ldots(90\text{·}5).$$

The supposed "mass-luminosity" law mentioned in § 76 has now disappeared completely. The luminosity of a star of given mass M can be anything from zero to infinity, and a star can adjust its configuration to any given emission of radiation by selecting a suitable surface temperature. Low surface temperature accompanies low luminosity and *vice versa*.

The constant on the right of equation (90·5) admits of evaluation in terms of constants of nature and the coefficients in Kramers' opacity-law. Eddington[*] found that when the constant was evaluated in this way, the absolute magnitudes given by the formula did not agree with those of observed stars. The average error proved to be about $2\tfrac{1}{4}$ magnitudes, the stars having only about an eighth of the luminosity that Kramers' law would require if they were purely gaseous stars built on the model $l = \tfrac{1}{2}$ or $k\bar{G} = constant$. For a star of given mass and luminosity, the average error in $\log T_e$ is of the order of $1\tfrac{1}{4}$, so that the stars have some 13 times the effective temperatures and so only about a two-hundredth part of the radius that Eddington's discussion would assign to them; on his model, a star of the mass and luminosity of the sun would be as big as Betelgeux.

Eddington accordingly treated the constant on the right of equation (90·5) as adjustable[†], selecting its value so that the formula gave the right absolute magnitude for Capella. Using this value for the constant, and taking $\mu = 2\text{·}11$, the formula was found to shew a fair agreement with observation which we shall discuss later (§ 118).

Russell first drew attention to a general objection affecting not only Eddington's model, for which $l = \tfrac{1}{2}$, but all stellar models for which l has a positive value.

If $l = \tfrac{1}{2}$, as in Eddington's model, then $k\bar{G}$ is constant, whence we readily find that

$$G = \frac{1}{\rho r^2} \frac{d}{dr} \left(\frac{1}{k} \int_0^r \rho r^2 dr \right) \times a \ constant.$$

Inserting Kramers' value for k, we obtain, since ρ/T^3 is constant on this model,

$$G = \frac{1}{\rho r^2} \frac{d}{dr} \left(T^{\tfrac{1}{2}} \int_0^r \rho r^2 dr \right) \times a \ constant.$$

In the outer regions of a star ρ is small and $T^{\tfrac{1}{2}}$ decreases rapidly with r, so that G assumes a large negative value. Thus the model demands a very

---

large positive generation of energy in the star's central regions, and a very large negative generation of energy in its outer regions. Since $\bar{G}$ becomes very small near the star's surface, these almost balance out, leaving only a small residue of net generation of energy.

This objection applies to all positive, but not to negative, values of l. Our model with l negative will give a true picture of the effect of a deficiency of energy-generation at the star's centre, and since there is continuity in passing through $l = 0$, small positive values of l will represent the general tendency of a central condensation of the generation of energy.

Solutions when $j = 0$ and $l = 0$.

91. We next consider solutions with $j = l = 0$, representing stars with uniform effective molecular weight, generation of energy uniformly distributed throughout the mass of the star, and opacity coefficient strictly proportional to $\mu\rho T^{-3\cdot5}$

The simplest case occurs when the star's mass is so small that λ may be treated as a large quantity. The value of n is then 3·25 and $B = 358$. Equation (89·1) now assumes the form

$$\lambda_c \mu^4 M^2 = 358 \quad \dots\dots\dots\dots\dots(91\cdot1),$$

while equation (88·6) becomes

$$T_c = 22\cdot4 \times 10^6 \frac{M\mu}{r} \quad \dots\dots\dots\dots\dots(91\cdot2).$$

These formulae make it very easy to calculate the values of λ_c and T_c for stars of given mass and radius, provided only this mass is so small that λ_c comes out to be a fairly large number.

When the masses are not so small as this, the procedure is more complicated. We start with a series of values of n and calculate the corresponding values of λ_c from equation (85·3). Table X next gives the value of B and with these values of B and λ_c known, equation (89·1) at once gives the value of $\mu^2 M$. Thus for any assigned value of μ, we have a system of values of M corresponding to different values of n. By interpolation we find the value of n, and hence the whole solution, corresponding to any given mass.

The following table shews the result of such calculations. No special value of μ has been assumed, but the table is arranged so as to facilitate calculation for the value $\mu = 2\cdot5$. In this case the factor $(\mu/2\cdot5)$ of course becomes equal to unity, and the fourth column gives the mass directly. For other values of μ it is necessary to calculate $M(\mu/2\cdot5)^2$ before commencing to use the table.

The last column of the table gives values of the central temperature T_c, or rather of the quantity $T_c r$, where r is measured in terms of the radius of the sun; these are calculated directly from equation (88·6).

Table XI. *Solutions for stars of different masses.*

κ	n	$B (\odot = 1)$	λ_c	$M\left(\dfrac{\mu}{2\cdot5}\right)^2$	$T_c r\left(\dfrac{\mu}{2\cdot5}\right)^{-1}$
1·308	3·250	358	∞	0	0
1·307	3·254	360	230	0·20	11×10^6
1·307	3·256	361	150	0·25	14×10^6
1·306	3·272	363	42	0·50	28×10^6
1·301	3·325	375	12·2	1·00	55×10^6
1·300	3·333	375	11·0	1·06	58×10^6
1·286	3·500	400	3·50	2·49	126×10^6
1·267	3·750	446	1·62	5·44	264×10^6
1·250	4·000	500	1·00	10·12	490×10^6
1·222	4·500	614	0·50	29·13	1900×10^6
1·204	4·900	758	0·32	65·25	15200×10^6
1·200	5·000	799	0·29	78·78	∞
1·141	7·000	—	0·00	∞	∞

As examples of the use of this table and Table X, Table XII gives calculated values of the central density and temperatures of actual stars, calculated on the supposition of a uniform effective molecular weight $\mu = 2\cdot5$.

Table XII. *Solutions for actual stars.*

(Solutions with $j = l = 0$ calculated for eff. mol. weight $\mu = 2\cdot5$.)

Star	Mass $(\odot = 1)$	λ_c	n	$\bar{\rho}$	ρ_c	$r\,(\odot = 1)$	T_c
B.D. 6° 1309	[78]	0·29	5·00	0·14	[∞]	—	[∞]
H.D. 1337 A	36·3	0·45	4·59	0·004	40	23·8	200×10^6
V Puppis A	19·2	0·67	4·27	0·06	120	7·6	160×10^6
u Herculis A	7·6	1·20	3·90	0·14	70	4·6	100×10^6
Sirius A	2·45	3·53	3·50	0·93	140	1·58	80×10^6
Sun	1·00	12·2	3·33	1·42	140	1·00	55×10^6
60 Kruger A	0·25	150	3·26	9·6	860	0·33	42×10^6
” 　 B	0·20	230	3·25	60	5300	0·17	65×10^6
Betelgeux	[40]	0·42	4·65	2×10^{-6}	[0·04]	300	10×10^6
Capella A	4·18	2·08	3·65	0·004	1	11	19×10^0
a Cent. B	0·97	12·9	3·32	0·76	70	1·2	45×10^6

92. In § 90 we saw that when the mass M of a star is given, and the effective molecular weight μ of the matter of which it is composed, it is possible to calculate the value of λ_c, the ratio of gas-pressure to pressure of radiation at its centre. It is also possible to calculate n. Thus all stars for which M and μ are specified have the same values of λ_c and n. They are therefore (since n is given) all built on the same model, and at corresponding points radiation-pressure has the same importance relative to gas-pressure.

It has also been found possible to calculate $T_c r$, the product of the central temperature and the radius of the star, but it has not been found possible to calculate T_c and r separately. Thus all stars which have given values of M and μ lie on a certain homologous series along which $T_c r$ is constant.

The question arises as to what further physical conditions determine which particular configuration will be assumed by an actual star. The analysis of § 90 supposed l and j (cf. § 79) both to be zero, and so supposed μ and $cF\bar{G}$ to have constant values throughout the star. We have specified the value of u, but have so far not found it necessary to specify any value for $cF\bar{G}$.

At the centre of the star we have, from equation (81·3),

$$\lambda_c(\lambda_c+1) = \frac{12\pi CR\gamma}{cF\bar{G}am} T_c^{\frac{1}{2}} \quad\dots\dots\dots\dots\dots(92\cdot1).$$

In § 77 we obtained the value of cF on Kramers' theory of absorption, in the form

$$cF = 4\cdot46 \times 10^{25} \left(\frac{N^2}{A}\right) \quad\dots\dots\dots\dots\dots(92\cdot2).$$

Inserting this value for cF, and using the known numerical values $C = 3 \times 10^{10}$, $R/m = 8\cdot26 \times 10^7$, $\gamma = 6\cdot66 \times 10^{-8}$, $a = 7\cdot64 \times 10^{-15}$, equation (92·1) becomes

$$\lambda_c(\lambda_c+1)\frac{N^2}{A}\bar{G} = 18\cdot3\, T_c^{\frac{1}{2}} \quad\dots\dots\dots\dots\dots(92\cdot3).$$

All stars which have the same mass M and the same effective molecular weight μ have the same values of λ_c. Equation (92·3) now shews that $T_c^{\frac{1}{2}}$ is proportional to $\left(\frac{N^2}{A}\right)\bar{G}$. As we pass along the homologous series of configurations T_c varies, and the actual position which a given star will assume on this series is determined jointly by the values of (N^2/A) and $\bar{G}$.

It will be remembered that N is the atomic number and A the atomic weight of the atoms of which the star is formed. For a given star these will be fixed, and treating N^2/A as a constant, equation (92·3) shews that T_c varies as $(\bar{G})^2$. Thus there will be one and only one configuration appropriate to a given value of $\bar{G}$, and if a star's mean generation of energy $\bar{G}$ is given, the star can always find a configuration of equilibrium by shrinking or expanding until its central temperature T_c has the value appropriate to the given value of $\bar{G}$. The question of the stability of this configuration will be considered later. Let us for the moment suppose it to be stable, then, since $\bar{G}$ is proportional to the luminosity of the star, we see that:

In gaseous stars which have the same given values for M, μ and N^2/A— i.e. stars whose mass and composition is fixed—T_c is proportional to the square of the luminosity of the star.

Since $T_c r$ is constant it follows that *the radius of the star is inversely proportional to the square of its luminosity.*

These are the laws which would be obeyed if all the quantities which we have assumed to be uniform were actually uniform and if the gas-laws were accurately obeyed throughout stellar interiors.

Observation reveals no tendency for these laws to be obeyed. As we shall see below (§ 167) the least luminous stars of a given mass generally have the smallest radii and the highest central temperatures. We shall see at once that the laws are not obeyed, since the atomic weights of stellar atoms, calculated on the supposition that the gas-laws are obeyed, will be found to vary widely from one star to another.

The Atomic Weight of Stellar Matter.

93. Equation (92·3) may be written in the form

$$\frac{N^2}{A} = \frac{18\cdot3\, T_c^{\frac{1}{2}}}{\lambda_c\,(\lambda_c+1)\,\bar{G}}\dots\dots\dots\dots\dots\dots(93\cdot1),$$

and so provides a means of determining N^2/A for the atoms of which actual stars are composed. Indeed, we have already calculated T_c and λ_c for a number of stars, and since $\bar{G}$ is readily calculated from the star's luminosity, the value of N^2/A is obtained at once.

It must of necessity be possible to determine N^2/A from observations on a star's structure, because the coefficient of opacity, by which the star's whole structure is determined, is proportional to N^2/A. If we cut every atomic nucleus in a piece of matter into two equal halves, we halve both N and A and so also N^2/A, with the result that the substance becomes twice as transparent as before. This is the theoretical basis of the well-known physical fact that a large clot of matter in the form of a massive nucleus is far more effective in absorbing X-radiation than a large number of small clots of equal total mass. It is for this reason that the physicist and surgeon both select lead as the material with which to screen their X-ray apparatus; a ton of lead is far more effective in stopping unwanted X-rays than a ton of wood or of iron. If we knew the strength of an X-ray apparatus, and the total weight of shielding material round it, we could form a very fair estimate of the atomic weight of the shielding material by measuring the amount of X-radiation which escaped through it.

In using formula (93·1) we are in effect using just such a method to determine the atomic weight of the atoms of which the stars are composed. A star is in effect nothing but a huge X-ray apparatus. We know the total mass of many stars, and we can readily calculate the rate at which they are generating X-rays—it is merely the rate at which they are radiating energy away into space. If we could shut a Maxwell demon inside a star and make him cut each atomic nucleus in half, keeping the star's mass and rate of generation unaltered, we should, as we have seen, halve the coefficient of opacity of the star. This would necessitate a change in the star's build: in actual fact its radius would increase fourfold while its surface temperature would be halved. We could follow the progress of the demon's work by watching the changes in either the radius or the surface-temperature of the star. Hence from the observed surface-temperature or the calculated radius of any star the mass and luminosity of which are known, it must be possible to estimate the atomic weight of the atoms of which the star is composed. Formula (93·1) provides the means.

Table XIII gives the values of N^2/A calculated from this formula. The third column contains the bolometric luminosity in terms of the sun's bolometric luminosity as unity, and the fourth contains the value of $\bar{G}$ measured in ergs per gramme per second.

Table XIII. *Atomic Weights of Stellar Matter.*

(Calculated on the supposition that the gas-laws are obeyed.)

Star	Mass ($\odot = 1$)	Bolom. Lum. ($\odot = 1$)	$\bar{G}$	λ_c	T_c	$\dfrac{N^2}{A}$
V Puppis	19·2	11,000	1,100	0·67	160×10^6	190
u Herculis	7·6	1,250	300	1·20	100×10^6	210
Sirius A	2·45	38	29	3·53	80×10^6	350
Sun	1·00	1·00	1·90	12·2	55×10^6	440
60 Kruger A	0·25	0·01	0·06	150	42×10^6	100
,, B	0·20	0·003	0·021	230	65×10^6	130
Betelgeux	[40]	6,000	[300]	0·42	10×10^6	320
Capella A	4·18	105	48	2·08	19×10^6	250
a Cent. B	0·97	0·46	0·90	12·9	45×10^6	750

The values of N^2/A in the last column cannot lay claim to any high degree of accuracy. There are two principal sources of error, namely the value assumed for $\bar{G}$ and the value assumed for λ_c in equation (93·1).

In the case of stars of very high or very low temperature, the value of $\bar{G}$ cannot be determined at all exactly. It is determined from the bolometric absolute magnitude of the star, and even if the visual absolute magnitude is known, the bolometric correction is generally very uncertain. An error of only half a magnitude in the bolometric correction would throw the value of N^2/A into error by a factor of 1·58.

This source of error almost vitiates the entries for V Puppis, u Herculis, 60 Kruger and a Orionis, but hardly affects those for such stars as the Sun and Capella, for which the bolometric correction is small, and the absolute emission of radiation is well known. But the entries for these stars are affected by errors in the value of λ_c. For stars of the mass of our sun, equation (91·1) shews that λ_c varies inversely as μ^{-4} so that $\lambda_c(\lambda_c + 1)$ varies approximately as μ^{-8}. As the value of T_c also varies approximately as μ (equation (90·2)) it follows that the calculated value of N^2/A varies approximately as $\mu^{3\cdot5}$. For stars of large mass the variation is different but of comparable amount. Thus a small error in the value assumed for μ, will result in a large error in the values calculated for N^2/A.

With this in mind let us examine the significance of the values for N^2/A shewn in Table XIII.

94. Atoms of moderate weight have an atomic weight A which is approximately twice their atomic number N, so that A is approximately four times N^2/A. In more massive atoms the factor is larger; for uranium $N = 92$ and $A = 238$, so that $N^2/A = 35\cdot5$ and $A = 6\cdot7$ (N^2/A). Thus all the entries in the last column represent atomic weights far higher than that of uranium, and in most cases atomic weights of thousands at least.

These values of N^2/A have been calculated relative only to a very special model of stellar structure. If we had complete confidence that this model gave a true picture of the structure of a star, and if we further had complete confidence in Kramers' theory of the absorption of X-radiation and of the use we have made of it, we could only conclude that stellar matter shewed a stopping power for radiation far beyond that of any known terrestrial atoms, and hence that it consisted of atoms far heavier than uranium.

There is no à priori objection to supposing that stellar matter consists of atoms heavier than uranium; indeed we shall see later that such a possibility must be taken very seriously into consideration and that there is much to be said for it. The only atoms with which the chemist is acquainted are the atoms which occur in the outer surfaces of the stars and near the surface of the earth, which in turn has probably at one time been formed out of the surface layers of the sun. Thus the atoms of chemistry are "surface-atoms" and there is no valid reason for expecting them to form a fair sample of the atoms of the universe as a whole. It is à priori far more probable that their very lightness has caused them to float to the surfaces of astronomical masses, and that concealed in the far depths of these bodies are atoms of far higher atomic weight, whose extreme weight has caused them to gravitate to the centres of the stars. If so terrestrial chemistry may properly be described as " surface-chemistry," and cosmical chemistry, of which terrestrial chemistry is a branch, deals with a wider range of elements.

But the acceptance of this hypothesis does not make it possible to accept the values of N^2/A shewn in the last column of Table XIII. These figures have been calculated for an assumed molecular weight $\mu = 2\cdot5$. Now even completely broken up uranium has an effective molecular weight $2\cdot56$ which is greater than this; atoms of atomic weight far above uranium would in all probability have still higher values of μ even if they were completely broken up, and our calculated central temperatures are inadequate to break up even atoms of uranium completely.

Thus our assumed value of $\mu = 2\cdot 5$ is too small to be consistent with the calculated values of N^2/A. Before we can accept these values we must amend our model by increasing the value of μ very substantially. The values of N^2/A must now be recalculated for this increased value of μ and since, as we have already seen, these values vary approximately as $\mu^{5\cdot5}$, the new values of N^2/A will be far greater than the old. This increase in the values of N^2/A demands a still further increase in μ and so on indefinitely. We know nothing about the relation between μ, N and A for our hypothetical heavy atoms, so that it is impossible to say whether or not the race will ever stop; it may be it is impossible to find values of N^2/A and μ which are consistent with one another and with the observed stopping-power of the stars for X-radiation on the model we have under discussion. What is abundantly clear is that if ever the race does stop, we shall by then have reached values of N^2/A far higher than those shewn in Table XIII, and shall be contemplating atoms of which the atomic weight is many thousands.

While no absolutely convincing reason can be assigned why such atoms should not exist in the stars, it will be generally felt that it is improbable that they do. If their existence is found to be a necessary consequence of our having assumed a special model for stellar structure, then it behoves us to look for other models which entail less improbable consequences.

95. Eddington, who found himself confronted with a similar difficulty in discussing his model of a star (§ 76), examined whether the difficulty could be avoided by concentrating the star's generation of energy near its centre *. If this is done, all the radiation has to pass through the whole radius of the star; on the model we have just been discussing, in which energy is generated uniformly throughout the star's interior, the radiation has, on the average, only to pass through about half of the star's radius before emerging into space. Eddington's plan, in effect, sets twice as many atoms at work to stop the radiation, so that each atom need only have half the stopping power it would otherwise require.

His actual calculation shews that in the extreme case in which the whole generation of energy is localised at the centre of the star, the observed facts can be explained by assuming a coefficient of opacity less by a factor of about $2\cdot5$ than would otherwise be required. Some later calculations of my own† confirm the general accuracy of this result, but apart from such detailed calculations it is in any case obvious, on the general grounds explained above, that the factor of reduction cannot be very far from 2.

Thus if we change our model and suppose its whole generation of energy to be localised at its centre, the values of N^2/A shewn in Table XIII may be divided by a factor of from 2 to $2\cdot5$. But when this is done the values of N^2/A are still of the order of 150, and such values correspond to atomic weights of the order of 1000 at least—more when we adjust for the increased value of μ. Thus redistribution of the generation of energy cannot solve, and hardly alleviates, the difficulty.

96. Another possible line of attack is to redistribute the atomic weights. Instead of supposing N^2/A to have a uniform value throughout, we imagine the heavier atoms to be concentrated near the centre of the star. In an actual star the heaviest atoms would naturally gravitate to the centre. This redistribution places the heaviest atoms where they are most effective in stopping radiation, and so, in a sense, increases the radiation-stopping efficiency of the stellar matter. I have examined the question in some detail*, and find it does not provide a solution of the problem. Redistribution enables us to reduce the average value of N^2/A throughout the star, but does not reduce the maximum value, which now occurs in the star's central regions, and the problem remains as acute as ever.

97. Still a third possibility is to suppose μ to vary inside the star. Detailed calculations which I have made†, shew that this can substantially lessen the values of N^2/A for stars of small mass. On the other hand, it makes no appreciable decrease in the values of N^2/A necessary in the most massive stars, and can hardly be regarded as providing a solution.

98. A far more drastic possibility remains. By formulae (72·3) and (77·3), the flow of radiant energy per unit area H is

$$H = -\frac{4aT^3C}{3k\rho}\frac{dT}{dr}$$

$$= -\frac{aC\mu T^{6\cdot5}}{3\cdot34 \times 10^{25}\rho^2(N^2/A)}\frac{dT}{dr} \quad\ldots\ldots\ldots\ldots(98\cdot1).$$

Near the surface of a star the value of H is a matter of direct observation. Our procedure has in effect been to calculate the values of T, ρ, and dT/dr for given stars, assuming that they conform to our particular model, and to examine what values of N^2/A are necessary to make the calculated value of H agree with the observed value. The necessary values of N^2/A have all been very improbably high.

We can reduce these values for N^2/A, while retaining the observed value of H, either by increasing some factor in the denominator of the fraction on the right of equation (98·1) or by decreasing some factor in the numerator. We cannot do much with ρ, since the mean density of the star is a matter of direct observation, and no possible change in μ can substantially alleviate the discrepancy. The only remaining factor is $T^{6\cdot5}dT/dr$, or

$$\frac{2}{15}\frac{d}{dr}(T^{7\cdot5}).$$

---

* *M.N.* LXXXVI. (1926), p. 570. The appropriate analysis is that given in the present chapter; we have to assign a positive value to l.

† *L.c.*, p. 571. To represent this possibility, we have to assign a negative value to j in the analysis of the present chapter.

Even a moderate decrease in temperature throughout the star lessens this factor enormously. Reducing the value of T uniformly to half its value, divides this factor by $(2)^{7 \cdot 5}$, which is equal to about 181. We can now divide our calculated values of N^2/A by 181, and find that, from being impossibly high, they have now become far too small to be at all plausible. Thus reducing T by something less than a half must lead to reasonable values of the atomic weights.

The pressure in a star's interior is fixed, as regards order of magnitude at least, by the circumstance that it has to support all the layers above against gravity. If the stellar matter obeys the laws of a perfect gas, this, in conjunction with the fixed mean density of the star, leaves no opportunity for substantial adjustment of the temperature so long as the gas-laws are obeyed.

Imagine that the temperature is artificially reduced to a certain uniform fraction θ of its value throughout one of the gaseous stars we have been considering, each particle of the star retaining its position, so that the density and the star's gravitational field remained unaltered. The gas-pressure is now reduced by a factor θ and the radiation-pressure by a factor θ^4, so that the total pressure is inadequate to support the star against its own gravitation. The star would start to collapse except that a new factor may immediately come into play. Let us suppose, to take a definite illustration, that the original temperature was so high as to keep the majority of the atoms ionised right down to their nuclei. The diminished temperature cannot maintain this high degree of ionisation, so that a certain number of atoms start to reform as far as their K-rings. Now at such densities as we have found would prevail at the centres of gaseous stars, K-ring atoms cannot be treated as mere points, so that the gas-laws will not be strictly obeyed. The new gas-pressure will accordingly be greater than θ times the old pressure, and may even exceed the old pressure if the deviations from the gas-laws are sufficiently large. If it should happen that the total new pressure is exactly equal to the total old pressure throughout the star, the star will be in equilibrium again. The pressure and density will have remained unaltered by the change, but the values of N^2/A which are necessary to maintain radiative equilibrium are reduced by a factor of $\theta^{7 \cdot 5}$.

It will of course be understood that this is a purely fictitious case; an actual star could not undergo the changes we have described without a good deal of internal readjustment. Let us nevertheless continue to use this purely fictitious case to examine what values of θ would be necessary to pass from an ideal gaseous star to an actual star.

We shall find later that the atomic numbers of stellar atoms are probably in the neighbourhood of $N = 95$, and so just higher than those of uranium for which $N = 92$. The value of μ for fully broken-up uranium is $2 \cdot 56$; for stellar atoms, which are not quite fully broken up, it will be about $2 \cdot 6$. The values of N^2/A given in table (134) were calculated for the value $\mu = 2.5$. To pass to the value $\mu = 2.6$, they must (§ 94) be multiplied by about $(2.6/2.5)^{87}$ or 1·40. When this is done we obtain the values:

Sun 620, Sirius A 490, α Centauri B 1050, Capella A 350, etc.

For uranium N^2/A is 35·5, so that for stellar atoms the actual values will be of the order of 37, and the values just calculated are about 16 times too large. The values of θ such that reduction by the requisite factor $\theta^{7.5}$ reduces them to the uniform value of 37 are:

Sun 0·69, Sirius 0·71, α Centauri B 0·64, Capella A 0·74.

For the average typical star we may suppose that the necessary reduction in temperature is one of 30 per cent. of the temperature in the gaseous state. At first sight this may appear too slight a change to produce any profound changes in our views of stellar mechanism. But the pressure of radiation $\frac{1}{3}aT^4$ is reduced by a factor of $(0.70)^4$ or 0·24, so that it becomes quite small, relative to the total pressure, even in quite massive stars, and quite insignificant in all others.

The free electrons and atomic nuclei are so small that after the temperature has been reduced they continue to exert pressure in accordance with Boyle's law. Their pressure is only 70 per cent. of what it originally was, so that for the star to remain in equilibrium, the atoms must make up the deficiency of 30 per cent. in the gas-pressure as well as the deficiency of 76 per cent. in the pressure of radiation. With an atomic number of 95, there are about 95 times as many free electrons as atoms, so that each atom must exert about $\frac{30}{70} \times 95$, or 40, times the pressure exerted by a free electron, and so about 40 times the pressure it would exert if Boyle's law were accurately obeyed.

To exert a pressure of this magnitude the atoms must be so closely packed as to be almost in contact, or rather so closely packed that their effective volumes occupy almost the whole of the available space. Such a condition may be properly described as a liquid, or semi-liquid, state.

Our main conclusion, then, is that if Boyle's law is assumed to be obeyed throughout the interior of a star, the observed capacity of the atoms for stopping radiation demands an impossibly high atomic weight. We can reconcile the observed opacity with reasonable values for the atomic weights by supposing the density to be so great that Boyle's law is not obeyed, but the deviations from Boyle's law have to be so great that the matter must be supposed to be in a liquid or semi-liquid state. This refers of course only to the central regions of the star; the outer layers must in any event be gaseous. We leave the discussion at this point for the moment, because an entirely independent chain of evidence, to be presented in the next chapter, points still more conclusively to the liquid or semi-liquid state, as does also a third line of evidence to be put forward in Chapter X.

CHAPTER IV

THE SOURCE OF STELLAR ENERGY

Inadequacy of Terrestrially-known Sources.

99. We have seen that each square centimetre of the sun's surface emits sufficient energy to drive an eight-horse-power engine continuously; the output from each square centimetre of an O or B type star, such as Plaskett's star or V Puppis, which is at least 200 times as great, is sufficient to drive an express locomotive at full speed year after year and century after century for millions of years. Since the full implications of the doctrine of conservation of energy have been understood, efforts have been made to discover the origin of the energy which is poured out with such terrific profusion by the sun and stars.

À priori there are two general possibilities open. Either the stream of energy liberated from a star's surface may be continually fed to the star from outside, or it may be generated in the star's interior, and driven out through its surface, as the only means of preventing an intolerable heating of the interior. An illustration of the former mode of liberation of energy is provided by a meteorite falling through the earth's atmosphere, the energy of its radiation being provided by the impact of molecules of air on its surface; an illustration of the latter is provided by an ordinary coal fire.

The only serious effort to explain the sun's energy as being supplied from outside was that of Robert Mayer, who conceived solar energy as arising from a continuous fall of meteors into the solar atmosphere. But simple calculations shew that a mass of meteors equal in weight to the earth would barely suffice to maintain the sun's radiation for a century, and that meteors sufficient to maintain the sun's radiation for only 30 million years would double its mass. As it is quite impossible to admit that the sun's mass can be increasing at such a rate, Mayer's explanation has to be abandoned. As no other way can be imagined by which energy of comparable amount can be brought in from outside, we are driven to regard the sun's generation of heat as taking place throughout its body.

The essential datum of the problem is no longer the energy discharged by a square centimetre of a star's surface, but the energy generated per gramme of its mass. As we have seen, the sun generates about 1·90 ergs per second for each gramme of its mass, while the corresponding figure for V Puppis is as high as 1100 ergs, and that for Pearce's star is probably of the order of 15,000.

100. Let us fix our attention on the special case of the sun, which is more typical of general stellar radiation than the two extreme cases just mentioned.

The radioactive contents of rocks provide what must apparently be accepted as quite conclusive evidence that the earth's crust has been solid for at least 1500 million years*. The age of the sun must, then, be greater than this. Speaking very loosely, we may say that the sun must have been generating energy for at least 1500 million years at a rate comparable with its present rate of 1·9 ergs per gramme per second. This represents a total energy generation of about 9×10^{16} ergs for each gramme of the sun's mass.

It has long been realised that combustion of matter, and chemical action in general, are quite inadequate to provide any such amount of energy. The formation of one gramme of carbon dioxide by the union of carbon and oxygen sets free 2140 calories or 9×10^{10} ergs, which is only about a millionth part of the required output of 9×10^{16} ergs per gramme. Some chemical reactions provide rather more energy per gramme than that just mentioned, but no known reaction provides ten times as much, or 9×10^{11} ergs per gramme, so that chemical action is unable to provide even one part in 100,000 of the energy which the sun has radiated away during the life of the earth.

Heat energy is equally insignificant. Even in the most favourable case in which matter is entirely broken up into its constituent electrons and protons, the heat energy of a gramme of matter at a temperature of 30 million degrees is only $1·5 \times 10^{15}$ ergs. Thus if the sun were merely radiating stored heat energy, the 9×10^{16} ergs per gramme which it must have radiated during the life of the earth, would require it to have had a temperature of about 1800 million degrees when the earth was born. The calculations of stellar temperatures which were given in the last chapter enable us to dismiss any such temperature as impossibly high.

In 1854 Helmholtz put forward his famous contraction-hypothesis, according to which the shrinkage of the sun under its own gravitational attraction, with the accompanying loss of gravitational potential energy, sets free the energy which appears as the sun's radiation. At a point in the sun's interior, the gravitational potential is of the order of magnitude of $\gamma M/r$, where M is the mass and r the radius of the sun, and γ is the gravitation constant. On inserting numerical values, this quantity is found to be of the order of 2×10^{15} in C.G.S. units, so that even if the whole mass of the sun had fallen in from infinity, each gramme could only provide about 2×10^{15} ergs of energy, and it appears that the Helmholtz contraction-hypothesis cannot account for more than about two per cent. of the energy which has been radiated by the sun during the earth's life.

Thus neither chemical energy, heat energy, nor gravitational energy are anything like adequate to account for the sun's emission of radiation even during the existence of the earth. The only remaining sources of energy are of sub-atomic nature, and of these radioactivity provides the only example of which we have any direct experience.

* A. Holmes, *The Age of the Earth.*

Radium, the most energetic of the radioactive elements, generates about 1,600,000 ergs of radiation per gramme per second, so that the sun's present radiation could be accounted for by supposing one part in 800,000 of its mass to be radium. But as the average life of radium is only about 2800 years, such a sun would be extinguished after a few thousand years. During its whole life a gramme of radium emits about 1.4×10^{17} ergs of radiation, so that a sun which consisted initially of pure radium would have sufficient capacity for generating energy to maintain the sun's radiation at its present rate for about 5000 million years. For uranium the corresponding figures are 2.2×10^{17} ergs of radiation per gramme and 8000 million years.

These figures shew that, contrary to a prevalent impression, radioactivity is capable of providing for the total amount of energy radiated by the sun since the earth solidified. On the other hand, no known combination of radioactive elements is capable of forming a sun which would radiate at the required rate for the required time—in brief, uranium is too slow, radium is too rapid, and all the other radioactive transformations have too little energy. Moreover we shall find evidence later that the ages of the stars are incomparably greater than the 1500 million years which we have taken to be the age of the solid earth, and radioactive substances of the kinds known on earth are found to be utterly inadequate to provide stellar radiation throughout the whole lives of the stars.

SUB-ATOMIC SOURCES OF ENERGY.

101. Thus we have to conclude that no source of energy known to us on earth is anything like adequate to account for the radiation of the sun and stars·

We can make progress only by striking the problem from an entirely new angle. Instead of surveying the sources of energy known to us on earth, we survey the general laws of physics which we believe to hold sway in the sun and stars as well as on earth, and examine whether any energy-transformation known to them, but not to us, is capable of providing adequate energy for the radiation of the sun and stars. In approaching this question it must be emphasised that the energy which is radiated away must have resided somewhere before it was radiated; it cannot have been created, as needed, out of nothing. Its total amount is so great that if it resided in the star, it must have resided in a very concentrated form indeed. Now the only places known to physics in which energy can reside in a concentrated form are the nuclei and electrons of which matter is formed, and the electromagnetic fields in their immediate neighbourhood.

It is well known that a moving electron has a greater mass than a stationary electron; more force is required to deflect it from its path or to produce a specified acceleration. If m_0 is its mass when at rest, its mass m when moving with a velocity v is given by

$$m = m_0 \left(1 - \frac{v^2}{C^2}\right)^{-\frac{1}{2}} \quad \dots\dots\dots\dots\dots\dots(101.1),$$

where C is the velocity of light. Its kinetic energy, when moving with velocity v, is

$$m_0 C^2 \left[\left(1 - \frac{v^2}{C^2} \right)^{-\frac{1}{2}} - 1 \right] \quad \dots\dots\dots\dots\dots(101\cdot2),$$

which, in virtue of the relation (101·1), can be written in the form

$$(m - m_0) C^2 \quad \dots\dots\dots\dots\dots\dots(101\cdot3).$$

Suppose that the electron, while moving with velocity v, is suddenly checked and brought to rest. Its kinetic energy $(m - m_0) C^2$ is transformed into radiation which travels off into space. At the same time the mass of the electron is reduced by an amount $m - m_0$. The conservation of mass, however, persists through this process, so that the mass $m - m_0$ is not lost, but must represent the mass of the radiation whose energy is $(m - m_0) C^2$. In this special instance the energy of the radiation is seen to be C^2 times its mass, and the emission of radiation involves a loss of mass to the emitting system equal to $1/C^2$ times the energy of the radiation.

This is no isolated special instance. In 1905 Einstein shewed it to be a general consequence of the theory of relativity that a change of energy δE in any system involved a change in its mass equal to $\delta E/C^2$. Thus the emission of radiation by a star or any other body at a rate E ergs per second necessarily involves a loss to the mass of the radiating body at the rate of E/C^2 grammes per second. The sun's present radiation of $3\cdot8 \times 10^{33}$ ergs per second involves a diminution of the sun's mass at the rate of $\dfrac{1}{C^2} (3\cdot8 \times 10^{33})$ grammes a second.

Putting $C = 3 \times 10^{10}$ we find that the sun's mass is diminishing at the rate of $4\cdot2 \times 10^{12}$ grammes per second, or about four million tons a second. In the 1500 million years which have elapsed since the earth solidified, the sun's mass must have diminished by 2×10^{29} grammes, or one part in ten thousand. The question arises as to the form in which this additional 2×10^{29} grammes resided in the sun 1500 million years ago.

102. It has long been known that the atomic weights of the elements cannot be expressed as exact whole numbers. If that of oxygen is taken to be 16, helium has atomic weight 4·00216, while that of hydrogen is 1·00778 *. If four hydrogen atoms each of weight 1·00778 combine to form one helium atom, the system experiences a loss of mass 0·029 (the unit being a sixteenth of the mass of the oxygen atom), and this must be the mass of radiation discharged into space when, if ever, helium is built up out of hydrogen. In 1919 Perrin † suggested that processes of the general type just described may be the origin of the radiation of the sun and stars, a suggestion which Eddington‡ repeated in 1920 in a more precise form. If, for instance, a star

---

* F. W. Aston, *Proc. Roy. Soc.* cxv. (1927), p. 510.

† *Annales de Physique*, ii. (1919), p. 89; and *Revue du Mois*, xxi. (1920), p. 113.

‡ British Association Cardiff (1920) Address to Section A.

consisted originally of pure hydrogen, the total transformation of this into heavier elements would involve a loss of mass equal to about 0·0072 gramme per gramme of the star's mass, so that the transformation would result in a total radiation of energy of amount $0·0072C^2$ or $6·5 \times 10^{18}$ ergs per gramme. This is ample to account for the sun's radiation throughout the whole of the earth's life and through epochs far beyond. But it is not adequate to account for the radiation of the stars through the lengths of life which, as we shall see later, are assigned to them by dynamical considerations, and this alone probably compels its abandonment. If the source of energy suggested by Perrin and Eddington were the true one, it would follow that young stars contained a larger proportion of hydrogen than old stars; there is no evidence of this being the case, although this can hardly be regarded as a fatal objection. A more serious objection is that if stellar energy were produced by the transformation of hydrogen into heavier elements, the rate of radiation of any particular star would be proportional to the frequency with which groups of hydrogen atoms got into the position appropriate for the formation of atoms of helium and heavier elements. A comparison of the rates of radiation of different stars does not suggest that they can be determined by any such law as this. Furthermore we shall see later that a star in which the energy was generated in the manner suggested by Perrin would be dynamically unstable, and this reason, apart from all others, would appear to compel the rejection of Perrin's scheme.

THE ANNIHILATION OF MATTER.

103. Fifteen years before Perrin had suggested that stellar energy might originate in the formation of heavy atoms out of simpler ones, I had pointed out* that an enormous store of energy could be derived out of the total annihilation of matter, positive and negative charges rushing together and neutralising and so annihilating one another, the resulting energy being set free as radiation. In 1918† I calculated the amount of energy which would thus be set free and the length of life which this source of energy allotted to the stars.

The principle of this calculation is very simple. When a proton and an electron of masses m, M neutralise one another in the sun and disappear, the sun loses material mass of amount $m + M$. A certain amount of radiation results and to satisfy the principle of conservation of mass, the mass of this radiation must be $m + M$. Hence, in accordance with Einstein's principle already explained, the energy of the radiation must be $(m + M) C^2$. In general the annihilation of a gramme of matter must set free C^2, or 9×10^{20}, ergs of energy.

The energy set free by the annihilation of matter is enormous compared with that which the same amount of matter can be made to yield in any

* *Nature*, LXX. (1904), p. 101.
† *Problems of Cosmogony and Stellar Dynamics*, p. 287.

other way. Some 3000 tons of oil must be burned to drive a liner across the Atlantic; the same amount of energy could be provided by the annihilation of about one-eightieth of an ounce of oil. Over four million tons of coal a week are raised to provide for the heating, lighting, power and transport of Great Britain alone; the annihilation of a single ton of this coal would provide for all these services for a century. And, to carry on the story, the total radiation emitted by the sun during the 1500 million years of the earth's existence could be provided by the annihilation of one ten-thousandth part of its mass, the result already mentioned in § 101; while the annihilation of the whole of its mass would provide radiation at the present rate for 15 million million years. This last result not only shews that the annihilation of matter provides an adequate source of stellar energy; it also makes it almost certain, as we shall now see, that it provides the actual source.

A mass of dynamical evidence, which will be brought forward later, indicates that the stars as a whole must have existed for millions of millions of years. The most direct evidence is perhaps provided by the orbits of binary stars. We shall see that, as a consequence of the manner of their formation, newly formed binary stars have circular, or nearly circular, orbits. Every gravitational pull on a circular orbit tends to make it more elliptical, so that the older a binary system is, the more elliptical its orbit ought to become. This is found by observation to be the case. But our knowledge of the density with which the stars are scattered in space gives us the means of calculating the actual rates at which the ellipticities of the orbits of binary systems must increase, so that from the observed ellipticities of orbits it is possible to calculate the ages of the binary systems. And the answer comes out in millions of millions of years.

We can calculate the total amount of radiation which a star has emitted during its life of millions of millions of years. Except in the case of the youngest stars, it is found that the total mass of the emitted radiation is far greater than the present mass of the star. The original mass of the star must have been the sum of the star's present mass and the mass of all the emitted radiation, so that the star must originally have been many times as massive as it now is. Indeed we shall shortly find observational evidence that young stars, as a class, are many times more massive than old stars.

The older views of stellar radiation regarded a star's gravitational potential energy and the heat and chemical energy of its molecules as reservoirs from which a star's radiation was drawn. When we look at the matter in terms of a time-scale of millions of millions of years, we see that the capacity of all these reservoirs is quite negligible; the reservoir in which the star's future radiation is stored is the star's mass. The time-scale of millions of millions of years requires that the energy stored in each gramme of a star's mass shall be of the order of magnitude of 9×10^{20} ergs, and we know of no way in which such a great concentration of energy can be stored except in the actual matter of electrons and protons.

We are not in a position to deny absolutely that the energy may be stored in other ways. Our knowledge of physics is derived from a study of molecules which are not liberating any appreciable amount of energy (§ 104), and this may be because they have no stored energy to liberate. A search for the mechanism of storing energy ought, if possible, to take place in regions where large amounts of energy are known to be stored, and we have to admit that a study of molecular physics in the Sun might disclose molecular mechanisms for storing energy, as so also of course mass, which are unknown to terrestrial physics merely because terrestrial molecules have no energy to store.

But so far as we can judge from terrestrial physics, the obvious place for storing the energy and mass in the enormous quantities needed is in the existence of electrons and protons, so that it seems reasonable to suppose that the liberation of energy arises from the annihilation of electrons and protons.

Thus we suppose that as a star ages, its atoms and electrons must undergo annihilation, their imprisoned energy being set free in the form of radiation. Coal, which has been picturesquely described as bottled sunshine, might more accurately be described as re-bottled sunshine. The bottles in which sunshine and all the radiation of the stars were first imprisoned, were the atoms and electrons of matter long since annihilated; the breakage of these bottles set free the radiation which warms and lights our earth and makes it a possible abode of life.

Those who feel that this solution of the problem of the source of stellar radiation is ultra-modern, and therefore under suspicion, may perhaps find comfort in the following transcript from Newton's *Opticks** (1704):

Query 30. Are not gross bodies and light convertible into one another; and may not bodies receive much of their activity from the particles of light which enter into their composition?

The changing of bodies into light, and light into bodies, is very conformable to the course of Nature, which seems delighted with transmutations. Water, which is a very fluid, tasteless salt, she changes by heat into vapour, which is a sort of air; and by cold into ice, which is hard, pellucid, brittle, fusible stone; and this stone returns into water by heat, and vapour returns into water by cold....Eggs grow from insensible magnitudes, and change into animals; tadpoles, into frogs; and worms, into flies. All birds, beasts and fishes, insects, trees, and other vegetables, with their several parts, grow out of water and watery tinctures and salts; and by putrefaction, return again into watery substances. And water, standing a few days in the open air, yields a tincture, which (like that of malt) by standing longer yields a sediment and a spirit; but before putrefaction is fit nourishment for animals and vegetables. And among such various and strange transmutations, why may not Nature change bodies into light, and light into bodies?

---

* I am indebted to Sir J. J. Thomson for bringing this to my notice.

Yet the consideration which will carry conviction to most minds that stellar radiation arises out of the annihilation of stellar matter is neither the fact of its being required by the laws of physics, in so far as we understand them, nor the fact of its having seemed to Newton two centuries ago, to be "very conformable to the course of Nature," but the general evidence of astronomy that young stars are uniformly more massive than old stars. The evidence for this will appear as our book proceeds.

THE CONDITIONS OF LIBERATION OF STELLAR ENERGY.

104. Although all available evidence points to annihilation of matter as being the source of stellar energy, we shall not find it necessary to assume this to be the source in the following investigation, or indeed in most of the investigations or discussions which follow.

Whatever the source of stellar energy may be, we wish to know in what way the liberation of energy takes place, and why it is proceeding so much more rapidly in some stars than others. Why should each gramme of the sun's mass liberate on the average 1·9 ergs per second, and why should each gramme of V Puppis average 1100 ergs per second? If, for instance, radiation is produced by the annihilation of matter, why is the sun transforming its matter into radiation so much more slowly than V Puppis? And, if it is a general property of matter to transform itself into radiation, why is this process so little in evidence on the earth?

If the earth generated no energy at all, its surface would assume a steady average temperature such that the energy radiated away by its area of $5\cdot1 \times 10^{18}$ square centimetres was exactly equal to the energy it received from the sun. This latter amount of energy is $1\cdot7 \times 10^{24}$ ergs per second, and the surface-temperature requisite to radiate this away is readily calculated to be about 276° abs. or 3° C. If the earth's surface is not treated as a perfect radiator, the form of argument is slightly different, but the final result is precisely the same. Suppose, however, that the earth's mass generated energy at an average rate even of a ten-thousandth part of an erg per gramme per second. A total internal generation of energy of $0\cdot6 \times 10^{24}$ ergs per second would now have to be added to the $1\cdot7 \times 10^{24}$ ergs received from the sun, and to radiate away the total of $2\cdot3 \times 10^{24}$ ergs per second would need a surface temperature of 29° C., or higher in so far as the earth's surface is not a perfect reflector. This is far above the average temperature of the earth's surface, so that we may be confident that if the earth's mass is generating energy at all, it does not do so at a rate approaching $\frac{1}{10000}$ erg per gramme per second. Thus the generation of energy which is proceeding in the sun and stars is almost or completely absent from the earth. Whatever causes are liberating energy in the sun are out of action on earth.

This leads us to inquire what determines the rate at which a star liberates its energy. It might at first sight seem a quite hopeless task to try to probe so recondite a matter as the physical details of a process of which we have no direct knowledge, but good luck comes to our rescue, and the one available method proves to be so unexpectedly powerful, that in actual fact the problem admits of almost complete solution.

The method is that of examining the dynamical stability of the star. It is *à priori* likely that changes in the physical condition of a star's interior will have some effect on the rate at which its radiation is generated and liberated by the star. If we double its temperature and compress it into one-eighth of its original volume, the star is hardly likely to go on generating energy at precisely the same rate as before, as though nothing had happened —so at least it would seem natural to conjecture. We can imagine all kinds of laws connecting a star's rate of liberation of energy with the density and temperature of its interior. But on discussion the exceedingly useful fact emerges that almost all these laws would give an unstable star; under some laws the star might expand or contract indefinitely through the rate of liberation of energy in the star's interior failing to keep pace with the rate at which energy was radiated away from its surface, while under other laws the star might transform its whole mass explosively into radiation. The range of laws under which the stars can continue to exist as stable structures is very limited indeed, and the undoubted fact that the stars actually do continue to exist limits the possibilities to this quite small range.

The Stability of Stellar Structures.

105. In the preceding chapter it was found that two independent equations were necessary to express that a star could rest in equilibrium:

(1) a dynamical equation expressing that the total pressure at any point is just adequate to support the total weight of stellar matter above that point, and

(2) an equation of radiative equilibrium, expressing that the generation of energy in any small volume of the star is just adequate to make good the net loss by radiation.

These equations can be written in the forms (cf. equations (75·2) and (75·1)),

$$\frac{1}{\rho}\frac{d}{dr}(p_G + \tfrac{1}{3}aT^4) = -\frac{\gamma M_r}{r^2} \quad \dots\dots\dots\dots\dots(105\cdot1),$$

$$\frac{1}{r^2}\frac{d}{dr}(r^2 H) = \rho G \quad \dots\dots\dots\dots\dots\dots(105\cdot2),$$

where G is the rate of generation of energy per unit mass, H is the flow of radiant energy per unit area at the point in question, p_G is the pressure of the stellar material at the point, whether this is gaseous or not, and M_r is the mass of the star which lies inside a sphere drawn through this point.

In the same way, when the star is in motion, two equations determine that motion:

(1) a dynamical equation expressing that any pressure in excess or defect of the amount requisite for equilibrium produces accelerations in the star, and

(2) a radiative equation expressing that any excess in the generation of energy over the net amount radiated away produces an increase in the energy of the stellar matter.

These equations are easily found to be [*]

$$\frac{d^2 r}{dt^2} = -\frac{1}{\rho}\frac{d}{dr}(p_G + \tfrac{1}{3}aT^4) - \frac{\gamma M_r}{r^2} \quad \dots\dots\dots\dots(105{\cdot}3),$$

$$\rho C_v \frac{dT}{dt} - (p_G + \tfrac{1}{3}aT^4)\frac{1}{\rho}\frac{d\rho}{dt} = \rho G - \frac{1}{r^2}\frac{d}{dr}(r^2 H) \quad \dots\dots\dots(105{\cdot}4),$$

where ρC_v denotes the specific heat of matter and radiation per unit volume.

106. The equations are too complicated for discussion in the most general case imaginable. For the present we shall discuss only a special model of stellar structure, or rather a structure about which special assumptions are made. We suppose that the star already has such stability that a local disturbance spreads in the form of waves until it is uniformly distributed throughout the star, and proceed to consider the stability, or reverse, of a star affected by a disturbance which is already uniformly spread throughout the star. To represent this we suppose the star's motion to be such that r changes proportionately throughout the star, so that the motion at each instant is one of uniform contraction or expansion. We also suppose T to change proportionately throughout the star, so that at every instant it is being uniformly heated or cooled. When these assumptions are made, the changes in any one shell or element of the star are characteristic of the whole star.

Let the equilibrium values of $T, \rho, G, H \dots$ for any shell of the star be $T_0, \rho_0, G_0, H_0 \dots$, and in the displaced configuration let their values be $T_0 + \delta T, \rho_0 + \delta\rho, \dots$. Throughout the motion we must have $\rho r^3 = \rho_0 r_0^3$.

We have already been led to suspect (§ 98) that the ordinary gas-laws will not be accurately obeyed in stellar interiors. Without in any way assuming this, we shall find it convenient to allow for the possibility from the outset of our present investigation by supposing the material pressure p_G to be connected with the density by a relation of the form

$$p_G \propto T\rho^{1+s} \quad \dots\dots\dots\dots\dots\dots(106{\cdot}1).$$

It is not necessary to suppose that this is the true relation between p_G and ρ in an actual star. Whether it is or not, we can always obtain the exact results appropriate to a purely gaseous star by putting $s = 0$, and we can obtain some indication of the general effect of a departure from the gas-laws by assigning a small value to s, but it cannot be supposed that results obtained by giving finite values to s are accurately true.

[*] I gave these equations in *M.N.* LXXXV. (1925), p. 917, but with $\tfrac{1}{3}aT^4$ in place of $\tfrac{1}{3}aT^4$ on the left-hand side of the second equation. My error was pointed out by Vogt (*Ast. Nach.* 232 (1928), No. 5545).

Using this relation, the dynamical equation of motion (105·3) assumes the form

$$\frac{d^2r}{dt^2} = -\frac{Tr_0^{1+3s}}{T_0 r^{1+3s}}\left(\frac{1}{\rho}\frac{dp_G}{dr}\right)_0 - \frac{T^4 r^2}{T_0^4 r_0^2}\left(\frac{d}{dr}(\tfrac{1}{3}aT^4)\right)_0 - \frac{\gamma M_r}{r_0^2}\left(\frac{r_0}{r}\right)^2.$$

If λ denotes the ratio of material pressure p_G to pressure of radiation, $\tfrac{1}{3}aT^4$, this equation can be expressed in the form

$$\frac{d^2}{dt^2}(\delta r) = \frac{\gamma M}{r_0^2}\left[\frac{\lambda+4}{\lambda+1}\left(\frac{\delta T}{T_0}+\frac{\delta r}{r_0}\right) - \frac{3s\lambda}{\lambda+1}\frac{\delta r}{r_0}\right]\quad\ldots\ldots\ldots(106\cdot2).$$

A displaced configuration will be one of dynamical equilibrium if the right-hand member of this equation vanishes. Putting the right-hand equal to zero and integrating, we see that from any equilibrium configuration r_0, T_0, a whole series of other equilibrium configurations can be derived, in which r and T are connected by the relation

$$r^{1-\frac{3s\lambda}{\lambda+4}}T = r_0^{1-\frac{3s\lambda}{\lambda+4}}T_0 \quad\ldots\ldots\ldots\ldots\ldots(106\cdot3).$$

These may be regarded as forming a series of homologous configurations. When the gas-laws are obeyed, so that $s = 0$, relation (106·3) assumes the form

$$rT = r_0 T_0$$

which merely expresses Lane's law, the series of homologous configurations specified by equation (106·3) now becoming identical with the homologous configurations of a sphere of perfect gas already discussed in § 60.

107. Let us suppose the coefficient of opacity k to be given by

$$k = \frac{c\mu\rho}{T^{3+n}} \quad\ldots\ldots\ldots\ldots\ldots\ldots(107\cdot1),$$

so that we pass to Kramers' opacity-law on taking $n = \tfrac{1}{2}$. The flux of energy per unit area, H, is now given by

$$H = -\frac{4aT^3C}{3k\rho}\frac{dT}{dr} = -\frac{4aC}{3c\mu}\left(\frac{T^3}{\rho}\right)^2\frac{\partial T}{\partial r}\,T^n,$$

whence we readily obtain

$$\frac{1}{r^2}\frac{d}{dr}(r^2H) = \left(\frac{T}{T_0}\right)^{7+n}\left(\frac{r}{r_0}\right)^4\left[\frac{1}{r^2}\frac{d}{dr}(r^2H)\right]_0\quad\ldots\ldots\ldots(107\cdot2).$$

Suppose, as a possible law of liberation of energy, that G, the rate of generation of energy per unit mass, depends on both T and ρ, being proportional to $\rho^\alpha T^\beta$. When the density and temperature of any element of the star undergo slight increments δT, $\delta\rho$, the rate of generation of energy G experiences an increment δG given by

$$\frac{\delta G}{G_0} = \alpha\frac{\delta\rho}{\rho_0} + \beta\frac{\delta T}{T_0}$$

$$= -3\alpha\frac{\delta r}{r_0} + \beta\frac{\delta T}{T_0} \quad\ldots\ldots\ldots\ldots(107\cdot3).$$

Using these relations, equation (105·4) may be put in the form

$$C_v \frac{d}{dt}(\delta T) + \frac{3p_G + 4aT_0^4}{\rho r}\frac{d}{dt}(\delta r) = G_0\left(-3\alpha\frac{\delta r}{r_0} + \beta\frac{\delta T}{T_0} - \frac{(7+n)\,\delta T}{T_0} - \frac{7\delta r}{r_0}\right)$$

...(107·4).

This and equation (106·2) together determine the motion of the star throughout any small displacement from its equilibrium configuration.

108. To examine the stability of the star, we eliminate δT from these two equations, thus obtaining an equation in δr alone. After some simplification this equation becomes

$$\frac{d^3}{dt^3}(\delta r) + \frac{(7+n-\beta)\,G_0}{C_v T_0}\frac{d^2}{dt^2}(\delta r) + \frac{\gamma M_r}{r_0^3}\left[\frac{\lambda+4}{\lambda+1}\left(\frac{3p_G+4aT_0^4}{\rho C_v T_0}-1\right)+\frac{3s\lambda}{\lambda+1}\right]\frac{d}{dt}(\delta r)$$

$$+\frac{\gamma M_r}{r_0^3}\frac{G_0}{C_v T_0}\left[\frac{\lambda+4}{\lambda+1}(3\alpha+\beta-n)+\frac{3s\lambda}{\lambda+1}(7+n-\beta)\right](\delta r)=0 \quad (108\cdot1).$$

Let us write this in the form

$$\frac{d^3}{dt^3}(\delta r) + B\frac{d^2}{dt^2}(\delta r) + C\frac{d}{dt}(\delta r) + D(\delta r) = 0 \quad\text{........}(108\cdot2),$$

where B, C, D merely represent the last three coefficients in equation (108·1). In virtue of the assumptions on which we are working, we must regard B, C and D as having constant values throughout the star, so that the solution will be of the form

$$\delta r = \sum_{s=1,2,3} A_s e^{\theta_s t},$$

where $\theta_1, \theta_2, \theta_3$ are the roots of the equation

$$f(\theta) \equiv \theta^3 + B\theta^2 + C\theta + D = 0 \quad\text{.................}(108\cdot3).$$

We proceed to examine the nature of these three roots.

As θ decreases from $+\infty$ to 0, $f(\theta)$ changes from $+\infty$ to D. If D is negative the equation has at least one real positive root for θ, and the star is unstable. Thus a first condition for stability is that D shall be positive.

As θ increases from $-\infty$ to 0, $f(\theta)$ changes from $-\infty$ to D, so that with D positive there must be at least one negative real root $\theta = -\epsilon$, and the equation assumes the form

$$(\theta^2 + 2P\theta + Q)(\theta + \epsilon) = 0 \quad\text{.................}(108\cdot4),$$

in which Q, as well as ϵ, is positive, since their product ϵQ, which is equal to D, is positive. Apart from the root $\theta = -\epsilon$, the two remaining roots are

$$\theta = -P \pm (P^2 - Q)^{\frac{1}{2}} \quad\text{......................}(108\cdot5).$$

If P^2 is greater than Q, $(P^2-Q)^{\frac{1}{2}}$ is real and is numerically less than P so that both roots are real and of the same sign as $-P$. If P^2 is less than Q, both roots are imaginary and have $-P$ for their real part. In either event, the star is stable if P is positive and is unstable if P is negative.

Comparing coefficients in the equivalent equations (108·3) and (108·4),

$$B = 2P + \epsilon, \quad C = 2P\epsilon + Q, \quad D = \epsilon Q \quad\text{............}(108\cdot6).$$

Since ϵ and Q are both positive, we see that if $BC < D$, P is negative and the star is unstable. So long as D is kept positive, a change from stability to instability can only occur when P vanishes, and this occurs only when $BC = D$. Hence the necessary and sufficient condition for stability is that D and $BC - D$ shall both be positive, or that

$$BC > D > 0 \quad \dots\dots\dots\dots\dots\dots\dots(108\cdot7).$$

To assign a physical interpretation to these conditions, let us first examine the critical case of $P = 0$. The roots for θ are now

$$\theta = \pm iQ^{\frac{1}{2}}, \, -\epsilon,$$

where, from equations (108·6),

$$\epsilon = B = \frac{G_0}{C_v T_0}(7 + n - \beta) \quad \dots\dots\dots\dots(108\cdot8),$$

$$Q = C = \frac{\gamma M_r}{r_0^3}\left[\frac{\lambda + 4}{\lambda + 1}\left(\frac{3p_G + 4aT_0^4}{\rho C_v T_0} - 1\right) + \frac{3s\lambda}{\lambda + 1}\right] \quad \dots\dots(108\cdot9).$$

We see that ϵ is of the order of magnitude of $G_0/C_v T_0$. Now G_0 is the rate at which energy is generated per unit mass, and so is equal to the rate per gramme at which the star radiates away energy when in equilibrium; $C_v T_0$ is equal to the heat-content of the star per gramme when in equilibrium. Thus $C_v T_0/G_0$ is the time during which the star's store of heat would last if it were radiated away without being replenished by any fresh generation of energy. This time is of the order of thousands of years for a giant star, and of millions of years for a dwarf star. We see that a time factor of $e^{-\epsilon t}$ represents a gradual shrinkage or expansion whose duration is thousands or millions of years.

Equation (108·9) shews that Q is of the order of magnitude of $\gamma M_r/r_0^3$, or of $\frac{4}{3}\pi\gamma\bar{\rho}$. For a fairly compact star $\bar{\rho}$ is of the order of magnitude of unity and $\gamma = 6\cdot66 \times 10^{-8}$, so that Q is of the order of magnitude of 3×10^{-7}. A time factor $e^{\pm iQ^{\frac{1}{2}}}$ is now seen to represent a pulsation whose period is of the order of $2\pi(3 \times 10^{-7})^{-\frac{1}{2}}$ seconds or a few hours or days.

Returning to the general equations (108·6), we see that $Q^{\frac{1}{2}}$ is in general of the order of $(\gamma\bar{\rho})^{\frac{1}{2}}$, while P and ϵ are of the order of the far smaller quantity $G_0/C_v T_0$. Neglecting P and ϵ in comparison with $Q^{\frac{1}{2}}$, the vibrations given by equation (108·5) have time factors

$$e^{-Pt}\cos Q^{\frac{1}{2}}t, \quad e^{-Pt}\sin Q^{\frac{1}{2}}t$$

and so represent vibrations of a few hours or days period, the amplitude of which increases or decreases exponentially in periods of thousands or millions of years.

109. We now see that the inequality $BC > D$ expresses the condition that small pulsations of the type just described, if once set up, shall not increase indefinitely in amplitude. Restoring the values of B, C and D, this condition is found to be

$$3\alpha + \beta - n < (3\alpha + 7) \frac{\dfrac{3p_G + 4aT_0^4}{\rho C_v T_0} - 1}{\dfrac{3p_G + 4aT_0^4}{\rho C_v T_0}} \quad \dots\dots\dots\dots(109\cdot1).$$

The remaining inequality $D < 0$, which expresses the condition that there shall be no unstable expansion or contraction, similarly becomes

$$3\alpha + \beta - n > -(3\alpha + 7) \frac{\dfrac{3s\lambda}{\lambda + 4}}{1 - \dfrac{3s\lambda}{\lambda + 4}} \quad \dots\dots\dots\dots(109\cdot2),$$

in which the denominator is always to be taken numerically positive.

We can best understand the meaning of these equations if we regard them as limiting the value of β. If β is positive and very great, heat generates more heat so violently that the star explodes. Generally an additional generation of energy increases the pressure and so causes the star to expand, and if β is not too great the consequent cooling may neutralise the excess generation of heat; the maximum value of β for which this can happen is given by equation (109·1).

Similarly if β is negative and very great, any excess of heat entirely inhibits the generation of energy, causing the star to contract under its own gravitation as in the Helmholtz-Kelvin contraction theory. This contraction heats the star to even higher temperatures, so that the capacity for generating energy is never regained and the star contracts indefinitely. The limiting value of β beyond which this happens is given by equation (109·2).

Whatever the values of α and β may be, the range of values of $3\alpha + \beta - n$ which are consistent with stability is limited in both directions. By subtraction of the right-hand members of the inequalities (109·1) and (109·2), we find as the length of this range

$$\frac{3\alpha + 7}{\dfrac{3p_G + 4aT_0^4}{\rho C_v T_0}\left(1 - \dfrac{3s\lambda}{\lambda + 4}\right)} \left[\frac{3p_G + 4aT_0^4}{\rho C_v T_0} + \frac{3s\lambda}{\lambda + 4} - 1 \right] \quad \dots(109\cdot3).$$

Should this expression vanish or become negative for any particular star, the two ranges of instability would have met and crushed all stable configurations out of existence. Hence expression (109·3) must be positive for every type of star which continues to exist stably in the sky.

110. Let there be N atoms, electrons or other separate units of the stellar matter per unit volume, and let each of these be supposed to have q degrees of freedom in addition to those represented by its freedom to move in space. Then the energy per unit volume is

$$\tfrac{1}{2}(q + 3) NRT + aT^4 \quad \dots\dots\dots\dots\dots(110\cdot1),$$

and the ratio of the specific heats γ is, as usual,

$$\gamma = 1 + \frac{2}{q+3}.$$

By differentiation of formula (110·1) with respect to T, we find for the specific heat,

$$\rho C_v = \tfrac{1}{2}(q+3)NR + 4aT^3 \quad\ldots\ldots\ldots\ldots\ldots(110\cdot 2),$$

so that

$$\rho C_v T_0 = \frac{NRT_0}{\gamma - 1} + 4aT_0^4.$$

Let us allow for deviations from the gas-laws by supposing the gas-pressure to be $(1 + \xi)NRT$, where ξ is a positive quantity which vanishes when Boyle's law is obeyed. Then the total pressure p is given by

$$p = (1 + \xi)NRT + \tfrac{1}{3}aT^4\ldots\ldots\ldots\ldots\ldots(110\cdot 3).$$

Thus

$$(1 + \xi)NRT_0 = \tfrac{1}{3}\lambda aT_0^4$$

and

$$3p_G + 4aT_0^4 = 3(1 + \xi)NRT_0 + 4aT_0^4.$$

With these values for ρC_v and p, the length of the range of stability as given by formula (109·3) becomes, after some simplification,

$$(3\alpha + 7)\lambda\left[\frac{3\gamma - 4 + 3\xi(\gamma - 1)}{3(\gamma - 1)(1 + \xi)(\lambda + 4)} + \frac{3s}{\lambda + 4 - 3s\lambda}\right]\quad\ldots(110\cdot 4),$$

in which the denominator $\lambda + 4 - 3s\lambda$ is to be taken always positive.

For a perfect gas, $s = \xi = 0$, and this reduces to

$$(3\alpha + 7)\frac{\lambda(3\gamma - 4)}{3(\gamma - 1)(\lambda + 4)}\quad\ldots\ldots\ldots\ldots(110\cdot 5).$$

Let us notice in passing that when pressure of radiation is neglected ($\lambda = \infty$), formula (110·5) vanishes when $\gamma = \tfrac{4}{3}$. Thus the range of stability vanishes when $\gamma = \tfrac{4}{3}$, and if $7 + 3\alpha$ is positive there will be no range of stability when $\gamma < \tfrac{4}{3}$. This includes as a special case the result first obtained by Ritter[*] that an ordinary sphere of gas in which the generation and radiation of energy are disregarded is dynamically unstable if $\gamma < \tfrac{4}{3}$.

111. For ordinary stellar matter, γ must be very approximately equal to $1\tfrac{2}{3}$. If γ has any value greater than $1\tfrac{1}{3}$, expression (110·4), which gives the length of the range of stable configurations, has the same sign as $(3\alpha + 7)$. Thus for any stable configurations to exist at all, $(3\alpha + 7)$ must be positive. This condition merely rules out large negative values of α, which general physical principles would in any case shew to be highly improbable.

With $(3\alpha + 7)$ positive, the length of range of stable configurations, as given by expression (110·4), is positive for all values of λ, but it becomes

* *Anwendungen der Mechanische Wärmetheorie auf Kosmologische Probleme* (Leipzig, 1882).

very restricted for gaseous stars of large mass, and finally disappears altogether for gaseous stars of infinite mass for which $\lambda = 0$. For such stars formulae (109·1) and (109·2) shew that stability is impossible unless $3\alpha + \beta - n$ has the precise value

$$3\alpha + \beta - n = 0 \dots \dots (111\cdot1).$$

As α and β are constants of nature and cannot vary from star to star, it is infinitely improbable that they have the exact values necessary to satisfy the equation. Thus it is *à priori* infinitely improbable that stars of enormously great mass, which continue to exist stably, are wholly gaseous.

The larger component of Plaskett's star has at least 80 times the sun's mass, while if Otto Struve is right in assigning 950 times the sun's mass to the quadruple system 27 Canis Majoris, the most massive star of this system must have at least 240 times the sun's mass. For gaseous stars of 80 and 240 times the sun's mass, the values of λ are 0·29 and 0·15 respectively, and the corresponding lengths of range of values for $3\alpha + \beta - n$, as determined by formula (110·5), are $0·034(3\alpha + 7)$ and $0·018(3\alpha + 7)$. The numerical coefficients are so small that it is extremely improbable that the natural values of α and β will be such as to make $3\alpha + \beta - n$ lie within these small ranges, and on this evidence alone we might conjecture with fair certainty that neither Plaskett's star nor 27 Canis Majoris can be wholly gaseous.

Let us, however, examine the conditions for stability in more detail. On putting $\gamma = 1\frac{3}{5}$, and introducing the values of ρC_v, $3p_G + 4aT_0^4$ and λ already given, the condition that there shall be no explosive vibrations, as expressed by inequality (109·1), becomes

$$3\alpha + \beta - n < (3\alpha + 7)\frac{(\xi + \frac{1}{2})\lambda}{(1 + \xi)(\lambda + 4)} \dots \dots (111\cdot2).$$

For a purely gaseous star, $\xi = 0$, and the condition reduces to

$$3\alpha + \beta - n < (3\alpha + 7)\frac{\lambda}{2(\lambda + 4)} \dots \dots (111\cdot3)$$

while, to secure dynamical stability, condition (109·2) requires that $3\alpha + \beta - n$ shall be positive.

For a gaseous star of the mass of Pearce's star, $\lambda = 0·45$, so that inequality (111·3) becomes

$$3\alpha + \beta - n < 0·050(3\alpha + 7) \dots \dots (111\cdot4)$$

and, since large values of α are out of the question, we see that $3\alpha + \beta - n$ must be quite small.

We have, however, just seen that the most massive stars of all are very unlikely to be gaseous, and at the end of the last chapter (§ 98) we found reasons for suspecting that the stars in general are not wholly in the gaseous state. When deviations from the gas-laws are admitted, the range of values of $3\alpha + \beta - n$ can be extended almost indefinitely; in the extreme case in which the deviations from the gas-laws are made infinite, ξ and λ both become infinite, independently of the mass of the star, and condition (111·2) is satisfied by all values such that $\beta - n < 7$.

Let us consider the effect of the deviations considered at the end of the last chapter, these being such as to make the observed data consistent with an atomic number of about 95, or a value of N^2/A equal to about 37. The value of N^2/A for Pearce's star calculated on the supposition that it is wholly gaseous is only 27, which would of itself suggest that the star is in the gaseous state. As already explained, the data are very uncertain and the value 27 is probably too small, but even after allowing for all sources of error, it seems unlikely that the star can be very far removed from the gaseous state. Thus the fact that it continues to exist stably shews that $3\alpha + \beta - n$ cannot be much greater than right-hand member of inequality (111·4), and so must be quite small. It would be gratifying if we could confirm the result from other stars, but adequate data for 27 Canis Majoris and Plaskett's star are wanting. For V Puppis the inequality (111·2) is found to assume the form

$$2\cdot1\alpha + \beta - n < 2\cdot2.$$

Since α and β are constants of nature, they may be determined from any one star, and the evidence of Pearce's star is that $3\alpha + \beta - n$ must be fairly small or negative. As n is not greatly different from $\frac{1}{2}$, this involves that $3\alpha + \beta$ must be fairly small or negative. If $3\alpha + \beta$ were greater than about unity Pearce's star, and probably also the more massive stars 27 Canis Majoris and Plaskett's star, would be unstable through a tendency to develop explosive pulsations.

The Mechanism of Generation of Energy.

112. We have seen that the generation of stellar energy must be a sub-atomic process, but we have so far obtained no indication on physical or astronomical grounds of the mechanism by which this generation of energy takes place.

The one familiar instance of sub-atomic generation of energy is radio-activity. Here the generation of energy by an atom is "spontaneous" in the sense that it is not influenced by interaction with other atoms or electrons; by analogy with monomolecular reactions in chemistry, we may describe it as a "mono-atomic" process. But we have no more right to assume that all sub-atomic generation of energy is "mono-atomic" than we should have to assume that all chemical reactions are monomolecular.

Alternative possibilities are that sub-atomic generation of energy may result from the interaction between an atom and other atoms, or free electrons or possibly radiation. The condition just obtained enables us to rule out all these possibilities in turn.

The rate of generation of energy per unit mass has been supposed to be proportional to $\rho^\alpha T^\beta$. If the mechanism of generation were set into operation by collisions of any kind, the rate of generation would be proportional to the number of collisions per unit time and possibly also to a further factor, representing the energy of the collisions which would increase with increasing temperature. The number of collisions per unit mass per unit time being proportional to $\rho T^{\frac{1}{2}}$, we should have $\alpha = 1$, $\beta \geqslant \frac{1}{2}$, and hence $3\alpha + \beta \geqslant 3\cdot5$.

If the mechanism of generation of energy were set into operation by the proximity of other bodies, the rate would be proportional to ρ, or to ρ raised to some power higher than unity, so that we should have $\alpha \geqslant 1$, $\beta = 0$ and hence $3\alpha + \beta \geqslant 3\cdot0$.

If the mechanism of generation were set into operation by ordinary temperature radiation, the rate would be proportional to the density of this radiation, and so to T^4, so that we should have $\alpha = 0$, $\beta = 4$ and hence $3\alpha + \beta = 4$.

All these values of $3\alpha + \beta$ are too large to be reconciled with the restriction on the value of $3\alpha + \beta$ just obtained, so that the corresponding possibilities may be dismissed. It must of course be admitted that radiation of precisely the right wave-length to stimulate the sub-atomic reaction must operate in the direction of increasing the rate of generation of energy, but we shall find that the increase produced in this way is quite insignificant.

113. The condition that $3\alpha + \beta$ must be small would also seem to rule out the suggestion of Perrin and Eddington as to energy being generated by atoms of the lighter elements combining to form heavier atoms. For before, for example, four atoms of hydrogen can combine into an atom of helium, it is necessary that the four atoms of hydrogen should come together, and the frequency with which this would happen per unit mass of hydrogen would be proportional to the third power of the density. This would give $3\alpha + \beta = 9$, a value which is far too high for stability.

114. The same condition would seem also to rule out a scheme of generation of stellar energy proposed by H. N. Russell[*], according to which the rate of generation of energy remains insignificant until the stellar matter attains a certain critical temperature T_c, after which generation proceeds vigorously. This scheme demands a rate of generation which may be represented in its main essentials by a law of the type

$$G \propto \left(\frac{T}{T_c}\right)^\beta,$$

where β is large. Our discussion has shewn that large values of β would render the star unstable, so that Russell's scheme would seem to be untenable.

---

[*] See § 155 below; also H. N. Russell, *Nature*, August 8, 1925, or Russell, Dugan and Stewart, *Astronomy*, Chapter xxvi.

115. We are thus led to the conclusion that the generation of energy must be a "mono-atomic" process; the atom by which the energy is generated is alone involved. Since the concepts of density and temperature have no meaning with reference to single atoms, it follows that the process is independent both of density and temperature. The rate of generation must proceed at a rate determined solely by the constants which specify the structure of the atoms; the process is spontaneous in the way in which the emission of energy by radioactive substances is spontaneous.

The quantum theory provides a reason why the emission of energy by radioactive substances cannot be either inhibited or expedited by such changes of temperature as are possible in the physical laboratory. A sub-atomic generation of energy, as Einstein has shewn, can occur in either of two ways—either spontaneously (i.e. through the lapse of time) or through the stimulus of incident radiation. In this latter case the wave-length of the incident radiation must be such that the energy of one quantum is equal to the energy set free by the sub-atomic change in question. This makes it easy to calculate the temperature at which the second process becomes operative. The temperature necessary to expedite the disintegration of uranium is in this way found to be of the order of thousands of millions of degrees, and it at once becomes clear why warming up uranium in the laboratory does not speed up its disintegration.

A similar calculation gives the temperature necessary to influence the rate of annihilation of matter. The process of annihilation of which the energy is lowest is the simple annihilation of an electron and a proton. The total mass annihilated in this process is $1\cdot66 \times 10^{-24}$ grammes, so that the total energy transformed in the process is C^2 times this, or $0\cdot0015$ ergs. Only quanta of radiation having this energy can have any effect on the rate of annihilation of matter, and a simple calculation shews that the number of such quanta is quite inappreciable until the temperature is of the order of a million million degrees. It follows that the temperature necessary to influence the rate of sub-atomic annihilation of matter is of the order of a million million degrees. It may be argued that a lower temperature, although not adequate to bring about the actual annihilation of matter, might set up sub-atomic processes of adequate intensity. This is true as regards a star's momentary radiation, but such processes cannot provide an adequate duration for the radiation. All processes which are affected by temperatures of less than about a million million degrees leave the total number of electrons and the total number of protons in a star unaltered, whereas the evidence of astronomy is that the number of electrons and protons in a star continually decreases.

With this figure before us, it is clear that the comparatively feeble stellar temperatures of less than a thousand million degrees must be quite inoperative in regard to the main generation of stellar energy; indeed the heat of the hottest of stellar interiors can have no more influence on the rate of annihilation of matter than a warm summer's day has on the rate of disintegration of uranium. We can now see why we were of necessity led to the conclusion that the rate of generation of energy cannot be influenced by changes of temperature. Any other conclusion would have been a violation of the laws of physics. With this result before us it would seem to be abundantly clear that what is annihilating the matter of the stars is neither heat nor cold, neither high density nor low, but merely the passage of time.

116. This conclusion has been reached from two abstract theoretical discussions, to which observational astronomy has hardly made any contribution at all. Thus observational astronomy is perfectly free either to confirm or to deny the conclusion we have reached. But before appealing to observational astronomy, one further abstract consideration must be put forward.

We may imagine a star to consist of a mixture of chemical atoms of different types, and from analogy with the radioactive elements, we should expect these to generate radiation at very different rates. But if a gramme of a special type of atom generates energy at a high rate, it is merely because the atoms are dissolving into energy at a high rate. The elements which generate energy at a great rate are those whose atoms are short-lived. Indeed if the whole atom is annihilated, it is readily seen that the rate of generation of energy per unit mass varies inversely as the average life of the atom.

As a star ages those elements which are most short-lived disappear first, while the most permanent atoms survive the longest. This is equivalent to saying that its most energetic generators of radiation disappear first, while the atoms which persist into the old age of the star are but feeble generators of radiation. It follows that as a star ages, its average rate of generation of energy per unit mass decreases. And, of course, since its radiation is produced at the expense of the matter it contains, its mass also decreases. Thus if we are right in believing that it is merely old age that annihilates matter, we ought to find that the stars which radiate most energetically per unit mass are the youngest stars, and so also the most massive stars; they ought not to be the stars of highest or lowest internal temperature or of highest or lowest central density.

117. This consequence of our theory is fully confirmed by observational astronomy. The stars which radiate most energetically are neither the hottest nor the densest; they are the most massive and hence the youngest. Tables and diagrams have been prepared by Hertzsprung*, by Russell, Adams and Joy†, and by Eddington‡, in which a star's luminosity is given in terms of its mass, and each shews a very strong correlation between mass and

---

* *Bull. Astr. Inst. Netherlands*, No. 43 (1923).
† *Publications Ast. Soc. Pac.* xxxv. (1923), p. 189.
‡ *M.N.* LXXXIV. (1924), p. 308.

luminosity; not only does a star's luminosity depend very closely on its mass, but so also does the star's luminosity per unit mass, which gives a measure of the average rate at which it is generating energy per unit mass. Stars of great mass, which our theory requires to be the youngest stars, shew high luminosity per unit mass, while stars of small mass, which are old stars, shew low luminosity per unit mass.

No such correlation can be discovered between a star's luminosity and its internal temperature or density. Certainly there is no correlation in the sense of the hottest or densest stars radiating more per unit mass than other stars; indeed many of the hottest and densest stars are entirely put to shame in the matter of radiation by very cool stars of low density such as Antares and Betelgeux. If we arrange the stars in order of radiation per unit mass, we shall find that we have arranged them neither in order of temperature nor of density, but very approximately in order of mass and age; the most massive and the youngest stars radiate most energetically regardless of their interior temperatures and density; the older stars appear to be tired out.

The general tendency is shewn in the following table:

Table XIV. *Generation of Energy by Typical Stars.*

Star	Generation (ergs per gramme)	Central Temperature	Central Density	Mass	Age (years)
B.D. 6° 1309 *A*	(11,000)	Very high	Very great	(78)	
H.D. 1337 *A*	15,000	200,000,000	40	36·3	less than 10^{11}
V Puppis *A*	1,100	160,000,000	120	19·2	
Betelgeux	(300)	10,000,000	(0·04)	(40)	
Capella *A*	48	19,000,000	1	4·18	less than 10^{12}
Sirius *A*	29	80,000,000	140	2·45	10^{12}
Sun	1·90	55,000,000	140	1·00	7×10^{12}
a Centauri *B*	0·90	45,000,000	70	0·97	7×10^{12}
60 Kruger *B*	0·02	65,000,000	5300	0·20	2×10^{14}

The rates of generation of energy which occupy the second column are those already given in Table IX (p. 59). The central temperatures and densities are taken from Table XII (p. 100). We subsequently found that the central temperatures needed a reduction of the order of 30 per cent. to allow for deviations from the gas-laws, but an all round reduction of about this amount would leave the general run of the figures unaltered. The central densities probably need adjustment for the same reason, but these differ so widely between themselves that no adjustment of this type is likely to change the order much. The masses shewn in the fifth column give a rough indication of the star's age, but the result of a more exact calculation is shewn in the last column.

The Ages of the Stars.

118. The method by which these ages are calculated needs explanation. The observed correlation between a star's mass and its total rate of radiation (or bolometric luminosity) is so marked that it is possible to say, within a not very great limit of error, that a star of known mass M has a definite bolometric luminosity L, which depends only on M.

To a fairly good approximation, the emission of radiation by different stars is found to be proportional to the cubes of their masses, so that

$$L = M^3 \quad \dots\dots\dots\dots\dots\dots\dots\dots\dots(118{\cdot}1),$$

where L is the star's bolometric luminosity and M is its mass, both being measured with reference to the sun as unity. Taking logarithms of both sides we deduce that

$$m = 4{\cdot}85 - 7{\cdot}5 \log M \quad \dots\dots\dots\dots\dots\dots(118{\cdot}2),$$

where m is the absolute bolometric magnitude.

Examples of the use of this formula, which for the present is best regarded as purely empirical, are shewn in the following table:

Table XV. *Mass-Luminosity Relation.*

Star	Mass	Absolute bolometric magnitude (calc.)	Absolute bolometric magnitude (obs.)	(obs. – calc.) (118·2)	(obs. – calc.) (Eddington)
B.D. 6° 1309 A	(75·6 +)	− 9·24 +	− 9·31	− 0·07 +	− 1·42 +
H.D. 1337 A	36 3	− 6·85	− 8·82	− 1·97	− 2·28
V Puppis A	19·2	− 4·77	− 5·26	− 0·5	+ 0·4
Capella A	4·18	0·20	− 0·20	− 0·4	—
,, B	3·32	0·94	0·22	− 0·7	+ 0·20
Sirius	2·45	1·93	1·00	− 0·9	+ 0·30
Sun	1·00	4·85	4·85	—	+ 0·33
a Centauri A	1·14	4·42	4·7	+ 0·3	+ 0·37
,, B	0·97	4·95	5·7	+ 0·7	+ 0·15
60 Kruger A	0·25	9·36	10·0	+ 0·6	− 1·41
,, B	0·20	10·1	11·5	+ 1·4	− 1·33

It will be noticed that the errors (obs. − calc.) in the fifth column are fairly small, shewing that L is fairly close to M^3. There is a certain amount of systematic deviation in the sense of L being greater than M^3 for stars of largest mass, but it should be noticed that if the mass of B.D. 6° 1309 A is much greater than the minimum value 75·6, L is less than M^3 for this star. Except for this the agreement could be greatly improved by making L proportional to a somewhat higher power of M. For instance the law $L = M^{3\cdot4}$ gives errors which average only $\pm 0{\cdot}4$ magnitudes for all stars except the first.

For comparison the last column shews the errors given by the mass-luminosity law discussed in § 90* which Eddington has derived from a

---

* *M.N.* LXXXIV. (1924), p. 308.

consideration of his special model. We see that the very simple law $L = M^3$ gives errors which are, on the whole, rather less than those given by the far more complicated mass-luminosity relation of Eddington.

As a result of its emission of radiation, a star's mass necessarily decreases at a rate which is proportional to the bolometric luminosity L of the star, and so, assuming formula (118·1), to M^3. Hence we must have a general relation

$$\frac{dM}{dt} = -\alpha M^3 \quad \ldots\ldots\ldots\ldots\ldots\ldots(118\cdot3),$$

where α is the same for all stars. Inserting the values appropriate to the sun ($M = 2 \times 10^{33}$ grammes, and $\dfrac{-dM}{dt} = 4\cdot2 \times 10^{12}$ grammes per sec.), we find that

$$\alpha = 5\cdot2 \times 10^{-66}.$$

Solving equation (118·3) we find that the time t required for a star to shrink from a mass M' to a smaller mass M is

$$t = \frac{1}{2\alpha}\left(\frac{1}{M^2} - \frac{1}{M'^2}\right) \quad \ldots\ldots\ldots\ldots\ldots(118\cdot4).$$

For instance, we find that if ever the sun had four times its present mass, the interval which has elapsed since this period must be about $7\cdot1 \times 10^{12}$ years. The time since the sun had an enormously great mass would be approximately the same, for while a star is very massive it radiates its mass away at such a great rate that this stage of its existence does not last for long. Putting $M' = \infty$ in formula (118·4) we obtain $t = 7\cdot6 \times 10^{12}$ years as the interval since the sun was, theoretically, of infinite mass.

The age of any other star calculated as the interval since it was of very great mass, is given by the formula

$$t = \frac{7\cdot6 \times 10^{12}}{M^2} \text{ years,}$$

where M is the star's mass in terms of that of the sun. These are the ages entered in the last column of Table XIV; they provide upper limits to the ages of the stars, except for a reservation to be made below (§§ 127, 167).

The Chemical Evolution of the Stars.

119. The mass of the star V Puppis is at present 19·2 times that of the sun, while Plaskett's star has a mass at least 76 times that of the sun, and probably greater still. After an interval of the order of 7×10^{12} years, each of these stars will be reduced to a mass about equal to that of the sun.

If radiation results from an actual annihilation of matter, it follows that about 99 per cent. of the electrons and protons now in Plaskett's star and about 95 per cent. of those now in V Puppis will have been annihilated during this period, their energy having been transformed into radiation. As we have already seen, the evidence of the opacity of the stars shews that there is no very great range in the average atomic weights of the atoms in different stars. Thus our statement may be put in the alternative form that some 99 per cent. of the atoms in Plaskett's star and some 95 per cent. of the atoms in V Puppis are fated to undergo annihilation within the next 7×10^{12} years or so.

We have no evidence as to whether stellar atoms are annihilated instantaneously or through successive stages by a gradual reduction of atomic weight and atomic number. But an essential feature of the process must in either case be the coalescence of electrons with the protons of the nucleus, since there is no other means of annihilating the protons. The electrons which fall into the nuclei might *à priori* either be free electrons rushing about through the star, or bound electrons describing orbits about the nuclei in question under the laws of quantum-dynamics.

If free electrons alone were liable to annihilation, the rate of annihilation, and so of generation of energy, would be in some way dependent on the frequency of collisions or of proximity between free electrons and nuclei. The discussion of § 112 enables us to rule out these possibilities, and we conclude that the process of annihilation consists mainly or wholly of the falling of bound electrons into the nucleus.

120. One conclusion appears to follow. A star whose atoms are stripped bare right down to their nuclei, ought to radiate little or no energy, for it has no bound electrons which can be annihilated, except perhaps a few in its surface layers. In looking for examples of such stars we naturally turn to the white dwarfs, the abnormally high density of these stars providing a positive guarantee that the majority of their atoms are stripped bare of electrons, since if even the K-rings were left the densities would be less than those actually observed. The predictions of theory are found to be confirmed by observation, the white dwarfs emitting abnormally little radiation. The faint companion to Sirius, of mean density about 50,000, radiates only 0·007 ergs per gramme as against the 1·9 ergs per gramme radiated by the sun. The faint component of o_2 Eridani, with a mean density of about 100,000, radiates about 0·002 ergs per gramme. The spectral type of the companion to Procyon is unknown. If it were of type $M6$ its visual luminosity would give it a surface only one-eightieth of that of 60 Kruger A, and so a radius equal to about a twenty-seventh of that of the sun, or about equal to that of Sirius B. A higher surface-temperature would give a still smaller radius. Thus, whether we call the star a white dwarf or not, its density is so great that we must suppose most of its atoms to be stripped bare of electrons. Its radiation of energy is only about 0·0005 ergs per gramme. Finally van Maanen's star (0 h. 43·9 m., + 4·55), an undoubted white dwarf with a mean density of hundreds of thousands, radiates only about 0·001 ergs per gramme.

More generally any stripping of the electrons from the atoms of stellar matter ought to inhibit the energy-generating capacity of the matter, since it increases the number of free electrons which are immune from annihilation. Broadly speaking, the atoms near the centre of a star are more highly ionised than those in its outer regions, and this must produce a tendency to inhibit the generation of energy in the central regions of a star and to throw the generation of energy into the star's outer layers. Other tendencies may of course operate to neutralise this, wholly or in part. We shall return to this question later.

121. The conclusion reached in § 115 that the rate of annihilation of matter is unaffected by changes of density and temperature, was based on the supposition that the matter to be annihilated retained the same constitution throughout all changes of temperature and density. Our conclusion implied, for instance, that a mass of matter in which each atom is ionised down to its K-ring but not below, will generate energy at precisely the same rate no matter how its temperature and density are changed, so long as these changes do not affect the ionisation of the matter. As soon as the ionisation is affected, the problem of the dependence of energy-generation on temperature and density becomes identical with that of the dependence of ionisation on temperature and density.

An increase of temperature increases the degree of ionisation of stellar atoms because the quanta become more energetic; this in turn inhibits the generation of energy. On the other hand, an increase of density decreases the degree of ionisation because there are now fewer quanta per atom, and this in turn increases the generation of energy. Thus the rate of generation of energy per unit mass must decrease with temperature and increase with density. We have tentatively supposed this rate to be proportional to $\rho^\alpha T^\beta$; it now appears that β must be negative and α positive. But a brief calculation, based on the known formulae connecting ionisation with temperature and density shews that α and β are both comparatively small. To a preliminary approximation which is adequate for most purposes, we may disregard both α and β, and suppose the rate of generation of energy to be a constant for a given type of matter.

122. Stellar radiation must either orginate in types of matter known to us on earth or else in other and unknown types. When once it is accepted that high temperature and density can do nothing or almost nothing to accelerate the generation of radiation by ordinary matter, it becomes clear that stellar radiation cannot originate in types of matter known to us on earth. Terrestrial matter as we have seen (§ 104) generates almost no energy, and as increasing its temperature to stellar temperatures and decreasing its density to stellar densities could not increase its generating capacity, we must conclude that other types of matter are responsible for the radiation of the sun and stars. The only possible reservation to this statement arises in connection with the radioactive elements. We have, however, seen that even if the sun were built of pure uranium, its radiating power would be only about one-half of that observed, and would only last for a minute fraction of what is believed to have been the sun's life. A sun of pure radium would radiate more than enough for the moment, but its life would be limited to a few thousand years. No possible combination of terrestrial elements can give the combination of high radiation and of staying power which is observed in the sun and stars.

According to practically all theories of the origin of the solar system, the earth and the other planets must have been formed out of the body of the sun at some epoch which we may date at approximately 1500 million years ago. The question then arises why the atoms of the sun, which are generating energy at an average rate of some 1·9 ergs a second, should be of different type from those of the earth, which are certainly generating less than 0·0001 ergs per second.

If they are of different type now, they must have been of different type when the earth was born out of the sun. The sun's present rate loss of mass by radiation indicates that the present average life of solar atoms is of the order of 15×10^{12} years. Thus the chemical composition of the sun, or of a fair sample of solar atoms, can hardly have changed appreciably in the 15×10^8 years of the earth's existence. The earth's atoms cannot have formed a fair sample of the sun's atoms when the earth was born, otherwise they would still do so, and the phenomenon of energy-generation shews that they do not.

Thus we must suppose that the atoms in the outer layers of the sun, from which the earth was formed, do not constitute a fair sample of the sun as a whole. This supposition can hardly be considered unreasonable, any more than the supposition that the earth's atmosphere is not a fair sample of the earth as a whole. Whatever mixture of elements may constitute the sun, the heavier elements are likely to sink to its far interior*, so that the atoms in its outer layers are selected for their light atomic weight. If for simplicity we regard the sun as containing only two kinds of atoms—light elements which rise to its surface, and heavy elements which sink into its far interior—then the light elements, being similar to those found on earth, have practically no capacity for generating energy, so that the sun's generation of energy must originate in the heavy elements which reside in its interior regions. But if, as we have supposed (§ 120), ionisation of stellar atoms partially or wholly inhibits their generation of energy, then the problem of the distribution of this energy generation becomes very intricate.

123. More generally we may suppose that the matter of the sun and stars consists in its earliest state of a mixture of elements of different atomic

* See § 87 (p. 93) above.

weights, those elements whose atomic weights are highest having the greatest capacity for the spontaneous generation of radiation by annihilating themselves, and, in consequence, having the shortest lives. Such elements will necessarily be the first to disappear as the star ages, their disappearance reducing not only the mean atomic weight in the star but also the mean rate of radiation per unit mass, since these heavy elements are the most energetic radiators. Just as, on the coast, the hardest rocks survive for longest the disintegrating action of the sea, so in a star the lightest elements survive for longest the disintegrating action of time, with the result that ultimately the star contains only the lightest elements of all and so has lost all radiating power. Our terrestrial elements have so little capacity for spontaneous transformation that they may properly be described as "permanent." The result of our previous calculations of § 104 may be stated in the form that if the terrestrial elements underwent any appreciable transformation in periods comparable with a period of 10^{17} years, the resulting generation of heat by the earth's mass would make the earth too hot for human habitation. Again the radioactive elements must be mentioned as an exception; they probably represent the last surviving vestiges of more vigorous primeval matter, thus forming a bridge between the inert permanent elements and the heavier and shorter-lived elements of the stars.

The half-period of uranium (5×10^9 years) is so short that we must suppose that the supply of uranium in the sun is being continually replenished. Otherwise, as Lindemann* has remarked, an interval of 5×10^{11} years would reduce the amount of uranium in the sun by a factor of 10^{-30}. Even if the sun had consisted wholly of uranium at the beginning of this period there would be less than 2 kilograms left at the end, and more than this exists now on the earth alone.

It is hardly likely that this replenishment can occur through the synthesis of lighter elements. We have seen that a temperature of the order of a thousand million degrees is necessary to effect the disintegration of uranium and a similar temperature would, in all probability, be needed to produce uranium by the synthesis of its components.

Thus we are led to suppose that the sun's store of uranium is continually being formed by the disintegration of a parent element of higher atomic weight and far longer half-period. The amount of uranium would in this case be kept constant if the amounts of the two substances were proportional to their half-periods, and a half-period of the order of 10^{13} years for the parent element removes all difficulties associated with the time-scale without calling for any improbably large amount of this element.

THE PROCESS OF ANNIHILATION.

124. We are now in a position to form some sort of a picture of the mechanism by which a star generates and radiates its energy.

* *Nature*, cxv. (1925), p. 229.

We must suppose that one of the heavy atoms in a star's interior first begins to change into radiation through one of its bound electrons falling into the nucleus, and coalescing with one of the nuclear protons so that both are annihilated. It is immaterial whether the whole atom changes into radiation at once or through a succession of comparatively slow changes. In either case the process of annihilation is likely to consist of a series of events in each of which a single proton and a single electron are annihilated simultaneously.

As we have seen, the energy set free by the annihilation of a proton of mass M and an electron of mass m is $(m + M)C^2$, which is equal to 0.0015 ergs. In accordance with general quantum principles each such annihilation must result in the production of a single quantum of radiant energy of frequency ν given by

$$h\nu = 0.0015 \text{ ergs},$$

so that the frequency ν is 2.3×10^{23} and the wave-length is 1.3×10^{-13} cms.

Each time a proton and electron are annihilated a splash of radiant energy of this wave-length and of total energy 0.0015 ergs is produced, and sets off to travel through the star until, after innumerable absorptions and re-emissions, it reaches the star's surface and wanders off into space. Except for being many thousands of times more powerful, each splash is similar to the splashes produced by radioactive material in the spinthariscope. The great energy of the splashes is to some extent counterbalanced by their rarity. In the sun, for instance, only about one atom in every 10^{17} annihilates itself each hour.

As this very high-frequency radiation travels through a star, it may be either scattered or absorbed when it meets an atom. Absorption can only be by complete quanta; the absorption of a quantum ejects an electron with a velocity representing kinetic energy of 0.0015 ergs, and so equal to 0.99999985 times the velocity of light. When this electron strikes an atom a new quantum of radiation is emitted whose energy, and therefore also wave-length, is equal to that of the original radiation. The hardness of the radiation is thus unaffected by absorption and re-emission. The scattering of the radiation, on the other hand, is readily shewn to produce a softening of its quality, just as in the ordinary Compton effect, and a succession of such scatterings will increase the wave-length of the radiation until it becomes indistinguishable from ordinary temperature radiation.

Newly generated radiation, in spite of its extreme hardness, will not penetrate far through the interior of a star without being changed in this way, so that we should expect the radiation emitted from the surface of a star to be ordinary temperature radiation, retaining no traces of its origin as radiation of extremely short wave-length.

On the other hand, astronomical bodies exist which are transparent to ordinary light and so, *à fortiori*, must be transparent to this high-frequency radiation. The irregular nebulae and the outer shells of the planetary nebulae form obvious instances, but the most important for our present purpose are the spiral and other extra-galactic nebulae. All the radiation generated in the transparent parts of these nebulae ought to pass without appreciable absorption or softening into outer space, so that we should expect to find space filled with high-frequency radiation of wave-length of the order of $1\cdot3 \times 10^{-13}$ cms. Here and there this radiation may devastate isolated atoms, ejecting 940-million volt electrons as it passes, but the greater part of it will pass through space unhindered until it meets a medium of substantial absorbing powers. Thus we should expect the atmospheres of the stars, sun and earth, and even the solid body of the earth, to be under continual bombardment by highly penetrating radiation of nebular origin.

Highly Penetrating Radiation.

125. Such radiation has been detected in the earth's atmosphere by McLennan, Burton, Cook, Rutherford, Kolhörster, Millikan, and many others. It seems to be satisfactorily established that the radiation is of extra-terrestrial origin, since it does not decrease in intensity with increasing height as it would if, as at first was thought, it originated in the earth's radioactivity. Indeed, by sending up balloons, Kolhörster, and later Millikan and Bowen, have shewn that the intensity of the radiation is substantially greater at high altitudes than at low, proving clearly that the radiation comes into the earth's atmosphere from outside. It does not come from the stars, for if it did the main part would come from the sun, and the amount received by day would be far greater than that received at night; this is not found to be the case. Thus the radiation must originate in nebulae or cosmic masses other than stars. Kolhörster and von Salis have recently found* that the intensity of the radiation received at any point on the earth's surface varies with the orientation of the earth in space in a way which indicates that the radiation is received largely from regions near the Milky Way, especially the regions of Andromeda and Hercules.

126. This is entirely consistent with the radiation being the direct product of the annihilation of matter in the transparent parts of the great nebulae. If g ergs per second are generated in any small region of space, the flux of radiation per square cm. at a distance r will be $g/4\pi r^2$. If a shell of matter surrounding the earth and having radii r_1, r_2 generates G ergs per gramme per second, the amount falling on a square cm. of the earth's atmosphere will be

$$\frac{1}{3}\int_{r_1}^{r_2} \frac{G\rho 4\pi r^2\, dr}{4\pi r^2} = \tfrac{1}{3}G\rho\,(r_2 - r_1)\dots\dots\dots\dots\dots(126\cdot1).$$

We can form a rough estimate of the actual flux of this highly penetrating radiation through the earth's atmosphere. At sea-level it is found to produce $1\cdot4$ ions per cubic centimetre per second, so that if it underwent no absorption

* *Nature*, Oct. 9, 1926.

at all in the earth's atmosphere the flux of energy per square centimetre
would produce over a million ions per second, representing a flux of energy of
about 0·00003 ergs per second. The actual flux must be far greater than this,
partly because the figure of 1·4 ions per second represents the activity of
the radiation after it has suffered absorption by the earth's atmosphere, and
partly because the main part of this radiation must expend its energy in other
ways than in the ionisation of molecules. To allow for absorption we may
multiply our estimate by about 10, increasing it to about 0·0003 ergs per
second; but it seems impossible to estimate the second source of error.

According to Hubble (§ 17) the extra-galactic nebulae are so evenly
spaced that we can regard them, to a first approximation, as forming a distribution of uniform density round the earth, with a density of matter, at least up
to about 100 million light years, of about $1·5 \times 10^{-31}$ grammes per cubic cm.

Giving ρ this value in equation (126·1), we find that a flux of 0·0003 ergs
per square cm. requires that

$$G (r_2 - r_1) = 6 \times 10^{27}.$$

If we put the outer radius r_2 equal to the 140 million light years distance
of the furthest visible nebulae ($r_2 = 1·4 \times 10^{26}$) we obtain $G = 43$. Thus we
can account for the amount of highly penetrating radiation received by the
earth's atmosphere, by supposing all the nebulae within 140 million light years
to be generating energy at the rate of 43 ergs per gramme per second. The
calculation is of course very rough; the energy actually received is greater
than 0·0003 ergs per second for the reasons already stated; against this no
doubt a large amount of highly penetrating radiation reaches us from nebulae
at distances greater than 140 million light years.

Setting these various corrections off against one another, a rate of generation
of energy of $G = 43$ ergs per gramme per second would seem to be of the right
order of magnitude to account for the observed reception of highly penetrating
radiation; we notice at once that it is of the same order of magnitude as the
rate of generation of fairly young stars—for Sirius, for instance, $G = 29$; for
Capella, $G = 50$. We shall return to a fuller discussion of this fact later; for
the present we note that the agreement as regards order of magnitude confirms
our conception of stellar energy originating in the annihilation of matter.

The wave-length of the observed radiation might be expected to disclose
its source. If the radiation originated in the annihilation of matter, the
minimum wave-length ought to be $1·31 \times 10^{-13}$ cms., the energy of a quantum
being 0·00150 ergs, and that of an electron 940 million volts. But the only
quantity susceptible of direct measurement is the penetrating power of the
radiation, and the theoretical relation between this and wave-length is still
uncertain. The original formula of Compton and a later formula of Dirac and
Gordon agreed in assigning a wave-length far greater than $1·31 \times 10^{-13}$ cms.
to the observed radiation, but a more recent investigation of Klein and Nishina[*]

* *Nature*, cxxii. (1928), p. 399.

takes account of the relativity-correction and leads to very different results. Gray has found that the available experimental evidence on the whole supports the new formula. And the observed penetrating power of the radiation agrees excellently with that which it assigns to radiation of 940-million volts energy, such as would be produced by the annihilation of protons and electrons†.

Rosseland‡ has suggested that bombardment by this radiation may be the cause of the observed bright lines in stellar spectra; I had previously§ suggested a similar origin for the luminosity of the irregular nebulae.

Finally we may notice that, as a gramme of matter contains 9×10^{20} ergs of energy, our estimate of $G = 43$ ergs per gramme per second for the rate at which energy is generated by the annihilation of matter throughout the universe, assigns to the matter of the universe an average expectation of life of 2×10^{19} seconds, which is only about 600,000 million years. But our whole estimate is of so rough a character that this figure can claim at most to be accurate as regards order of magnitude.

SUMMARY.

127. The general impression produced by our analysis of observational facts is that of a universe slowly but inexorably dissolving into radiation. The process may not properly be compared to that of ice dissolving into water, a process whose speed is governed by the rate at which heat is supplied from outside, nor to gunpowder dissolving into smoke, a process whose rapidity is checked only by mechanical conditions extraneous to the process itself; it should rather be compared to the dissolving of uranium into lead, a process which pursues its steady course uninfluenced by all external factors.

One reservation only need be made. We have found reasons for supposing that the complete ionisation of stellar matter inhibits the process of transformation into energy, and it seems likely that partial ionisation may produce a partial inhibition of this process. The chemist who is given a sample of rock in which lead and uranium are mixed can estimate the age of the rock by measuring the proportion of uranium which remains. The astronomer who is given a sample of stellar matter cannot estimate its age in a similar way; he must first know for how long it has been protected from dissolution by being in a state of complete or partial ionisation. The ages which we have calculated for the stars may need to be substantially increased for this reason, and what appear to be the youngest of the stars may conceivably be constituted of atoms which are as old as any in the universe but have been protected by ionisation from dissolution throughout the greater part of their lives.

† See a general discussion by Sir E. Rutherford, *Proc. Roy. Soc.* 122ᴀ (1929), p. 15.

‡ *Astro. Journ.* May, 1926.

§ *Nature*, December 12, 1925.

CHAPTER V

LIQUID STARS

THE GENERAL CONDITION OF LIQUID STARS.

128. IN Chapter III we investigated the internal equilibrium of the stars on the supposition that they were masses of gravitating gas, in which the gas-laws were obeyed throughout. The investigation was abandoned when it was found to lead to impossibly high values for the atomic weights of the stellar atoms. This created a suspicion that the hypothesis on which it was based was unfounded, and that the gas-laws are not obeyed in stellar interiors.

The last chapter provided further evidence to the same effect. We there investigated the mode of generation of stellar energy, using the guiding principle that all modes of generation of energy which make the stars dynamically unstable can be ruled out of the list of practical possibilities. We found that when the gas-laws are supposed to be obeyed, no possibilities remain for stars of enormously great mass. Further the only mode of generation of energy which was both physically acceptable and consistent with the stability of actual stars proved to be one in which the rate of generation of energy is uninfluenced by changes of density and temperature, as in radioactive substances, and this, as we shall see at once (§ 134), requires substantial deviations from the gas-laws in stars of all masses.

We now rediscuss the problem of the physical constitution of the stars, and examine the form it assumes when the gas-laws are no longer obeyed.

129. With Kramers' law for the coefficient of opacity, the flow of radiant energy per unit area is, as in equation (98·1),

$$H = - \frac{aC\mu T^{6\cdot5}}{3\cdot34 \times 10^{25}\rho^2(N^2/A)} \frac{dT}{dr} \quad \dots\dots\dots\dots(129\cdot1),$$

where N, A are the atomic number and weight of the stellar atoms. This equation is true independently of whether the gas-laws are obeyed or not.

Let us start from a standard configuration in which the star is in equilibrium with a specified rate of generation of energy for each element. We can obtain a succession of configurations suitable to stellar matter of different atomic weights by keeping every element of the star's mass in its original position, thus maintaining the star's distribution of density and its gravitational field unaltered, but varying N^2/A and T throughout the star. If we change N^2/A and T in such a way as to keep the ratio of N^2/A to $T^{7\cdot5}$ unaltered throughout the star, equation (129·1) will still be satisfied throughout the star with the same value of H, so that if the generation of energy is maintained unaltered, the new configuration will all be in radiative equilibrium. Since H retains the same value throughout the star, the star's luminosity remains unaltered.

Let us suppose that the standard configuration from which we started was one in which the stellar matter was wholly in the gaseous state, and that the gas-laws were obeyed throughout the star. Let us further suppose the standard configuration to be one of dynamical, as well as radiative, equilibrium, so that equation

$$\frac{dp}{dr} = -g\rho \dots\dots\dots\dots\dots\dots(129\cdot2)$$

is satisfied throughout, where p is the total pressure, including pressure of radiation, calculated on the supposition that the gas-laws are obeyed. Equation (129·2) is the equation of dynamical equilibrium for all the other configurations also, and since we are supposing g and ρ to remain the same in all these configurations, the equation will be satisfied if p retains its original values throughout the star.

When we assumed the star to be wholly gaseous we found that the values of N^2/A which were necessary to give the true values of H for actual stars came out uniformly something like 16 times too high (§ 98). This suggested that the gaseous configuration is not the true configuration for actual stars, and we found that to obtain possible configurations we must pass along the series just described until we come to configurations in which T has about seven-tenths of its value in the gaseous state.

For the new configuration to be in dynamical equilibrium the total pressure p must be equal to the total pressure p in the gaseous configuration. The pressure of radiation $\frac{1}{3}aT^4$, having only $(0\cdot7)^4$ times, or about 24 per cent. of, its value in the old configuration, is negligible for all except the most massive stars, and the new gas-pressure must shoulder the whole burden previously carried by gas-pressure as well as three-quarters of that previously carried by the pressure of radiation.

130. Incidentally, Eddington * and myself† have investigated the relation between stellar masses and luminosities on the supposition that the gas-laws are obeyed, and the pressure of radiation played a fundamental part in both of our discussions. On the hypothesis of liquid stars we both assumed values for the pressure of radiation that were something like four times too high, so that both our discussions fail entirely and any apparent success achieved either by Eddington's mass-luminosity law, or my mass-luminosity-temperature relation, must have been fortuitous; indeed we should have obtained far more accurate results by disregarding the pressure of radiation entirely. This probably explains why, as shewn in Table XV (p. 130), the errors in Eddington's

* *M.N.* LXXXIV. (1924), p. 308, and *The Internal Constitution of the Stars.* See also § 90 above.
† *Ibid.* LXXXV. (1925), pp. 196 and 394.

law become large for stars of great mass, where the pressure of radiation in gaseous stars is large, and no doubt my own mass-luminosity-temperature relation would shew similar errors *.

This estimate of the importance of radiation-pressure can hardly be much affected by errors of observation, since all stars agree in telling practically the same story. The estimate depends directly on the numerical coefficient F by which we multiplied Kramers' opacity formula in § 77. Probably our estimate of $F = 20$ was, if anything, on the high side, but lowering the value of F would depress the importance of radiation-pressure still lower.

131. In place of the vanished or insignificant pressure of radiation a new pressure assumes importance. In the actual configuration of equilibrium, the free electrons and bare atomic nuclei probably, on account of their smallness, still exert pressure in accordance with Boyle's law. Thus the deficiency of about 30 per cent. in the gas-pressure and of about 76 per cent. in the pressure of radiation must be made good by the deviations from the gas-laws, which arise from the finite sizes of the atoms. To make good this deficiency, we have seen that each atom must exert about 40 times the pressure which it would exert if Boyle's law were actually obeyed. This requires the atoms to be so closely jammed together that the condition of the stellar material may properly be described as liquid or semi-liquid.

When the atoms are jammed as closely as they can be packed, a limiting density ρ is reached (cf. § 144 below), and we can obtain a general idea of the conditions in the central parts of a star by studying the ideal case in which all the matter in the central regions of the star is compressed to this limiting density. This of course represents an extreme case, and the actual truth will lie somewhere intermediate between this extreme and the other extreme of a purely gaseous star already discussed in Chapter III.

132. In general the flux of energy across a sphere of radius r drawn round the centre of a star is $4\pi r^2 H$, where H is given by equation (129·1). It is also $\frac{4}{3}\pi r^3 \overline{\rho G}$ where $\overline{\rho G}$ denotes the mean value of ρG, estimated by volume, inside the sphere of radius r. Hence

$$H = \tfrac{1}{3} r \overline{\rho G} \quad \dots\dots\dots\dots\dots\dots\dots\dots(132\cdot1)$$

or, by reference to (129·1)

$$\frac{\partial}{\partial r}(T^{7\cdot5}) = - B\rho^2 r \overline{\rho G}$$

where B is a constant. On integration from $r = 0$ outwards, this gives

$$T^{7\cdot5} = T_c^{7\cdot5} - B\int_0^r \rho^2 \overline{\rho G}\, r\, dr \quad \dots\dots\dots\dots\dots(132\cdot2),$$

* For a comparison between the two relations, see Ohlsson, *Charlier Festschrift* (Lund, 1927), p. 51.

where T_c is the temperature at the centre of the star. Taking the integration over the whole radius of the star, we obtain

$$T_c^{7\cdot5} = B \int_0^r \rho^2 \rho G r \, dr$$

and this gives the central temperature corresponding to any given rate of generation of energy. For stars built on the same model, we see that

$$T_c^{7\cdot5} \propto E M^2 R^{-7}$$

where E is the emission of energy $\frac{4}{3}\pi R^3 \overline{\rho G}$. On inserting numerical values for actual stars it appears that the central temperatures of the white dwarfs must be enormously high, while those of giant stars of large radius must be comparatively low; in each case the factor R^{-7} preponderates in importance over $E M^2$. As a very rough approximation indeed we may neglect variations in $E M^2$ in comparison with those in R^{-7} on account of the high index of the latter, and find that the central temperatures of the stars must vary something like inversely as their radii. Two white dwarfs are shewn in fig. 5 (p. 61) as

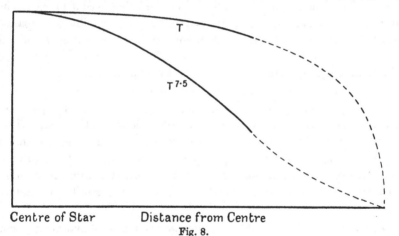

Centre of Star   Distance from Centre
Fig. 8.

having radii of about a hundredth of that of the sun, so that their central temperatures must be of the general order of a hundred times that of the sun, and so must be measured in thousands of millions of degrees.

Let us pass to the consideration of stars in which the atoms near the centre are so closely packed that ρ has an approximately uniform value up to a distance r from the centre of the star. Let us further suppose that G, the rate of generation of energy per unit mass is also uniform within the same range. Then equation (132·2) becomes

$$T^{7\cdot5} = T_c^{7\cdot5} - \tfrac{1}{2} B \rho^3 G r^2 \quad\quad\quad\quad\quad (132\cdot3).$$

The graph of $T^{7\cdot5}$ against r given by this equation is a parabola as shewn by the lower curve in fig. 8. But with this value for $T^{7\cdot5}$ the graph of T is the much flatter upper curve; the largeness of the index 7·5 secures that the temperature is approximately constant throughout a large distance from the star's centre.

Thus we may regard the central regions of a star of the type we are now discussing as being at approximately uniform temperature and density. Outside these lie a region of transition in which the temperature and density fall rapidly, the gas-laws being partially obeyed, and outside this a further region in which the gas-laws are obeyed entirely. The discussion of § 86 has suggested that in very massive stars this outermost region may be of very great extent in comparison with the two inner regions.

STABILITY.

133. With this general, and necessarily very vague, picture of a stellar interior before us, we may resume the discussion of stellar stability at the point at which we abandoned it in the last chapter.

We had found that two conditions were necessary to ensure the stability of a star, the first ensuring that the star should not be liable to explosive vibrations, and the second that it should not be liable to continuous unchecked contraction or expansion. Strictly speaking, it is impossible to separate the parts played by dynamics and thermodynamics in determining the stability or instability of a star, but we may conveniently refer to the two conditions for stability as the thermodynamical and the dynamical conditions respectively.

In the last chapter we hypothetically assumed the rate of generation of energy per gramme of the stellar matter to depend on the temperature and density of the gramme in question through a factor of the form $\rho^{\alpha}T^{\beta}$. The condition for thermodynamical stability was then found to require that α and β should be small. To a first approximation it was possible to put them both equal to zero, so that the star's generation of energy became independent of its conditions of temperature and pressure, and so similar to radioactive generation.

We found that there is only one adequate source of stellar energy, namely, the annihilation of stellar matter, electrons and protons coalescing and destroying one another, setting free their energy in the process in the form of radiation.

We found, however, that free electrons must be immune from annihilation, since if they were liable to annihilation the resulting values of α and β would be so large that every star would be thermodynamically unstable. Annihilation can only overtake electrons which are bound and are describing orbits about the nuclei either of complete atoms or of partial atoms from which a number of electrons have been stripped. In such atoms annihilation has been seen to be a spontaneous process, the likelihood of which cannot be affected by changes of density or temperature. Nevertheless an increase of temperature or a decrease of density, by increasing the degree of ionisation of stellar matter, lessens the number of electrons liable to annihilation and so indirectly lessens the rate of generation of energy per unit mass of the stellar matter. This requires us to give a small positive value to α and a small negative value to β.

134. The second condition of stability, the dynamical condition, is that D shall be positive in equation (108·2) or that

$$(3\alpha + \beta - n) + \frac{3s\lambda}{\lambda + 4}(7 + n - \beta) > 0 \quad \ldots\ldots\ldots\ldots(134\cdot1).$$

Here α and β have the meanings just explained, n is defined through the opacity formula, the coefficient of opacity being supposed proportional (§ 107) to $\mu\rho T^{-(3+n)}$; λ is the ratio of material pressure to gas-pressure in the central regions of the star, and s is a quantity which exists only when there are deviations from the gas-laws, the material pressure p_G being supposed proportional to $\rho^{1+s}T$.

To our first approximation $3\alpha + \beta$ is zero; calculation shews that even to the second approximation, in which changes of ionisation are taken into account, $3\alpha + \beta$ is, generally speaking, very small and negative—i.e. the effect of changes of temperature outweighs that of changes of density, so that $3\alpha + \beta$ takes the sign of β. With Kramers' formula for the opacity $n = \frac{1}{2}$, and in general n may be assumed positive.

Thus the first term on the right of the above inequality is negative, and since $7 + n - \beta$ is in any case positive, the inequality can only be satisfied by assigning a positive value to s. This merely reiterates our former conclusion that stability requires deviations from the gas-laws, but we now see that such deviations are necessary whatever the mass of the star, and relation (134·1) enables us to estimate the amount of the necessary deviations.

To a sufficiently good approximation for our present purpose, we may neglect α and β and put $n = \frac{1}{2}$. Relation (134·1) now assumes the form

$$-\tfrac{1}{2} + \frac{3s\lambda}{\lambda + 4}(7\cdot5) > 0,$$

and we find as condition for stability

$$s > \frac{1}{45}\left(1 + \frac{4}{\lambda}\right) \quad \ldots\ldots\ldots\ldots\ldots\ldots\ldots\ldots(134\cdot2).$$

For stars of moderate mass, λ is large, and the condition for stability is that s must be greater than $\frac{1}{45}$; for stars of large mass λ is comparatively small, and s must have a value substantially higher than $\frac{1}{45}$.

Values of s in the neighbourhood of $s = \frac{1}{45}$ look small until we make numerical calculations of what they involve. The calculations of Chapter III have shewn that the density in the central regions of a gaseous star may well be about 100 times the mean density of the star, and so more than 100 times that in the outer regions of the star in which Boyle's law is obeyed. Thus with $s = \frac{1}{45}$ the additional factor ρ^s in the pressure requires a pressure in the centre of the star of at least $(100)^{\frac{1}{45}}$ or 1·107 times that given by Boyle's law.

The total pressure, it is true, is only increased by 10·7 per cent., but this increase must be contributed solely by the atomic nuclei with their bound electrons, the free electrons being so minute that their pressure will always obey Boyle's law. Even if we suppose that there are only 90 free electrons to each atomic nucleus (which will soon prove to be an under-estimate), the atoms will contribute only one-ninety-first part of the total pressure when Boyle's law is obeyed, so that their additional contribution of 10·7 per cent. of the whole pressure is $91 \times 10\cdot7$ per cent. of their own pressure as given by Boyle's law, or say 10 times this pressure. Thus even a value $s = \frac{1}{45}$ requires the nuclei to be so closely packed at the centre of the star that the pressure they exert is about 11 times that given by Boyle's law. This is what we may describe as a semi-fluid state.

This represents the minimum deviation from Boyle's law which is adequate to ensure stability. Our analysis has shewn that unless the atoms in the star's central regions are packed so close as to provide a firm unyielding base of the kind just described, the star will be liable to start contracting or expanding, this contraction or expansion continuing unchecked until a firm base is formed at its centre.

In the average stable star the deviations from Boyle's law must naturally be more than the minimum; the discussion of § 131 suggested that in actual stars the pressure of the nuclei may be about 40 times that given by Boyle's law, and so nearly 4 times the minimum required for stability.

135. The thermodynamical stability criterion discussed in the last chapter (cf. formula (109·1)) did not involve s at all; s enters only in the second (dynamical) criterion which has just been discussed. As, however, it is clear that the fictitious assumed law $p \propto \rho^{1+s}$ cannot represent the actual facts of deviations from Boyle's law with any accuracy, it becomes important to examine what form is assumed by the second stability criterion when this very special law is no longer assumed to hold.

Let a star's emission of radiation be plotted against its radius as in fig. 9. We have already seen how a star of given mass can assume configuration of different radii, in which the star will emit radiation at different rates. Each of these configurations will be represented by a single point in the diagram. Let the curved line MM' be supposed to represent, in a purely diagrammatic way, the various configurations of equilibrium which can be assumed by a star of specified mass M as its rate of internal generation of energy changes.

Let the internal rate of generation of energy at every point of a star be supposed to depend quite generally on the temperature T and density ρ at that point, so that the total rate of generation of energy G of the star will be represented by an integral of the type

$$G = \int 4\pi\rho r^2 f(T, \rho)\, dr \quad \dots\dots\dots\dots\dots(135\cdot1),$$

where $f(T, \rho)$ depends not only on T and ρ, but also on the particular type of matter of which the star is made.

From this formula we can calculate the value of G for each configuration on the line MM' and can represent each value of G, measured as an emission of radiation, by a point P' either vertically above or vertically below the point P to which it refers. Corresponding to any one specified type of matter, the various points P' will form a line such as AA' (which, for diagrammatic simplicity, has been drawn as a straight line). We may call AA' a "generation line" corresponding to the "emission line" MM'. There is a different "generation line" for each type of stellar material, but the

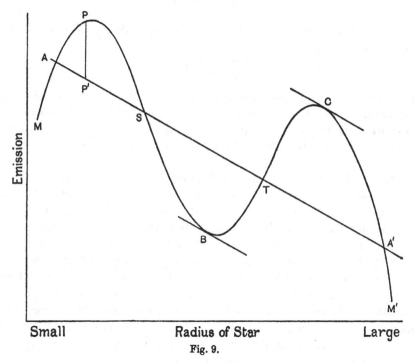

Fig. 9.

"emission line" is fixed by the mass of the star, except for small variations of the type we had under consideration in §§ 95–97, such as might arise from a redistribution within the body of the star of its energy-generating material or its radiation-stopping atoms. We must be content to disregard such small variations in the present investigation.

The intersections ST of a generation line AA' and the corresponding emission line MM' represent configurations at which $E = G$, and so determine the possible configurations of equilibrium for a star of given mass, made of that special type of matter to which the line AA' refers.

As a star ages and its more readily transformable material becomes exhausted, the line AA' will move downwards. Corresponding to certain ages of a star the intersections ST may coalesce in two adjacent positions of equilibrium as at B or C. When such a situation arises, the star can move from one of these configurations to the adjacent one without any forces of restitution coming into play, so that B and C represent configurations of neutral equilibrium.

If G denote the rate of generation of energy of the whole star, and E its total rate of emission in any configuration, the tangents of the slopes of the lines of generation and emission in fig. 9 are dG/dR and dE/dR. In a configuration of neutral equilibrium these slopes are the same, so that

$$\frac{d}{dR}(E - G) = 0.$$

A transition from dynamical instability to stability or *vice versa* accordingly occurs whenever

$$\frac{d}{dR}(E - G)$$

passes through a zero value, and it only remains to determine which sign of this quantity corresponds to stable and which to unstable configuration. We can easily do this by considering the special model already discussed in which the pressure and density are connected by the law $p \propto \rho^{1+s}T$.

136. With reference to the special model considered in the last chapter, in which the pressure is determined by the law $p \propto \rho^{1+s}$, the stability criterion which we now have under discussion requires that D in equation (108·2) shall be positive. When D is negative the star becomes unstable through the time-factor for the corresponding expansion or contraction of the star assuming the form $e^{\theta t}$ with θ positive. Changes from stability to instability or *vice versa* occur in configurations for which D vanishes; when $D = 0$, $\theta = 0$ and the star is in neutral equilibrium. We have just seen that configurations of neutral equilibrium are precisely those at which two intersections of the emission and generation lines such as S, T coalesce in our diagram. In other words, they are the configurations such as B and C in which the tangents to the lines of generation and of emission coincide.

When the rate of generation of energy G is assumed to be proportional to $\rho^a T^\beta$, we have, at each point of the star,

$$\frac{1}{G}\frac{dG}{dR} = \frac{\alpha}{\rho}\frac{d\rho}{dR} + \frac{\beta}{T}\frac{dT}{dR} \quad\dots\dots\dots\dots\dots(136\cdot1),$$

where d/dR denotes differentiation with respect to the different values of R in the various configurations which are possible for a star of given mass. These configurations have been seen to fall into homologous series, along each of which

$$R^{1 - \frac{3s\lambda}{\lambda+4}}T = \text{constant}.$$

By differentiation of this equation we obtain

$$\frac{1}{T}\frac{dT}{dR} = -\left(1 - \frac{3s\lambda}{\lambda + 4}\right)\frac{1}{R},$$

while the relation $R^3\rho = $ constant gives

$$\frac{1}{\rho}\frac{d\rho}{dR} = -\frac{3}{R}.$$

Equation (136·1) now becomes

$$\frac{1}{G}\frac{dG}{dR} = -\frac{3\alpha}{R} - \frac{\beta}{R}\left(1 - \frac{3s\lambda}{\lambda + 4}\right) \quad\dots\dots\dots\dots(136\text{·}2).$$

The total outward flow of radiation across a sphere of radius r in the interior of the star is

$$E = 4\pi r^2 \frac{4aC}{3c\mu}\left(\frac{T^3}{\rho}\right)^2 \frac{\partial T}{\partial r} T^n \quad\dots\dots\dots\dots(136\text{·}3),$$

in which $\frac{\partial}{\partial r}$ denotes differentiation along the radius of the star. From this we readily obtain

$$\frac{1}{E}\frac{dE}{dR} = \frac{7+n}{T}\frac{dT}{dR} - \frac{2}{\rho}\frac{d\rho}{dR} + \frac{1}{R}$$

$$= -\frac{7+n}{R}\left(1 - \frac{3s\lambda}{\lambda + 4}\right) + \frac{7}{R} \quad\dots\dots\dots\dots(136\text{·}4).$$

When the star is in equilibrium, $E = G$, so that equations (136·2) and (136·4) give

$$\frac{d}{dR}(E - G) = \frac{E}{R}\left[(3\alpha + \beta - n) + \frac{3s\lambda}{\lambda + 4}(7 + n - \beta)\right]\dots\dots(136\text{·}5).$$

Our second stability criterion $(D > 0)$ merely expressed that the term in square brackets on the right of this equation must be positive for stability, or, what is the same thing, that the right-hand member of the equation must be negative. But as we have seen that, quite apart from this special model, the stability criterion depends only on the sign of $\frac{d}{dR}(E - G)$, it is clear that, in general, the second stability criterion takes the form that

$$\frac{d}{dR}(E - G) \text{ must be positive for dynamical stability.}$$

Since R increases as we pass to the right of the diagram shewn in fig. 9, the stable ranges will be those in which E increases more rapidly than G as we pass to the right.

To a first approximation the first (thermodynamical) stability criterion required that G should be independent of changes in the density and temperature of the star. If we neglect the dependence of G on density and temperature, dG/dR is zero, and the condition for stability assumes the simple form that dE/dR must be positive—for a stellar configuration to be stable, the emission of radiation must decrease as the star contracts. This is the reverse of the relation which obtains in a gaseous star, in which we have seen (§ 92) that the emission of radiation increases as the star contracts. Thus to secure stability the deviations from the gas-laws must be of such amount as to reverse the sign of dE/dR.

To discuss the general relation between E and R we need a knowledge of the deviations from the gas-laws, and this in turn demands a knowledge of the state of ionisation of the stellar matter. We have already seen (§ 132) that throughout the centre of a liquid star both ρ and T will have tolerably uniform values, so that the degree of ionisation, which depends only on T and ρ, will also be tolerably uniform. Thus it will be legitimate to discuss only the ionisation of a single element fairly near to the centre of the star, and treat this as typical of the whole of the central parts of the star.

Ionisation.

137. The degree of ionisation of any element of stellar matter is determined by a formula obtained by R. H. Fowler* as an amplification of an earlier very important formula of Saha†. On inserting numerical values for well-known constants, Fowler's formula becomes‡

$$\log \left(\frac{x}{1-x} \rho \right) = -\frac{69000\,N^2}{T\,(\tau+1)^2} + \tfrac{3}{2}\log T + \log \left(\tfrac{1}{4}\mu\sigma \right) - 7{\cdot}790$$

$$\dots\dots(137{\cdot}1),$$

where all logarithms are to the base 10; here ρ, T, μ denote as usual the density, temperature and effective molecular weight of the stellar material, x is the proportion of atoms which are ionised down to their τ-quantum ring of electrons, and σ is the number of electrons which describe orbits in the $(\tau+1)$ quantum ring in an unbroken atom.

If ρ is kept constant, an increase of temperature increases both the first two terms of the right, and so increases x, the degree of ionisation, although complete ionisation ($x=1$) is not attained until T has an infinite value. Similarly a decrease of density, T remaining constant, leaves the whole left-hand side of the equation unaltered, and so increases x, the degree of ionisation, although again complete ionisation is not attained at any finite temperature until ρ vanishes.

138. We proceed to apply this formula to actual stellar matter in which T and ρ change simultaneously in the manner determined by the dynamics of the star.

Let λ be defined by the equation

$$\frac{R}{m\mu}\rho T = \tfrac{1}{3}a\lambda T^4 \dots\dots\dots\dots\dots\dots(138{\cdot}1).$$

* R. H. Fowler, *M.N.* LXXXIII. (1923), p. 407.

† M. N. Saha, *Phil. Mag.* XL. (1920), pp. 472 and 809; *Proc. Roy. Soc.* 99 A (1921), p. 135.

‡ J. H. Jeans, *M.N.* LXXXV. (1925), p. 928.

When the gas-laws are obeyed, λ becomes the ratio of gas-pressure to pressure of radiation, and so becomes identical with the λ of which the value was discussed in Chapter III. When the gas-laws are not obeyed, λ becomes a generalisation of our former λ, being proportional to $\rho/\mu T^3$.

Taking logarithms to the base 10, and inserting numerical values, we obtain from (138·1)

$$\log \rho = 3 \log T + \log (\lambda\mu) - 22\cdot511 \dots\dots\dots\dots(138\cdot2).$$

Using this value for $\log \rho$, equation (137·1) assumes the form

$$\log \left(\frac{x}{1-x}\right) = - \frac{69000N^2}{T(\tau+1)^2} - \tfrac{3}{2} \log T + \log \left(\frac{\sigma}{4\lambda}\right) + 14\cdot721$$
$$\dots\dots(138\cdot3).$$

Let us denote the right-hand member of this equation by ϕ, so that

$$1 - x = \frac{1}{1 + 10^\phi} \dots\dots\dots\dots\dots\dots(138\cdot4).$$

If all the atoms have the same atomic number N, then $1-x$ is the proportion of atoms at any point of the star which retain some at least of the atoms of their $(\tau+1)$ quantum ring. The total volume occupied by such atoms in a unit volume of the star will be proportional to $(1-x)\rho$. We may legitimately disregard by comparison the space occupied by more highly ionised atoms, so that $(1-x)\rho$ will give a measure of the total space occupied by all the atoms in unit volume being, roughly speaking, proportional to the b of Van der Waals' equation; it will, therefore, give a rough measure of the deviations of the stellar matter from the gas-laws.

From equations (138·2) and (138·4) we obtain

$$\log [(1-x)\rho] = 3 \log T + \log (\lambda\mu) - 22\cdot511 - \log (1+10^\phi)\dots(138\cdot5).$$

As the star moves through a series of different configurations, its density ρ and temperature T will change. The effective molecular weight μ will change slightly as the degree of ionisation varies, but the changes will be so small that they may be treated as negligible. Finally λ will change. So long as the gas-laws are approximately obeyed, λ depends almost entirely on the mass of the star so that its changes will be slight. When the gas-laws are appreciably departed from, there will be greater variations in λ. Finally, when the atoms are so closely packed that changes in ρ may be disregarded by comparison with changes in T, equation (138·1) shews that λ varies as $1/T^3$, so that

$$\frac{\partial\lambda}{\partial T} = - \frac{3\lambda}{T}\dots\dots\dots\dots\dots\dots\dots(138\cdot6).$$

As the deviations from the gas-laws increase, $\partial\lambda/\partial T$ changes from zero to $-3\lambda/T$.

So long as λ remains finite, equation (138·3) shews that x is zero both at zero and at infinite temperature. At zero temperature x is zero because the quanta of radiation are inadequate in strength to ionise the atoms. At infinite temperature, x is zero because the quanta of radiation are inadequate in number to ionise the atoms; the density being infinite, the number of atoms per cubic centimetre is infinite, while the number of quanta per cubic centimetre remains finite. The latter condition could not of course be reached in an actual gas, since the gas-laws would fail long before infinite density was reached, and the density cannot exceed a certain limiting value.

Similarly equation (138·5) shews that so long as λ remains finite, $(1-x)\rho$ increases from 0 to $+\infty$ while T increases from 0 to ∞, although again conditions at $T=\infty$ could not occur in an actual star.

139. It will be convenient to examine the changes in $(1-x)\rho$ by studying its maxima and minima. Treating μ but not λ as constant, we obtain by differentiation of equation (138·5),

$$\frac{\partial}{\partial T}\log\left[(1-x)\rho\right] = \left(\frac{1}{\lambda}\frac{\partial\lambda}{\partial T}+\frac{3}{T}\right)0\cdot4343$$
$$-\frac{10^{\phi}}{1+10^{\phi}}\left[\frac{69000N^2}{T^2(\tau+1)^2}-\left(\frac{3}{2T}+\frac{1}{\lambda}\frac{\partial\lambda}{\partial T}\right)0\cdot4343\right] \quad\ldots\ldots(139\cdot1).$$

The maxima and minima of $(1-x)\rho$, if any exist, will occur when the right-hand member of this equation vanishes, and this is when

$$\frac{3}{T}\left[1+10^{\phi}\left(\frac{3}{2}-\frac{53100N^2}{T(\tau+1)^2}\right)\right]+\frac{1}{\lambda}\frac{\partial\lambda}{\partial T}(1+2\times10^{\phi})=0 \ldots(139\cdot2).$$

If we put

$$\frac{53100N^2}{T(\tau+1)^2}=y \quad\ldots\ldots\ldots\ldots\ldots\ldots\ldots(139\cdot3),$$

and restore the value of ϕ from (138·3) this becomes

$$y^{\frac{3}{2}}\left(y-\frac{3}{2}-\frac{2T}{3\lambda}\frac{\partial\lambda}{\partial T}\right)10^{7\cdot635-1\cdot303y}=\frac{4\lambda N^3}{\sigma(\tau+1)^3}\left(1+\frac{T}{3\lambda}\frac{\partial\lambda}{\partial T}\right)\ldots(139\cdot4).$$

140. Let us first discuss the problem on the supposition that the gas-laws are so nearly obeyed that $\partial\lambda/\partial T$ may be neglected. Equation (139·4), which gives the occurrence of maxima and minima of $(1-x)\rho$, now assumes the form

$$y^{\frac{3}{2}}\left(y-\frac{3}{2}\right)10^{7\cdot635-1\cdot303y}=\frac{4\lambda N^3}{\sigma(\tau+1)^3} \quad\ldots\ldots\ldots\ldots(140\cdot1).$$

The left-hand member of this equation is a function of y which is positive only when y is greater than $\frac{3}{2}$. It is readily found, by elementary methods, that it attains a single maximum when $y=1\cdot95$, its value at this maximum being $1\cdot52\times10^5$.

Since its left-hand member can never exceed $1\cdot52\times10^5$, equation (140·1) can have no roots at all if

$$\frac{4\lambda N^3}{\sigma(\tau+1)^3}>1\cdot52\times10^5 \quad\ldots\ldots\ldots\ldots\ldots(140\cdot2).$$

In such a case $(1-x)\rho$, which measures the extent of the deviations from the gas-laws, passes through neither maximum nor minimum values but increases steadily from 0 to ∞ as T increases from 0 to ∞, at any rate until the gas-laws fail.

On the other hand, if $\dfrac{4\lambda N^3}{\sigma(\tau+1)^3} < 1\cdot 52 \times 10^5$ ....(140·3),

equation (140·1) has two roots for y. As T starts from zero temperature and increases, $(1-x)\rho$ also increases at first. It reaches a maximum for some value of y intermediate between ∞ and 1·95, after which it starts decreasing and declines to a minimum which occurs for some value of y between 1·50 and 1·95; after this it again increases and would attain an infinite value at $T = \infty$ except that the gas-laws necessarily fail before this value of $(1-x)\rho$ is reached.

The value of y at which $(1-x)\rho$ is a minimum is always intermediate between 1·50 and 1·95. By equation (139·3), the corresponding temperature T is given by

$$T(\tau+1)^2 = QN^2 \qquad\qquad ....(140·4),$$

where Q lies between 27200 and 35400. To within an error of 14 per cent. in T or of 7 per cent. in N, we may replace Q by the geometric mean of its extreme values and write equation (140·4) in the form

$$T(\tau+1)^2 = 31000N^2 \qquad ....(140·5).$$

This equation gives the values of T at which $(1-x)\rho$ has successive minimum values corresponding to the ionisation of the successive quantum rings defined by the different values of $\tau+1$, calculated on the assumption that the gas-laws are obeyed at these minima.

141. The foregoing analysis has shewn what is obvious on general principles, that as a star steadily contracts along a homologous series of equilibrium configurations, the deviations from Boyle's law in its central regions may fluctuate through a succession of maxima and minima as the different rings of electrons surrounding the atoms are ionised in turn. The contraction of the star is accompanied by an increase in its temperature and, by ionising one ring after another of electrons, this causes the atoms to diminish in size. The star and its atoms contract together but the star contracts steadily while the atoms, so to speak, contract by jerks. There will be times when the contraction of the star has rather outstripped that of the atoms, so that the atoms are jammed together. The ionisation of a new ring of electrons may now relieve the congestion and set the whole structure free again.

Equation (140·5) gives the approximate temperatures at which the deviations from Boyle's law are a minimum, calculated on the supposition that at these minima the deviations are slight. If the deviations from Boyle's law, even at the minima, may not legitimately be treated as slight, equation (140·5) must be corrected accordingly, but unless the deviations are quite large, the necessary correction will be slight as compared with the whole value of T, so that this equation will in any case give approximate values of T at the minima.

The general nature of the results we have obtained may be represented graphically as in fig. 10, in which the ordinate represents $(1-x)\rho$, or the amount of deviation from Boyle's law, and the abscissa represents the central temperature of the star, high temperatures being represented on the left of the diagram.

The values of $(1-x)\rho$ corresponding to different degrees of ionisation of any one ring of electrons are represented by a curve such as $OABC$, B representing the temperature at which the deviations from the gas-laws are a minimum.

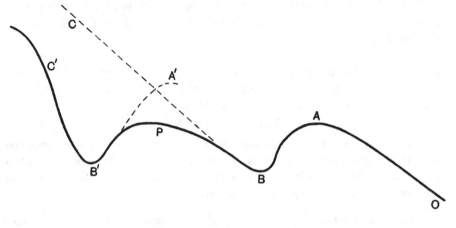

High   Temperature   Low

Fig. 10.

As a star contracts, the deviations from the gas-laws, which were considerable at A, become slight at B, and then begin to increase again. Our theoretical formula would make them increase without limit. But before the star has moved far along the branch BC, the deviations from the gas-laws again become appreciable, so that the theoretical curve BC, determined on the supposition that the gas-laws are obeyed, begins to fail.

Corresponding to a given temperature and pressure, the density ρ is less than it would be if Boyle's law were obeyed, with the result that $(1-x)$ is also less, as is shewn by formula (137.1), and consequently $(1-x)\rho$ is also less. Thus the curve BC gives values for $(1-x)\rho$ which are too large, and must be replaced by a lower curve such as BP. At very high temperature very few atoms would be left unionised, so that the ionisation of the next ring becomes of importance; the values of $(1-x)\rho$ produced by the imperfect ionisation of this ring will be represented by a curve $A'B'C'$ similar to the original curve ABC, but shifted to a higher temperature. On joining up the relevant parts of these curves by a curve such as that drawn thick in the diagram, we get a graph of the actual values of $(1-x)\rho$.

142. We have seen that $(1-x)\rho$ will not pass through a minimum resulting from the ionisation of the $(\tau+1)$ quantum ring, unless

$$\frac{4\lambda N^3}{\sigma(\tau+1)^3} < 1\cdot52 \times 10^5 \quad\ldots\ldots\ldots\ldots\ldots(142\cdot1).$$

As a star of given mass contracts and the various rings of electrons are ionised in turn, λ and N^3 remain approximately the same, but σ and $\tau+1$ decrease with each successive ionisation. Thus a time may come, according to the values of λ and N, when a new ionisation does not produce a minimum of $(1-x)\rho$ at all, but instead $(1-x)\rho$ increases continuously and without limit. When this ring is reached, the graph of $(1-x)\rho$ rushes upwards without limit until deviations from the gas-laws occur. As before, the effect of these is to check the upward rush of $(1-x)\rho$, and ultimately the graph of $(1-x)\rho$ must become asymptotic to a curve giving values of $(1-x)\rho$ at which the atoms are jammed so close together that no further compression is possible.

Precise calculations given later (p. 162) will shew that the atomic numbers of actual stellar atoms are in the neighbourhood of 95. With this value for N, equation $(142\cdot1)$ shews that the ionisation of the τ-quantum ring will give a maximum and a minimum to $(1-x)\rho$ only if

$$\lambda < 0\cdot444\sigma(\tau+1)^3.$$

For the M-ring, $\tau+1=3$ and $\sigma=18$ so that M-ring ionisation will give a minimum value to $(1-x)\rho$ only if $\lambda<21\cdot6$, a condition satisfied by all stars whose mass is greater than about three-quarters of the sun's mass.

For the L-ring, $\tau+1=2$ and $\sigma=8$, so that L-ring ionisation gives a definite minimum only if $\lambda<2\cdot9$, a condition which is satisfied only by stars more massive than Sirius.

For the K-ring, $\tau+1=1$ and $\sigma=2$, so that K-ring ionisation gives a definite minimum only if $\lambda<0\cdot09$, a condition which is not satisfied by even the most massive of known stars. Thus actual stars can shew no minimum for K-ring ionisation, at any rate until the density is so great that our analysis has failed through the deviations from the gas-laws becoming excessive.

143. To examine the state of ionisation in a star in which the deviations from the gas-laws are great, we must replace equation $(140\cdot1)$ which we have so far had under discussion by the more general equation $(139\cdot4)$, namely

$$y^{\frac{3}{2}}\left(y-\frac{3}{2}-\frac{2T}{3\lambda}\frac{\partial\lambda}{\partial T}\right)10^{7\cdot835-1\cdot30\omega y} = \frac{4\lambda N^3}{\sigma(\tau+1)^3}\left(1+\frac{T}{3\lambda}\frac{\partial\lambda}{\partial T}\right)\ldots(143\cdot1).$$

The two terms in $\partial\lambda/\partial T$ in this equation represent the effect of deviations from the gas-laws. It is readily seen* that the term in $\partial\lambda/\partial T$ on the left-

* *M.N.* LXXXVII. (1927), p. 781.

hand side of the equation has no very great effect on the equation as a whole, the graph of the left-hand side being, in its general nature at least, similar to what it is when this term is absent.

On the other hand, the term in $\partial\lambda/\partial T$ on the right-hand side of the equation has a profound effect, with the result that the equation always has a root even for stars of the smallest mass.

For by the definition of λ (equation (138·1)),

$$\rho = \lambda T^3 \times \text{a constant},$$

so that, by logarithmic differentiation,

$$\frac{1}{\rho}\frac{\partial\rho}{\partial T} = \frac{3}{T}\left(1 + \frac{T}{3\lambda}\frac{\partial\lambda}{\partial T}\right).$$

The effect of introducing the term in $\partial\lambda/\partial T$ is accordingly to replace λ by $\dfrac{\lambda T}{3\rho}\dfrac{\partial\rho}{\partial T}$, and this is proportional to $\dfrac{1}{T^2}\dfrac{\partial\rho}{\partial T}$.

As the deviations from the gas-laws become greater, $\partial\rho/\partial T$ falls steadily below the value it would have if the gas-laws were obeyed, and would finally vanish in a state in which the density increases no further, heating now being accompanied by expansion as in a solid or liquid. But the two sides of equation (143·1) will have become equal at some point before this state is reached, so that the deviations from the gas-laws again diminish until $(1 - x)\rho$ reaches a minimum value.

Thus we see that for every star, whatever its mass, and for every ring of electrons, $(1 - x)\rho$ must in time pass through a minimum. But the more the mass of the star falls short of the critical limit calculated in § 140, the greater the deviations from the gas-laws must be before $(1 - x)\rho$ reaches a turning point. The occurrence of the turning point in an actual star is of course dependent on deviations of this magnitude being attained; it may well be that such deviations are not attained in a given star through the next ring of electrons being ionised first. This reservation cannot apply to K-ring ionisation since there is no further ring of electrons to be ionised and calculation shews that nuclear disintegration is non-existent, or at any rate unimportant, even at the central temperatures of the white dwarfs. Thus the K-ring ionisation must reach a turning point in every star, and pass through maxima and minima in succession. But as the masses of actual stars are all far below the critical value ($\lambda = 0·09$) calculated in § 140, it appears that there must always be very great departures from the gas-laws before the turning point for K-ring ionisation occurs.

144. We proceed to consider the stability of the various types of configuration which have been under discussion.

If for the moment we neglect all dependence of the rate of generation of energy G on physical conditions of density, temperature and ionisation, a given configuration will be dynamically stable if, and only if,

$$\frac{dE}{dR} > 0 \quad \dots\dots\dots\dots\dots\dots\dots\dots(144\cdot1),$$

where E is the rate of emission of energy and R is the radius, of the star.

If the gas-laws are obeyed throughout the star, and the coefficient of opacity is given by Kramers' law, E varies as $R^{-\frac{1}{2}}$ (§ 92) and all configurations are unstable, as has already been seen. We have seen that we can pass from a configuration in which the gas-laws are obeyed to one of the same mass, density and radius in which they are not obeyed, by a process of lowering the temperature throughout. Since the emission E for a star of given mass, density and radius is proportional to $T^{7\cdot5}$, this depresses the value of E also. Thus the rhythmical variations which have been found to occur in $(1 - x)\rho$, which measures the deviations from the gas-laws, will shew themselves as rhythmical variations in the emission E.

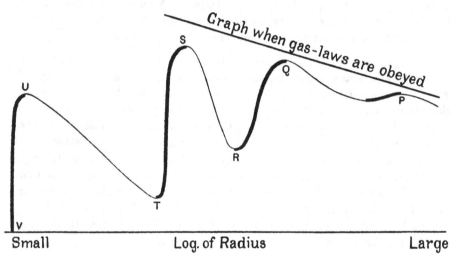

Fig. 11.

In fig. 11 let the ordinate represent $\log E$ and the abscissa $\log R$, so that stars of low density and low temperature are to the right. If the gas-laws are obeyed, the relation between E and R is $E \propto R^{-\frac{1}{2}}$, and the graph of E is a straight slant line such as the one at the top of the diagram. To take account of variations from the gas-laws we must depress the value of T and so also of E. The rhythmical fluctuations in $(1 - x)\rho$ which result from the ionisation of successive rings of electrons will shew themselves by the graph hanging below this line in a series of festoons. The minimum deviations whose positions are determined by equations (140·4) or (140·5) must coincide very approximately with the highest points of these festoons. As we pass to the left of the diagram, the general magnitude of the deviations from Boyle's law tends to increase, so that the festoons fall deeper and deeper below the slant line which gives the graph of E when Boyle's law is obeyed.

Let the festoon PQ represent deviations from Boyle's law arising from the congestion of atoms ionised down to their M-rings. At Q the M-rings become ionised and the gas-laws are nearly obeyed. The next festoon QRS represents deviations caused by L-ring congestion, and at S the L-ring is ionised. Further contraction of the star causes the K-ring atoms to become congested, the long nearly vertical stretch ST representing configurations in which the deviations from the gas-laws become ever greater, and finally become very great indeed for a star of moderate mass. At last the congestion is relieved through the increase in $(1 - x)\,\rho$ ceasing. The majority of atoms now become ionised down to their nuclei, but with further contraction, even the nuclei, together with the fraction of K-ring atoms which are still ionised, become jammed together, so that the departures from the gas-laws again increase, and the graph begins to form a new festoon UV, a festoon which can shew no subsequent upward turn since there is no further ring of electrons whose ionisation can relieve the congestion.

The approximately vertical parts of the graph represent configurations in which the atoms are packed together approximately as closely as they will go, changes in temperature producing only very slight changes in density. Strictly speaking any change in temperature must produce a change in the degree of ionisation, and the density must change simply because the number of atoms which are available for close-packing changes. But calculations based on equation (137·1) shew that the importance of this effect is quite slight numerically, and to a legitimate approximation we may suppose that there are vertical stretches of the graph such as ST, UV, in which the atoms are packed together approximately as closely as they will go, and the density approaches asymptotically to a definite minimum. This provides the physical justification for the analysis given at the end of §132.

Those parts of the graph of E for which dE/dR is positive are drawn thick. So long as variations in G are neglected, these represent stable configurations, by relation (144·1), while all others are unstable.

When variations in G are taken into account, the stability condition (144·1) must be replaced by the more complete condition

$$\frac{dE}{dR} > \frac{dG}{dR} \qquad\qquad\qquad\qquad\text{.........................(144·2)}.$$

We have seen that increasing ionisation must diminish, or at any rate cannot increase, G, so that dG/dR is zero or positive, with the result that the ranges of stability are somewhat more restricted than those which are drawn thick in fig. 11.

COMPARISON WITH OBSERVATION.

145. The graph of E shewn in fig. 11 is drawn for a star of given constant mass. On drawing a number of such graphs in a diagram in which $\log E$ and $\log R$ are taken as coordinates, we obtain a complete map of the configurations possible for stars of all masses. We have already noticed that, since $E = 4\pi R^2 \sigma T_e^4$,

$$\log E = 2 \log R + 4 \log T_e + \text{a cons.} \quad \dots\dots\dots(145 \cdot 1).$$

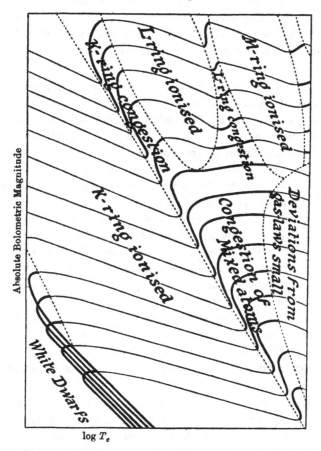

Fig. 12. Division of the Temperature-Luminosity diagram into Stable (thick) and Unstable (thin) configurations.

Instead of taking $\log E$ and $\log R$ as coordinates, it is more convenient to take $\log E$ and $\log T_e$, the transformation between the two sets of co-ordinates being made by equation (145·1). In this way we obtain a diagram of the kind already introduced in § 56 (fig. 5). On changing coordinates in this way and drawing graphs of $\log E$ for a number of stars of different masses we obtain a diagram of the general character shewn in fig. 12.

This is of course directly comparable with the diagram already given in fig. 5 (p. 61), this latter presenting the results of observation while fig. 12 presents the predictions of theory. It is also very nearly, but not quite, directly comparable with the diagram shewn in fig. 6, in which 2100 stars are plotted in respect of absolute magnitude and spectral type.

146. Theory predicts that only those configurations are stable which lie in parts of the diagram (fig. 12) in which the lines are drawn thick. Thus, unless theory is at fault, stars ought to occur in the corresponding regions in the observational diagrams shewn in figs. 5 and 6, and in no others; the regions in which the lines are drawn thin ought to be untenanted by stars since the corresponding configurations are unstable, to be quitted with all possible speed.

A comparison of the theoretical and observational diagrams shews that the predictions of theory are borne out at least to a sufficient extent to suggest that the theory is on the right general lines. This being so, it is possible to identify the various areas of stable configurations predicted by theory with the areas found observationally, and thus to specify the physical structure of the stars which inhabit the various areas in the diagrams. The identification is shewn in fig. 13, in which the diagram of 2100 stars already shewn in fig. 6 (p. 62) is parcelled out into stable and unstable regions in accordance with the requirements of theory. The main sequence is seen to consist of stars whose atoms are ionised down to their K-rings and jammed together. On the left-hand edge of the sequence the deviations from the gas-laws are so great that the atoms are packed almost as closely as they will go. The giant branch has a similar interpretation with the difference that the atoms are ionised only down to their L-rings. In the white dwarfs the atoms are mainly ionised down to their nuclei, but a few K-ring atoms remain and these, although few in number, probably occupy the major part of the available space; it is their jamming, rather than that of the nuclei, which results in the departures from the gas-laws which ensure the stability of the star.

The theoretical diagram suggests that the white-dwarf series ought to extend upwards right into the earliest spectral types. I have suggested* that the so-called "Dwarf Wolf-Rayet" stars (O-type stars of small radius) may be found to occupy the upper half of this series; if this conjecture is confirmed the O-type stars ought to be found to be divided sharply into stars of small and of fairly large radius.

147. As we have seen, the general similarity of the observational and theoretical diagrams makes it possible to superpose them in such a way that the areas of stability coincide fairly well. In fig. 13 the diagrams are shewn

---

* *M.N.* LXXXVII. (1927), p. 412.

superposed in such a way as to make these areas fit as well as possible. In choosing this particular superposition we have in effect assigned values to various adjustable constants, and it·is necessary to examine what these constants are and what values have been assigned to them in fig. 13.

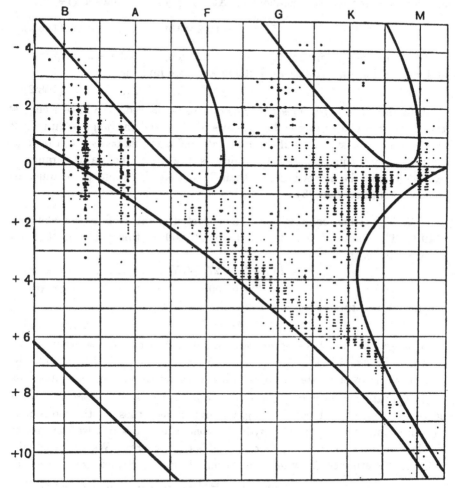

Fig. 13. The partition of the Russell diagram into regions of Stable and Unstable configurations.

Apart from details, the different ways of superposing the two diagrams depend in the main on the numerical values of four quantities representing the orientation, scale and the two coordinates of any one selected point in one diagram relative to the other.

Atomic Numbers.

148. As selected point it is convenient to choose the inner fork of the junction between the giant branch and the main sequence. Fig. 11 suggests that at the right-hand edges of the areas of stability the gas-laws are approximately obeyed, so that there will be no very great error involved in supposing the gas-laws to be obeyed at this point. The point is generally agreed to be about absolute bolometric magnitude $+1$ and spectral type F. The value of λ for stars at this point is about 3. At this point by equation (140·2),

$$\frac{4\lambda N^3}{\sigma(\tau+1)^3} = 1\cdot52 \times 10^5 \quad\ldots\ldots\ldots\ldots\ldots\ldots(148\cdot1).$$

The point is a minimum for L-ring ionisation, so that we must put $\sigma = 8$ and $\tau + 1 = 2$ in this equation. Putting $\lambda = 3$ we find that

$$N = 93\cdot9 \quad\ldots\ldots\ldots\ldots\ldots\ldots\ldots\ldots(148\cdot2).$$

Equation (148·1) assumes the gas-laws to be obeyed. We have seen that the effect of deviations from the gas-laws is to extend the range within which the minimum exists. Thus if the true value of λ at this point is 3, the value which satisfies equation (148·1), which is the value before the range is extended, must be somewhat less than 3, so that the value of N must be somewhat higher than 93·8.

149. The other coordinate of the selected point is fixed by equation (140·4). The value of Q is known to be 27,200 at this point, so that the equation takes the form

$$T(\tau+1)^2 = 27,200 N^2 \quad\ldots\ldots\ldots\ldots\ldots\ldots(149\cdot1),$$

and on putting $\tau + 1 = 2$, and $T = 60,000,000$, this being the approximate central temperature of stars at this point, we find

$$N = 93\cdot2\ldots\ldots\ldots\ldots\ldots\ldots\ldots\ldots(149\cdot2).$$

The agreement between these two values of N is entirely satisfactory, but it cannot be claimed that the value of N can be determined with anything like the accuracy suggested by small differences between these determinations.

It appears, however, that the atomic numbers of stellar atoms are in the neighbourhood of, and possibly slightly higher than, those of the radioactive elements, and this fits in well with the conclusions to which we were led in the last chapter as to the generation of stellar energy. We could equally well have determined N by using equation (149·1) in its more general form

$$T(\tau+1)^2 = QN^2 \quad\ldots\ldots\ldots\ldots\ldots\ldots(149\cdot3),$$

and applying it at a point higher up the main sequence. We have already seen that as we advance up the main sequence the values of T which are calculated from observation increases slightly, while theory compels the value of Q to increase slightly. Actually the value of T appears to increase rather more rapidly than that of Q, so that the values of N calculated from equation (149·3) would increase as we pass up the main sequence—i.e. N would be greater for younger stars than for older. This again is entirely in agreement with our conclusions as to the generation of stellar energy, but the increases in N are too slight for us to lay much stress on them.

Stellar Temperatures.

150. This last feature indicates that the orientation of our diagrams has been accurately arranged, and the only outstanding question is that of scale.

Equation (149·3) shews that the minima of ionisation of L, M and N-ring electrons ought to occur at temperatures which are approximately in the ratio of

$$\frac{1}{2^2} : \frac{1}{3^2} : \frac{1}{4^2},$$

or $36 : 16 : 9$. Thus the central temperatures of stars on the extreme right-hand edges of the main sequence, the giant branch and the yet further branch ought to be in approximately this ratio. Calculated central temperatures seem to shew a rather greater range than this theoretical ratio would indicate, but the data for almost all stars except possibly those on the main sequence are so uncertain that we cannot attach much weight to the apparent discrepancy.

Apart from this, our calculations have proceeded on the supposition that the ionisations of the separate rings of electrons do not overlap, so that one is completely ionised before the next begins to be ionised at all. A quite simple calculation shews that this is not a wholly legitimate assumption, so that inferences based on it are not likely to be fulfilled with high accuracy.

Atomic Diameters.

151. If it is granted that the two diagrams have been superposed in a legitimate way, one further test remains which the theory must survive if it is to be considered at all tenable. This consists in estimating the actual diameters of atoms ionised down to their K, L and M-rings, and examining whether matter formed of such atoms and compressed to the densities which prevail at the centres of the stars would shew deviations from the gas-laws of amount adequate to ensue the stability of the star.

This test is as difficult in practice as it is simple in principle. We have no means of calculating the effective diameters of these highly ionised atoms, and can only attempt an estimate from the known diameters of their outermost electronic orbits. According to Bohr's theory the diameter of the hydrogen atom, in which only the K-ring exists, is $1·08 \times 10^{-8}$ cms., while the diameter of an atom of atomic number N ionised down to its K-ring is $1·08 \times 10^{-8}/N$, and that of the same atom ionised down to its L-ring is four times this or $4·32 \times 10^{-8}/N$.

Unfortunately these diameters give only the slightest indication of the spaces occupied by the atoms themselves. In liquid or solid helium each atom occupies a sphere of diameter 4×10^{-8} cms., whereas the calculated diameter of the outermost ring of electrons in the helium atom is only $0·54 \times 10^{-8}$ cms. If theory provides no means of estimating even the effective diameter of the electrically neutral helium atom to better accuracy than this, it becomes almost impossible to predict with any accuracy the effective diameters of highly ionised stellar atoms which are surrounded by powerful electric fields.

Let us, however, attempt the calculation for the right-hand (low density) edge of the main sequence at which the deviations from the gas-laws ought to be small, having the lowest amount consistent with stability. Inserting the very rough values $N = 100$, $T = 10^6$, equation (137·1) gives as the proportion x of ionised atoms,

$$\log \left(\frac{x}{1-x} \rho \right) = -\frac{6·9}{(\tau+1)^2} + \log \left(\tfrac{1}{4} \mu \sigma \right) + 4·210 \quad \ldots\ldots(151·1).$$

The value of μ is approximately 2·5. For L-ring ionisation we put $\sigma = 8$, $\tau + 1 = 2$ and obtain

$$\frac{x}{1-x} \rho = 1528.$$

As the central density of stars in this region is about 100, x is about 0·938, so that about fifteen out of every sixteen atoms are ionised down to their K-rings.

The diameters of atoms of atom number 100 ionised down to their K-rings and L-rings are given by Bohr's theory as $1·08 \times 10^{-10}$ and $4·32 \times 10^{-10}$ cms. respectively. At a density of 100, each cubic centimetre will contain $6·06 \times 10^{23}$ atoms, and if fifteen-sixteenths of these are ionised to their K-rings and the remainder to their L-rings, the total volume occupied, if their effective diameters were those just stated, would be about 0·000004 cu. cms.

We have already noticed that the effective diameter of the helium atom in the liquid state is 7·4 times the diameter calculated from Bohr's theory. If we increase the calculated diameters of the K-ring and L-ring atoms by a similar factor, the volume just calculated must be increased by a factor of $(7·4)^3$ and becomes 0·001 cu. cms. This ought to represent the value of Van der Waals' b in configurations in which the deviations from the gas-laws are just adequate, and no more than adequate to ensure stability. It is undoubtedly too small, but it is not of a hopelessly wrong order of magnitude, and possibly this is the best we have any right to expect in view of our almost complete ignorance of the effective dimensions of ionised atoms.

A large accumulation of evidence, especially in connection with the fission of stars into binary systems (Chap. x, below) seems to suggest that the effective radii of highly-ionised atoms must be very much greater than the radii of the last surviving ring of electrons as calculated on Bohr's theory.

SUMMARY.

152. To sum up, we have found that considerations of stability demand that all stars should be in a state in which the deviations from the gas-laws are appreciable, while actually the majority are found to be in states in which these deviations are so large, that their central regions may properly be described as in the liquid state.

We have seen that the configurations which theory predicts to be stable, group themselves in respect of radius, luminosity and spectral type in the way in which observed stars are found actually to be grouped.

This has enabled us to assign physical meanings to the various groups of observed stars such as the white dwarfs, the giants and the main-sequence stars. To secure quantitative agreement between this interpretation and observation, we have found that the atomic numbers of stellar atoms must be in the neighbourhood of 95. This fits in well with the conclusions reached in the last chapter as to the source of stellar energy. It was there found that, for the stars to be stable structures, their energy must be generated by a sort of generalised radioactivity, the stellar atoms not merely undergoing transformation into other atoms, as in ordinary radioactivity, but undergoing complete annihilation, and setting free the stored up energy of their masses as radiation. As this is not a property of terrestrial atoms, we were led to suppose that the stellar atoms which experience annihilation are unknown on earth and therefore of atomic number higher than 92.

The estimates of atomic number which have been made in the present chapter fit in with this view, as does also the fact that the estimated atomic numbers shew a tendency, although not very marked, to decrease as a star gets older.

CHAPTER VI

THE EVOLUTION OF THE STARS

General Principles.

153. The early spectroscopists believed that the spectrum of a star provided a sure indication of the star's age. Huggins and Lockyer had found, for instance, that the spectrum of Sirius exhibited hydrogen lines very strongly and calcium lines rather weakly; in the solar spectrum the relative strength of these two sets of lines was reversed, calcium being strong and hydrogen weak. They concluded that hydrogen was specially prominent in the constitution of Sirius and calcium in that of the sun. Believing that Sirius must one day develop into a star similar to our sun, they conjectured that its substance must gradually change from hydrogen into calcium and other more complex elements, thus finding support for the long-established hypothesis that the more complex elements were formed by gradual evolution out of the simplest. In this way they were led to regard a star's spectrum as an index to its age.

As we have seen, the true interpretation of these observations is merely that the surface of Sirius is at a temperature at which hydrogen is specially active in emitting and absorbing radiation, while the sun's surface is at a lower temperature at which hydrogen is comparatively inert, while calcium, iron, etc., have become active in its place. Just as the laboratory physicist can produce different spectra from the same vacuum tube by varying the mode and conditions of excitation, so Nature produces different spectra from the same stellar material by varying its temperature. The linear sequence into which the spectra of stars fall is merely one of varying surface temperature.

Clearly this circumstance robs stellar spectra of all direct evolutionary significance. The spectra of the stars merely inform us as to their present surface temperatures, so that even if we could arrange the stars in order of age, a comparison of their spectra would only shew whether their surfaces were becoming hotter or cooler; it would give no information as to any evolutionary or chemical changes occurring in their substance.

To obtain evidence as to evolutionary changes in a star, we must probe deeper into the star than we can by a study of the star's surface spectrum and try to obtain information as to changes in progress in the star's interior.

We have already been driven to look to the annihilation of matter for the source of stellar energy. It necessarily follows that the primary evolutionary change in a star is decrease of mass—the older a star gets, the less massive it becomes. Other evolutionary changes follow as an inevitable corollary. The elements which have the greatest energy-generating capacity are necessarily the shortest lived, and so disappear most rapidly. As they disappear, the star's generation of energy per unit mass necessarily declines, so that as a star ages its generation of energy in ergs per gramme must decline.

Our sun radiates less energy per gramme than V Puppis because it is seven million million years older, and because in this interval most of the elements responsible for the present high radiation of V Puppis have disappeared from the sun through being transformed into radiation. If the white dwarfs are excluded from consideration (for reasons explained in § 120), the two quantities, mass and rate of generation of energy per gramme, change together. The results already mentioned (§ 117) of Hertzsprung, of Russell, Adams and Joy, and of Eddington, in which a star's luminosity is plotted against its mass, have disclosed a very strong correlation between these two quantities. Not only does the luminosity increase with the star's mass, as might in any case be expected, but the generation of energy per unit mass also increases with the mass. Either of these quantities accordingly provides a rough indication of a star's age.

Although this is far less certain, it seems probable that the elements of highest capacity for generating energy will, on the whole, be the elements of highest atomic weight. If this is so, the early disappearance of these elements ought to cause the *average* atomic weight of the atoms in a given star to diminish as the star grew older. So far as it went, the discussion of the last chapter favoured this view, at least for main sequence stars, since the calculated values of the atomic numbers increased with the mass of the star. There is room for some doubt as to the reality of this support, since the calculated range of atomic numbers seems rather too large to correspond to actuality, but it is, nevertheless, reasonably probable that the elements which disappear first are those of high atomic weight, since those which have no capacity for annihilation—the permanent terrestrial elements—occupy all the places of low atomic weight, while the evidence that the elements in the stars have average high atomic weights is too strong to be disregarded.

This result, if established, has far-reaching implications. Contrary to the views of the early spectroscopists, and contrary to what is still probably the prevalent belief, it begins to look as though the atoms in a star become simpler as the star grows older; evolution appears to be from complex to simple, and not, as in biology, from simple to complex. There is at present no direct experimental evidence bearing on this question except that provided by radioactivity, where evolution is certainly from complex to simple, atoms of lower atomic weight being continually produced by the disappearance of atoms of higher atomic weight.

The evidence of physical astronomy, pointing to an evolution of matter in the same direction, suggests that the main evolution of matter in the universe may be of the same type as, but a generalisation of, the radioactive processes as they occur on earth. The circumstance that the atomic numbers calculated in the last chapter came out somewhere in the neighbourhood of the atomic numbers of the radioactive elements gave support to this view.

Nevertheless, something more than mere radioactivity is necessary to account for the radiation of the stars. So far as is at present known, every radioactive process increases the number of atoms in the universe, whereas stellar radiation can only result from a decrease; it calls for actual annihilation of atoms, a process of which terrestrial radioactivity knows nothing.

It would obviously be rash to base any very definite conclusions on the very uncertain calculations of atomic weight given in the last chapter. It is entirely possible that the true atomic weights of stellar matter are all less than that of uranium, and that the atoms which undergo annihilation in the stars are merely isotopes of the heavier of the terrestrial elements.

Russell's two Theories of Stellar Evolution.

154. The possibility that the observed spectral sequence corresponded to evolutionary development in the chemical composition of the stars, was in effect ruled out when Hertzsprung discovered in 1905 that red stars fell into the two distinct classes which he designated as giants and dwarfs. The giant red stars, of far higher luminosity than Sirius, shewed the lines of heavy elements in their spectra as well as conspicuous bands which were identified with the bands of titanium oxide. If the chemical evolution of the stars was from hydrogen to more complex elements, it was absurd to find titanium, of atomic number 22, figuring prominently in the atmospheres of the youngest of the stars, not to mention strontium (37), yttrium (38) and barium (56).

Russell emphasized this in 1913, when he published his first diagram of the distribution of spectral types by absolute magnitude, and based a theory of stellar evolution upon it[*].

In place of the reversed γ formation which characterises the true Russell diagram (cf. fig. 6, p. 62), Russell's first diagram seemed to shew the formation exhibited in fig. 14. The most luminous stars were distributed over all spectral types along the range PQ, with the least luminous at R.

This led Russell to propound a theory that the curve PQR represented an evolutionary sequence, the spectrum of the normal star first advancing from type M to type B or A and then receding again to type M.

It was easy to find a theoretical justification for this hypothesis. We have seen how a star's density can be calculated from its observed luminosity and surface temperatures. Giants of types M, K and G are found to have mean densities of the order of 0·000002, 0·0005 and 0·004 respectively. In general

---

[*] *American Association for the Advancement of Science*, 1913 meeting, and *Nature*, xciii. (1914), pp. 227, 252.

the calculated mean densities of the giant stars were so small as to suggest that the stars must be constituted of gas of such low density that the ordinary gas-laws would be approximately obeyed. The densities of dwarf stars proved to be so high that they might be either gaseous or liquid or solid, but, if gaseous, they were so dense as to necessitate wide deviations from the gas-laws, at any rate, on the supposition, on which the whole theory was based, that stellar molecules or atoms remained intact and so occupied the same volume of space as terrestrial atoms and molecules. It was now easy to see why, in the giant stars, increase of temperature and density go together; this was merely a consequence of Lane's law. But the dwarfs, so Russell thought, might be

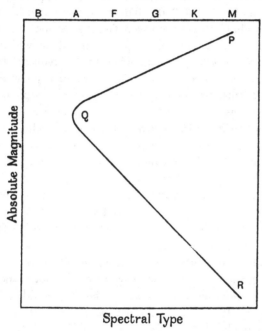

Fig. 14. Russell's first Theory of Stellar Evolution.

more properly compared to solid bodies, in which a loss of heat does not result in contraction and heating, but rather in cooling without much change of volume. The point Q in fig. 14 at which the sequence begun to turn was accordingly interpreted as the point at which the star's density became so great that the stellar matter ceased to behave like a gas.

155. Gradually the growth of knowledge of stellar interiors made this view untenable. The concept which I introduced in 1917 of stellar matter consisting of highly-ionised atoms and free electrons left it without any sure theoretical foundation, since it became an open question whether the much-ionised atoms retained sufficient size to cause the gas-laws to fail. Finally the discussions of observational material* by Hertzsprung, by Russell, Adams

* See above § 117.

and Joy, and by Eddington revealed no abrupt break in the correlation between mass and luminosity such as might have been expected to occur at a transition from the gaseous to the solid state, while, as already mentioned, Eddington met with a substantial measure of success in explaining the mass-luminosity correlation in terms of a theoretical formula which supposed all stars, including dense dwarfs such as Kruger 60, to obey the gas-laws.

This led Russell to propound an alternative theory of stellar evolution in 1925*, based on an acceptance of the hypothesis that the source of stellar energy was the transformation of matter into radiation.

Russell believed that all main-sequence stars have approximately the same central temperature of about 30,000,000 degrees, and based his theory on this supposed fact. He obtained his supposed fact by calculating all his central temperatures in terms of Eddington's special model for which $n = 3$ (§ 90). We have seen that when n is given the proper value required by the mass of the star and by Kramers' opacity law, the values of n generally differ very appreciably from 3, with the result that the central temperatures of main-sequence stars vary considerably, although still not enormously, *inter se*. The supposed constancy of the central temperatures led Russell to conjecture that the transformation of matter into radiation could not occur, at any rate in main-sequence stars, until the matter reached a temperature of 30,000,000 degrees, after which energy could be supplied to a practically unlimited extent. If for instance the central temperature of the sun fell at any instant to below 30,000,000 degrees, the transformation of matter into radiation would instantly cease. The sun would continue to radiate energy from its surface, and as this energy would not be replaced, the sun would contract, getting hotter throughout in so doing. In time the central temperature would again reach 30,000,000 degrees, after which the generation of energy would restart, and so on. If too much energy were generated at any instant, the star would expand until the central temperature fell to below the critical 30,000,000 degrees, when the cycle would repeat itself.

One cannot but admire the ingenuity of this theory, but our analysis has shewn (§ 114) that it comes into fatal conflict with dynamical principles. We have seen that oscillations of the type described would increase indefinitely in amplitude, so that the star would be violently unstable.

Apart from this the theory meets very grave difficulties in explaining the giant branch. As a matter of calculation the central temperatures on this are well below 30,000,000 degrees, and shew no very marked approximation to constancy. Russell accordingly found it necessary to postulate that in addition to normal matter which begins to be transformed at 30,000,000 degrees, there must be innumerable other types of matter which transform themselves into radiation at the various temperatures of the centres of the giant stars. The

* *Nature*, cxvi. (1925), p. 209.

necessity of postulating these manifold types of matter rather destroys the simplicity of the main conception, with the result that the whole hypothesis begins to look somewhat artificial. If there are about as many types of matter as there are giant stars, the uniformity of central temperature on the main branch becomes meaningless and we might as well postulate a different type of matter for each central temperature on the main sequence as well. Indeed Russell's own calculations compel us to do this. These shew that the central temperatures of main sequence stars are uniform if the stars are built after Eddington's model, so that they would not be uniform for stars built on any other model. Russell's scheme requires a star's whole generation of energy to be concentrated at its centre, and calculation shews that main sequence stars, built on this model, would not have anything like uniform central temperatures.

The highly penetrating radiation which has already been mentioned, presents a further and no less serious difficulty. Its penetrating powers though great are inadequate to carry it through more than an exceedingly small fraction of the radius of a star. In whatever bodies this radiation originates, it must originate so near to their surfaces that no great optical thickness of matter lies between the point of origin and outer space. Thus the temperature of the matter in which the radiation originates cannot be at all comparable with that of stellar centres; something of the order of 50,000 degrees would appear to be an upper limit. In view of this, it is difficult to believe that the normal radiation of the stars cannot be generated at temperatures lower than about 30,000,000 degrees.

The main reason which compels the abandonment of this rather fascinating theory is, however, that mentioned first of all, namely, that stars functioning in the way imagined by the theory would be violently unstable. Their thermodynamical properties would be those of gunpowder at its flash point, and gunpowder heated to its flash point does not shine with the steady light of the stars; it explodes.

156. The two theories of stellar evolution just considered both recognised that certain parts only of the Russell diagram are tenanted by stars, and both tried to interpret the tenanted parts as an evolutionary sequence; these parts were supposed to form a system of roads along which the stars march as they age.

In the last chapter, however, we found this feature of the Russell diagram to be adequately explained by simple considerations of stability. Some of the possible equilibrium configurations for a star are unstable, some are stable. Those parts of the Russell diagram which represent unstable configurations are naturally untenanted by stars (or possibly are sparsely populated by stars which are not in stable equilibrium); those parts which represent stable configurations alone ought to be occupied, and these have in actual fact been found to coincide very closely with those parts of the diagram in which observation places stars.

157. The problem of stellar evolution is now seen to be quite distinct from that of explaining the distribution of stars in the Russell diagram, and, furthermore, the problem can expect no assistance from the observed distribution of stars in this diagram. The problem of the evolution of the stars must be attacked anew, starting from first principles.

Fortunately these principles are perfectly clear cut. As a star ages, its mass decreases and its rate of generation of energy also decreases. Fig. 12 (p. 159) maps out the configurations possible for stars of given masses and given rates of generation of energy: the problem of following a star's evolution simply reduces to that of calculating the path that a star will follow, in this or any other diagram, as its mass and rate of generation of energy change with the passage of time.

It is convenient to divide the problem into two parts. In the first we shall consider what would be the star's evolutionary path if the gas-laws were exactly obeyed throughout its whole life; in the second we examine in what way this evolutionary path is modified by deviation from the gas-laws.

THE EVOLUTION OF GASEOUS STARS.

158. We must first find the relation between a star's mass, luminosity and surface temperature when the gas-laws are supposed to be obeyed throughout the star's interior.

From equation (92·3) of Chapter III, we obtain

$$\lambda_c(\lambda_c + 1)\frac{N^2}{A}\bar{G}R^{\frac{1}{2}} = 18\cdot3\,(T_c R)^{\frac{1}{2}} \quad \ldots\ldots\ldots\ldots(158\cdot1),$$

where R is the radius of the star. Assuming G to be constant throughout the star, $E = M\bar{G}$, so that

$$ER^{\frac{1}{2}} = \frac{18\cdot3\,M\,(T_c R)^{\frac{1}{2}}}{\left(\dfrac{N^2}{A}\right)\lambda_c(\lambda_c + 1)} \quad \ldots\ldots\ldots\ldots\ldots(158\cdot2).$$

In our previous discussion of the configurations of gaseous stars, $T_c R$ and λ_c were found to depend only on M, the mass of the star, and their values were tabulated in Table XI. Thus the right-hand member of equation (158·2) depends only on the mass of the star, and, so long as the mass of a star remains unaltered, E varies as $R^{-\frac{1}{2}}$. When the mass of the star is allowed to vary, equation (158·2) gives $ER^{\frac{1}{2}}$ as a function of M. Eliminating the star's radius R between this equation and

$$E = 4\pi\sigma R^2 T_e,$$

where T_e is the star's effective temperature, we find that

$$E = f(M)\,T_e^{0\cdot8} \quad \ldots\ldots\ldots\ldots\ldots\ldots(158\cdot3),$$

where

$$f(M) = (4\pi\sigma)^{\frac{1}{5}} \left[\frac{18\cdot3\, M\, (T_e R)^{\frac{1}{4}}}{(N^2/A)\, \lambda_c\, (\lambda_c + 1)} \right]^{\frac{4}{5}} \quad \ldots\ldots\ldots(158\cdot4).$$

This would give the relation between a star's mass, bolometric luminosity and effective temperature if the gas-laws were obeyed.

We could of course exhibit the relation in a diagram of the type shewn in fig. 12, in which $\log E$ and $\log T_e$ are taken as ordinate and abscissa respectively. The lines of constant mass would have equations of the type

$$E \propto T_e^{0\cdot8},$$

and so would be a series of straight parallel slant lines.

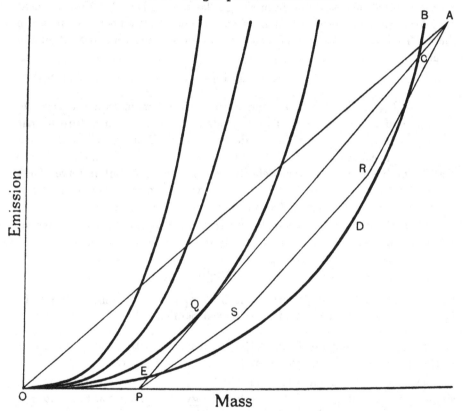

Fig. 15. Effective temperature (or spectral type) as a function of mass and luminosity.

It will, however, be more convenient to exhibit the relation in a different form. Equation (158·3) shews that a star of any given mass M and any given bolometric luminosity E can always find a value of T_e such that it will be in equilibrium provided the gas-laws are obeyed throughout its mass. Thus we may think of a gaseous star's effective temperature T_e as being determined by its mass M and its bolometric luminosity, or total rate of generation of energy E, and we may exhibit the values of T_e in a diagram in which E and M are taken as ordinate and abscissa, as in fig. 15.

The curves along which T_e has constant values are found to lie somewhat as shewn in fig. 15. No two of these curves can ever intersect since the value of T_e is uniquely determined by equation (158·3). It is readily verified that as M increases, $f(M)$ increases more rapidly than M, so that the curves are convex to the axis of M at every point.

A general idea of the shape of the curve is obtained by returning to the rough approximation $E \propto M^3$, which we found (§ 118) to represent the relation between E and M with tolerable accuracy for actual stars. This gives a moderately good approximation to the form of $f(M)$ but gives a better approximation still to the luminosities of actual stars, so that it already takes partial account of deviations from the gas-laws. The law $E \propto M^3$ was tested only for stars on or near the main sequence, and so contains no temperature factor. To extend it to stars of all effective temperatures, we write it in the form

$$E = \alpha M^3 \left(\frac{T_e}{T_{em}} \right)^{0·8} \quad\dots\dots\dots\dots\dots\dots(158·5).$$

Here T_{em} is the effective temperature of the main-sequence stars of mass M, and α is a constant. The dependence of E on temperature is that given by equation (158·3) while the dependence of E on mass is that given by our empirical law $E \propto M^3$, so long as we limit ourselves to main-sequence stars. The law is neither very definite or very accurate, but it serves for a general discussion of the kind in which we are now engaged, in which simplicity is more important than precision.

When the emission is given by equation (158·5), the curves of constant effective temperature approximate closely to the cubical parabolas

$$\frac{E}{M^3} = \text{constant},$$

since the changes in $(T_{em})^{0·8}$ are slight in comparison with those in M^3. In fig. 15 the curves actually drawn are cubical parabolas.

159. We now attempt to trace the evolutionary path followed by a star in this diagram as the emission of radiation causes its mass to diminish.

Consider first the ideal case in which the star consists entirely of one single type of material which liberates energy at a given constant rate per gramme. In this case the star's emission E is always exactly proportional to its mass M. Thus throughout its evolution

$$\frac{E}{M} = \text{cons.} \quad\dots\dots\dots\dots\dots\dots(159·1).$$

The star's path in fig. 15 is determined by the condition that E/M shall remain constant, and so is a straight line through the origin such as AO. A glance at the diagram shews that the effective temperature of such a star would continually increase as the star aged, its spectrum passing through the types M, K, G, F, A, B, O, ... in this order. The hottest stars would thus be the least massive, which is not what is found in nature.

160. Coming nearer to reality, let us next suppose that a star is formed of two kinds of matter. Let it contain a mass M_0 of matter of permanent type which does not transform itself into radiation at all, and a mass $M - M_0$ of another type which produces radiation spontaneously at a given rate. The star's emission of radiation is no longer proportional to M, but to $M - M_0$, and the star's evolutionary path in the diagram will have as its equation

$$\frac{E}{M - M_0} = \text{cons.} \quad\dots\dots\dots\dots\dots\dots(160\cdot1).$$

This is a straight line such as AP which does not pass through the origin but meets the axis of M in the point $M = M_0$; the mass does not continually diminish down to $M = 0$, but only to the limiting mass $M = M_0$, after which no further decrease of mass occurs.

Let the curve $BCDE$ in our diagram correspond to the effective temperature at which a star's surface is so cool as to be only just visible, so that stars are only visible when their representative points are above the curve $BCDE$. The hypothetical star we are now considering would first become visible at C, after which its effective temperature would increase until it attained a maximum at Q, thereafter decreasing until the star again lapsed into invisibility, as a star of very small mass, at E. Its sequence of spectral types would be M, K, G, F, ..., F, G, K, M. This is precisely the sequence predicted by Russell's 1913 theory of stellar evolution (§ 154), although it has been derived from utterly different premises. It is now clear that a star might well be urged, merely by loss of mass consequent on the passage of time, to pursue a course along the giant branch and then down the dwarf half of the main sequence, the path which Russell regarded as the normal evolutionary path for a star.

161. Passing to higher degrees of generality, let us consider a star consisting of any number of different kinds of matter annihilating themselves at different rates. To be specific, let us suppose that a star originally contains masses M_1, M_2, M_3, ... of different kinds of matter which break down at rates represented by moduli of decay κ_1, κ_2, κ_3, ...

After a time t the surviving amounts of these different types will be $M_1 e^{-\kappa_1 t}$, $M_2 e^{-\kappa_2 t}$, $M_3 e^{-\kappa_3 t}$, ..., so that the total mass of the star will be

$$M = M_1 e^{-\kappa_1 t} + M_2 e^{-\kappa_2 t} + M_3 e^{-\kappa_3 t} + \dots \quad\dots\dots\dots\dots(161\cdot1).$$

The star's rate of emission of energy E is equal to its rate of generation of energy $- C^2 \dfrac{dM}{dt}$. Obtaining the value of this by differentiation of equation (161·1), we find

$$E = C^2 \left(\kappa_1 M_1 e^{-\kappa_1 t} + \kappa_2 M_2 e^{-\kappa_2 t} + \dots \right) \quad\dots\dots\dots\dots(161\cdot2).$$

These two equations determine the values of M and E which correspond to any values of t. The succession of values obtained on allowing t to increase determines the evolutionary course of the star in the diagram shewn in fig. 15. If the values of κ_1, κ_2, ... are very unequal, the path will be a broken line such as $ARSP$; this represents the extreme case in which the different types of matter are used up in turn, each type practically disappearing before the next and far slower type of matter has experienced any substantial loss. If the values of κ_1, κ_2, κ_3 are less unequal than this, the evolutionary path will be a curve of the same general type but with the corners rounded off.

When these results are translated into the ordinary Russell diagram they are found to give a normal evolutionary curve of the general type already shewn in fig. 14.

The Influence of Fission on Evolution.

162. So far we have considered the evolution of a single star. Let us now examine how the course of evolution is affected if the star happens to break up by fission and form a binary system. We shall consider the dynamical

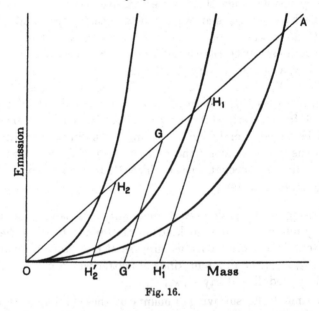

Fig. 16.

details of this process in a later chapter: at the moment we are only concerned with the final result which is a binary star—two detached masses describing orbits about their common centre of gravity.

Imagine first that the parent star divides into two parts of equal mass, with the various constituent types of matter divided equally between them, so that each component has precisely half the energy-generating capacity of the original star. If the parent star was of mass M and luminosity L, each component of the binary has mass $\frac{1}{2}M$ and luminosity $\frac{1}{2}L$. If A in fig. 16

represents the mass and luminosity of the original star, G, the middle point of OA represents the mass and luminosity of each of the components. Whatever the position of A, the effective temperature of G is necessarily higher than that of A, so that fission into two equal and similar components must always increase the effective temperature of a star.

Our approximate formula (158·5), namely,

$$E = \alpha M^3 \left(\frac{T_e}{T_{em}}\right)^{0·8}$$

provides the means of calculating the order of magnitude, at least, of the amount of this increase. If the process of fission halves both E and M, it must increase T_e/T_{em} in the ratio $4^{1·25}$, so that T_e must be increased about sixfold. Thus if an M-type star of effective temperature 3000° breaks up by fission in two equal and similar halves, each constituent will have an effective temperature of about 18,000°, and so be of spectral type $B2$ or $B3$.

This at once suggests an explanation in general terms of why newly formed binaries are generally observed to be of spectral types O, B or A. It also calls into being the giant half of the main sequence, since if the parent star was of large mass, the two components would both lie on this giant half. That we are on the right lines is suggested by the observed fact that very young binaries mainly lie on this giant half of the main sequence.

163. Nevertheless the course of nature is not quite so simple as we have supposed, and the two components of observed binary stars are usually unequal both in mass and in luminosity.

If the radiation-producing capacity of the parent star is divided between the two components in the ratio of their masses, each component will have the same value of E/M as the parent star, so that the representative points H_1, H_2 will still lie on the line OA joining the origin to the parent star. In such a case the less massive of the two components has the higher effective temperature, the effective temperatures of both being necessarily higher than that of the parent star.

The evolutionary paths of the two components will be two lines $H_1 H_1'$, $H_2 H_2'$, both parallel to the path of parent star before fission. Thus when the less massive component has any mass M_1 it will be of greater luminosity and of higher effective temperature than the more massive component will be when it has shrunk to mass M_1, and both components when they reach mass M_1 will be of greater luminosity and of higher effective temperature than the parent would have been had it shrunk to mass M_1 without fission taking place.

This scheme is still too simple to fit the observed facts. Observation of newly formed binaries does not indicate any tendency for E/M to be the same for both components. It rather shews that E/M, which is proportional to the rate of generation of energy per gramme, is different for the two components, being generally less for the less massive component.

To explain the observed facts we have to suppose that when a star breaks up by fission the less massive component is formed mainly from the outer layers of the parent star. There must have been some tendency at least for the heavier atoms, in which we believe that main capacity for generating energy resides, to have sunk towards the centre of the parent star, so that when fission occurs, an undue proportion of these heavy atoms are likely to stay with the more massive component. This accordingly starts life with a value of E/M which is greater than that either for the parent star or for the less massive component. Menzel[*] has found that, as compared with the average star, the brighter component of a visual binary system usually has an abnormally high effective temperature, while that of the fainter component is generally abnormally low, but it is doubtful how far this evidence is relevant since we shall find later (Chap. XII) that probably only a small number of visual binary systems have been formed by fission.

The extent to which the heavier atoms have sunk to the centre of the parent star must depend on the age of the star. A rough calculation[†] suggests that the time necessary for the elements of different atomic weights to separate out under gravity is of the order of 10^{13} years. Thus the process might be well advanced in a dwarf star, but could hardly have more than started in a giant star, at any rate if the atoms were thoroughly well mixed at the birth of the star. More probably, however, a newly born star is a mixture of masses of gas, some being of high molecular or atomic weight, and some of low. The former will rapidly find their way to the central regions of the star by convection, just as water falls through oil or mercury through water, so that we should expect to find a certain preponderance of heavy atoms at the star's centre even in its earliest stages. There ought to be a still greater preponderance in old stars in which the atoms have had time to become thoroughly sifted out under gravity.

It follows that at fission the more massive component ought always to have the greater value of E/M, but that the disparity in the values of E/M for the two components ought to be greater in old stars than in young. This may offer a partial explanation at least of the circumstances noticed by Leonard[‡] and others, that in giant binaries the brighter and so presumably more massive component is usually of later spectral type than the fainter component, while in dwarf binaries the reverse usually seems to hold.

164. So long then as we think of the stars as purely gaseous structures, the normal evolutionary track of a star must be thought of as a curved line of ascending and descending temperature of the general type indicated on

---

[*] *Science*, May 6, 1927.

[†] *M.N.* LXXXVI. (1926), p. 561.   [‡] *Lick Observatory Bulletin*, No. 343 (1923).

a Russell diagram in fig. 14. We may think of the stars as a vast army marching through the Russell diagram, each individual taking his marching orders from the rate of generation of energy in his interior. If all individuals of the same age were precisely similar, the army would of course march in Indian file through the diagram, the position of any individual being entirely fixed by his age. As individuals of the same age do not appear to differ very widely from one another, we may suppose that the line of march of the army forms a fairly clearly marked and well-trodden track. The main army marches down the giant branch in the upper half of the diagram to the point where this joins the main sequence and then wheeling left, marches down the main sequence. At intervals, when a member of the army breaks up by fission and forms a binary system, this routine is departed from. Both components of the new system are transported over to the left of the diagram and start independent evolutionary marches from there.

THE EVOLUTION OF LIQUID STARS.

Giant and Main-sequence Stars.

165. The scheme just described undergoes substantial modification when deviation from the gas-laws are taken into account, and the diagram is divided up into stable and unstable regions.

The evolution of the stars must no longer be compared to the steady march of an army through a perfectly flat featureless plain, but rather to the movements of an army scrambling down, and possibly at times up, a succession of terraces. The different terraces are the bands of stable configurations which correspond to the jamming of atoms ionised down to different rings of electrons. A star stands for a time on the terrace corresponding to one ring, and then, stepping on to the slippery unstable region between this terrace and the next, drops down to the next lower terrace—from the giant branch to the main sequence, let us say.

Apart from the considerations advanced in § 143, which we shall disregard for the moment and return to later, the lowest terrace of all, the main sequence, does not lead to a further drop down. It is not bounded on its further side by a slippery unstable slope but by an impenetrable barrier formed by configurations in which the atoms lie as close together as they can be packed. In course of time most stars reach this barrier but cannot cross it, and sidle along it indefinitely. This explains the great concentration of stars along the left-hand edge of the main sequence, against which the stars seem to press like flies against a window-pane.

When a star breaks up by fission into a binary system we have seen that both components of the binary are instantaneously transferred a long way over to the left of the diagram. If the effective temperature of the parent star was fairly high, both components will be thrown right over to the left-hand edge of the main sequence, against the impenetrable barrier. They cannot cross, so that their evolutionary progress consists in a march down the main sequence. This explains why binary stars, and newly formed binaries in particular figure so prominently on the left-hand edge of the main sequence. So much is this the case that perhaps the best way of mapping out the edge of the main sequence is through a succession of binaries which appear to have recently broken up by fission—Pearce's star, V Puppis, u Herculis, β Aurigae, etc.

As a consequence the majority of newly formed binaries are of the early spectral types O, B and A. The astronomer of only a decade ago explained this by saying that stars of types O and B were specially liable to fission; he thought of the spectral type as something so permanently attached to the constitution of the star that even a cataclysm like fission could not alter it. Our explanation is rather that giant F, G, K and M stars shew a certain tendency to fission, but that as soon as fission has taken place the star ceases to be an F, G, K or M-type star; its components change their spectral type and become O or B-type stars. In brief O and B-type stars do not shew a tendency to fission, but to have fissioned.

We can trace this tendency to its origin. Fission, as we shall see in a subsequent chapter, results from a star having more angular momentum than it can carry without bursting. As a star's dimensions shrink, its angular momentum remains constant except for a slow loss in the form of radiation, but its capacity for carrying angular momentum diminishes. If a sudden shrinkage occurs, the star's angular momentum remains constant while its capacity for carrying it undergoes a sudden decrease, and this may result in fission. Such a sudden shrinkage occurs when a star reaches the unstable edge of the giant branch and suddenly drops down to the main sequence. Even if fission does not occur, such a star becomes a main sequence, and so probably an early type, star—this explains the existence of O and B-type stars which are not binaries—but if fission occurs the star not only becomes an early type star, but its components are thrown over to the extreme left-hand edge of the main sequence.

The same process can of course occur in other parts of the diagram. For instance a giant star ionised only down to its M-ring may suddenly reach the limits of stability and undergo fission in falling down to the giant branch of L-ring stars. Such considerations probably explain the origin of giant binaries such as Capella and W Crucis.

White Dwarfs.

166. For the sake of simplicity in exposition, we have spoken of the left-hand edge of the main sequence as forming an impenetrable barrier. Actually, as we have seen, the barrier is not absolutely impenetrable. The more a star presses against the barrier the greater (in general terms) become the density and the star's centre and the consequent deviations from the gas-laws. We have seen (§ 143) that when the deviations from the gas-laws become sufficiently great the barrier is no longer impenetrable. If the star presses sufficiently hard against the barrier a door suddenly opens, the star falls through to a still lower, and very much lower, terrace and finally regains stability as a white dwarf.

In a white dwarf the great majority of the atoms are stripped bare down to their nuclei and so are immune from annihilation and have almost no capacity for generating radiation. As a consequence stars on the white dwarf terrace are only very feebly luminous, and the fall to the white dwarf branch is the end of a star's career as a brilliant luminary. It is slightly disconcerting to find that our sun's position in the temperature-luminosity diagram suggests that it is pressing with perilous force against the dangerous edge of the main sequence, so that its collapse into a feebly luminous white dwarf may commence at any moment.

The white dwarf state represents the utmost limit of contraction which is observed. No further rings of electrons remain to be ionised, and even the central temperatures of the white dwarfs are insufficient to produce any appreciable nuclear disintegration. As regards further evolution, the white dwarf state is one of almost complete stagnation, changes of mass and luminosity being practically inappreciable within periods comparable with the whole life of ordinary stars. In the next 7×10^{12} years or so, each component of Plaskett's star is likely to change into a star very similar to our sun; in the same interval the only change in Sirius B will be a loss of one part in 700 of its present mass, probably accompanied by a quite inappreciable change in its luminosity.

Eddington[*] has pointed out that interesting and somewhat delicate questions arise as to the ultimate state of the white dwarfs, to which R. H. Fowler[†] has given an answer in terms of the Fermi-Dirac mechanics. In brief, Fowler finds that in its last stage, all the nuclei and electrons of a white dwarf may be regarded as forming one gigantic molecule which is in its lowest quantum-state. In this state it emits no radiation and its energy cannot be further diminished.

We can see in a general way what type of star is most likely to become a white dwarf. In brief, of course, it is the type of star which presses most forcibly against the not quite impenetrable barrier formed by the left-hand edge of the main sequence. This again is the type which would wander furthest over to the left-hand of the diagram if the barrier were not there— i.e. if the gas-laws were obeyed.

In discussing stars in which the gas-laws were obeyed, we saw that the process of fission placed both components well over to the left of the diagram. In a young star in which the atomic ingredients are fairly well mixed, the

---

[*] *The Internal Constitution of the Stars*, § 117.   [†] *M.N.* LXXXVII. (1926), p. 114.

smaller component will go further to the left of the diagram, and the smaller this component is, the further it will go. Thus if the mass-ratio is very uneven when fission takes place, there is a grave danger of the smaller component becoming a white dwarf. The two systems of well-determined mass-ratio in which the ratio is furthest from equality are* Sirius ($m'/m = 0.29$) and Procyon ($m'/m = 0.33$). The less massive component of Sirius is certainly a white dwarf and that of Procyon is in all probability one. Even if fission does not cause the less massive component to fall into the white dwarf state a second fission may do so, since the secondary small components so formed tend to be thrown very forcibly against the barrier. Such consideration may explain why o_2 Eridani *B* (of mass only 0.22 times the sun) is a white dwarf, although again we cannot press these interpretations since it is very uncertain (Chap. XII) whether these systems have originated by fission. In any case, a number of single stars of exceptionally small mass and originally of high luminosity may press so hard against the barrier as to fall through and end as white dwarfs.

167. Let us next consider the evolution of a star in which no fission occurs.

In any case this evolution must be in the direction of decreasing mass; if we neglect the probable inhibition of energy-generation caused by ionisation, it must also be in the direction of continually decreasing bolometric luminosity.

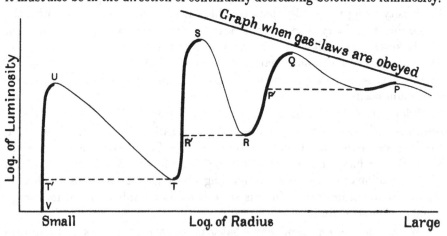

Fig. 16*a*.

The curve shewn in fig. 16*a* reproduces that already shewn in fig. 11 for the possible configurations of a star of given mass. As before, stable configurations are drawn thick and unstable configurations are drawn thin.

When a contracting star reaches a configuration such as *R*, its stable track comes to an end, and it must drop on to the next stable branch *ST*. It cannot

---

* Aitken, *Binary Stars*, pp. 205 and 216.

start at S, since it has no means of increasing its luminosity to the value appropriate to S. It accordingly starts at R', a configuration of the same mass and luminosity as R. Thus ranges such as SR', UT' are never traversed by stars in their normal course of evolution; they can only be occupied by binaries which have evolved by fission.

Thus the configurations open to a system of stars contracting without fission consist of a series of branches such as $P'R$, $R'T$, $T'V$, etc., each being lower than the preceding. The distinctness of these branches results from the distinctness between K-ring, L-ring and M-ring stars. If we disregard this distinctness, and look at the problem in a more "coarse-grained" manner, the configurations open to a star of given mass M form roughly a slant line stretching across the diagram, the direction of slant being such that high luminosity goes with low surface-temperature and conversely.

On the hypothesis of gaseous stars the configurations open to a star of given mass M also formed a slant line stretched across the diagram—but the slant is in the opposite direction (§ 158), high luminosity accompanying high surface temperature. Thus a diagram shewing the observed configurations of stars of given masses in the sky ought to provide a test between the two theories.

Such a diagram has been obtained by Seares* and is shewn in fig. 16 b. As this diagram was obtained by purely statistical methods, it is not "fine-grained" enough to shew any clear distinction between K-ring, L-ring and M-ring stars. Moreover it omits white dwarfs entirely.

The general direction of slant is unmistakably in the sense of high luminosity accompanying low surface-temperature. That is to say, the slant is in the direction needed to confirm the hypothesis of liquid stars, and is antagonistic to the hypothesis of gaseous stars. The direction of slant is reversed in the region (approximately) of G type stars. The diagram, however, represents visual absolute magnitudes whereas theory is concerned only with bolometric absolute magnitudes. The effect of applying the bolometric correction, so as to reduce visual magnitudes to bolometric, is to raise all the curves in the diagram throughout the regions of G, K and M spectral type; the topmost chain line in the diagram shews the corrected curve for stars of mass ten times that of the sun. When all the curves are similarly treated the reversal of slant and corresponding minimum luminosity at about type K 0 disappears from all curves except those representing stars of the highest mass, and Seares considers that the results for these stars are relatively uncertain, being obtained only by extrapolation of observational data. As a general statement we may say that the curves as a whole slant from high luminosity and low surface-temperature to low luminosity and high surface-temperature, as is required by the hypothesis of liquid stars.

---

* *Astrophys. Journ.* LV. (1922), p. 412.

It now appears that the diagram we used to discuss the course of stellar evolution in § 158 needs amendment. It was drawn on the supposition that high surface-temperature was associated with high luminosity, and it now appears that, except where fission enters into the problem, the exact reverse is the case. To represent the configurations of simple stars, the lines *OB*, *OQ*, etc., which represented stars at given surface-temperatures, must be replaced

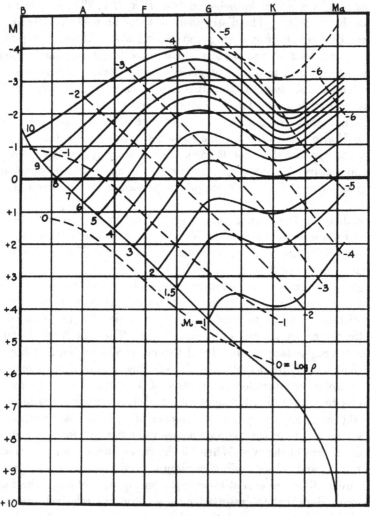

Fig. 16 b.

by bands of finite width representing white dwarfs, *K*-ring stars, *L*-ring stars, etc. And the white dwarf band must come lowest, in some position such as *OB* in fig. 15, followed in succession by bands representing *K*-ring, *L*-ring stars, etc.

In terms of this amended diagram we can discuss evolution much as before. The evolutionary path of a star will be of the general nature of the line $AQEP$ in fig. 15.

We notice in the first place that every star must necessarily end its career as a white dwarf. Also a star may begin as a white dwarf (e.g. at A) and, as it follows such a path as AQP, it may rise from being a white dwarf to being a main-sequence star, or even a giant M star, and then sink back to the white dwarf state. Whereas Russell's theory, discussed in § 154, gave an evolutionary sequence of cool-hot-cool configurations, the present sequence is hot-cool-hot, in which, however, the word "hot" refers less to the surface-temperature than to that of the interior; the final "hot" state will in general be that of a white dwarf with the surface-temperature falling gradually to zero.

This concept has important cosmogonical consequences in that it makes it possible to suppose that very massive stars have ages comparable with those of ordinary stars. In the white dwarf stage of a star's existence, whether this comes at the end or at the beginning, the atoms are ionised down to their nuclei, and, as we have seen, observation and theory agree that such a star should emit very little radiation. Plaskett's star, which at present emits about 15,000 ergs per gramme, cannot, from considerations of mass, have radiated at its present rate for more than about 10^{11} years. The present theory makes it possible to extend the star's age far beyond 10^{11} years by supposing that it may have had a previous existence as a white dwarf, in which, although its mass was of course very great, it emitted but little radiation.

Such a concept removes the otherwise almost insuperable difficulty of explaining how the most massive stars can be such very recent creations. It opens up the way to supposing that all the stars of the galactic system may have come into being at approximately the same time regardless of the great differences in their apparent ages as calculated by the methods of § 118.

We shall leave this question an open one for the present, to return to it in § 352.

We have now sketched in general terms the view of stellar evolution to which we are led by the twin hypotheses of the annihilation of matter as the source of stellar energy and of high ionisation as the state of stellar interiors. Apart from the questions discussed in the present section, it may almost be said that there has been no true choice of roads at any point of our journey, unless indeed we have dashed past some bifurcation of roads without noticing it; what may at first sight have appeared to be alternative routes have proved, on further investigation, to be barred by the well-established laws of dynamics, thermodynamics and physics.

Happily the conclusions to which we have been led accord tolerably well with the conclusions reached by astronomical observations. It is fortunate that it is so, for had there been any serious divergence we should hardly have known how to retrace our steps and take a different road in the hope of coming to some other conclusion. On the other hand, our conclusions suffer from a good deal of vagueness. If we liked to introduce further definite assumptions we could make the picture as much more definite as we wished. Some assumptions would give results which would agree with observations, while others naturally would not. By a skilful choice of assumptions we could doubtless both add precision to our picture and accentuate its agreement with observation. But the procedure is too risky to be profitable. We might find an assumption which would intensify the agreement with observation enormously, but we should remain in ignorance of perhaps ten others that would do the same thing equally well. We should start on our journey with odds of ten to one that we were on the wrong track and that all our labour would be wasted. Considerations of this nature suggest that it is more prudent to leave the theory in its present vague form than to gain additional precision by introducing special assumptions of a speculative or semi-speculative kind.

CHAPTER VII

NON-SPHERICAL MASSES—DYNAMICAL PRINCIPLES

CONFIGURATIONS OF EQUILIBRIUM.

168. THE last four chapters have been devoted to a discussion of the commonest object in the sky—the simple star. Its mechanism and build were discussed on the supposition that it was of spherical shape.

We turn now to the discussion of non-spherical masses. A star which was initially spherical may assume other shapes as the result of being set into rotation, of coming within the field of gravitational force arising from some other body, or in other ways. Our main problem is to calculate the shapes assumed by astronomical bodies under such circumstances, the ultimate aim being, of course, to compare the calculated shapes with the observed shapes of astronomical objects. The discussion will no longer be limited to stars. The present chapter will contain an exposition of dynamical principles which we shall subsequently apply to nebulae, stars, the sun, the earth and the planets, although the principles are so general as to apply to all dynamical systems and not merely to astronomical bodies.

169. Let the configuration of a dynamical system at any instant be fixed by a number of co-ordinates

$$\theta_1, \theta_2, \theta_3, \dots \theta_n \quad \dots\dots\dots\dots\dots\dots(169\cdot1),$$

while its motion is specified in terms of the corresponding velocities

$$\dot{\theta}_1, \dot{\theta}_2, \dot{\theta}_3, \dots \dot{\theta}_n \quad \dots\dots\dots\dots\dots\dots(169\cdot2).$$

The potential energy W of the system is a function of the co-ordinates of position only, say

$$W = f(\theta_1, \theta_2, \dots \theta_n) \quad \dots\dots\dots\dots\dots\dots(169\cdot3),$$

while the kinetic energy T is a function both of the co-ordinates of position and of the velocities, say

$$T = F(\theta_1, \theta_2, \dots \theta_n, \dot{\theta}_1, \dot{\theta}_2, \dots \dot{\theta}_n) \quad \dots\dots\dots\dots(169\cdot4),$$

this function being of the second degree in the velocities $\dot{\theta}_1, \dot{\theta}_2, \dots \dot{\theta}_n$.

The motion of the system will be determined by the usual Lagrangian equations

$$\frac{d}{dt}\left(\frac{\partial T}{\partial \dot{\theta}_s}\right) - \frac{\partial T}{\partial \theta_s} = -\frac{\partial W}{\partial \theta_s} + F_s \quad (s = 1, 2, \dots n) \quad \dots\dots\dots(169\cdot5)$$

where $F_1, F_2, \dots F_n$ are the "generalised forces" applied from outside.

In a great number of cosmogonical problems we are concerned with astronomical masses which are either in a state of equilibrium or whose motion is so slow that their kinetic energy is negligible. For such configurations we may put $T = 0$, so that equations (169·5) reduce to

$$\frac{\partial W}{\partial \theta_1} = 0, \quad \frac{\partial W}{\partial \theta_2} = 0, \quad \frac{\partial W}{\partial \theta_3} = 0, \text{ etc. } \quad \ldots\ldots\ldots\ldots(169\cdot6).$$

These may be regarded either as equations of equilibrium or as equations determining the configurations of a very slowly changing mass. Regarded as equations in $\theta_1, \theta_2, \theta_3, \ldots$, there will be a number of solutions of which a typical one may be taken to be

$$\theta_1 = \Theta_1, \quad \theta_2 = \Theta_2, \text{ etc. } \quad \ldots\ldots\ldots\ldots\ldots(169\cdot7).$$

In this solution the quantities $\Theta_1, \Theta_2, \ldots$ depend on the constants which specify the potential energy W in terms of $\theta_1, \theta_2, \ldots$, as given by equation (169·3). In problems of cosmogony in which changes of a secular or evolutionary nature occur, these constants must themselves be supposed to vary; they are better spoken of as parameters than as constants. When equations such as (169·6) are satisfied, an astronomical mass is momentarily in a position of equilibrium, but if the physical conditions change in the course of time, this particular configuration of equilibrium will give place to another. We may represent this process by supposing slow changes to occur in the parameters which enter into the specification of W by equation (169·3).

Linear Series.

170. Let us consider in detail the changes produced in $\Theta_1, \Theta_2, \ldots$, the co-ordinates of a configuration of equilibrium, as one of the variable parameters, say μ, is allowed slowly to vary.

A slight change from μ to $\mu + d\mu$ in the value of μ will alter the values of $\Theta_1, \Theta_2, \ldots$ by quantities which will in general be small quantities of the same order of magnitude as $d\mu$. On making such a small change in μ, a configuration of equilibrium such as that given by equations (169·7) gives place to an adjacent configuration of equilibrium. On continually varying μ we pass through a whole series of continuous configurations of equilibrium, which form what Poincaré has called a "linear series[*]."

Suppose for the moment that our dynamical system is specified by only two co-ordinates θ_1, θ_2, as, for instance, the co-ordinates of a particle moving on a curved surface. We may in imagination construct a three-dimensional space having

$$\theta_1, \theta_2, \mu$$

as co-ordinates. Any one plane $\mu = \text{cons.}$ will be suitable for the representation of all the configurations which are possible for one value of μ, and therefore for all which are possible for one definite physical state of the system. The

---

[*] Poincaré, *Acta Math.* VII. (1885), p. 259, or *Figures d'équilibre d'une masse fluide.* (Paris, 1902.) See also Lamb, *Hydrodynamics*, p. 680.

particular points in this plane determined by equations such as (169·7) will represent the configurations of equilibrium in this physical state.

From its meaning the function W must be a single valued function of θ_1, θ_2, and μ, so that the surfaces $W = $ cons. in the three-dimensional space are necessarily non-intersecting surfaces. The condition that a configuration shall be one of equilibrium, as expressed by equations (169·6), is exactly identical with the condition that the tangent to the surface $W = $ cons. shall be perpendicular to the axis of μ. If we make the axis of μ vertical, the configurations of equilibrium are represented by the points at which the tangents to the surfaces $W = $ cons. are horizontal. We may speak of these as "level points." Each level point represents a configuration of equilibrium corresponding to one value of the parameter μ. On joining up a succession of level points we get a line such that points on it represent configurations of equilibrium for different values of μ. Such a succession of configurations forms a "linear-series."

We have illustrated the meaning of our terms by using a simple system with only two co-ordinates θ_1, θ_2, but the method is quite general. If the system has n co-ordinates we represent its configurations, as μ changes, in an imaginary space of $(n + 1)$ dimensions, and soon reach the same result.

As the value of any parameter μ changes, the configuration of the system will change, and so long as the system remains in equilibrium, its various configurations will lie on a linear series.

Points of Bifurcation.

171. As we pass along a linear series, the regular succession of points representing configurations of equilibrium may be broken in various ways. One obvious way is by a change in the direction of curvature of the W-surfaces, resulting in the formation of a kink, such as is shewn occurring at the point

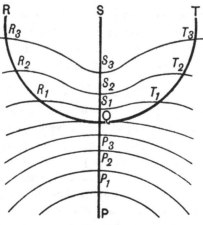

Fig. 17.

Q in fig. 17. On any surface on which this formation has just occurred, there will be three adjacent level points such as R_1, S_1, T_1, in the figure. The original linear series PQ will accordingly become replaced by three linear series such as QR, QS and QT as soon as we pass above the point Q at which the kink first forms. It is readily seen that at Q two of the series QR and QT must run continuously into one another, and so in effect form a single new series, while the series QS may be regarded as a continuation of PQ. We may accordingly suppose that there are two linear series PQS and RQT crossing one another at the point Q. Poincaré calls a point such as Q a "point of bifurcation."

The succession of level points can also be broken—or rather deviated—as shewn in fig. 18. In this case, as μ increases, two linear series such as $P_1 P_2 Q$ and $U_1 U_2 Q$ approach one another and finally coalesce in the point Q and then disappear. A point such as Q in this figure may conveniently be described as a "turning point."

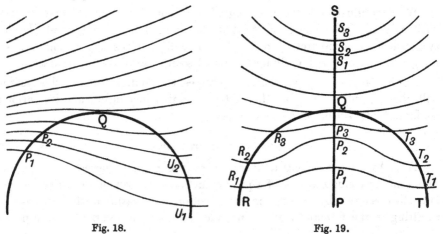

Fig. 18.   Fig. 19.

A third possibility, shewn in fig. 19, is only a variant of that shewn in fig. 17, and again leads to two linear series crossing one another in a point of bifurcation Q. Other minor variations may occur, but the principle possibilities are those shewn in figures 17, 18 and 19.

Stability and Instability.

172. Every point on a linear series is a configuration of equilibrium; a question which is of the utmost importance in cosmogonical problem is whether this equilibrium is stable or unstable. Confining our attention to one particular state of the system, and so to one of the planes $\mu = \text{cons.}$ the condition that a particular configuration of equilibrium in this plane shall be stable is that the value of W at the point in question shall be a minimum. Hence the configuration represented at any point on a linear series will be stable if the concavities of the different vertical sections of the W-surface through this point are all turned in the same direction, this direction being that of W-decreasing.

Suppose, for instance, that in fig. 17 W increases as we pass upwards, and that the concavities for all sections of the W-surface through P_1 are turned in the same direction as that shewn in the diagram. Then the configuration represented by the point P_1 will be one of stable equilibrium.

On passing along a series such as PQS in fig. 17 or 19, one of the sections must clearly change the direction of its concavity as we pass through the point Q at which a kink is first formed on the W-surfaces. Thus configurations which were initially stable give place to unstable configurations on passing through points such as Q. Thus we see that a principal series such as PQS loses its stability on passing through a point of bifurcation.

If P_1, P_2, P_3 represent stable configurations in fig. 17, the concavities of all the curvatures at these points must be turned downwards. The same is then true at the points R_1, R_2, R_3 and T_1, T_2', T_3, so that the configurations represented by R_1, R_2, R_3 and T_1, T_2, T_3 will also be stable. Thus stability, which leaves the principal series PQS at Q, may be thought of as passing to the branch series RQT. Thus we see that there is an *exchange of stabilities* at the point of bifurcation Q.

In fig. 19, on the other hand, we find that if the configurations represented by P_1, P_2, P_3 are stable, then those represented by R_1, R_2, R_3 and T_1, T_2, T_3 will be unstable, in addition to those represented by S_1, S_2, S_3. In this case there is a *disappearance of stability* at the point of bifurcation Q.

In fig. 18, it is clear that if $P_1, P_2, \ldots$ are stable, then $U_1, U_2, \ldots$ must be unstable; while conversely if $U_1, U_2, \ldots$ are stable, then $P_1, P_2, \ldots$ must be unstable. Thus in moving along a linear series a loss or gain of stability occurs on passing through a point such as Q at which μ is a maximum. But in a physical problem, μ will continually change in the same direction, and the physical phenomenon which will accompany the passing of μ through its value at Q will be a complete disappearance of two sets of equilibrium configurations.

These results are shewn diagrammatically in the following figures, in which

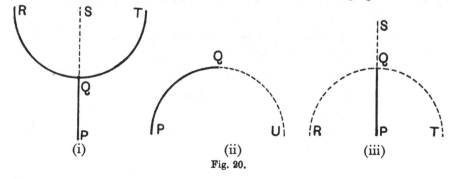

(i) (ii) (iii)

Fig. 20.

thick lines represent series of stable configurations, and thin lines series of unstable configurations, the series PQ being assumed to be stable in every case

173. Suppose that in any physical problem μ changes very slowly, the direction of change being that represented by an upward movement in our diagrams. From what has already been said, it is clear that the following rule will trace out the sequence of stable states which will be followed by the system as μ varies.

Start from a configuration in the diagram which is known to be stable and follow a path along linear series of equilibrium so as always to move upwards, and so as always to cross over from one series to another at a point of bifurcation. So long as we can do this we are following a sequence of configurations which is always stable. When it becomes impossible to do this any longer, a value of μ has been reached beyond which no stable configurations exist, and if the physical conditions continue to change so that μ attains to a still higher value, the statical problem gives place to a dynamical one; it is no longer a question of tracing out a sequence of gradual secular changes, but of following up a comparatively rapid motion of a cataclysmic nature.

At each point of bifurcation there is necessarily a certain amount of indefiniteness in the path which will actually be followed. For instance, in fig. 20 (i), the system on arriving at Q may proceed either along QT or along QR, both being equally consistent with the maintenance of stability, and so far as can be seen equally likely.

This complication causes no difficulty in actual problems. It arises from the obvious circumstance that a general discussion of stability, although competent to determine when stability ceases, cannot in general determine what will happen after stability has ceased. A general discussion of stability will readily shew that a top spinning slowly on its point is in unstable equilibrium, but it cannot determine in which precise direction the top will first fall to the ground.

174. In his classical paper[*] in which the theory of linear series and points of bifurcation was first developed, Poincaré used analytical methods to obtain results identical with those just given.

If μ is the only parameter which can vary, the potential energy W of the system may be written in the form

$$W = f(\theta_1, \theta_2, \ldots \theta_n, \mu)$$

and the configurations of equilibrium are given by the equations

$$\frac{\partial}{\partial \theta_1} f(\theta_1, \theta_2, \ldots \theta_n, \mu) = 0, \text{ etc.} \quad \ldots\ldots\ldots\ldots(174\cdot1).$$

[*] l.c. *ante.*

As in §169, let $\Theta_1, \Theta_2, \ldots$ be a configuration of equilibrium corresponding to this given value μ of the parameter, so that in this configuration

$$\frac{\partial W}{\partial \theta_1} = \frac{\partial W}{\partial \theta_2} = \frac{\partial W}{\partial \theta_3} = \ldots = 0 \quad \ldots \ldots \ldots (174\cdot2).$$

In any adjacent configuration $\Theta_1 + \delta\theta_1, \Theta_2 + \delta\theta_2, \ldots \mu + \delta\mu$, the value of W may be expressed in the form

$$W + \tfrac{1}{2}(\delta\theta_1)^2 \frac{\partial W}{\partial\theta_1^2} + (\delta\theta_1)(\delta\theta_2)\frac{\partial^2 W}{\partial\theta_1\partial\theta_2} + \ldots$$
$$+ \delta\mu\frac{\partial W}{\partial\mu} + \tfrac{1}{2}(\delta\mu)^2\frac{\partial^2 W}{\partial\mu^2} + (\delta\mu)(\delta\theta_1)\frac{\partial^2 W}{\partial\mu\partial\theta_1} + \ldots \quad (174\cdot3)$$

and the condition that this new configuration shall be one of equilibrium is, from equations (174·2),

$$\delta\theta_1\frac{\partial^2 W}{\partial\theta_1^2} + \delta\theta_2\frac{\partial^2 W}{\partial\theta_1\partial\theta_2} + \ldots + \delta\theta_n\frac{\partial^2 W}{\partial\theta_1\partial\theta_n} + \delta\mu\frac{\partial^2 W}{\partial\theta_1\partial\mu} = 0 \quad \ldots(174\cdot4)$$

and similar equations. Writing W_{12} for $\partial^2 W/\partial\theta_1\partial\theta_2$ and so on, the solution of these equations is

$$\frac{\delta\theta_1}{\begin{vmatrix} W_{12}, & W_{13}, \ldots & W_{1\mu} \\ W_{22}, & W_{23}, \ldots & W_{2\mu} \\ \ldots & & \ldots \end{vmatrix}} = \frac{\delta\theta_2}{\begin{vmatrix} \ldots \end{vmatrix}} = \ldots = \frac{\delta\mu}{\Delta} \quad \ldots\ldots(174\cdot5),$$

where Δ is the Hessian of W with respect to the variables $\theta_1, \theta_2, \ldots \theta_n$, given by

$$\Delta = \begin{vmatrix} W_{11}, & W_{12}, \ldots & W_{1n} \\ W_{21}, & W_{22}, \ldots & W_{2n} \\ \ldots & & \ldots \end{vmatrix} \quad \ldots\ldots\ldots\ldots(174\cdot6).$$

The values of the ratios $\delta\theta_1 : \delta\theta_2 : \ldots : \delta\mu$ determine the ratios of the small changes in $\Theta_1, \Theta_2, \ldots \mu$ as we pass along a linear series. At points such as Q in figs. 20 (i) and 20 (iii) one or more of these ratios must become indeterminate, so that we must have (say) $\delta\mu/\delta\theta_s = 0$. At a point such as Q in fig. 20 (ii) we must have (say) $\delta\mu/\delta\theta_r = 0$. Thus the three points Q in fig. 19 are all determined by the single condition

$$\Delta = 0 \quad \ldots\ldots\ldots\ldots\ldots\ldots\ldots(174\cdot7).$$

175. If μ is kept constant, the change of potential energy corresponding to changes $\delta\theta_1, \delta\theta_2, \ldots$ in the values of $\Theta_1, \Theta_2, \ldots$, will be given by

$$\delta W = \tfrac{1}{2}(\delta\theta_1)^2 W_{11} + (\delta\theta_1)(\delta\theta_2)W_{12} + \ldots \quad \ldots\ldots\ldots(175\cdot1)$$

in which no terms of degree beyond the second need be written down, since $\delta\theta_1, \delta\theta_2, \ldots$ are supposed small.

Let the co-ordinates $\delta\theta_1, \delta\theta_2, \ldots$ in this quadratic expression for δW be changed by a linear transformation to new co-ordinates $\phi_1, \phi_2, \ldots$, such that δW becomes a sum of squares, say

$$\delta W = \tfrac{1}{2}(b_1\phi_1^2 + b_2\phi_2^2 + \ldots + b_n\phi_n^2) \quad \ldots\ldots\ldots(175\cdot2),$$

the modulus of transformation being λ.

Since the discriminant remains invariant through all linear transformations, we have

$$\begin{vmatrix} b_1, & 0, & 0, & \ldots \\ 0, & b_2, & 0, & \ldots \\ & \ldots & \ldots & \end{vmatrix} = \lambda \begin{vmatrix} W_{11}, & W_{12}, & \ldots & W_{1n} \\ W_{21}, & W_{22}, & \ldots & W_{2n} \\ & \ldots & \ldots & \end{vmatrix}$$

or

$$b_1, b_2, \ldots b_n = \lambda \Delta \qquad \ldots\ldots\ldots\ldots\ldots\ldots\ldots(175\cdot3).$$

The condition that the configuration $\Theta_1, \Theta_2, \ldots$ under discussion shall be one of stable equilibrium is that δW shall be positive for all values of $\delta\theta_1, \delta\theta_2, \ldots \delta\theta_n$, or again that expression (175·2) shall be positive for all values of $\phi_1, \phi_2, \ldots \phi_n$. This condition is that $b_1, b_2, \ldots b_n$ shall all be positive. If any one of these quantities becomes negative, δW can become negative for a small displacement, so that the configuration has become unstable.

The coefficients $b_1, b_2, \ldots b_n$ are called by Poincaré "coefficients of stability." A change from stability to instability occurs when any one of these coefficients vanishes, and the values of μ for which this occurs are, from equation (175·3) given by

$$\Delta = 0 \qquad \ldots\ldots\ldots\ldots\ldots\ldots\ldots\ldots(175\cdot4).$$

Combining this with the result obtained in the last section, it appears that a change of stability occurs at every point of bifurcation, and at every point on a linear series at which μ passes through a maximum or a minimum value. This agrees precisely with the result obtained by other means in §§ 171 and 172. The stability or instability of the branch series at a point of bifurcation is most readily determined by the method already adopted in § 173; with the conventions there used, it appears that the branch series will be stable if it turns upwards from the point of bifurcation, and unstable if it turns downwards.

ROTATING SYSTEMS.

176. We have so far discussed the stability of statical systems only. The stability of motion of a dynamical system is a much more complicated question, but assumes a comparatively simple form when the motion consists mainly of a uniform rotation. We proceed to discuss the stability of such a system.

Let the system be referred to axes rotating in space with any velocity ω about the axis of z in the direction from Ox to Oy. Let x, y, z be the co-ordinates of any point referred to these axes, and let $\dot{x}, \dot{y}, \dot{z}$ denote their rates of increase. The components of velocity in space are then given by

$$u = \dot{x} - y\omega, \quad v = \dot{y} + x\omega, \quad w = \dot{z} \qquad \ldots\ldots\ldots\ldots\ldots(176\cdot1),$$

so that the kinetic energy T is given by

$$T = \tfrac{1}{2}\Sigma m (u^2 + v^2 + w^2)$$
$$= \tfrac{1}{2}\Sigma m (\dot{x}^2 + \dot{y}^2 + \dot{z}^2) + \omega\Sigma m (x\dot{y} - y\dot{x}) + \tfrac{1}{2}\omega^2\Sigma m (x^2 + y^2) \ldots(176\cdot2).$$

The total moment of momentum **M** about the z-axis is given by

$$\mathbf{M} = \Sigma m (xv - yu)$$
$$= \Sigma m (x\dot{y} - y\dot{x}) + \omega\Sigma m (x^2 + y^2) \qquad \ldots\ldots\ldots\ldots(176\cdot3).$$

Let us put

$$T_R = \tfrac{1}{2}\Sigma m\, (\dot{x}^2 + \dot{y}^2 + \dot{z}^2) \quad \dotfill (176\cdot4),$$
$$U = \Sigma m\, (x\dot{y} - y\dot{x}) \quad \dotfill (176\cdot5),$$
$$I = \Sigma m\, (x^2 + y^2) \quad \dotfill (176\cdot6),$$

so that T_R is the kinetic energy relative to the rotating axes, U is the moment of momentum relative to the moving axes, and I is the moment of inertia. Then the values of T and $\mathbf{M}$ just obtained assume the forms

$$T = T_R + \omega U + \tfrac{1}{2}\omega^2 I \quad \dotfill (176\cdot7).$$
$$\mathbf{M} = U + \omega I \quad \dotfill (176\cdot8).$$

The elimination of U gives

$$T = T_R + \omega\mathbf{M} - \tfrac{1}{2}\omega^2 I \quad \dotfill (176\cdot9).$$

177. The position of a rotating system may be supposed defined by a co-ordinate ψ fixing the position of the axes, such that $\dot{\psi} = \omega$, and $n - 1$ other co-ordinates $\theta_1, \theta_2, \ldots \theta_{n-1}$ fixing the configuration of the system relative to the axes, so that the system has n co-ordinates in all.

The equations of motion (169·5) take the form

$$\frac{d}{dt}\left(\frac{\partial T}{\partial \omega}\right) - \frac{\partial T}{\partial \psi} = G \quad \dotfill (177\cdot1),$$

$$\frac{d}{dt}\left(\frac{\partial T}{\partial \theta_s}\right) - \frac{\partial T}{\partial \theta_s} = -\frac{\partial W_s}{\partial \theta_s} + F_s \qquad (s = 1, 2, \ldots n-1) \ (177\cdot2),$$

in which G is the generalised force corresponding to the co-ordinate ψ, and so is the couple about the axis of z which acts upon the system.

With the value of T given by equation (176·7), we have $\partial T/\partial \psi = 0$, and $\partial T/\partial \omega = \mathbf{M}$, so that equation (177·1) reduces to

$$\frac{d\mathbf{M}}{dt} = G \quad \dotfill (177\cdot3),$$

which merely expresses that the rate of increase of the moment of momentum $\mathbf{M}$ is equal to the couple G.

If a mass is rotating freely in space, $G = 0$, and $\mathbf{M}$ remains constant. If a mass is constrained to rotate at a constant angular velocity while $\mathbf{M}$ changes, a couple G will be necessary to maintain the uniformity of rotation, and the amount of this couple will be determined by equation (177·3).

MASS CONSTRAINED TO ROTATE WITH CONSTANT ANGULAR VELOCITY.

178. Let us first consider the problem when ω is kept constant. The value of any co-ordinate of position x will be a function of the $n - 1$ co-ordinates $\theta_1, \theta_2, \ldots \theta_{n-1}$, so that

$$\frac{dx}{dt} = \Sigma\, \frac{\partial x}{\partial \theta_s}\, \frac{\partial \theta_s}{\partial t},$$

and hence

$$\frac{\partial \dot{x}}{\partial \theta_s} = \frac{\partial x}{\partial \theta_s}.$$

With the value of U given by equation (176·5) we have

$$\frac{\partial U}{\partial \theta_s} = \Sigma m \left(x \frac{\partial \dot{y}}{\partial \dot{\theta}_s} - y \frac{\partial \dot{x}}{\partial \dot{\theta}_s} \right) = \Sigma m \left(x \frac{\partial y}{\partial \theta_s} - y \frac{\partial x}{\partial \theta_s} \right),$$

so that

$$\frac{d}{dt} \left(\frac{\partial U}{\partial \dot{\theta}_s} \right) = \Sigma m \left(\dot{x} \frac{\partial y}{\partial \theta_s} - \dot{y} \frac{\partial x}{\partial \theta_s} \right) + \Sigma m \left[x \frac{d}{dt} \left(\frac{\partial y}{\partial \theta_s} \right) - y \frac{d}{dt} \left(\frac{\partial x}{\partial \theta_s} \right) \right].$$

Also

$$\frac{\partial U}{\partial \theta_s} = \Sigma m \left(\frac{\partial x}{\partial \theta_s} \dot{y} - \frac{\partial y}{\partial \theta_s} \dot{x} \right) + \Sigma m \left[x \frac{d}{dt} \left(\frac{\partial y}{\partial \theta_s} \right) - y \frac{d}{dt} \left(\frac{\partial x}{\partial \theta_s} \right) \right],$$

so that

$$\frac{d}{dt} \left(\frac{\partial U}{\partial \dot{\theta}_s} \right) - \frac{\partial U}{\partial \theta_s} = 2\Sigma m \left(\dot{x} \frac{\partial y}{\partial \theta_s} - \dot{y} \frac{\partial x}{\partial \theta_s} \right)$$

$$= 2\Sigma m \left[\Sigma \left(\frac{\partial x}{\partial \theta_r} \frac{\partial y}{\partial \theta_s} - \frac{\partial y}{\partial \theta_r} \frac{\partial x}{\partial \theta_s} \right) \dot{\theta}_r \right].$$

If we put

$$\beta_{rs} = 2\Sigma m \left(\frac{\partial x}{\partial \theta_r} \frac{\partial y}{\partial \theta_s} - \frac{\partial y}{\partial \theta_r} \frac{\partial x}{\partial \theta_s} \right),$$

so that $\beta_{rs} = -\beta_{sr}$ and $\beta_{rr} = 0$, we obtain

$$\frac{d}{dt} \left(\frac{\partial U}{\partial \dot{\theta}_s} \right) - \frac{\partial U}{\partial \theta_s} = \beta_{1s} \dot{\theta}_1 + \beta_{2s} \dot{\theta}_2 + \dots \quad \dots\dots\dots(178\cdot1).$$

Using the value of T given by equation (176·7) and keeping ω constant,

$$\frac{d}{dt} \left(\frac{\partial T}{\partial \dot{\theta}_s} \right) - \frac{\partial T}{\partial \theta_s} = \frac{d}{dt} \left(\frac{\partial T_R}{\partial \dot{\theta}_s} \right) - \frac{\partial T_R}{\partial \theta_s} + \omega \left[\frac{d}{dt} \left(\frac{\partial U}{\partial \dot{\theta}_s} \right) - \frac{\partial U}{\partial \theta_s} \right] - \tfrac{1}{2}\omega^2 \frac{\partial I}{\partial \theta_s},$$

so that the equations of motion (177·2) become

$$\frac{d}{dt} \left(\frac{\partial T_R}{\partial \dot{\theta}_s} \right) - \frac{\partial T_R}{\partial \theta_s} + \omega \left(\beta_{1s} \dot{\theta}_1 + \beta_{2s} \dot{\theta}_2 + \dots \right) = -\frac{\partial}{\partial \theta_s} \left(W - \tfrac{1}{2}\omega^2 I \right) + F_s$$

$$\dots\dots\dots(178\cdot2).$$

These are the equations of motion relative to rotating axes. They differ from the simpler equations appropriate to the case of $\omega = 0$ in two respects; first by the presence of what we may call "gyroscopic" terms such as $\beta_{1s} \omega \dot{\theta}_1$ and second, by $W - \tfrac{1}{2}\omega^2 I$ replacing the potential energy W of the simpler equations.

179. The system will be in equilibrium relative to the moving axes if

$$\dot{\theta}_1 = \dot{\theta}_2 = \dots = 0,$$

so that, from equation (178·2) the configurations of relative equilibrium are determined by the equations

$$\frac{\partial}{\partial \theta_s} \left(W - \tfrac{1}{2}\omega^2 I \right) = F_s, \text{ etc. } \quad \dots\dots\dots\dots(179\cdot1).$$

When there are no externally applied forces, these reduce to

$$\frac{\partial}{\partial \theta_s} \left(W - \tfrac{1}{2}\omega^2 I \right) = 0 \quad \dots\dots\dots\dots(179\cdot2).$$

These equations only differ from the simpler ones for a system at rest in that W has become replaced by $W - \tfrac{1}{2}\omega^2 I$. The configurations of relative equilibrium may accordingly be found just as though the system were at rest under a potential $W - \tfrac{1}{2}\omega^2 I$, and these configurations will fall into linear series as before.

Small Oscillations.

180. To discuss the small oscillations of such a system, we have to return to the equations of motion (177·2). Suppose we are considering the oscillations of a configuration which is one of equilibrium under no applied forces, say

$$\theta_1 = \Theta_1, \text{ etc.}$$

Let the co-ordinates be replaced by $\theta_1 - \Theta_1$, etc., so that the new values of θ_1, θ_2, ... all vanish in the configuration of equilibrium. The values of $W - \tfrac{1}{2}\omega^2 I$ and of T_R for any small displacement may now be expressed in the forms

$$2T_R = a_{11}\dot{\theta}_1{}^2 + 2a_{12}\dot{\theta}_1\dot{\theta}_2 + \dots \quad \dots\dots\dots\dots(180\cdot1),$$

$$2(W - \tfrac{1}{2}\omega^2 I) = b_{11}\theta_1{}^2 + 2b_{12}\theta_1\theta_2 + \dots \quad \dots\dots\dots\dots(180\cdot2),$$

the preliminary condition that equations (179·2) shall be satisfied in the configuration of equilibrium requiring the omission of terms of first degree in θ_1, θ_2, ... By a linear transformation, T_R and $W - \tfrac{1}{2}\omega^2 I$ may be simultaneously reduced to a sum of squares, so that we may legitimately assume the still simpler forms

$$2T_R = a_1\dot{\theta}_1{}^2 + a_2\dot{\theta}_2{}^2 + \dots \quad \dots\dots\dots\dots(180\cdot3),$$

$$2(W - \tfrac{1}{2}\omega^2 I) = b_1\theta_1{}^2 + b_2\theta_2{}^2 + \dots \quad \dots\dots\dots\dots(180\cdot4),$$

With these values for T_R and $W - \tfrac{1}{2}\omega^2 I$, the equation of motion (178·2) assumes the form

$$a_s\ddot{\theta}_s + b_s\theta_s + \omega(\beta_{s1}\dot{\theta}_1{}^2 + \beta_{s2}\dot{\theta}_2 + \dots) = F_s, \text{ etc. } \quad \dots\dots(180\cdot5).$$

Had the system been at rest, these equations would have reduced to

$$a_s\ddot{\theta}_s + b_s\theta_s = F_s, \text{ etc. } \quad \dots\dots\dots\dots\dots(180\cdot6),$$

a system in which each equation depends on only one co-ordinate. From the form of these equations it follows that the different co-ordinates θ_1, θ_2,... execute entirely independent vibrations; the changes in any one co-ordinate θ_1 does not influence, and is not influenced by, whatever changes may be occurring in the other co-ordinates θ_2, θ_3,.... These are of course the well-known properties of "principal co-ordinates."

But a glance at equations (180·5) and (180·6) will shew that these properties no longer persist when the system is in rotation. A disturbance in which θ_1 at first exists alone will soon set up oscillations in which θ_2, θ_3, ... have finite values, and the co-ordinates θ_1, θ_2, ... no longer correspond to independent vibrations.

Since these equations are linear with constant coefficients, it is clear that there will be systems of separate free vibrations. These may be found by putting $F_1 = F_2 = \ldots = 0$, and assuming $\theta_1, \theta_2, \ldots$, each proportional to the same time-factor $e^{\lambda t}$. The equations then reduce to

$$(a_1 \lambda^2 + b_1)\, \theta_1 + \omega\lambda\beta_{12}\theta_2 + \omega\lambda\beta_{13}\theta_3 + \ldots = 0,$$

$$\ldots \qquad \ldots \qquad \ldots$$

$$\omega\lambda\beta_{s1}\theta_1 + \omega\lambda\beta_{s2}\theta_2 + \ldots + (a_s\lambda^2 + b_s)\,\theta_s + \ldots = 0 \quad \ldots\ldots(180\cdot7).$$

Eliminating the θ's, we find as the equation giving λ,

$$\begin{vmatrix} a_1\lambda^2 + b_1, & \omega\lambda\beta_{12}, & \omega\lambda\beta_{13}, & \ldots \\ \omega\lambda\beta_{21}, & a_2\lambda^2 + b_2, & \omega\lambda\beta_{23}, & \ldots \\ \omega\lambda\beta_{31}, & \omega\lambda\beta_{32}, & a_3\lambda^2 + b_3, \ldots \\ \ldots & \ldots & \ldots \end{vmatrix} = 0.$$

Since $\beta_{rs} = -\beta_{sr}$, this equation is unchanged when the sign of λ is changed. The equation is therefore an equation in λ^2, just as when the system is at rest. But the roots in λ^2 are not necessarily all real as they are for a system at rest; in general they will occur in pairs of the form $\lambda^2 = \rho \pm i\sigma$, and these will lead to roots for λ of the form

$$\lambda = \pm q \pm ip,$$

so that the complete time-factor for an oscillation is found to be of the form

$$A e^{qt} \cos{(pt - \epsilon)} + B e^{-qt} \cos{(pt - \eta)}.$$

If q is different from zero for any vibration, the amplitude of this vibration must continually increase owing to the presence of the factors $e^{\pm qt}$, and the system will be unstable. Thus the condition for stability is that q shall be zero for every vibration, and this in turn requires that all the roots in λ^2 shall be real and negative—a condition which is the same in form as that for the stability of a non-rotating system.

Stability and Instability.

181. Multiplying the general equations of motion (178·2) by $\dot{\theta}_1, \dot{\theta}_2, \ldots$ and adding corresponding sides, we obtain

$$\frac{d}{dt}(T_R + W - \tfrac{1}{2}\omega^2 I) = F_1\dot{\theta}_1 + F_2\dot{\theta}_2 + \ldots \quad \ldots\ldots(181\cdot1).$$

When $F_1 = F_2 = \ldots = 0$, so that no forces act except the couple G necessary to maintain the rotation constant, the equation has the integral

$$T_R + W - \tfrac{1}{2}\omega^2 I = \text{constant} \quad \ldots\ldots\ldots\ldots(181\cdot2).$$

For equilibrium we have seen that $W - \tfrac{1}{2}\omega^2 I$ must be stationary. Consider first what kind of equilibrium obtains when $W - \tfrac{1}{2}\omega^2 I$ is an absolute minimum. When any small displacement of the system occurs, $W - \tfrac{1}{2}\omega^2 I$ is necessarily increased, so that the constant value of $T_R + W - \tfrac{1}{2}\omega^2 I$ is greater than its value when at rest in the equilibrium configuration by a small amount c. Throughout the subsequent motion the value of T_R can never increase beyond c, so that the motion is absolutely stable. This argument cannot however be reversed, and the system is not necessarily unstable if $W - \tfrac{1}{2}\omega^2 I$ is not an absolute minimum.

Secular Instability.

182. Let us next suppose that F_1, F_2, ... do not vanish, but represent dissipative forces such as viscosity. The special characteristic of these forces is that they always act against the velocity in such a way as to compel the system to work against them; in brief, they resist the motion. Thus, whatever the algebraic signs of $\dot{\theta}_1$, $\dot{\theta}_2$, ... may be in equation (181·1), the signs of the terms $F_1\dot{\theta}_1$, $F_2\dot{\theta}_2$, ... on the right will always be negative unless they are zero. Thus the right-hand member of equation (181·1) will be negative except when the system is relatively at rest, so that $T_R + W - \frac{1}{2}\omega^2 I$ must decrease indefinitely. If $W - \frac{1}{2}\omega^2 I$ is an absolute minimum in the position of equilibrium, this condition can only be satisfied by T_R being reduced to zero, and the system coming to rest in its position of equilibrium. But if $W - \frac{1}{2}\omega^2 I$ is not an absolute minimum in the configuration of equilibrium, there will be a possible motion in which $W - \frac{1}{2}\omega^2 I$ continually decreases while T_R remains small at first, but may increase beyond limit when $W - \frac{1}{2}\omega^2 I$ is sufficiently decreased. The system is now in a restricted sense unstable.

Instability of the kind just discussed is called "secular instability." The conception of "secular instability" was first introduced by Thomson and Tait[*]. It has reference only to rotating systems or systems in a state of steady motion; for systems at rest secular stability becomes identical with ordinary stability.

So long as $W - \frac{1}{2}\omega^2 I$ remains an absolute minimum the system is stable both secularly and ordinarily. At the moment at which $W - \frac{1}{2}\omega^2 I$ ceases to be an absolute minimum, secular instability sets in, although the system may remain ordinarily stable. It follows that a system which is ordinarily stable may or may not be secularly stable, but a system which is ordinarily unstable is necessarily secularly unstable.

Further, we see that as the physical conditions of a system gradually change, secular instability necessarily sets in before ordinary stability. Thus, for problems of cosmogony it is secular instability alone which is of interest. A system never attains to a configuration in which ordinary instability comes into operation, since secular instability must always have previously intervened.

MASS ROTATING FREELY IN SPACE.

183. We have so far considered only the case of a mass constrained to rotate with a constant angular velocity ω. Schwarzschild[†] has shewn that the conditions of secular stability assume a somewhat different form for a mass rotating freely in space. In this case the rate of rotation is not constant,

---

[*] *Nat. Phil.*, 2nd ed., vol. I, p. 391.

[†] See Schwarzschild: "Die Poincaré'sche Theorie des Gleichgewichts." *Neue Annalen d. Sternwarte München*, III. (1897), p. 275, or *Inaugural Dissertation*, München (1896).

but changes as the moment of inertia of the mass changes; if the motion is referred to axes rotating with a uniform velocity, the rotation of the freely rotating mass may lag behind that of the axes and the relative co-ordinates x, y, z may increase without limit although the configuration remains stable. It is therefore important to express the conditions of stability in a form which does not involve the constancy of ω.

When the mass is rotating freely in space, $G = 0$ in equation (177·3) so that $\mathbf{M}$ is constant. The elimination of ω from equations (176·7) and (176·8) leads to

$$T = T_s + \frac{\mathbf{M}^2}{2I},$$

where

$$T_s = T_R - \frac{U^2}{2I}.$$

Using the values of T_R, U and I already obtained in equations (176·4) to (176·6), we find that

$$2I T_s = [\Sigma m_1 (x_1^2 + y_1^2)] [\Sigma m_2 (\dot{x}_2^2 + \dot{y}_2^2 + \dot{z}_2^2)]$$
$$- [\Sigma m_1 (x_1 \dot{y}_1 - y_1 \dot{x}_1)] [\Sigma m_2 (x_2 \dot{y}_2 - y_2 \dot{x}_2)]$$
$$= \Sigma \Sigma m_1 m_2 [(x_1 \dot{x}_2 + y_2 \dot{y}_1)^2 + (x_2 \dot{x}_1 + y_1 \dot{y}_2)^2$$
$$+ (x_1 \dot{y}_2 - x_2 \dot{y}_1)^2 + (x_2 \dot{y}_1 - x_1 \dot{y}_2)^2 + \dot{z}_1^2 (x_2^2 + y_2^2) + \dot{z}_2^2 (x_1^2 + y_1^2)].$$

This expression, being a sum of squares, is necessarily positive. Thus, since I is necessarily positive, T_s is always positive. Since I does not involve $\dot{x}$, $\dot{y}$, $\dot{z}$, T_s of course is quadratic in $\dot{x}$, $\dot{y}$, $\dot{z}$.

The equation of energy, $T + W = \text{cons.}$, now assumes the form

$$T_s + W + \frac{\mathbf{M}^2}{2I} = \text{cons.} \quad \dots\dots\dots\dots\dots(183\cdot1).$$

This is of the same form as the former equation (181·2), T_s replacing T_R and $W + \mathbf{M}^2/2I$ replacing $W - \frac{1}{2}\omega^2 I$. By the argument already used in § 181, it appears that configurations for which

$$W + \frac{\mathbf{M}^2}{2I} \quad \dots\dots\dots\dots\dots\dots(183\cdot2)$$

is an absolute minimum ($\mathbf{M}$ being kept constant) will be thoroughly stable, while configurations for which this expression is not an absolute minimum will be secularly unstable, and may or may not be ordinarily unstable.

184. The last two sections have shewn that as we pass along a linear series of configurations of equilibrium of a rotating system, starting from a part of the series which is known to be stable, the configurations will become secularly unstable as soon as

$$W - \tfrac{1}{2}\omega^2 I \qquad (\omega = \text{constant}) \dots\dots(184\cdot1)$$

or

$$W + \tfrac{1}{2}\mathbf{M}^2/I \qquad (\mathbf{M} = \text{constant}) \dots\dots(184\cdot2)$$

ceases to be an absolute minimum, the former expression referring to the case in which the mass is compelled by external forces to rotate at a constant rate ω, while the latter refers to the case in which the mass is rotating freely in space.

The theory of linear series and stability developed in §§ 170—173 will be exactly applicable to the problem of the secular stability of a rotating mass, provided W is replaced in the argument of those sections by the appropriate one of expressions (184·1) or (184·2). Secular stability is lost at a "turning point" or "point of bifurcation." At a turning point stability is lost entirely; at a point of bifurcation it may be lost or may be transferred to the branch series through the point according as the branch series turns downwards or upwards in the appropriate diagram. And, finally, to determine the positions of "points of bifurcation" and of "turning points" we need only express the appropriate one of the two above expressions as a sum of squares, and the "points of bifurcation" and "turning points" occur whenever one of the coefficients vanishes.

For instance, to determine at what stage a mass rotating freely in space becomes secularly unstable, we calculate $W + \frac{1}{2}\mathbf{M}^2/I$ for any configuration which is arrived at by a small displacement from a configuration of relative equilibrium, $\mathbf{M}$ being kept constant throughout the displacement, and express the excess over the equilibrium value as a sum of squares in the form

$$\delta\left(W + \tfrac{1}{2}\mathbf{M}^2/I\right) = \tfrac{1}{2}\left(b_1\theta_1^2 + b_2\theta_2^2 + \ldots + b_{n-1}\theta_{n-1}^2\right)\ldots\ldots(184\text{·}3).$$

If the configuration under consideration is secularly stable, $W + \frac{1}{2}\mathbf{M}^2/I$ must be an absolute minimum, so that $b_1, b_2, \ldots b_{n-1}$ must all be positive. As the various parameters which determine the physical state of the system change, the values of $b_1, b_2, \ldots$ change. Secular stability persists so long as $b_1, b_2, \ldots$ all remain positive, but is lost as soon as one of them becomes negative.

EXAMPLES OF SECULAR INSTABILITY.

185. Some interesting examples of secular instability have been worked out by Lamb[*].

As a first instance consider a spherical bowl, which is made to rotate with

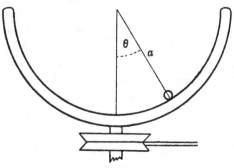

Fig. 21.

[*] *Proc. Roy. Soc.* A, lxxx. (1907), p. 170.

a constant angular velocity ω, while a small particle, such as a shot, is free to roll about inside. When the shot is in a position at an angle θ from the lowest point of the bowl, the potential energy W is equal to $-mga \cos \theta$. This has only one minimum, at $\theta = 0$, so that if the bowl is smooth, the particle will fall to the lowest point of the bowl and remain there; the rotation of the bowl is entirely irrelevant, and the equilibrium in this position is "ordinarily" stable.

To examine the question of secular instability, we have to consider the variations not of W but of $W - \frac{1}{2}\omega^2 I$ (formula 184·1). We have

$$W - \tfrac{1}{2}\omega^2 I = -mga \cos \theta - \tfrac{1}{2}m\omega^2 a^2 \sin^2 \theta \quad \ldots\ldots\ldots(185\cdot1).$$

Possible configurations of equilibrium occur when

$$\frac{\partial}{\partial \theta}\left(W - \tfrac{1}{2}\omega^2 I\right) = 0,$$

and this is when $\theta = 0$, or when

$$\cos \theta = \frac{g}{\omega^2 a} \quad \ldots\ldots\ldots\ldots\ldots\ldots\ldots\ldots(185\cdot2).$$

The root $\theta = 0$, of course, represents the position of equilibrium at the lowest point of the bowl, while the second root represents equilibrium in which the particle rotates with the bowl, at such a height up the face of the bowl that the tangential component of centrifugal force $m\omega^2 a \sin \theta \cos \theta$ is precisely equal to the tangential component of gravity $mg \sin \theta$. The second configuration only exists when ω^2 is greater than g/a. If ω^2 has a value less than g/a, equation (185·2) cannot be satisfied, since $\cos \theta$ is necessarily less than unity.

To examine the secular stability of the position of equilibrium at the lowest point of the bowl, we expand $W - \frac{1}{2}\omega^2 I$ in the neighbourhood of this position in powers of displacements from the position, and obtain

$$W - \tfrac{1}{2}\omega^2 I = -mga + \tfrac{1}{2}m\theta^2\left(ga - \omega^2 a^2\right).$$

Comparing this with the general formula (184·3), we see that the coefficient of stability corresponding to the single variable co-ordinate θ has the value $m\left(ga - \omega^2 a^2\right)$. This is positive so long as ω^2 is less than g/a, and for such values of ω^2 the equilibrium at the lowest point of the bowl is secularly stable. As soon as ω^2 exceeds this critical value, the coefficient of stability $m\left(ga - \omega^2 a^2\right)$ becomes negative, and equilibrium at the lowest point becomes secularly unstable. The particle now slips away from this configuration, the value of θ oscillating about a mean value which recedes further and further from the original equilibrium value $\theta = 0$.

The initial path of the particle is readily calculated. Let us take horizontal rectangular co-ordinates Ox, Oy passing through the lowest point of the bowl, and rotating with the bowl. Let us assume the frictional force acting on the particle to be $-k$ times the velocity of the particle relative to the bowl. Then, so long as x, y are small, the equations of motion of the particle are

$$\ddot{x} - 2\omega\dot{y} - \omega^2 x = -k\dot{x} - \frac{gx}{a},$$

$$\ddot{y} + 2\omega\dot{x} - \omega^2 y = -k\dot{y} - \frac{gy}{a}.$$

Multiplying corresponding sides by 1, i and adding, we obtain

$$\left[\frac{d^2}{dt^2} + (2i\omega + k)\frac{d}{dt} + \left(\frac{g}{a} - \omega^2\right)\right](x + iy) = 0,$$

of which the solution is

$$x + iy = A_1 e^{\lambda_1 t} + A_2 e^{\lambda_2 t},$$

where

$$\lambda_1 = -i\omega + i\left(\frac{g}{a}\right)^{\frac{1}{2}} - \tfrac{1}{2}k\left[1 - \omega\left(\frac{a}{g}\right)^{\frac{1}{2}}\right],$$

$$\lambda_2 = -i\omega - i\left(\frac{g}{a}\right)^{\frac{1}{2}} - \tfrac{1}{2}k\left[1 + \omega\left(\frac{a}{g}\right)^{\frac{1}{2}}\right].$$

Thus

$$x + iy = e^{-i\omega t}\left\{A_1 e^{i\left(\frac{g}{a}\right)^{\frac{1}{2}}t} e^{-\frac{1}{2}k\left[1-\omega\left(\frac{a}{g}\right)^{\frac{1}{2}}\right]t} + A_2 e^{-i\left(\frac{g}{a}\right)^{\frac{1}{2}}t} e^{-\frac{1}{2}k\left[1+\omega\left(\frac{a}{g}\right)^{\frac{1}{2}}\right]t}\right\}.$$

To refer to axes fixed in space instead of rotating with the bowl, we need merely multiply $x + iy$ by $e^{i\omega t}$. This neutralises the factor outside the curled brackets. We are left with two terms in A_1, A_2, representing circular motions of period $2\pi\left(\frac{a}{g}\right)^{\frac{1}{2}}$ in opposite directions in space, their amplitudes being proportional to $e^{-\frac{1}{2}k\left[1-\omega\left(\frac{a}{g}\right)^{\frac{1}{2}}\right]t}$ and to $e^{-\frac{1}{2}k\left[1+\omega\left(\frac{a}{g}\right)^{\frac{1}{2}}\right]t}$ respectively. The latter amplitude rapidly diminishes to zero; the former does the same if $\omega\left(\frac{a}{g}\right)^{\frac{1}{2}}$ is less than unity, but if $\omega\left(\frac{a}{g}\right)^{\frac{1}{2}}$ is greater than unity it increases indefinitely until x and y are no longer small, when our analysis no longer applies. Thus the initial path of the particle is a spiral of ever-increasing radius.

To examine the secular stability of the upper configuration of equilibrium given by $\cos\theta = g/\omega^2 a$, we take a new co-ordinate $z = a\cos\theta$, so that z is the distance of the particle below the centre of the bowl. Equation (185·1) now becomes

$$W - \tfrac{1}{2}\omega^2 I = -mgz - \tfrac{1}{2}m\omega^2 a^2 + \tfrac{1}{2}m\omega^2 z^2$$

$$= \tfrac{1}{2}m\omega^2\left(z - \frac{g}{\omega^2}\right)^2 - \tfrac{1}{2}m\left(\frac{g^2}{\omega^2} + \omega^2 a^2\right).$$

The coefficient of stability is thus seen to be $m\omega^2$ which is positive. Thus the configuration of equilibrium is secularly stable for all values of ω for which it exists, but it disappears altogether when $\omega^2 < g/a$. So long as $\omega^2 > g/a$, the particle can rest in an equilibrium which is secularly stable at the position given by $\cos\theta = g/\omega^2 a$.

We can represent the configurations we have been discussing in linear series, after the manner adopted in § 172. The system has only one co-ordinate θ, and ω^2 may be taken to be the varying parameter. If ω^2 is taken as ordinate and θ as abscissa, the diagram lies as in fig. 22.

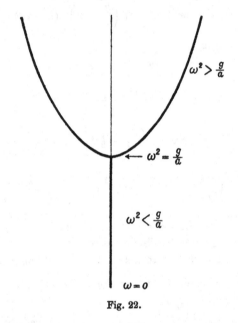

Fig. 22.

The configurations drawn thick are stable; there is an exchange of stability at the point of bifurcation at which $\omega^2 = g/a$, stability here passing from the series of configurations $\theta = 0$ to the series $\theta = \cos^{-1}(g/\omega^2 a)$. If the particle is placed at the bottom of the bowl while the bowl is at rest, and the bowl is then caused to rotate with a gradually increasing speed, the particle will lie at the bottom of the bowl until ω^2 reaches the value g/a. After this it will begin to ascend the side of the bowl, by spiral paths of ever increasing radius, ultimately clinging to a point whose elevation is given by $\cos\theta = g/\omega^2 a$.

186. As a second example, Lamb considers the problem of a pendulum symmetrical about a vertical axis which is hung by a Hooke's joint from a vertical shaft which can be made to rotate with any desired velocity. The position of the pendulum can be determined by two angles, θ, ϕ, where θ is the inclination to the vertical and ϕ is an azimuth. If A, B, C are the principal moments of inertia of the pendulum, and h is the depth of its centre of gravity below the point of suspension, we find

$$W - \tfrac{1}{2}\omega^2 I = - mgh \cos\theta - \tfrac{1}{2}\omega^2 (A \sin^2\theta + C \cos^2\theta).$$

On differentiating with respect to θ, there are found to be two linear series given by $\theta = 0$ and

$$\cos\theta = \frac{Mgh}{(A - C)\,\omega^2} \quad \ldots\ldots\ldots\ldots(186\cdot1).$$

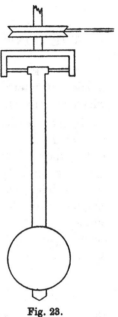

The first is found to be secularly stable until the second comes into existence, when the stability passes to it. The general form of the diagram is precisely that already shewn in fig. 22. Thus, if the shaft is set rotating at an increasing speed, the pendulum will hang vertically until the speed of rotation reaches the value $\omega^2 = Mgh/(A - C)$, after which it will rotate as a conical pendulum at an angle given by equation (186·1).

In the next chapter further instances of the principles just explained will be provided by problems which have reference to actual astronomical conditions.

Fig. 23.

CHAPTER VIII

THE CONFIGURATIONS OF ROTATING LIQUID MASSES

187. WE proceed to apply the general principles just explained to the problem of determining the shapes assumed by astronomical or other masses which are subjected to forces such as to make the spherical shape impossible.

In the present chapter we deal with the shapes of rotating bodies which are acted on by no forces but their own gravitation, confining ourselves to the simplest case in which the matter is supposed to be homogeneous and incompressible; the far more difficult problem presented by masses which are compressible, and so are not of uniform density, is reserved for the next chapter. Even the simplified problem has taxed the skill of some of the most eminent mathematicians. Among those who have specially worked at it may be mentioned Maclaurin, Jacobi, Lord Kelvin, Poincaré, Sir George Darwin and Liapounoff.

GENERAL EQUATIONS.

188. If a homogeneous mass of matter of uniform density ρ is rotating about the axis of z, the conditions that it shall rotate in relative equilibrium as a rigid body rotating with angular velocity ω are expressed by the system of equations

$$\frac{\partial p}{\partial x} = \rho \frac{\partial V}{\partial x} + \omega^2 \rho x \quad \dots\dots\dots\dots\dots\dots(188\cdot1),$$

$$\frac{\partial p}{\partial y} = \rho \frac{\partial V}{\partial y} + \omega^2 \rho y \quad \dots\dots\dots\dots\dots\dots(188\cdot2),$$

$$\frac{\partial p}{\partial z} = \rho \frac{\partial V}{\partial z} \quad \dots\dots\dots\dots\dots\dots(188\cdot3).$$

Here p is the pressure at the point x, y, z, while V is the gravitational potential. The equations are of course merely the ordinary hydrostatic equations we have used before, modified by the additional terms $\omega^2\rho x$, $\omega^2\rho y$, which represent the effect of centrifugal force. It is not necessary to specify the mechanism by which the pressure p is exerted. It is the total pressure, including pressure of radiation if this is appreciable, and there is no need to pecify it further.

Since ρ is supposed to be constant throughout the mass, the three equations have the common integral

$$\frac{p}{\rho} = V + \tfrac{1}{2}\omega^2(x^2 + y^2) + \text{cons.} \quad \dots\dots\dots\dots(188\cdot4).$$

The pressure p, however it arises, must be zero over any free surface of an astronomical body. If this condition is satisfied, equation (188·4) fixes the value of p at every point inside the body, and the three equations of equilibrium (188·1) to (188·3) are then automatically satisfied. Thus the necessary and sufficient condition that any configuration shall be a possible figure of equilibrium for a homogeneous mass of density ρ rotating with uniform angular velocity ω is that

$$V + \tfrac{1}{2}\omega^2 (x^2 + y^2) \quad \dotfill (188\cdot5)$$

shall be constant over its surface, where V is the gravitational potential of the body.

If $\omega = 0$, the condition becomes simply that V shall be constant over the surface, so that the configuration must be spherical.

If ω is small, although not actually zero, a spherical surface does not satisfy the condition, the term $\tfrac{1}{2}\omega^2 (x^2 + y^2)$ destroying the spherical symmetry. In this case, as we shall see almost immediately, the configuration is that of an oblate spheroid of small ellipticity—a shape similar to that of the earth.

ELLIPSOIDAL CONFIGURATIONS.

189. Instead of attacking the problem piecemeal, we shall proceed at once to consider under what conditions the ellipsoid

$$\frac{x^2}{a^2} + \frac{y^2}{b^2} + \frac{z^2}{c^2} = 1 \quad \dotfill (189\cdot1)$$

can be a configuration of equilibrium for a homogeneous rotating mass of density ρ. In dealing with points inside or outside the boundary, it is convenient to think of the boundary as constituting the surface $\lambda = 0$ in the family of confocal ellipsoids

$$\frac{x^2}{a^2 + \lambda} + \frac{y^2}{b^2 + \lambda} + \frac{z^2}{c^2 + \lambda} = 1 \quad \dotfill (189\cdot2).$$

When this is done, we shall write for brevity

$$a^2 + \lambda = A, \quad b^2 + \lambda = B, \quad c^2 + \lambda = C$$

$$[(a^2 + \lambda)(b^2 + \lambda)(c^2 + \lambda)]^{\tfrac{1}{2}} = (ABC)^{\tfrac{1}{2}} = \Delta \quad \dotfill (189\cdot3).$$

The potential V_0 of this mass at any external point x, y, z is, by a well-known formula,

$$V_0 = -\pi\gamma\rho abc \int_\lambda^\infty \left(\frac{x^2}{A} + \frac{y^2}{B} + \frac{z^2}{C} - 1 \right) \frac{d\lambda}{\Delta} \quad \dotfill (189\cdot4),$$

in which the lower limit of integration λ is the root of equation (189·2) and so is the parameter of the confocal ellipsoid on which the point x, y, z lies.

The potential V_i of the mass at an internal point x, y, z is

$$V_i = -\pi\gamma\rho abc \int_0^\infty \left(\frac{x^2}{A} + \frac{y^2}{B} + \frac{z^2}{C} - 1 \right) \frac{d\lambda}{\Delta} \quad \dotfill (189\cdot5),$$

and so is a quadratic function of x, y, z. If we write

$$\int_0^\infty \frac{d\lambda}{\Delta} = J,$$

and more generally put

$$\int_0^\infty \frac{d\lambda}{A^m B^n C^p \Delta} = J_{A^m B^n C^p} \qquad \text{...............}(189\cdot6),$$

this equation assumes the form

$$V_i = -\pi\gamma\rho abc\,(J_A x^2 + J_B y^2 + J_C z^2 - J) \text{...........}(189\cdot7).$$

It is easily verified that

$$J_A + J_B + J_C = \frac{2}{abc} \qquad \text{...................}(189\cdot8),$$

as can also be seen from the circumstance that $\nabla^2 V_i$ must be equal to $-4\pi\gamma\rho$. We may also note the formulae

$$J_B - J_A = (a^2 - b^2)\,J_{AB} \qquad \text{.................}(189\cdot91),$$

$$J_{A^m B^{n+1} C^p} - J_{A^{m+1} B^n C^p} = (a^2 - b^2)\,J_{A^{m+1} B^{n+1} C^p} \text{......}(189\cdot92),$$

$$J_{A^n B} + J_{A^n C} + (2n+1)\,J_{A^{n+1}} = \frac{2}{a^{2n} abc} \qquad \text{.........}(189\cdot93),$$

all of which are easily verified by algebraic transformations.

190. The necessary and sufficient condition that the standard ellipsoid (189·1) shall be a figure of equilibrium for a homogeneous mass of density ρ freely rotating with angular velocity ω is that

$$V_i + \tfrac{1}{2}\omega^2\,(x^2 + y^2) \qquad \text{.....................}(190\cdot1)$$

shall be constant over the boundary, V_i being given by equation (189·7).

Now consider the function

$$V_i + \tfrac{1}{2}\omega^2\,(x^2 + y^2) + \theta\pi\gamma\rho abc\left(\frac{x^2}{a^2} + \frac{y^2}{b^2} + \frac{z^2}{c^2} - 1\right) \quad \text{......}(190\cdot2),$$

where θ is a constant, as yet undetermined. Operating with ∇^2, we find that this function will be a spherical harmonic if

$$-4\pi\gamma\rho + 2\omega^2 + 2\theta\pi\gamma\rho abc\left(\frac{1}{a^2} + \frac{1}{b^2} + \frac{1}{c^2}\right) = 0,$$

a condition which can be satisfied by assigning to θ the value

$$\theta = \frac{2\left(1 - \dfrac{\omega^2}{2\pi\gamma\rho}\right)}{abc\left(\dfrac{1}{a^2} + \dfrac{1}{b^2} + \dfrac{1}{c^2}\right)} \qquad \text{...................}(190\cdot3).$$

With this value for θ, expression (190·2) becomes harmonic. The necessary and sufficient condition that the standard ellipsoid (189·1) shall be a figure of equilibrium is that this function shall have a constant value over the boundary. Since the function is harmonic, this is equivalent to the condition that the function shall have a constant value throughout the interior of the ellipsoid. We must accordingly have

$$-\pi\gamma\rho abc\,(J_A x^2 + J_B y^2 + J_C z^2 - J) + \tfrac{1}{2}\omega^2\,(x^2 + y^2)$$

$$+ \theta\pi\gamma\rho abc\left(\frac{x^2}{a^2} + \frac{y^2}{b^2} + \frac{z^2}{c^2} - 1\right) = \text{cons.},$$

where θ is given by equation (190·3). Equating coefficients of x^2, y^2 and z^2, this equation is seen to be equivalent to the three separate equations

$$J_A - \frac{\omega^2}{2\pi\gamma\rho abc} = \frac{\theta}{a^2} \quad \dots\dots\dots\dots\dots(190\text{·}4),$$

$$J_B - \frac{\omega^2}{2\pi\gamma\rho abc} = \frac{\theta}{b^2} \quad \dots\dots\dots\dots\dots(190\text{·}5),$$

$$J_C \qquad\quad = \frac{\theta}{c^2} \quad \dots\dots\dots\dots\dots(190\text{·}6).$$

By addition of corresponding sides we again obtain equation (190·3) which gives the value of θ. Thus the three equations just written down contain within themselves the necessary and sufficient conditions that the ellipsoid

$$\frac{x^2}{a^2} + \frac{y^2}{b^2} + \frac{z^2}{c^2} = 1,$$

shall be a figure of equilibrium under the rotation ω.

191. Subtracting corresponding sides of the first two equations, we obtain

$$(a^2 - b^2)\, J_{AB} = (a^2 - b^2)\frac{\theta}{a^2 b^2},$$

and the elimination of θ between this and the third equation gives

$$(a^2 - b^2)\,[a^2 b^2 J_{AB} - c^2 J_C] = 0.$$

Thus the necessary equations can be satisfied in two ways, either by taking

$$a^2 = b^2 \qquad \dots\dots\dots\dots\dots\dots(191\text{·}1),$$

or by taking

$$a^2 b^2 J_{AB} = c^2 J_C \qquad \dots\dots\dots\dots\dots\dots(191\text{·}2).$$

The first condition requires that the body shall be spheroidal, the cross sections by the axis of rotation (Oz) being circles. If, however, the second condition is satisfied, a and b are not in general equal (as we shall shortly see), and the configurations are ellipsoids with three unequal axes.

Clearly the two equations (191·1) and (191·2) represent two linear series of equilibrium configurations, which are spheroidal and ellipsoidal respectively. The configurations which form the first series are commonly known as Maclaurin's spheroids; those which form the second as Jacobi's ellipsoids, their existence having first been demonstrated by Jacobi[*] in 1834. We shall examine these two linear series in turn.

Maclaurin's Spheroids.

192. When $a = b$ equations (190·4) and (190·5) become identical, and the three equations (190·4) to (190·6) reduce to two:

$$J_A - \frac{\omega^2}{2\pi\gamma\rho a^2 c} = \frac{\theta}{a^2} \quad \dots\dots\dots\dots(192\text{·}1),$$

$$J_C \qquad\quad = \frac{\theta}{c^2} \quad \dots\dots\dots\dots(192\text{·}2).$$

---

[*] *Pogg. Ann.* xxxiii. (1834), p. 229.

Elimination of θ gives

$$\frac{\omega^2}{2\pi\gamma\rho} = c\,(a^2 - c^2)\int_0^\infty \frac{\lambda\,d\lambda}{\Delta AC} \quad\ldots\ldots\ldots\ldots(192\cdot3).$$

Since ω^2 must be positive, a^2 must be greater than c^2, so that the spheroids are all oblate. When $a = b$, the integral on the right admits of evaluation in finite terms, the equation then assuming the form

$$\frac{\omega^2}{2\pi\gamma\rho} = \frac{3 - 2e^2}{e^3}(1 - e^2)^{\frac{1}{2}}\sin^{-1}e - 3\left(\frac{1}{e^2} - 1\right)\quad\ldots\ldots\ldots(192\cdot4),$$

where e is the eccentricity of the spheroid, defined by

$$e^2 = \frac{a^2 - c^2}{a^2}.$$

We notice that the eccentricity depends only on the ratio $\omega^2/2\pi\gamma\rho$. Further, when $\omega^2/2\pi\gamma\rho$ is given, the eccentricity is fixed, but not the size of the spheroid; a spheroid of any size is a possible figure of equilibrium if its eccentricity has the value given by equation (192·4). The following table of values of $\omega^2/2\pi\gamma\rho$ and e is compiled from values given by Lamb*, Darwin† and Thomson and Tait‡.

Table XVI. *Maclaurin's Spheroids.*

e	a/r_0	c/r_0	$\dfrac{\omega^2}{2\pi\gamma\rho}$	Ang. Momentum$/M^{\frac{3}{2}}r_0^{\frac{1}{2}}$
0	1·0000	1·0000	0	0
·1	1·0016	·9967	·0027	·0255
·2	1·0068	·9865	·0107	·0514
·3	1·0159	·9691	·0243	·0787
·4	1·0295	·9435	·0436	·1085
·5	1·0491	·9068	·0690	·1417
·6	1·0772	·8618	·1007	·1804
·7	1·1188	·7990	·1387	·2283
·8	1·1856	·7114	·1816	·2934
·81267	1·1972	·6977	·18712	·30375
·9	1·3189	·5749	·2203	·4000
·91	1·341	·5560	·2225	·4156
·92	1·367	·5355	·2241	·4330
·93	1·396	·5131	·2247	·4525
·94	1·431	·4883	·2239	·4748
·95	1·474	·4603	·2213	·5008
·96	1·529	·4280	·2160	·5319
·97	1·602	·3895	·2063	·5692
·98	1·713	·3409	·1890	·6249
·99	1·921	·2710	·1551	·7121
1·00	∞	0	0	∞

The values of e range from 0 to 1·00, so that every oblate spheroid is a configuration of equilibrium for some value of $\omega^2/2\pi\gamma\rho$. On the other hand, the values of $\omega^2/2\pi\gamma\rho$ never exceed 0·225, so that there are no spheroidal configurations of equilibrium when $\omega^2 > 0\cdot225 \times 2\pi\gamma\rho$.

* *Hydrodynamics* (4th Ed.), p. 673.
† *Proc. Roy. Soc.* XLI. (1887), p. 319, or *Coll. Works*, III. p. 119.   ‡ *Nat. Phil.* § 772.

Jacobi's Ellipsoids.

193. Let us now examine the configurations which constitute the second linear series, determined by equation (191·2). In these configurations a is no longer equal to b, and the integrals do not admit of integration in finite terms. They have been discussed by C. O. Meyer*, and also reduced to elliptic integrals and treated numerically by Darwin†.

The ellipsoids are found to form one single continuous series. The maximum value of $\omega^2/2\pi\gamma\rho$ is found to be 0·18712; this occurs for the particular ellipsoid which $a = b = 1·7161\ c$. This configuration is also of course a Maclaurin spheroid, and so forms a point of bifurcation on this latter series. It is the configuration printed in heavy type in the table above.

As we pass along the Jacobian series, the ratio a/b may be supposed to vary continuously from 0 to ∞, and the point of bifurcation occurs when $a = b$. The two halves of the series are, however, exactly similar, either one changing into the other on interchanging a and b, so that we may legitimately confine our attention to one half, let us say that for which $a > b$. We now regard the series of Jacobian ellipsoids as starting at the value $a = b$ (the point of bifurcation), and the ratio a/b continually increases from 1 to ∞ as we pass along the series. The following numerical values are given by Darwin‡:

Table XVII. *Jacobian Ellipsoids.*

$\dfrac{a}{r_0}$	$\dfrac{b}{r_0}$	$\dfrac{c}{r_0}$	$\dfrac{\omega^2}{2\pi\gamma\rho}$	Angular Momentum
1·1972	1·1972	·6977	·18712	·30375
1·216	1·179	·697	·1870	·304
1·279	1·123	·696	·186	·306
1·3831	1·0454	·6916	·1812	·3134
1·6007	·9235	·6765	·1659	·3407
1·88583	**·81498**	**·65066**	**·14200**	**·3898**
1·899	·8111	·6494	·1409	·3920
2·346	·7019	·6072	·1072	·4809
3·1294	·5881	·5434	·0661	·6387
5·0406	·4516	·4393	·0259	1·0087
∞	0	0	0	∞

194. We have now mapped out the various configurations on the two linear series of ellipsoidal configurations—the Maclaurin spheroids and the Jacobian ellipsoids—and the stability of these configurations can be investigated by the methods already explained.

* *Crelle's Journ.* xxxiv. (1842).
† *Proc. Roy. Soc.* xli. (1887), p. 319, or *Coll. Works*, iii. p. 118.
‡ *Coll. Works*, iii. p. 130.

STABILITY.

Stability when angular velocity is increasing, as in Plateau's experiments.

195. In 1842 Plateau devised an experiment in which he attempted to observe directly the sequence of configurations in a rotating mass of fluid, with a view to testing whether they were at all similar to those assumed by Laplace as the basis of his nebular hypothesis. Plateau mixed water and alcohol until they were of just the right density to float freely in olive oil. A globule of this mixture was then set in rotation in the oil by spinning a wire through its centre, the globule being kept in position on the wire by a disc round which it clustered. As the speed of rotation increased, the globule was observed to flatten itself more and more until finally a dimple formed at the centre, and the globule detached itself from the disc in the form of a perfect ring. The conditions of this experiment were very different from those which prevail in astronomical masses, since the globule was not held

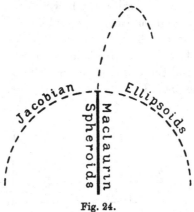

Fig. 24.

together by its own gravitational attractions, but by surface tensions, and was not made to shrink while moving freely in space, but had its angular velocity mechanically increased by the medium of the wire and disc.

As a problem suggested by Plateau's experiments, let us examine what would be the sequence of configurations assumed by a mass of gravitating matter which had its angular velocity continually increased by some mechanical means such as the spinning at an ever increasing rate of a pole through its centre.

The possible configurations of equilibrium are those already discussed; so long as the mass is constrained to remain ellipsoidal, they consist of Maclaurin spheroids and Jacobian ellipsoids. To examine the stability of these figures we draw a diagram in which the angular velocity ω is the vertical co-ordinate as in fig. 24, using the values of ω provided by Tables XVI and XVII.

The stability of the Maclaurin spheroids is seen to be terminated by the occurrence of a point of bifurcation when $\omega^2/2\pi\gamma\rho = 0\cdot18712$, the branch series through this point being the Jacobian ellipsoids. As this latter series turns downwards from the point of bifurcation, the Jacobian ellipsoids also are unstable. Thus there are no stable configurations of equilibrium for a rotation greater than that given by $\omega^2/2\pi\gamma\rho = 0\cdot18712$. When the rotation exceeds this amount, the problem ceases to be a statical one and becomes a dynamical one; here we shall not attempt to follow it.

Stability when the angular momentum is increased.

196. The problem just considered is of interest as illustrating the theory of points of bifurcation, but it entirely fails to represent the conditions which prevail in astronomical events. To represent natural conditions the mass must be supposed to rotate freely in space so that its angular momentum

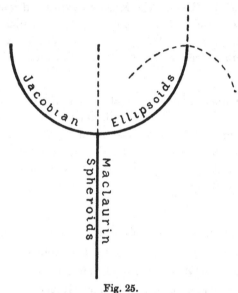

Fig. 25.

remains constant. If the rotating body shrinks, its density will increase, and this may or may not result in an increase of angular velocity. To study the problem by the most direct method, we should have to look for series of configurations of constant angular momentum and varying density. It is, however, a convenience to suppose that the density remains constant while the angular momentum increases, and it is easily seen that this leads to exactly the same mathematical problem. We accordingly proceed to study the stability of the Maclaurin and Jacobian series, supposing ρ to remain constant while the angular momentum is made continually to increase.

In this problem the angular momentum is given in the last columns of Tables XVI and XVII (pp. 210, 211), and in a diagram in which the angular momentum is taken for ordinate, the series will be found to be as in fig. 25. Clearly the Maclaurin spheroids will be stable up to the point at which they meet the Jacobian ellipsoids. At this point of bifurcation they lose their stability, and since the series of Jacobian ellipsoids turns upward at this point, it follows that stability passes to them.

If the mass is constrained to remain ellipsoidal there is no further point of bifurcation on the Jacobian series, and, as the angular momentum continually increases along this series, it follows that all configurations on it are stable. But it will be found later (§ 203) that when the constraint to remain ellipsoidal is removed, the Jacobian series loses its stability at a certain stage by meeting a series of non-ellipsoidal (pear-shaped) configurations. This has been anticipated in fig. 25.

We have of course been concerned only with secular stability or instability. The conditions of ordinary stability are entirely different; for instance, as has been shewn by G. H. Bryan *, the Maclaurin spheroid remains "ordinarily" stable until its eccentricity is $e = 0{\cdot}9529$. But its secular stability has disappeared long before this eccentricity is reached, so that, as has already been noticed to be generally true, it is secular stability alone which is of interest in cosmogony.

197. Before we proceed further into abstract mathematical discussion, let us attempt to understand the physical significance of the results so far obtained.

Our investigation of the configurations of equilibrium of a rotating incompressible mass has been based upon the equations of equilibrium of the separate elements of the mass, but the same results could have been obtained from the general equations

$$\frac{\partial}{\partial \theta_s}(W - \tfrac{1}{2}\omega^2 I)$$

of § 179. These again may be put in either of the forms

$$\delta(W - \tfrac{1}{2}\omega^2 I) = 0 \quad (\omega = \text{cons.}) \dots\dots\dots\dots(197{\cdot}1),$$
$$\delta(W + \tfrac{1}{2}\mathbf{M}^2/I) = 0 \quad (\mathbf{M} = \text{cons.}) \dots\dots\dots\dots(197{\cdot}2),$$

and either equation would have given precisely the same configurations of equilibrium as those already obtained.

We have found, for instance, that an incompressible mass in slow rotation will assume a spheroidal shape. Thus $W - \tfrac{1}{2}\omega^2 I$ must have a stationary value in this configuration, and since the configuration has been found to be stable, the stationary value is a minimum. Thus any displacement which causes a decrease in W must necessarily cause a decrease of at least equal amount in $\tfrac{1}{2}\omega^2 I$. It is easy to find displacements which decrease W, as, for instance, by displacing the surface until it becomes more spherical, but this will increase $W - \tfrac{1}{2}\omega^2 I$, and the displaced mass will oscillate stably about its position of equilibrium.

Any vibration of an incompressible mass may be regarded loosely as a system of surface waves, and the distance from one point of zero displacement

* *Phil. Trans.* 190 A (1887), p. 187.

of the surface to the next may be regarded as a sort of wave-length of the vibration. The stability or instability of a vibration depends on which is the greater—the gain in $\frac{1}{2}\omega^2 I$ or the gain in gravitational energy when the vibration takes place. But as between a vibration of great wave-length and one of short wave-length there is this important distinction: for vibrations of equal amplitude the gravitational disturbance caused by the vibration of great wave-length is much greater than that caused by the vibration of short wave-length, since the elements of the latter very largely neutralise one another. Thus the change in gravitational energy is enormously the greater for vibrations of great wave-length, while it is easily seen that the changes in $\frac{1}{2}\omega^2 I$ are approximately the same. It follows that when a rotating mass first becomes unstable, instability will set in through a vibration of the greatest possible wave-length.

Any vibration or disturbance on the surface of an ellipsoid may be analysed into series of ellipsoidal harmonics. The harmonics of order n have wave-lengths, in the sense loosely defined above, of the order of $2\pi a/n$, where a is the semi-major-axis of the ellipsoid. Thus the vibrations of greatest wave-length are those in which the surface-displacements are proportional to harmonics of the lowest orders. Thus we should expect a rotating incompressible mass first to become unstable through harmonic displacements of the lowest order which is physically possible.

Displacements of order $n = 1$ are not physically possible, for they displace the centre of gravity of the mass off the axis of rotation and so are prohibited from the outset.

The lowest order of harmonics available, then, is the second. The effect of superposing a second order displacement onto a spheroid is to transform it into an ellipsoid of three unequal axes. Hence we should anticipate that if the series of Maclaurin spheroids ever becomes unstable, it will be by deformation into an ellipsoid. This is precisely what we have found, the ellipsoidal figures in question being the Jacobian ellipsoids.

When we apply the same train of thought to the Jacobian ellipsoids, we find that there must be a possibility of these becoming unstable in turn through the deformation of greatest wave-length which is possible for them. This is represented by a third harmonic, since a second harmonic deformation merely changes an ellipsoid into another ellipsoid and so only represents a step along the Jacobian series of ellipsoids. Our analysis so far has been definitely restricted to the consideration of configurations which cannot leave the ellipsoidal shape, and so has not been capable of disclosing an instability which enters through a third harmonic displacement.

Stability of the Jacobian Ellipsoids.

198. The problem of determining the stability of the Jacobian ellipsoids was undertaken by Poincaré*, who proved rigorously that instability first

---

* *Acta Math.* VII. (1885), p. 259, and *Phil. Trans.* 198 A (1902), p. 333.

entered through a displacement which was everywhere proportional to the third zonal harmonic. The onset of instability is thus marked by a point of bifurcation, the second series through which is a series of the pear-shaped figures obtained by imposing a third zonal harmonic displacement upon an ellipsoid. On following the Jacobian series beyond this point of bifurcation, Poincaré found that a whole succession of further points of bifurcation were encountered, representing instabilities which entered respectively through harmonics of orders 4, 5, 6, ... All this is in entire agreement with the general principles just discussed.

Poincaré discussed the problem of stability in terms of ellipsoidal harmonic analysis, and in this he was followed by Darwin* and Schwarzschild†. I have found‡, however, that both this and more difficult questions to follow are more easily treated by using ordinary Cartesian co-ordinates x, y, z; we shall accordingly use these in the present discussion of the problem.

PEAR-SHAPED CONFIGURATIONS OF EQUILIBRIUM.

199. In discussing ellipsoidal configurations, it proved convenient to regard the standard ellipsoid

$$f_0 \equiv \frac{x^2}{a^2} + \frac{y^2}{b^2} + \frac{z^2}{c^2} - 1 = 0 \dots\dots\dots\dots(199\text{·}1),$$

which formed the boundary of the rotating mass, as the special member $\lambda = 0$ of the family of surfaces

$$f \equiv \frac{x^2}{a^2+\lambda} + \frac{y^2}{b^2+\lambda} + \frac{z^2}{c^2+\lambda} - 1 = 0 \dots\dots\dots(199\text{·}2).$$

We have now to consider surfaces which are not themselves ellipsoidal but are derived from ellipsoids by distortion. We shall take the boundary of the distorted surface to be

$$\frac{x^2}{a^2} + \frac{y^2}{b^2} + \frac{z^2}{c^2} - 1 + eP_0 = 0 \dots\dots\dots\dots(199\text{·}3),$$

where e is a small parameter which measures the amplitude of the distortion, while P_0 determines its distribution, and we shall find it convenient to regard this surface as the particular surface $\lambda = 0$ of the family of surfaces

$$\frac{x^2}{a^2+\lambda} + \frac{y^2}{b^2+\lambda} + \frac{z^2}{c^2+\lambda} - 1 + eP = 0 \dots\dots\dots(199\text{·}4).$$

To deduce the value of P from P_0, we write P_0 in the form

$$P_0 = F\left(\frac{x}{a^2}, \frac{y}{b^2}, \frac{z}{c^2}, a^2, b^2, c^2\right)$$

and the value of P is taken to be

$$P = F\left(\frac{x}{a^2+\lambda}, \frac{y}{b^2+\lambda}, \frac{z}{c^2+\lambda}, a^2, b^2, c^2\right).$$

---

* *Coll. Works*, III. Papers 10, 11, 12 and 13.

† *Neue Ann. d. Sternwarte*, München, III. (1897), p. 275, and *Inaug. Dissert.* München. 1896.

‡ *Phil. Trans. Roy. Soc.* 215 A (1915), p. 27.

Write ξ, η, ζ for $\dfrac{x}{a^2+\lambda}$, $\dfrac{y}{b^2+\lambda}$, $\dfrac{z}{c^2+\lambda}$, so that P is a function of ξ, η, and ζ, and introduce a differential operator D defined by

$$D=\left(\frac{1}{a^2}-\frac{1}{a^2+\lambda}\right)\frac{\partial^2}{\partial\xi^2}+\left(\frac{1}{b^2}-\frac{1}{b^2+\lambda}\right)\frac{\partial^2}{\partial\eta^2}+\left(\frac{1}{c^2}-\frac{1}{c^2+\lambda}\right)\frac{\partial^2}{\partial\zeta^2}.$$

Then I have shewn* that the gravitational potential of a uniform mass of density ρ bounded by the surface

$$\frac{x^2}{a^2}+\frac{y^2}{b^2}+\frac{z^2}{c^2}-1+eP_0=0 \quad\dots\dots\dots\dots\dots\text{(199·5)}$$

is

$$V=-\pi\gamma\rho abc\int_0^\infty \frac{f+e\phi}{\Delta}\,d\lambda \quad\dots\dots\dots\dots\text{(199·6)}$$

where $\phi = P-(\tfrac14 f)\,DP+\dfrac{1}{2^2}(\tfrac14 f)^2 D^2P-\dfrac{1}{2^2.3^2}(\tfrac14 f)^3 D^3P+\dots$

$$-\tfrac18 e[DP^2-(\tfrac18 f)\,D^2P^2+\tfrac13(\tfrac18 f)^2 D^3P^2-\tfrac{1}{18}(\tfrac18 f)^3 D^4P^2+\dots]$$

$$+\tfrac{1}{192}e^2[D^2P^3-\tfrac{1}{12}fD^3P^3+\tfrac{1}{384}f^2D^4P^3-\dots]+\dots \quad\dots\dots\dots\text{(199·7)}.$$

Here V gives the potential either at an internal point or at the boundary of the mass, Δ as before being given by

$$\Delta^2=(a^2+\lambda)(b^2+\lambda)(c^2+\lambda).$$

200. For the surface (199·5) to be a possible boundary for a mass of rotating liquid, it is necessary that

$$V+\tfrac12\omega^2(x^2+y^2)$$

should be constant over this surface.

If the term eP_0 represents a third zonal harmonic distortion, we may assume

$$P=\xi(\alpha\xi^2+\beta\eta^2+\gamma\zeta^2+\kappa)$$

so that

$$P_0=\frac{x}{a^2}\left(\frac{\alpha x^2}{a^4}+\frac{\beta y^2}{b^4}+\frac{\gamma z^2}{c^4}+\kappa\right) \quad\dots\dots\dots\dots\text{(200·1)}.$$

For small displacements from the ellipsoidal configuration we may neglect e^2, and obtain from equation (199·7)

$$\phi=\frac{x}{A}\left\{\alpha\frac{x^2}{A^2}+\beta\frac{y^2}{B^2}+\gamma\frac{z^2}{C^2}\right\}$$

$$-\frac12\left(\frac{x^2}{A}+\frac{y^2}{B}+\frac{z^2}{C}-1\right)\left(\frac{3\alpha\lambda}{a^2 A}+\frac{\beta\lambda}{b^2 B}+\frac{\gamma\lambda}{c^2 C}\right)\frac{x}{A}+\kappa\frac{x}{A} \quad\dots\dots\text{(200·2)},$$

so that we may write

$$\int_0^\infty \frac{\phi}{\Delta}\,d\lambda=x(a_1 x^2+a_2 y^2+a_3 z^2+a_4) \quad\dots\dots\dots\text{(200·3)}.$$

Thus the distortion introduces additional terms of degrees 3 and 1 into the potential.

---

* *Problems of Cosmogony and Stellar Dynamics*, chap. IV. or *Phil. Trans.* 217 A (1916), p. 7.

It is now clear that for $V + \frac{1}{2}\omega^2(x^2 + y^2)$ to be constant over the surface of the mass, we must have

$$V + \tfrac{1}{2}\omega^2(x^2 + y^2) = -\pi\gamma\rho\,abc\,\theta\left(\frac{x^2}{a^2} + \frac{y^2}{b^2} + \frac{z^2}{c^2} - 1 + eP_0\right) \quad \dots(200\text{·}4)$$

at all points of the boundary, θ being at our disposal. Equating coefficients we obtain

$$\left.\begin{aligned}
J_A - \frac{\omega^2}{2\pi\gamma\rho\,abc} &= \frac{\theta}{a^2} \\[1mm]
J_B - \frac{\omega^2}{2\pi\gamma\rho\,abc} &= \frac{\theta}{b^2} \\[1mm]
J_O &= \frac{\theta}{c^2}
\end{aligned}\right\} \quad \dots\dots\dots\dots\dots\dots(200\text{·}5),$$

which are identical with the equations already obtained in § 190, together with

$$\int_0^\infty \frac{\phi}{\Delta}\,d\lambda = \theta P_0.$$

On inserting the value of ϕ in this last equation and equating coefficients of x^3, xy^2, xz^2 and x, we obtain

$$\left.\begin{aligned}
a_1 &= \theta\,\frac{\alpha}{a^6} \\[1mm]
a_2 &= \theta\,\frac{\beta}{a^2 b^4} \\[1mm]
a_3 &= \theta\,\frac{\gamma}{a^2 c^4} \\[1mm]
a_4 &= \theta\,\frac{\kappa}{a^2}
\end{aligned}\right\} \quad \dots\dots\dots\dots\dots\dots(200\text{·}6).$$

If we introduce new quantities c_1, c_2, c_3 defined by

$$c_1 = \int_0^\infty \frac{\lambda\,d\lambda}{\Delta A BC}, \quad c_2 = \int_0^\infty \frac{\lambda\,d\lambda}{\Delta A^2 C}, \quad c_3 = \int_0^\infty \frac{\lambda\,d\lambda}{\Delta A^2 B}$$

then equations (200·6) are found to assume the form

$$\left.\begin{aligned}
\frac{\alpha}{2a^2}(c_2 + c_3) - \frac{\beta}{2b^2}c_3 - \frac{\gamma}{2c^2}c_2 &= \theta\,\frac{\alpha}{a^6} \\[1mm]
-\frac{3\alpha}{2a^2}c_3 + \frac{\beta}{2b^2}(3c_2 + c_1) - \frac{\gamma}{2c^2}c_1 &= \theta\,\frac{\beta}{a^2 b^4} \\[1mm]
-\frac{3\alpha}{2a^2}c_2 - \frac{\beta}{2b^2}c_1 + \frac{\gamma}{2c^2}(3c_2 + c_1) &= \theta\,\frac{\gamma}{a^2 c^4}
\end{aligned}\right\} \quad \dots\dots(200\text{·}7),$$

$$\frac{3\alpha}{2a^2}\int_0^\infty \frac{\lambda\,d\lambda}{\Delta A^2} + \frac{\beta}{2b^2}\int_0^\infty \frac{\lambda\,d\lambda}{\Delta AB} + \frac{\gamma}{2c^2}\int_0^\infty \frac{\lambda\,d\lambda}{\Delta AC} + \kappa\int_0^\infty \frac{d\lambda}{\Delta A} = \theta\,\frac{\kappa}{a^2} \quad (200\text{·}8).$$

201. The elimination of α/a^2, β/b^2 and γ/c^2 from the three equations (200·7) gives

$$a^2 b^2 c^2\left(\frac{3}{a^2} + \frac{1}{b^2} + \frac{1}{c^2}\right)(c_1 c_2 + c_1 c_3 + 3c_2 c_3)$$

$$-\frac{2\theta}{a^2}\left[c_1(b^2 + c^2) + c_2(3a^2 + c^2) + c_3(3a^2 + b^2)\right] + \left(\frac{2\theta}{a^2}\right)^2 = 0 \quad (201\text{·}1).$$

Adding corresponding sides of the three equations (200·6), we again obtain the value for θ already obtained in equation (190·3), namely,

$$\theta = \frac{2\left(1 - \frac{\omega^2}{2\pi\gamma\rho}\right)}{abc\left(\frac{1}{a^2} + \frac{1}{b^2} + \frac{1}{c^2}\right)} \qquad\qquad\qquad (201\cdot2),$$

and on inserting this value for θ, equation (201·1) becomes an equation in a, b, c alone. It is this equation which determines the first point of bifurcation on the Jacobian series of ellipsoids.

The numerical solution of this equation is an arduous piece of work, for the integrals c_1, c_2, c_3 cannot be evaluated in finite terms, and the solution of the equation can only proceed by trial and error. The necessary computations were carried through by Darwin*, who obtained the solution

$$\frac{a}{r_0} = 1\cdot885827, \quad \frac{b}{r_0} = 0\cdot814975, \quad \frac{c}{r_0} = 0\cdot650659 \qquad (201\cdot3)$$

where $r_0{}^3 = abc$, so that r_0 is the radius of the sphere of equivalent volume. The corresponding value of ω is found to be given by

$$\frac{\omega^2}{2\pi\gamma\rho} = 0\cdot141999 \qquad\qquad\qquad (201\cdot4).$$

Darwin's calculations were based on the equations he obtained by harmonic analysis, not on equations (201·1) and (201·2), but I have verified† that his solution satisfies these equations.

The shape of the critical ellipsoid defined by equations (201·3) is shewn

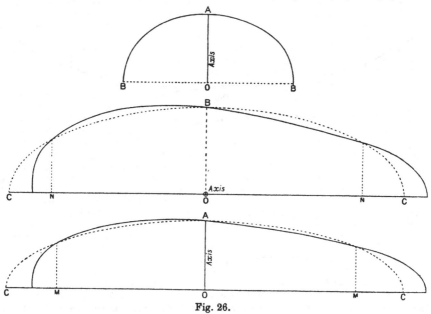

Fig. 26.

* *Coll. Works*, III. p. 288; or *Phil. Trans.* 198 A (1901), p. 301.   † *Phil. Trans.* 215 A (1915), p. 53.

in fig. 26. The dotted lines represent the surface distorted by the third zonal harmonic displacement through which instability first sets in.

202. Proceeding along the series of Jacobian ellipsoids from its junction with the Maclaurin spheroids, the configurations remain stable up to the critical ellipsoid defined by equations (201·3). Through this point two linear series pass, namely, the series of Jacobian ellipsoids and the series of pear-shaped configurations which are specified, in the immediate neighbourhood of the Jacobian series, by giving small values to e in equation (199·5). This specification of course breaks down as soon as the values of e^2 become appreciable.

It would be a feasible, although extremely laborious task, to use the formulae already given to calculate the whole series of pear-shaped configurations. I have calculated* these as far as terms in e^3, but find no new features are introduced beyond those shewn in fig. 30, until terms in e^4 and higher powers of e become appreciable and the calculation fails.

The corresponding problem in two dimensions—the determination of the analogous figures for infinitely long cylinders rotating under no forces beyond their own gravitational attractions—is incomparably simpler, because the formula for the gravitational potential of a two-dimensional cylinder is far simpler than that for a three-dimensional body. I have calculated† the two-

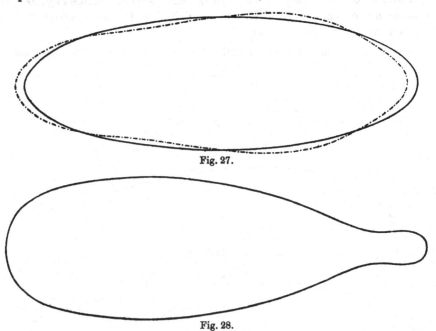

Fig. 27.

Fig. 28.

* *Phil. Trans.* 217 A (1916), p. 1.

† *Problems of Cosmogony and Stellar Dynamics*, p. 102, and *Phil. Trans.* 200 A (1902), p. 67. The original calculation in the *Phil. Trans.* contained a number of numerical errors, mainly quite unimportant, which were corrected in the later presentation in *Problems of Cosmogony*.

dimensional figures as far as e^6, and obtained the cross-sections shewn in figs. 27—30 for successively increasing values of e, the last one being largely conjectural.

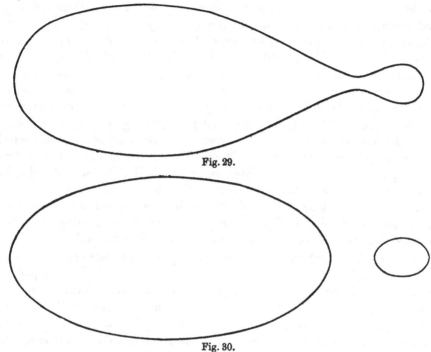

Fig. 29.

Fig. 30.

The two-dimensional figures are seen to end by fission into two detached masses, and there can be but little doubt, although this has never been definitely proved, that the three-dimensional figures do the same.

Stability of the Pear-shaped Configurations.

203. The series of Jacobian ellipsoids necessarily loses its stability at the point of bifurcation which occurs at its junction with the series of pear-shaped figures just discussed. The question arises whether stability is lost for good at this point, or is merely transferred to the pear-shaped figures in the way in which the stability of the Maclaurin spheroids is transferred to the Jacobian ellipsoids at the point of bifurcation of these two series. To settle this we have to examine whether the pear-shaped figures are stable or unstable.

The criterion of stability for these pear-shaped figures has already been given in § 173; if the angular momentum initially increases on passing along the series from the point of bifurcation, then the figures are stable; if on the other hand it is found initially to decrease, then the figures are unstable. As far as first order terms, the angular momentum will necessarily be the same as at the point of bifurcation, so that to apply this criterion, we must proceed to terms of higher order than the first in our determination of the series.

This problem formed the subject of a series of classical papers by Poincaré, Darwin and Liapounoff*. The general problem was first broached by Poincaré's memoir in Vol. 7 of the *Acta Mathematica* (1885), to which reference has already been made. The criterion of stability was inaccurately stated there, and the necessary modification was announced by Schwarzschild† in 1896. The accuracy of Schwarzschild's criterion of stability was admitted by Poincaré in a paper published in 1901‡; in this paper Poincaré also developed a method of carrying ellipsoidal harmonic potentials as far as the second order terms, and reduced the criterion of stability to an algebraic form, without, however, undertaking the necessary computations. At this stage the problem was taken up by Darwin, who, after preparing the ground by preliminary investigations§, published in 1902 a paper‖, "The Stability of the Pear-shaped Figure of Equilibrium of a Rotating Mass of Fluid." In the paper the equation of the pear-shaped figure was found, as far as terms of the second order, by a method which was substantially identical with that which Poincaré had developed the previous year; and the moment of momentum of the configuration was calculated. This was found to increase on passing along the series, so that the pear-shaped figure was announced to be stable.

Darwin's investigation had not long been published when doubt was cast on the accuracy of his conclusions. In 1905 Liapounoff published a paper¶ in which he stated that he could prove that the pear-shaped figure was unstable. Liapounoff's method was very different from that of Darwin, and a large part of his investigation appeared in the Russian language ; owing perhaps to these circumstances, neither investigator was able to announce the exact spot in which the error of the other lay, and the problem remained an open one for some years.

In 1915** I approached the problem from a new angle, using ordinary Cartesian co-ordinates, in the way already explained, in place of the ellipsoidal

---

* Poincaré's papers on this subject occupy 122 pages in the *Acta Mathematica*, and 41 pages in the *Phil. Trans.* Darwin's papers occupy 247 pages in the *Phil. Trans.* or 237 pages in his *Collected Works.* Liapounoff's papers occupy 750 pages in the *Memoirs* of the St Petersburg Academy. My own two papers referred to later occupy 86 pages in the *Phil. Trans.* Considerations of space make it impossible to give more than the barest outlines of this great mass of mathematical research. A fuller account of my own investigation will be found in my *Problems of Cosmogony and Stellar Dynamics*, pp. 87—102.

† K. Schwarzschild, *Münchener Inaug. Dissert.* (1896).

‡ "Sur la Stabilité de l'Equilibre des Figures Pyriformes affectées par une Masse Fluide en rotation," *Phil. Trans.* 197 A (1901), p. 333.

§ "Ellipsoidal Harmonic Analysis," *Phil. Trans.* 197 A (1901), p. 461; "On the Pear-shaped Figure of Equilibrium of a Rotating Mass of Liquid," *Phil. Trans.* 198 A (1901), p. 301.

‖ *Phil. Trans.* 200 A (1902), p. 251; see also papers in *Phil. Trans.* 208 A (1908), p. 1, and *Proc. Roy. Soc.* LXXXII A (1909), p. 188, all combined in one paper in *Collected Scientific Papers*, III. p. 317.

¶ "Sur un Problème de Tchebychef," *Mémoires de l'Académie de St Pétersbourg*, XVII. 3 (1905), and other papers published by the Academy.

** *Phil. Trans.* 215 A (1914), p. 27.

harmonic analysis which had been used by Poincaré, Darwin and Liapounoff. My results agreed with those previously given by Darwin up to a certain distance, and where they began to disagree I was able to discover* a quite simple error which invalidated Darwin's discussion of the problem. On carrying the discussion of the figure as far as the third order of small quantities†, I confirmed the conclusion already reached by Liapounoff, that the pear-shaped series was initially unstable, and was further able to shew that the correction of Darwin's error made the difference between stability and instability in his final result. When Darwin's error had been corrected in this way, the three investigations of Darwin, Liapounoff and myself agreed in finding the pear-shaped figure to be unstable. In 1920 H. F. Baker‡ expressed doubts (unsubstantiated by detailed calculation) as to the accuracy of my treatment of certain series, but withdrew his criticism five years later§, and expressed himself as satisfied that the expansion in series which I had used in my papers could be placed on a sure foundation.

There seems, then, to be little room for doubt that the series of pear-shaped configurations is initially unstable.

Sequence of Configurations of Rotating Liquid.

204. The equilibrium configurations of a mass of homogeneous rotating liquid accordingly fall into linear series which may be arranged diagrammatically as shewn in fig. 25, and also on the left of fig. 31 (the right-hand half anticipates results which will be obtained later). In this diagram the angular

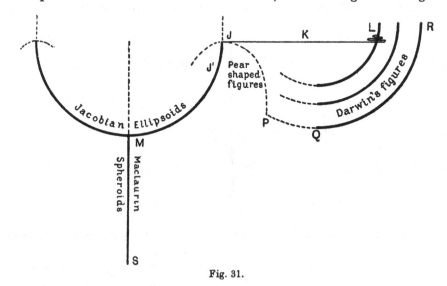

Fig. 31.

* *l.c.* p. 76.
† *Phil. Trans.* 217 A (1916), p. 1, and *Problems of Cosmogony and Stellar Dynamics*, p. 87.
‡ *Proc. Camb. Phil. Soc.* xx. (1920), p. 198.   § *Ibid.* xxiii. (1925), p. 1.

momentum is taken as the variable parameter, increase of angular momentum being represented by an upward motion.

The spherical configuration of no rotation, or zero angular momentum, is represented by the point S at the bottom of the diagram. As the angular momentum increases, the mass first moves along the series of Maclaurin spheroids SM until it comes to the point of bifurcation M at which the series intersects the series of Jacobian ellipsoids. At this point the Maclaurin spheroids lose their stability, and the motion proceeds along the series of Jacobian ellipsoids $MJ'J$ until the point of bifurcation J is reached. The Jacobian ellipsoids lose their stability here. The second series through J is, as we have seen, a series of pear-shaped figures such as JP in the diagram. The angular momentum of these figures decreases as we proceed along the series from J, so that the series is unstable and the curve JP turns downwards in the diagram after leaving J. Thus there is no stable configuration beyond J, and dynamical motion of some kind must occur as soon as shrinkage has proceeded so far that the angular momentum is greater than that represented by the point J.

Let us attempt to examine what type of dynamical motion is to be expected when a mass of fluid having the configuration of a Jacobian ellipsoid reaches the point at which secular instability sets in.

In fig. 32 let JJ' represent the series of stable Jacobian ellipsoids in the neighbourhood of the point of bifurcation J. For any configuration within the range JJ', the third harmonic (pear-shaped) vibration is stable both ordinarily and secularly. Thus if any small pear-shaped vibration is set up

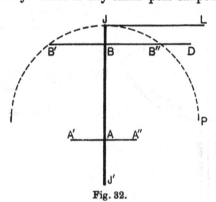

Fig. 32.

when the mass is in a configuration such as A, the representative point will oscillate backwards and forwards through some small range such as $A'AA''$ until the vibration is damped by viscosity. If the vibration is set up when the representative point is at some point B close to J, there may still be oscillation through a small range, but the motion can only be stable if this range is less than the range $B'B''$ in Fig. 32. For the point B'' represents a secularly unstable configuration, so that if the representative point once passes beyond B'', on the line $BB''D$, it will not return but will describe some path such as $BB''D$ in the plane through B.

As the point B approaches J, the range of vibration which is possible without instability setting in becomes smaller and smaller and finally vanishes altogether, so that in the limit any disturbance, no matter how slight, causes the representative point to move permanently away from the line $J'J$. The path of this point is necessarily in the horizontal plane through J, and we know that the direction of this path initially is that of the tangent JL at J to the pear-shaped series JB''. In other words, the motion is one in which a furrow of the type depicted in fig. 26 forms on the ellipsoid, and this furrow continually deepens.

It seems likely that the furrow will deepen until the mass divides into two parts. If so, the motion, which must be in the plane JL in fig. 32, will end by the representative point coming to rest at some point such as L in fig. 31 on a series of configurations representing two detached bodies revolving around one another. Such configurations have been studied in detail by Roche and Darwin. We shall refer to the study of these configurations as the Binary Star problem, from its obvious application to the dynamics of binary stars.

THE BINARY STAR PROBLEM.

205. We proceed to search for equilibrium configurations of two detached bodies of masses M, M' revolving round one another in such a way that the whole system remains at rest relative to a system of axes rotating with angular velocity ω in the plane of xy.

Let the centre of gravity of the mass M be taken as origin, the line joining the centres of gravity of the two masses being taken as axis of x, R being the distance between the two centres. Then the axis of rotation, which of course passes through the centre of gravity of the two bodies, is

$$x = \frac{M'}{M + M'} R, \quad y = 0.$$

The centrifugal forces acting on the mass M may accordingly be derived from a potential

$$\tfrac{1}{2}\omega^2 \left[\left(x - \frac{M'}{M + M'} R \right)^2 + y^2 \right] \quad \dots\dots\dots\dots(205{\cdot}1).$$

The remaining forces which have to be considered are the gravitational attractions of the masses, both on themselves and on one another. Our problem is to search for configurations in which both bodies can rotate in relative equilibrium under these attractions and the centrifugal forces resulting from their rotation.

ROCHE'S PROBLEM.

206. The simplest problem occurs when the mass M', which we shall describe as the secondary, may be treated as a rigid sphere; this is the special problem dealt with by Roche. In this simple case the gravitational forces acting on the body of mass M, which we shall call the primary, may be derived from a potential

$$\frac{\gamma M'}{R} + \frac{\gamma M'}{R^2}\, x + \frac{\gamma M'}{R^3}\,(x^2 - \tfrac{1}{2}y^2 - \tfrac{1}{2}z^2) + \dots \quad \dots\dots(206\cdot1).$$

We shall for the present be content to omit all terms beyond those written down. The correction required by the neglect of these terms will be discussed later, and will be found to be so small that the results now to be obtained are hardly affected.

Omitting these terms, and combining the two potentials (205·1) and (206·1), we find that, apart from its own gravitation, the primary may be supposed to be acted on by a total field of force derived from a potential

$$\frac{M'}{R^2}\, x\left(\gamma - \frac{\omega^2 R^3}{M + M'}\right) + \frac{\gamma M'}{R^3}(x^2 - \tfrac{1}{2}y^2 - \tfrac{1}{2}z^2) + \tfrac{1}{2}\omega^2(x^2 + y^2)$$

$$\dots\dots(206\cdot2).$$

Denoting this potential by V', the equations of equilibrium of an element of the primary are, as in § 188,

$$\frac{\partial p}{\partial x} = \rho\,\frac{\partial V}{\partial x} + \rho\,\frac{\partial V'}{\partial x}\ \text{etc.}$$

where V is the potential of the primary, and these have the common integral

$$\frac{p}{\rho} = V + V' + \text{cons.}$$

Thus the primary will be in equilibrium if $V + V'$ is constant over its surface.

Since V' is quadratic in x, y and z, it is at once clear that ellipsoidal configurations of equilibrium are possible, as in the former problem. The surface

$$\frac{x^2}{a^2} + \frac{y^2}{b^2} + \frac{z^2}{c^2} - 1 = 0 \quad \dots\dots\dots\dots\dots(206\cdot3)$$

will be a possible configuration of equilibrium if (cf. equation 189·7)

$$-\pi\gamma\rho abc\,(J_A x^2 + J_B y^2 + J_C z^2 - J)$$
$$+ \frac{M'}{R^3}\, x\left(\gamma - \frac{\omega^2 R^3}{M + M'}\right) + \frac{\gamma M'}{R^3}(x^2 - \tfrac{1}{2}y^2 - \tfrac{1}{2}z^2) + \tfrac{1}{2}\omega^2(x^2 + y^2)\dots\dots(206\cdot4)$$

is constant over its surface.

To make this expression constant over the surface, we must first remove the term in x by assigning to ω^2 the appropriate value

$$\omega^2 = \frac{\gamma(M + M')}{R^3} \quad \dots\dots\dots\dots\dots\dots(206\cdot5).$$

We next equate the expression, as in § 190, to

$$-\pi\gamma\rho abc\theta\left(\frac{x^2}{a^2} + \frac{y^2}{b^2} + \frac{z^2}{c^2} - 1\right),$$

and, on equating coefficients of x^2, y^2 and z^2, we obtain as the conditions of equilibrium,

$$J_A - \frac{\mu}{\pi\gamma\rho abc} - \frac{\omega^2}{2\pi\gamma\rho abc} = \frac{\theta}{a^2} \quad \dots\dots\dots\dots(206\cdot61),$$

$$J_B + \frac{\mu}{2\pi\gamma\rho abc} - \frac{\omega^2}{2\pi\gamma\rho abc} = \frac{\theta}{b^2} \quad \dots\dots\dots(206\cdot62),$$

$$J_C + \frac{\mu}{2\pi\gamma\rho abc} = \frac{\theta}{c^2} \quad \dots\dots\dots\dots\dots(206\cdot63),$$

where μ is written for $\gamma M'/R^3$. Putting $\mu = 0$ causes the secondary to go out of existence. The problem then reduces to that of a single mass rotating freely in space, the equations becoming identical with those already discussed.

It is convenient to put $M/M' = p$, so that

$$\omega^2 = (1 + p)\mu \quad \dots\dots\dots\dots\dots(206\cdot64).$$

The equations then reduce to

$$J_A - \frac{(3+p)\mu}{2\pi\gamma\rho abc} = \frac{\theta}{a^2} \quad \dots\dots\dots\dots(206\cdot71),$$

$$J_B - \frac{p\mu}{2\pi\gamma\rho abc} = \frac{\theta}{b^2} \quad \dots\dots\dots\dots(206\cdot72),$$

$$J_C + \frac{\mu}{2\pi\gamma\rho abc} = \frac{\theta}{c^2} \quad \dots\dots\dots\dots(206\cdot73).$$

The elimination of θ from the first two equations gives

$$(b^2 - c^2)\int_0^\infty \frac{\lambda\,d\lambda}{(b^2+\lambda)(c^2+\lambda)\Delta} = (b^2 p + c^2)\frac{\mu}{2\pi\gamma\rho abc}$$
$$\dots\dots(206\cdot8),$$

while similarly the elimination of θ from the first and third yields

$$(a^2 - c^2)\int_0^\infty \frac{\lambda\,d\lambda}{(a^2+\lambda)(c^2+\lambda)\Delta} = (a^2 p + c^2 + 3a^2)\frac{\mu}{2\pi\gamma\rho abc}$$
$$\dots\dots(206\cdot9).$$

Except for differences of notation, these two equations are identical with the two which Roche takes as the basis of his discussion [*].

207. It will be most convenient to examine the solution of these equations by a graphic method. We may suppose a, b, c to be connected by the usual relation $abc = r_0^3$, where r_0 remains constant throughout the changes of shape of a given mass. Thus the two quantities a, b specify the shape of the mass.

Let us take a and b as abscissa and ordinate, as in fig. 33. At each point in this diagram the values of p and μ may be uniquely determined from equations (206·8) and (206·9). If we map out a curve along which p or M/M' is constant, this curve will represent a linear series of configurations corresponding to different values of μ or $\gamma M'/R^3$, and so corresponding to different

---

[*] *Acad. de Montpellier (Sciences)*, I. (1850), p. 243. Our two equations (206·8) and (206·9) are identical with equations (4) and (5) (p. 247) of Roche's memoir.

distances between the two masses. Since the equations determine p uniquely in terms of a and b, none of the curves $p = $ cons. can ever intersect.

The special case of $p = \infty$ has already been fully investigated; when $p = \infty$, $M' = 0$ and the problem reduces to that of a single mass rotating freely in space under its own gravitation alone. Thus when $p = \infty$ the solution of the equations represents the two linear series of Maclaurin spheroids ($a = b$) and Jacobian ellipsoids. In fig. 33, let S represent the spherical configuration $a = b = r_0$, and let SM represent the series of Maclaurin spheroids ($a = b > c$). Let B represent the point of bifurcation with the series of Jacobian ellipsoids, and let JBJ' represent this latter series.

When $p = -1$ equations (206·7) shew that $b = c$. Thus the curve $p = -1$ is represented by the curve $ab^2 = r_0^3$; let this be the line $T'ST''$ in fig. 33.

All points which lie to the left of the median line OSM represent configurations for which $b > a$, and therefore configurations in which the primary is broadside on to the secondary. These configurations are unstable, for they would be unstable even if the primary were constrained to remain rigid. They need not trouble us further and we may confine our attention to the right-hand half of the diagram.

Linear series for all values of p pass through S. The series for $p = +\infty$ is the broken line SBJ, that for $p = -1$ is the line ST, while that for $p = -\infty$ is easily seen to be the line SO, along which $a = b < c$. Since two linear series cannot cross, it is clear that the series for a very large positive value of p must be asymptotic to the line SBJ. All the series from $p = +\infty$ to $p = -1$ accordingly lie within the small area bounded by the lines JB, BS, ST. The series in the area OST are of course series for which p is negative and numerically greater than 1, while those in the area MBJ are again series for which p is negative, a second series for $p = -\infty$ coinciding with the line MBJ.

Let us now confine our attention to the series which lie inside the area $JBST$, these being as we have seen the only ones of physical interest. Each series starts at S and ends at the point in which the lines BJ and ST ultimately meet at infinity. Thus each series begins with a sphere and ends with an infinitely long prolate spheroid. As we pass along any one of these series μ changes while p remains constant. The value of ω^2 which is given by equations (206·6) accordingly changes also, this giving the value of a real angular velocity when p is positive, and being regarded simply as an algebraic quantity when p is negative. The value of ω^2 vanishes only when $p = -1$ or when $\mu = 0$; consequently it vanishes at S, along the line ST, and at the points at which a is infinite, but nowhere else. It follows that ω^2 is positive everywhere inside the area $SBJT$.

Since ω^2 vanishes at both ends of every series, it follows that on passing along each series ω^2 at first increases, and then after passing a maximum value decreases. Roche *, treating equations (206·8) and (206·9) by a laborious method of numerical calculation, found that there is only one maximum on each series. On the series $SBJ (p = \infty)$ the maximum occurs as we have seen at B; the value of $\omega^2/2\pi\gamma\rho$ here is 0·18712. Roche calculated the maxima of $\omega^2/2\pi\gamma\rho$ on other series. On the series $p = 0$, the series of configurations in which the primary is infinitesimal, he finds the maximum value of $\omega^2/2\pi\gamma\rho$ to be 0·046, and the configuration at which this maximum occurs is that in which $a = 1·63r_0$, $b = ·81r_0$; this is represented by the point R'' in the diagram. When $p = 1$, the maximum value of $\omega^2/2\pi\gamma\rho$ is 0·072, and Roche finds that the value of $\omega^2/2\pi\gamma\rho$ at the various maxima increases continuously from $p = 0$ to $p = \infty$.

On connecting the points B, R'', T'' by a continuous line, we obtain the points at which ω^2 is a maximum on the various linear series.

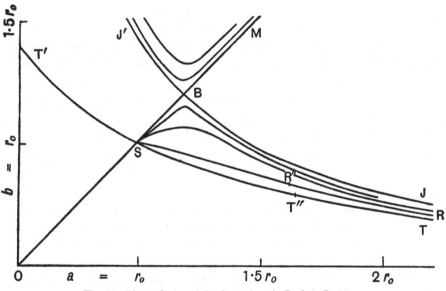

Fig. 33. Linear Series of Configurations in Roche's Problem.

Stability.

208. In a physical problem in which $\omega^2/2\pi\gamma\rho$ increased continuously, it would follow, from the principles already discussed, that all configurations on the left of the line $BR''T''$ would be stable, while all the configurations on the right would be unstable. The configurations on the left would of course only have been proved stable so long as the configurations were constrained to remain ellipsoidal, but it can be proved that this restriction makes no difference †.

In the natural double-star problem, the change in physical conditions is not adequately represented by making $\omega^2/2\pi\gamma\rho$ increase continuously. Even

* *l.c.* p. 251.   † *Problems of Cosmogony,* p. 85.

if both masses shrink, the rates of shrinkage, and consequent rates of increase in density, will in all probability be quite different for the two masses. We can, however, construct an artificial problem in which the density, if supposed uniform to begin with, remains uniform throughout the shrinkage, or in which the two densities, if not supposed equal to begin with, change so as always to retain the same ratio. The physical conditions are now represented by an increase in the absolute densities, while the moment of momentum remains constant and, exactly as in § 196, these conditions may equally be represented by supposing both densities to remain constant while the moment of momentum increases.

209. In the more general problem in which the secondary is not regarded merely as a point, the moment of momentum of the primary about the centre of gravity of the system is

$$M \left[k^2 + \left(\frac{M'}{M + M'} \right)^2 R^2 \right] \omega$$

where k is the radius of gyration of the primary. Adding the similar expression for the secondary, we obtain for the total moment of momentum **M** of the system

$$\mathbf{M} = \left(Mk^2 + M'k'^2 + \frac{MM'}{M+M'} R^2 \right) \omega \quad \ldots\ldots\ldots\ldots(209\text{·}1),$$

or, replacing R by its value $\gamma^{\frac{1}{3}} (M + M')^{\frac{1}{3}} \omega^{-\frac{2}{3}}$,

$$\mathbf{M} = (Mk^2 + M'k'^2) \omega + \frac{\gamma^{\frac{2}{3}} MM'}{(M+M')^{\frac{1}{3}}} \omega^{-\frac{1}{3}} \quad \ldots\ldots\ldots(209\text{·}2).$$

210. When the primary M is infinitely massive compared with the secondary M', the total moment of momentum **M** has the value $\mathbf{M} = Mk^2 \omega$, and the variation of **M** is precisely that of a freely rotating mass; it increases steadily from $\mathbf{M} = 0$ to $\mathbf{M} = \infty$ as we pass along the series SBJ in fig. 33.

For finite values of the ratio M/M', the value of **M** given by equation (209·1) becomes infinite when $\omega = 0$, i.e. at the two ends of the linear series of configurations shewn in fig. 33. Thus on leaving S, **M** decreases until a minimum is reached, and all configurations beyond this minimum will be unstable. Thus the curved line $BR''T'''$ which divided stable from unstable configurations in fig. 33 must now be replaced by another curved line passing through S.

It accordingly appears that when M/M' is large the linear series becomes unstable very near to S, the range of stability vanishing altogether when M/M' is infinite. If both masses are rigid, so that k^2 and k'^2 are constants, the limit of this range is easily found from equation (209·1) by making $\delta \mathbf{M} = 0$. The limit of stability is found to be given by

$$3\omega^{\frac{4}{3}} (Mk^2 + M'k'^2) = \frac{\gamma^{\frac{2}{3}} MM'}{(M+M')^{\frac{1}{3}}} \quad \ldots\ldots\ldots\ldots(210\text{·}1),$$

or, in terms of R (using equation (206·5)),

$$R^2 = \frac{3\,(M + M')}{MM'}\,(Mk^2 + M'k'^2) \quad \ldots\ldots\ldots\ldots(210\cdot2),$$

and the range of stability is again seen to be infinitesimal—i.e. limited to very great values of R—when M/M' is infinite.

The result shews that there can only be secular stability of a large and small mass rotating round one another when the smaller mass is at a very great distance from the larger*. We are dealing, of course, with secular stability only; the question means nothing except when dissipative forces are present. When there are no dissipative forces, as for instance if both bodies are perfectly rigid, a circular orbit of no matter what radius is thoroughly stable, the orbit $r = a$ giving place when slightly disturbed to the slightly elliptical orbit $r = a\,(1 - e\cos\theta)$. And when dissipative forces are introduced, as for instance by making the masses fluid, or by supposing the solid masses covered by shallow oceans, the instability is one of orbital motion only and not one of the configurations of the masses. When the secondary is supposed wholly fluid so that k'^2 is variable, the fluidity of the mass modifies the stage at which instability sets in, but introduces no new instability of its own. The mechanism by which this instability is set up has been studied by Darwin under the name of "Tidal Friction"†; it produces a secular change in the mean radius of the orbit which we shall study more fully below (Chap. XI).

It is important to notice that the case of $M/M' = \infty$, in which M' is of infinitesimal mass, is not, from our present point of view, identical with the case of $p = \infty$ in the diagram shewn in fig. 33, in which M' is supposed to disappear altogether. The former problem has one more degree of freedom than the latter, and this one degree of freedom happens to be secularly unstable for all finite values of R. In the latter problem, in which the system is supposed to reduce to a single rotating body, the angular momentum increases steadily from 0 to ∞ on passing along the path SBJ in fig. 33, so that the configurations on this path are all stable so long as the mass is constrained to remain ellipsoidal.

Roche's Limit.

211. A distinct problem arises when the rotation of the secondary is not affected by forces exerted on it by the primary. The primary, which is the body whose configurations and stability we are specially considering, may now be a small satellite rotating round a massive planet, which our choice of terms compels us to call the secondary. The term $M'k'^2\omega$ in equation (209·2) may now be replaced by $M'k'^2\omega'$, where ω' is the angular velocity of the secondary and neither this nor k'^2 is subject to variation. The angular momentum is accordingly

$$\mathbf{M} = Mk^2\omega + \frac{\gamma^{\frac{2}{3}}MM'}{(M + M')^{\frac{1}{3}}}\,\omega^{-\frac{1}{3}} + \text{cons.} \quad \ldots\ldots\ldots\ldots(211\cdot1).$$

* Cf. Sir G. Darwin, *Coll. Works*, III, p. 442.   † *Ibid.* II.

A similar problem occurs when the secondary is treated as a point so that $k'^2 = 0$. This leads back to Roche's problem discussed in § 206. The moment of momentum is again given by equation (211·1) in which the final constant now vanishes. Let us investigate the stability of systems in which the moment of momentum is given by equation (211·1). We shall consider in turn the cases of M being infinitesimal and of M being finite.

212. When the primary M is infinitesimal, $\mathbf{M}$ also becomes infinitesimal, but $\mathbf{M}/M$ remains finite, being given by

$$\frac{\mathbf{M}}{M} = (\gamma M')^{\frac{2}{3}} \omega^{-\frac{1}{3}} + \text{cons.}$$

The minimum value of M now coincides with the maximum value of ω. The series of configurations are those represented in the series $(p = 0)$ in fig. 33, and the minimum value of ω occurs at the point R''. Thus configurations on the branch SR'' are stable, while those beyond R'' are unstable.

The actual value of ω at the point R'' is given by

$$\frac{\omega^2}{2\pi\gamma\rho} = 0\cdot04503 \,*.$$

The general value of $\omega^2/2\pi\gamma\rho$ on this series is

$$\frac{\omega^2}{2\pi\gamma\rho} = \frac{M'}{2\pi\rho R^3} = \frac{2}{3}\left(\frac{\rho'}{\rho}\right)\frac{r_0'^3}{R^3},$$

so that in the critical configuration we have

$$R = 2\cdot4554 \left(\frac{\rho'}{\rho}\right)^{\frac{1}{3}} r_0' \quad\dots\dots\dots\dots\dots(212\cdot1).$$

Thus a small satellite rotating about a rigid primary of mass enormously greater than its own cannot be in equilibrium in any configuration whatever if its distance from the centre of its planet is less than $2\cdot4554\,(\rho'/\rho)^{\frac{1}{3}}$ radii of the planet. This distance is commonly spoken of as Roche's limit.

213. In the more general case in which M is not infinitesimal, and the angular momentum is given by the complete equation (211·1), the maximum value for ω is not so easily found since k^2 will vary with ω. It is, however, clear that $\mathbf{M}$ will be infinite when $\omega = 0$, and that ω will again increase to a maximum and again decrease, so that $\mathbf{M}$ will pass through a minimum value which will again divide stable from unstable configurations. Again there will be a limiting value of R similar to Roche's limit, and there will be no configurations of equilibrium at all for smaller values of R than this.

---

* Roche gave 0·046; both here and in equation (212·1) I quote the more accurate values deduced from Darwin's calculations, *Coll. Works*, III. p. 436.

Darwin's Problem.

214. The double-star problem has been discussed by Sir G. Darwin* in the most general case in which both masses are supposed fluid, so that each is subject to distortion from the tidal forces generated by the other. As in the simpler case, the discussion falls into two parts—the determination of figures of equilibrium and the determination of the stability or instability of these figures.

In Roche's problem the secondary was supposed to be a rigid sphere, so that its potential could be written down in the form

$$\frac{\gamma M'}{R} + \frac{\gamma M'}{R^2}\,x + \frac{\gamma M'}{R^3}\,(x^2 - \tfrac{1}{2}y^2 - \tfrac{1}{2}z^2) + \ldots \quad \ldots\ldots\ldots(214\cdot1).$$

In Darwin's problem, the shape of the secondary is determined by the mutual tidal actions between the two bodies, and an expansion such as the foregoing is no longer permissible. To a first approximation both bodies may be regarded as ellipsoids. To a second approximation, Darwin supposes the bodies to be distorted ellipsoids and expresses the distortions in terms of ellipsoidal harmonics. The amount of this distortion is found to be in every case quite small, so that the supposition that the figures are actually ellipsoidal is found to give a tolerably accurate solution. In illustration of this the following figures may be quoted from Darwin's paper†; they express the proportional increase $\delta a/a$ in the semi-major axis of the primary which would be produced by the removal of the ellipsoidal constraint when the masses are at the closest distance consistent with stability (cf. § 217 below).

$M/M' =$	...	...	...	0·4	0·7	1·0
$\delta a/a$ in direction towards secondary ...				$\frac{1}{17}$	$\frac{1}{20}$	$\frac{1}{23}$
$\delta a/a$ in direction away from secondary				$\frac{1}{25}$	$\frac{1}{33}$	$\frac{1}{39}$

The amount of these corrections is shewn by the dotted lines in figs. 37, 38, 39 (p. 241 below). When the masses are further apart than this minimum distance, the error in the ellipsoidal solution will of course be still less, so that the assumption that the figures are ellipsoidal is seen to give a very fair approximation.

215. Let the secondary now be assumed to be an ellipsoid of mass M' and semi-axes a', b', c'.

Let us momentarily take the centre of the secondary for origin, then its potential at any external point x', y', z', will be

$$V = -\tfrac{3}{4}\gamma M' \int_{\lambda'}^{\infty} \left(\frac{x'^2}{a'^2 + \lambda} + \frac{y'^2}{b'^2 + \lambda} + \frac{z'^2}{c'^2 + \lambda} - 1 \right) \frac{d\lambda}{\Delta'} \quad \ldots(215\cdot1)$$

---

* *Phil. Trans.* 206 A (1906), p. 161, or *Coll. Works*, III. p. 436. The actual paper occupies 88 pages in each place, so that it will be understood that only the merest outline of it is given here. And, to avoid the complicated methods of ellipsoidal harmonic analysis employed by Darwin, I have substituted a simpler discussion of the fundamental equations, deriving them in a form analogous to the equations of Roche already discussed.   † *l.c.* p. 510.

where the lower limit λ' is the root of

$$\frac{x'^2}{a'^2+\lambda} + \frac{y'^2}{b'^2+\lambda} + \frac{z'^2}{c'^2+\lambda} - 1 = 0 \quad \dots\dots\dots\dots(215\cdot2),$$

and Δ' stands for

$$[(a'^2+\lambda)(b'^2+\lambda)(c'^2+\lambda)]^{\frac{1}{2}}.$$

Differentiating, and bearing in mind that the lower limit is a function of x', y' and z', we obtain

$$\frac{\partial V}{\partial x'} = -\tfrac{3}{2}\gamma M'x' \int_{\lambda'}^{\infty} \frac{d\lambda}{(a'^2+\lambda)\Delta'},$$

$$\frac{\partial^2 V}{\partial x'^2} = -\tfrac{3}{2}\gamma M' \left[\int_{\lambda'}^{\infty} \frac{d\lambda}{(a'^2+\lambda)\Delta'} - \frac{2\left(\dfrac{x}{a'^2+\lambda'}\right)^2}{\left[\left(\dfrac{x}{a'^2+\lambda'}\right)^2 + \left(\dfrac{y}{b'^2+\lambda'}\right)^2 + \left(\dfrac{z}{c'^2+\lambda'}\right)^2\right]\Delta'} \right]$$

and similar equations give $\partial V/\partial y'$, $\partial^2 V/\partial y'^2$, etc.

To evaluate these quantities at a point on the axis of x', we put $y'=z'=0$, and $\lambda' = x'^2 - a'^2$. The equations become

$$\frac{\partial V}{\partial x'} = -\tfrac{3}{2}\gamma M'x \int_{\lambda'}^{\infty} \frac{d\lambda}{(a'^2+\lambda)\Delta'}; \quad \frac{\partial V}{\partial y'} = \frac{\partial V}{\partial z'} = 0 \quad \dots\dots(215\cdot3),$$

$$\left.\begin{aligned}
\frac{\partial^2 V}{\partial x'^2} &= -\tfrac{3}{2}\gamma M' \left[\int_{\lambda'}^{\infty} \frac{d\lambda}{(a'^2+\lambda)\Delta'} - \frac{2}{\Delta'} \right]\\
\frac{\partial^2 V}{\partial y'^2} &= -\tfrac{3}{2}\gamma M' \int_{\lambda'}^{\infty} \frac{d\lambda}{(b'^2+\lambda)\Delta'}, \text{ etc.}
\end{aligned}\right\} \quad \dots\dots\dots(215\cdot4).$$

We obtain the differential coefficients of the potential V at the centre of the primary on putting $x'=R$, and $\lambda'=R^2-a'^2$. If V denotes the value of the potential at this point, and $\partial V/\partial x$, etc., denote the value of differential coefficients at this point, the general value of the potential of the secondary, referred to the centre of the primary as origin, will be

$$V + x\frac{\partial V}{\partial x} + y\frac{\partial V}{\partial y} + z\frac{\partial V}{\partial z} + \frac{1}{2}\left(x^2\frac{\partial^2 V}{\partial x^2} + y^2\frac{\partial^2 V}{\partial y^2} + z^2\frac{\partial^2 V}{\partial z^2} \right) \dots(215\cdot5),$$

the terms in xy, yz, zx, vanishing on account of the symmetry of figure of the ellipsoid. To the approximation to which we are now working, the primary is supposed to remain of ellipsoidal shape, so that all tide-generating terms of degrees three and higher must be neglected. Or we may, if we please, regard the problem as one in which the primary is constrained to remain ellipsoidal, in which case the forces of constraint must be just sufficient to neutralise the omitted terms in formula (215·5).

The coefficients in this formula are precisely those which have been already evaluated in equations (215·3) and (215·4). For we have

$$\frac{\partial V}{\partial x} = -\frac{\partial V}{\partial x'}, \text{ etc., so that } \frac{\partial V}{\partial y} = \frac{\partial V}{\partial z} = 0, \text{ and similarly } \frac{\partial^2 V}{\partial x^2} = \frac{\partial^2 V}{\partial x'^2}, \text{ etc.}$$

If ξ, ξ' denote for the moment the distances of the centres of the primary and secondary from the axis of rotation, the primary must, as before, be in equilibrium under a statical field of force of potential $\frac{1}{2}\omega^2[(x-\xi)^2+y^2]$ or

$$\tfrac{1}{2}\omega^2(x^2+y^2)-\omega^2\xi x + \text{a constant}$$

and the condition for the equilibrium of the primary is, as in § 206, that

$$V_b + \frac{1}{2}\left(x^2\frac{\partial^2 V}{\partial x'^2} + y^2\frac{\partial^2 V}{\partial y'^2} + z^2\frac{\partial^2 V}{\partial z'^2}\right) + \tfrac{1}{2}\omega^2(x^2+y^2) - x\frac{\partial V}{\partial x'} - \omega^2\xi x = \text{a constant}$$

$$\ldots\ldots(215\text{·}6)$$

over the surface of the primary. The term in x on the right of this equation is removed if

$$\omega^2\xi = -\frac{\partial V}{\partial x'} = \tfrac{3}{2}\gamma M'R\int_{R^2-a'^2}^{\infty}\frac{d\lambda}{(a'^2+\lambda)\Delta'} \quad\ldots\ldots\ldots(215\text{·}71).$$

The corresponding equation for the secondary is

$$\omega^2\xi' = \tfrac{3}{2}\gamma MR\int_{R^2-a^2}^{\infty}\frac{d\lambda}{(a^2+\lambda)\Delta} \quad\ldots\ldots\ldots\ldots(215\text{·}72).$$

Since $\xi+\xi'=R$, these two equations suffice to determine ξ, ξ' and ω^2. The ratio ξ/ξ' is obtained at once by division of corresponding sides of the two equations.

These equations refer to masses which are constrained to remain ellipsoidal by the supposed application of small external forces. Had the bodies been rotating freely in space, the ratio ξ/ξ' would have been given directly by the condition that the centre of gravity of the two masses must be on the axis of rotation, namely

$$\frac{\xi}{M'} = \frac{\xi'}{M} = \frac{R}{M+M'} \quad\ldots\ldots\ldots\ldots\ldots(215\text{·}8).$$

The two values of ξ/ξ' obtained in these two different ways are not found to be identical, their difference giving a measure of the error introduced in supposing the masses to remain ellipsoidal. If we put

$$\int_{R^2-a^2}^{\infty}\frac{d\lambda}{(a^2+\lambda)\Delta} = I; \quad \int_{R^2-a'^2}^{\infty}\frac{d\lambda}{(a'^2+\lambda)\Delta'} = I',$$

then equation (215·8) gives $\xi/\xi' = M'/M$, whereas equations (215·7) give

$$\frac{\xi}{\xi'} = \frac{M'}{M}\times\frac{I'}{I}.$$

The difference between the two values of ξ/ξ' is seen to be represented by the multiplying factor I'/I.

The integral I is easily expanded in descending powers of R in the form

$$I = \frac{2}{3R^3} + \frac{1}{5R^5}(2a^2-b^2-c^2) + \ldots$$

while I' is obtained by replacing $2a^2-b^2-c^2$ by $2a'^2-b'^2-c'^2$.

If the two ellipsoids are equal, $I'=I$ and the factor I'/I is equal to unity. In general $2a^2-b^2-c^2$ and $2a'^2-b'^2-c'^2$ will both be small in comparison with R^2, so that both I and I' are nearly equal to $2/3R^2$. Let ζ be defined by

$$\tfrac{1}{2}(I+I')=\frac{2}{3R^2}(1+\zeta)$$

then no serious error will be involved in supposing both I and I' to be equal to the quantity on the right.

Equations (215·7) now become accordant with (215·8). By addition of (215·71) and (215·72) we obtain

$$\omega^2=\gamma\frac{(M+M')}{R^3}(1+\zeta) \quad\ldots\ldots\ldots\ldots(215\cdot9).$$

This determines ω^2. To balance centrifugal force, the gravitational attraction between the two masses must be $\omega^2 RMM'/(M+M')$, or, using the value just obtained for ω^2,

$$\gamma\frac{MM'}{R^2}(1+\zeta)$$

so that ζ measures the proportional increase in the gravitational attraction between the two bodies produced by their ellipsoidal forms.

216. The terms in ξ have now been made to disappear from equation (215·6) and the condition for equilibrium is that

$$V_b+\frac{1}{2}\left(x^2\frac{\partial^2 V}{\partial x'^2}+y^2\frac{\partial^2 V}{\partial y'^2}+z^2\frac{\partial^2 V}{\partial z'^2}\right)+\tfrac{1}{2}\omega^2(x^2+y^2)=\text{cons.} \quad\ldots(216\cdot1)$$

over the boundary.

Following the procedure of § 206, we find that equation (216·1) may be replaced by the three separate equations

$$J_A-\frac{M'}{M}\left(\int_{\lambda'}^{\infty}\frac{d\lambda}{(a'^2+\lambda)\Delta'}-\frac{2}{\Delta'}\right)-\frac{\omega^2}{2\pi\gamma\rho abc}=\frac{\theta}{a^2} \quad\ldots(216\cdot21),$$

$$J_B-\frac{M'}{M}\int_{\lambda'}^{\infty}\frac{d\lambda}{(b'^2+\lambda)\Delta'}-\frac{\omega^2}{2\pi\gamma\rho abc}=\frac{\theta}{b^2} \quad\ldots\ldots\ldots(216\cdot22),$$

$$J_C-\frac{M'}{M}\int_{\lambda'}^{\infty}\frac{d\lambda}{(c'^2+\lambda)\Delta'}=\frac{\theta}{c^2} \quad\ldots\ldots\ldots\ldots\ldots(216\cdot23),$$

in which the lower limit λ' is put equal to $R^2-a'^2$.

These are the equations of equilibrium for the primary, and there is a similar set for the secondary. This set of six equations contains the solution of the problem. Unfortunately the only method of solving these equations is by laborious numerical computations. But as they are of the same general form as equations (206·6) already discussed in connection with Roche's problem, it is readily seen that the general arrangement of figures of equilibrium must be the same as that already found in § 206 for Roche's problem.

Stability.

217. We now turn to the problem of stability. The angular momentum of the system is still given by equation (209·1), namely,

$$\mathbf{M}=\left(Mk^2+M'k'^2+\frac{MM'}{M+M'}R^2\right)\omega \quad\ldots\ldots\ldots\ldots(217\cdot1)$$

but the value of R is now given in terms of ω by

$$R = \gamma^{\frac{2}{3}} (M + M')^{\frac{1}{3}} (1 + \zeta)^{\frac{1}{3}} \omega^{-\frac{2}{3}} \quad \dots\dots\dots\dots(217\cdot2),$$

so that $\mathbf{M}$, expressed in terms of ω alone, becomes

$$\mathbf{M} = (Mk^2 + M'k'^2)\,\omega + \frac{\gamma^{\frac{2}{3}} MM'}{(M+M')^{\frac{1}{3}}}(1 + \zeta)^{\frac{1}{3}} \omega^{-\frac{1}{3}} \quad \dots\dots(217\cdot3).$$

This differs from the former value of $\mathbf{M}$ obtained in § 209 only through the occurrence of the factor $(1 + \zeta)^{\frac{1}{3}}$, and as this is never far from unity, the general discussion of stability given in §§ 210 and 211 remains valid, at least in its general features. Except in the special case of $p = \infty$, there is always a configuration in which $\mathbf{M}$ is minimum; starting out of this are two series of configurations, along each of which $\mathbf{M}$ increases indefinitely up to $\mathbf{M} = \infty$, the end configurations each being configurations of zero rotation ($\omega = 0$). One of these series ends in two spherical masses rotating infinitely slowly round one another at an infinite distance, and this series is stable throughout. At the end of the other series the primary is an infinitely elongated Jacobian ellipsoid, and this series is unstable throughout.

Instead of eliminating R from equations $(217\cdot1)$ and $(217\cdot2)$, and so obtaining $\mathbf{M}$ as a function of ω, we might equally well have eliminated ω from these equations and obtained $\mathbf{M}$ as a function of R, in the form

$$\mathbf{M} = \left[Mk^2 + M'k'^2 + \frac{MM'}{M + M'}\,R^2 \right] (1 + \zeta)^{\frac{1}{2}} (M + M')^{\frac{1}{2}} R^{-\frac{3}{2}}.$$

If M is not greatly different from M', the value of $\mathbf{M}$ reduces to its last term when R is large. Even for configurations in which the ellipsoids are almost in contact, it is readily seen that by far the greater part of the value of $\mathbf{M}$ comes from this last term, so that $\mathbf{M}$ varies approximately as $R^{\frac{1}{2}}$. It follows that the configuration for which $\mathbf{M}$ is a minimum nearly coincides with that for which R is a minimum, this latter being the configuration of closest approach of the centres of the ellipsoids.

Let R_0 be the distance of closest approach. For any value $R_0 + \delta R$ which is just greater than R_0, there will be two configurations; in one the ellipsoids are more elongated than in the configuration of closest approach, while in the other they are less elongated. In the former configuration, then, k^2, k'^2 and ζ are all greater than in the latter, so that, as the values of R are the same in both configurations, it follows that the more elongated configuration has the larger value of $\mathbf{M}$.

Now the diagram of configurations, drawn with $\mathbf{M}$ as vertical co-ordinate, lies as in fig. 34. Here O is the configuration of minimum angular momentum, the less elongated configurations OP are stable, while the more elongated configurations OP' are unstable. We have seen that at the configuration of closest approach R, increasing elongation goes with increasing angular momentum; it follows that R is on the unstable branch OP' of the series.

Thus the configuration of closest approach is always unstable; it is fairly near to O when M and M' are nearly equal, but is far removed from O in other cases. Passing to the limit of configurations of greater elongation, it is easily shewn that in the extreme configuration P' in which $\mathbf{M} = \infty$ and $\omega = 0$, the two bodies must overlap; thus this configuration, although satisfying the

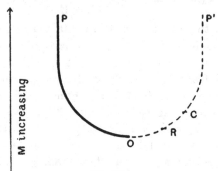

Fig. 34.

mathematical equations, is physically impossible. At some stage between R and P, there must be a configuration C, in which the bodies are just in contact, but without overlapping; this configuration, which we may call the contact configuration, is the last one which is physically possible. It is clear that all contact configurations are necessarily unstable.

Darwin calculates in detail the configurations C, R and O for the case in which the two masses are equal ($p = 1$ or $M = M'$). The calculation for configuration O is not very accurate, for his series do not give good approximations when the masses are in or close to actual contact.

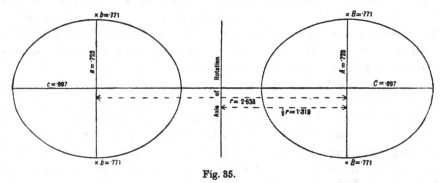

Fig. 35.

When $M = M'$, Darwin finds for the configuration O of limiting stability
$$a = a' = 0\!\cdot\!897, \quad b = b' = 0\!\cdot\!771, \quad c = c' = 0\!\cdot\!723, \quad r = 2\!\cdot\!638,$$
the unit being the radius of the sphere formed by rolling the two masses into one, and the cross-section is shewn in fig. 35 which is reproduced from Darwin's *Collected Works**.

---

* Vol. III. p. 513. I am indebted to the Syndics of the Cambridge University Press for permission to reproduce this figure, and also figs. 36—39 from the original blocks.

Again, when $M = M'$, Darwin gives the value of r for the configuration R of closest approach as 2·343, but he does not compute the axes.

For the contact configuration C in the same case of $M = M'$ Darwin finds *

$$r = 2·372, \quad a = a' = 1·186.$$

In these solutions the figures are assumed to be ellipsoidal; the harmonic deformations which have to be superposed will of course bring the vertices closer, so that actual contact will occur before the vertices of the ellipsoids touch. Darwin gives the following figure, which he describes as "highly conjectural" for such a case.

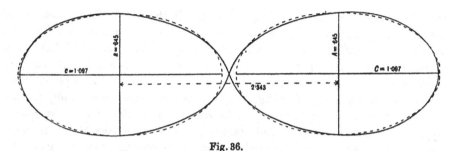

Fig. 36.

Darwin calculates the value of R in the configurations of limiting stability and of closest approach (i.e. the minimum possible value of R) in some other cases; from his results we can compile the following table:

$$M/M' = \quad 1, \qquad 0·8, \qquad 0·5, \qquad 0·4, \qquad 0.$$
$$R \text{ (limiting stability)} = 2·638 \qquad\qquad 2·574, \quad 2·59, \ \infty .$$
$$R \text{ (closest approach)} \ = 2·343, \quad 2·36, \qquad\qquad\qquad 2·457.$$

Partial Stability.

218. As has already been noticed, the entry $p = 0$, $R = \infty$ means that there can be no secular stability for an infinitesimal planet until it has been driven off to infinity. The agency by which this driving off is accomplished is tidal friction; the satellite M raises tides in the primary M'; the dissipation of energy in the tides provides the dissipation necessary for secular instability to come into operation, and the tidal forces result in an acceleration of the small body at the expense of the energy of rotation of the large, this process continuing until the bodies are infinitely far apart.

On the other hand, if the big body is regarded as a point or rigid sphere, so that tidal friction cannot operate, the problem becomes identical with Roche's problem already discussed. The value of R in limiting stability when $\rho = 0$ is no longer ∞, but $2·455r$. Thus tidal friction in the primary can increase the value of R from this value to infinity.

---

* Approximately: I have extrapolated to get initial contact in Darwin's table on p. 514 of *Collected Works*, III. Stress should not be laid on exact values, as Darwin specially draws attention to the bad convergence of the series used in this and similar calculations.

Darwin describes a system as "partially stable" when it is stable except for the tidal friction arising from the tides in the primary, remarking that, inasmuch as tidal friction is a slowly acting cause of instability, partial stability of this kind is from the point of view of cosmical evolution of even greater interest than the full secular stability of the system.

Darwin believes that the limit of partial stability of a series of configurations, such as that represented in fig. 34, can be found by discovering the value at which

$$\mathbf{M'} = \left(Mk^2 + \frac{MM'}{M+M'} R^2 \right) \omega$$

is a minimum, this value $\mathbf{M'}$ representing all that part of the moment of momentum which is liable to variation when tides cannot be raised in M'. A slight modification of the argument of § 217 will shew that the configuration of closest approach cannot be even "partially stable." It accordingly appears that the configuration of limiting "partial stability" must lie at a point intermediate between O and R in fig. 34. Darwin calculates some configurations of "partial stability," and gives the following table of values for R, the closest approach consistent with partial stability, and for the axes of the primary and secondary when in this critical position, a being the mean radius of the combined mass:

Table XVIII. *Axes of Binary Systems.* (Darwin.)

$\dfrac{M}{M'}$	$\dfrac{R}{a}$	Axes of primary			Axes of secondary			$\dfrac{\omega^2}{2\pi\gamma\rho}$
		a/a	b/a	c/a	a'/a	b'/a	c'/a	
0	2·457	$1 \div \infty$	$0·511 \div \infty$	$0·482 \div \infty$	1·030	1·030	0·942	0·0449
0·4	2·484	0·843	0·603	0·562	0·988	0·886	0·815	0·0435
0·5	2·485	0·870	0·642	0·597	0·979	0·860	0·792	0·0434
0·6	2·490	0·888	0·674	0·627	0·969	0·836	0·772	0·0432
0·7	2·497	0·901	0·701	0·652	0·958	0·815	0·753	0·0428
0·8	2·502	0·912	0·725	0·673	0·947	0·796	0·737	0·0426
0·9	2·508	0·921	0·744	0·691	0·937	0·778	0·722	0·0423
1·0	2·514	0·927	0·762	0·708	0·927	0·762	0·708	0·0420

Changes in the ratio of the masses are found to produce surprisingly little change in the critical value of R/a. Whatever the ratio of the masses, it remains approximately true that a satellite cannot rotate with partial secular stability round a planet if their centres are within a distance less than $2\frac{1}{2}$ radii of the combined mass, the densities being the same.

The critical figures for $M/M' = 0·4$, $0·7$ and $1·0$ are shewn in figs. 37, 38, 39, taken from Sir G. Darwin's *Collected Works*[*].

* III. pp. 508, 509.

219. The general nature of the motion which sets in when two masses of a binary system approach too close for stability, can be seen from considerations

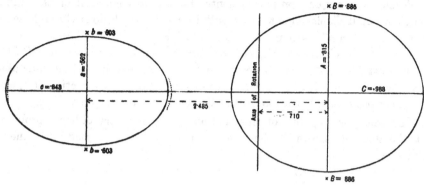

Fig. 37 $(M/M'=0.4)$.

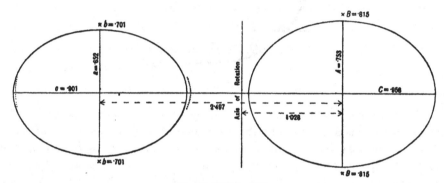

Fig. 38 $(M/M'=0.7)$.

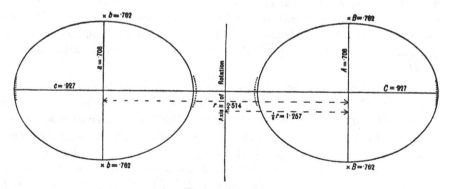

Fig. 39 $(M=M')$.

precisely similar to those brought forward in considering the dynamical motion in the rotational problem (§ 204).

Let POP' (fig. 40) represent the series of equilibrium configurations for the satellite already exhibited in fig. 34, the branch PO being stable, the branch OP' being unstable, and the point O·representing the configuration of limiting stability.

When the configuration is represented by a point such as A on the stable branch, a small displacement will result in stable oscillations through some small range $A'AA''$. When the representative point is at C the range of stable oscillations is very small, and an oscillation of range greater than $C'CC''$ will be unstable. Finally at O any oscillation at all will result in an unstable motion which will initially be represented by motion in a direction OO', and so will consist of an elongation of the ellipsoidal figure of the satellite.

The exact tracing out of this motion presents a very difficult problem, but the general nature of the motion is disclosed bv a study of general principles.

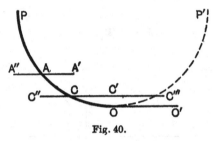

Fig. 40.

The radius of the orbit is determined by the same equation as it would be if the whole mass of the satellite were concentrated at its centre of mass. The satellite may be thought of as consisting of two halves H and H', the former being nearer to the primary than the centre of mass and the latter further away. If it were not for the presence of H', the half H would be too near the primary for a circular orbit to be possible under the prescribed rotation; equilibrium is maintained by the gravitational pull from H' which neutralises part of the attraction of the primary on H. Similarly it is only the gravitational attraction of H which makes a circular orbit possible for H'.

When the configuration reaches limiting stability at the point O, a rapid elongation of figure begins, and this lessens the gravitational attraction between H and H'. The immediate result is that H is drawn in closer to the primary, while H' is driven further away. At first this motion is only another representation of the elongation of the figure of the satellite, but it is clear that this elongation cannot continue for ever—a long thin filament of matter must be unstable under all conditions. Before long the satellite must break up into detached masses, the innermost of which will fall in towards the primary while the outermost will recede from it. If we think of these fragments as ultimately describing elliptic orbits, the point at which instability sets in will approximately coincide with the aphelia of the inner pieces and with the perihelia of the outer ones.

If no change of density takes place in the matter of the satellite, the orbits of the inner fragments will all be within the radius of limiting stability, so that for each fragment the same process must repeat itself indefinitely, a limit only being reached when the fragments are so small that their chemical cohesion is able to defy the disruptive effects of gravitation and rotation. The outer fragments, on the other hand, will describe orbits which will all lie outside the radius of limiting stability, and so they will not suffer further disintegration at first. But the perihelia of these orbits are already very close to the sphere of limiting stability, and if the agencies which drove the original satellite inside this sphere are still operative, it may be expected that before long the new satellites also will be driven in and broken up in turn.

220. If the matter of the satellite is even slightly compressible, and therefore liable to changes of density, an entirely new feature presents itself. For the initial elongation of the satellite when the configuration of limiting stability is reached will be accompanied by a rapid diminution of pressure in the interior of the satellite, and therefore by a rapid diminution of average density. The radius of the sphere of limiting stability, however, depends on the density ρ of the satellite, varying as $\rho^{-\frac{1}{3}}$. Thus the elongation of the satellite will be accompanied by a rapid expansion of the sphere of limiting stability; when the satellite breaks into fragments all these will be within the new sphere of limiting stability, and the process of breaking up will repeat itself indefinitely.

Whichever way we approach the problem, we see that the final result of the motion must be a ring of broken fragments, each fragment being so small that its forces of cohesion can resist the mechanical tendency to disintegration.

221. Roche has suggested that Saturn's rings were formed in this way.

It will be remembered that Roche's critical radius of stability was found to be equal to 2·45 radii of the primary when the densities of primary and satellite are the same. The following data accordingly seem to provide *prima facie* support for Roche's suggestion:

Radius of Saturn's outermost ring   = 2·30 radii of Saturn.
Radius of orbit of Saturn's innermost satellite = 3·11 radii of Saturn.
Radius of orbit of Jupiter's innermost satellite = 2·54 radii of Jupiter.
Radius of orbit of Mars' innermost satellite   = 2·79 radii of Mars.

But the figures have no significance unless we suppose the primaries and the satellites to have the same density. In actual fact the densities are likely to be so different, and always to have been so different, that it is difficult to attach much importance to the figures just mentioned.

222. Our study of the configurations of binary systems has shewn that a number of configurations of stable equilibrium exist in which the two bodies rotate about one another. When a rotating mass loses its secular stability on the pear-shaped series of configurations, we have seen that the furrow around its middle part is likely to deepen until it results in fission into two parts. The configurations just studied provide possible final configurations for the mass; they are the configurations which were tentatively inserted in fig. 31 as providing the most likely conclusion to the unstable motions of the pear-shaped figure.

We leave the theoretical problem at this stage to return to it in its more practical aspects in Chapter x. We now turn to study the theoretical problem of the configurations of compressible non-homogeneous masses.

CHAPTER IX

THE CONFIGURATIONS OF ROTATING COMPRESSIBLE MASSES

General Theory.

223. In the last chapter we discussed the configurations assumed by masses of homogeneous and incompressible matter in rotation. Our earlier discussion of the constitution of the stars had, however, shewn that actual stellar matter is far from being either homogeneous or incompressible. Thus it becomes of the utmost importance to inquire how far compressible matter behaves in the way in which incompressible matter has been found to behave.

In the present chapter we shall examine the general theory of the configurations of equilibrium of compressible masses, and shall attempt to discuss in a general way the effect of compressibility in introducing departures from the motion predicted by the incompressible model which we have so far had under consideration.

224. If p is the total pressure, including pressure of radiation at any point x, y, z of a mass rotating with uniform angular velocity ω about the axis of z, the equations of equilibrium are again, as in § 188,

$$\frac{\partial p}{\partial x} = \rho \frac{\partial V}{\partial x} + \omega^2 \rho x \dots\dots\dots\dots(224 \cdot 11),$$

$$\frac{\partial p}{\partial y} = \rho \frac{\partial V}{\partial y} + \omega^2 \rho y \dots\dots\dots\dots(224 \cdot 12),$$

$$\frac{\partial p}{\partial z} = \rho \frac{\partial V}{\partial z} \dots\dots\dots\dots(224 \cdot 13),$$

in which V is the potential of the whole gravitational field of force, including any tidal forces which may act on the rotating mass. With a view to the subsequent discussion of more general problems, let us write

$$V = V_M + V_T \dots\dots\dots\dots(224 \cdot 2),$$

where V_M is the gravitational potential of the rotating mass alone, and V_T is the potential of the tidal field.

Writing $\qquad \Omega = V + \tfrac{1}{2}\omega^2 (x^2 + y^2) \dots\dots\dots\dots(224 \cdot 3),$

these equations of equilibrium become

$$\frac{\partial p}{\partial x} = \rho \frac{\partial \Omega}{\partial x} \dots\dots\dots\dots(224 \cdot 41),$$

$$\frac{\partial p}{\partial y} = \rho \frac{\partial \Omega}{\partial y} \dots\dots\dots\dots(224 \cdot 42),$$

$$\frac{\partial p}{\partial z} = \rho \frac{\partial \Omega}{\partial z} \dots\dots\dots\dots(224 \cdot 43).$$

On equating the two values of $\partial^2 p/\partial y \partial z$ given by the last two of these equations, we obtain

$$\frac{\partial \rho}{\partial z}\frac{\partial \Omega}{\partial y} = \frac{\partial \rho}{\partial y}\frac{\partial \Omega}{\partial z},$$

so that

$$\frac{\dfrac{\partial \rho}{\partial x}}{\dfrac{\partial \Omega}{\partial x}} = \frac{\dfrac{\partial \rho}{\partial y}}{\dfrac{\partial \Omega}{\partial y}} = \frac{\dfrac{\partial \rho}{\partial z}}{\dfrac{\partial \Omega}{\partial z}} \quad \dotsc\dotsc\dotsc\dotsc\dotsc\dotsc(224 \cdot 5).$$

Thus the surfaces $p =$ cons. coincide with the equipotentials $\Omega =$ cons., and it further follows from equations (224·4) that these surfaces also coincide with the surfaces of constant pressure $p =$ cons. The boundary of the fluid must of course be one of this family of surfaces ($p = 0$), and the necessity for the condition that Ω shall be constant over the boundary, which has so far been used as the condition for equilibrium, becomes obvious.

The condition that Ω shall be constant over the boundary no longer suffices to ensure equilibrium; while still necessary, it is no longer sufficient, and equations of equilibrium must now be satisfied throughout the mass.

Masses of Uniform Composition.

225. The case which is most amenable to mathematical treatment is that in which the pressure depends on the density only, through a relation of the general type

$$p = f(\rho).$$

When a mass is of uniform composition throughout, the pressure will in general depend on both the density p and the temperature T, and it will only be when the temperature depends solely on the density that this relation is strictly satisfied.

We have, however, seen that the matter in a star at rest arranges itself in such a way that the pressure and density are approximately connected by a relation of the form

$$p \propto \rho^{\kappa},$$

where κ has values which we have calculated in detail for different stars. It has not been proved that this relation persists when the star is in rotation, but it is clear that in this case also the pressure and density must increase or decrease together following a relation of the same general type. In what follows we shall assume that, as an approximation at least, the pressure in the equilibrium configuration depends only on the density, being given by the relation $p = f(\rho)$.

With this value for p, equations (224·4) assume the form

$$\frac{\partial f(\rho)}{\partial \rho}\frac{\partial \rho}{\partial x} = \rho \frac{\partial \Omega}{\partial x}, \text{ etc.}$$

If $\phi(\rho)$ is defined by

$$\phi(\rho) = \int \frac{1}{\rho}\frac{\partial f(\rho)}{\partial \rho}\,d\rho \quad \dotsc\dotsc\dotsc\dotsc\dotsc\dotsc(225 \cdot 1),$$

these become $\dfrac{\partial \phi(\rho)}{\partial x} = \dfrac{\partial \Omega}{\partial x}$, etc.,

so that the equations of equilibrium have the common integral

$$\phi(\rho) = \Omega + C \dots\dots\dots\dots\dots(225\cdot2),$$

where C is a constant. At any point inside the mass, $\nabla^2 V_m = -4\pi\gamma\rho$ and $\nabla^2 V_T = 0$, so that, from equation (224·3)

$$\nabla^2 \Omega = 2\omega^2 - 4\pi\gamma\rho \dots\dots\dots\dots(225\cdot3).$$

Thus on operating on equation (225·2) with ∇^2, we obtain

$$\nabla^2 \phi(\rho) + 4\pi\gamma\rho = 2\omega^2 \dots\dots\dots\dots(225\cdot4).$$

This is the differential equation which must be satisfied by ρ for equilibrium to be possible. It of course includes the standard equation (62·2) of Chapter III as a special case, to which we pass by putting $\omega^2 = 0$ and $f(\rho) = K\rho^k$.

226. Let R denote the distance from any point P inside the mass to a variable point x, y, z inside the mass; let dS' be an element of surface of a small sphere S' surrounding P, and let dS be an element of surface of the boundary S. Then the value of V_M at the point P is

$$V_M = \gamma \iiint \frac{\rho}{R}\,dx\,dy\,dz = -\frac{1}{4\pi}\iiint \frac{\nabla^2 V}{R}\,dx\,dy\,dz.$$

Both these integrals may legitimately be supposed to extend throughout the region between S and S', so that V_M may be expressed as a surface integral in the form

$$V_M = -\frac{1}{4\pi}\iint \left\{\frac{\partial V}{\partial n}\left(\frac{1}{R}\right) - V\frac{\partial}{\partial n}\left(\frac{1}{R}\right)\right\}(dS + dS')$$

$$= V + \frac{1}{4\pi}\iint \left\{\frac{\partial V}{\partial n}\left(\frac{1}{R}\right) - V\frac{\partial}{\partial n}\left(\frac{1}{R}\right)\right\}dS.$$

Here V is the total potential, so that $V = V_M + V_T$, and hence

$$V_T = -\frac{1}{4\pi}\iint \left\{\frac{\partial V}{\partial n}\left(\frac{1}{R}\right) - V\frac{\partial}{\partial n}\left(\frac{1}{R}\right)\right\}dS.$$

The integral on the right is of course the potential of a Green's equivalent stratum; it is a known theorem that the potential of this together with that of external masses has a constant value inside the surface.

Since $V + \frac{1}{2}\omega^2(x^2 + y^2)$ is constant over the boundary in a configuration of equilibrium, this value of V_T becomes

$$V_T = -\frac{1}{4\pi}\iint \frac{\partial V}{\partial n}\left(\frac{1}{R}\right)dS - \frac{\omega^2}{8\pi}\iint (x^2 + y^2)\frac{\partial}{\partial n}\left(\frac{1}{R}\right)dS + \text{a cons.}$$

For a problem in which V_T, ω and the shape of the boundary are given, the values of

$$V_T \text{ and } \frac{\omega^2}{8\pi}\iint (x^2 + y^2)\frac{\partial}{\partial n}\left(\frac{1}{R}\right)dS$$

are given at every point inside the boundary. It follows that

$$\iint \frac{\partial V}{\partial n} \frac{1}{R} \, dS$$

is given at all points inside the boundary except for a constant. Also

$$\iint \frac{\partial V}{\partial n} \, dS$$

is given, being equal to 4π times the total mass of the rotating body, so that $\partial V/\partial n$ is uniquely determined at every point of the boundary.

It follows from equation (225·2) that ρ and $\partial\rho/\partial n$ are determined at every point of the boundary, and from this and equation (225·4) it can be seen that the solution for ρ is unique *.

The important result follows that: *when the pressure depends only on the density, configurations of equilibrium may be specified by their boundaries alone.*

An even more important result also follows. When V_T and ω^2 are given and the boundary is given, there will be an endless number of possible vibrations in which the internal particles move, while those at the boundary remain in position. The result just obtained shews that none of these can ever be of zero frequency, so that no points of bifurcation can occur. Thus: *the various internal vibrations, if stable in the initial configuration of the mass, must always remain stable.*

227. From the circumstance that configurations of equilibrium may be specified by their boundaries alone, the various configurations must fall into linear series much in the same way as the incompressible problem.

The configuration for no rotation and no tidal forces will of course be spherical. Starting out from this, there will obviously be a series in the rotational problem analogous to the Maclaurin spheroids, in which the boundary is a figure of revolution. The configurations near to the spherical one are spheroids of small ellipticity, but the series does not remain spheroidal throughout its length. But the far end of this series is again spheroidal, being in fact identical with the Maclaurin spheroid for a mass which has a uniform density equal to the density at the surface of the matter ($p = 0$). Just as in the incompressible problem, this is unstable for all displacements specified by sectorial harmonic deformations of its boundary†. It follows that, on the series we are considering, there must be points of bifurcation corresponding to all sectorial harmonics. The general physical principles explained in §197 lead us to expect with confidence that the first of these to occur will be that corresponding to the second harmonic. At this point the circular cross-section of the figure gives place at first to an elliptic cross-section of

* *Proc. Roy. Soc.* 93 (1917), p. 416. It is difficult to construct a rigorous proof, for complications of a mathematical nature arise.

† Cf. Poincaré, *Acta Math.* VII. 1885, p. 259.

small ellipticity, and the configurations on the new series are analogous to the Jacobian ellipsoids.

Further, the far end of the series analogous to the Jacobian series is again identical with that in the incompressible problem, both as regards configuration and stability, so that again this series must have the same points of bifurcation as the Jacobian series.

These statements obviously require modification in the extreme case in which the density of the matter falls to zero at the surface, but except for this case it is clear that the general arrangement of series and points of bifurcation will be very similar to that in the problem of the configurations assumed by an incompressible mass. It ought again to be possible to construct a diagram similar to that shown in fig. 25 (p. 213); the general arrangement will be the same, but the numerical values different, and the shape of the configurations will of course be different except at the extreme ends of the various series.

Figures of equilibrium which take the place of the spheroidal figures of the incompressible problem, whether rotational or tidal, may conveniently be referred to as " pseudo-spheroids." Similarly figures which take the place of ellipsoidal figures of equilibrium may be referred to as " pseudo-ellipsoids."

This general discussion does not touch the question of the stability of the various branch series; this can only be determined by detailed calculations in individual problems. Thus it is not possible to predict from a consideration of general principles whether the pseudo-ellipsoidal series of configurations for a rotating mass will initially be stable or unstable. If it is unstable in any particular case, cataclysmic motion will begin as soon as the first point of bifurcation on the pseudo-spheroidal series is reached. This motion will consist at first of an ellipsoidal elongation of the pseudo-spheroid, the circular cross-sections giving place to elliptical ones, and the points of bifurcation on the pseudo-ellipsoidal series will be replaced by " dynamical points of bifurcation " in this motion. If ever such a case occurs, it seems possible that the rotating mass may divide up into a number of detached masses, instead of into only two.

It must, however, be remembered that the angular momentum of the pseudo-ellipsoidal series is infinite at its far end, so that much the most likely event is that it increases all along the length of this series; in this event the pseudo-ellipsoidal series would initially be stable. But no such general consideration can be brought forward in the case of the pear-shaped series which branches off at the first point of bifurcation, and nothing justifies us in predicting whether this will in general be stable or unstable. Indeed, it appears to be at least possible that in some problems this series may be initially stable, a possibility which has been mentioned by Poincaré*.

* "Sur la Stabilité de l'Equilibre des Figures Pyriformes affectées par une Masse Fluide en Rotation," *Phil. Trans.* 198 A (1902), p. 335.

DISCUSSION OF SPECIAL MODELS.

228. Leaving the realm of general principles, we now turn to a discussion of the behaviour of particular models, conforming to special laws of compressibility. There are, of course, an infinite variety of arrangements of compressible matter possible, while the full discussion of even a single case presents a problem of considerable difficulty and complexity. It therefore behoves us to choose the special cases which we attempt to solve with skill and care, so as to economise labour as much as possible.

Compressibility of matter necessarily results in variations of density in the compressible mass, and the greater the compressibility of the matter, the greater these variations of density will be. In the last chapter we discussed the special case of a mass having no compressibility at all, and so having no variations of density.

In a sense this problem formed a limiting case of the problem of the motion of a compressible mass. At the other extreme there will be another limiting case in which the compressibility is so great that infinite variations of density may be expected. Mathematically this limiting case may be specified by the condition that the density is infinite or zero at different places. Physically, this limiting case proves not to be so artificial as its mathematical specification might lead us to suppose.

In a mass of gas at rest with the temperature uniform throughout (isothermal equilibrium), the density at great distances from the centre falls off as $1/r^2$. The complete law of density has been obtained by Darwin[*] and others[†]. But it is clear without detailed analysis that at a sufficient distance from the centre, the law of density must become [‡]

$$\rho = \rho_0 \left(a^2/r^2\right)$$

so that, when viewed from a very great distance, the density may be regarded as infinite at the centre and zero everywhere else. The total mass is, however, infinite, so that a finite mass of gas in isothermal equilibrium will be of zero density everywhere.

Similarly for a mass of gas in adiabatic equilibrium with the ratio of the specific heats κ equal to $1\frac{1}{5}$, the law of density is (formula 64·4)

$$\rho = \rho_0 \left(1 + r^2/a^2\right)^{-\frac{5}{2}}.$$

Again, when this mass of gas is viewed from a sufficient distance, the value of ρ becomes infinite at the centre and zero everywhere else. The same is true for any value of from 1 to $1\frac{1}{5}$. The mass is infinite when $\kappa < 1\frac{1}{5}$, but becomes finite when $\kappa = 1\frac{1}{5}$.

---

[*] "On the Mechanical Conditions of a Swarm of Meteorites and on Theories of Cosmogony." *Phil. Trans.* 180 A (1889), p. 1 and *Coll. Works*, IV. p. 362.

[†] For detailed references see Darwin's paper.

[‡] *l.c.* p. 377.

This same model, in which the density is infinite or very great over a point or small concentrated area but zero everywhere else, has been largely utilised by Roche* in his researches on cosmogony, and may suitably be called "Roche's model." Roche interpreted it physically as referring to a small and intensely dense solid nucleus surrounded by an atmosphere of negligible density. In Roche's model, the whole of the mass is supposed concentrated at the centre; in this respect it differs from a mass of gas in isothermal equilibrium, although giving a faithful representation of an adiabatic mass for which $\kappa = 1\frac{1}{3}$.

Roche's Model.

229. It is now clear that Roche's model and the incompressible model form the two limiting cases of the general compressible mass. The latter has already been studied in detail; it is natural to begin our investigation of the compressible problem with a discussion of the former.

In studying the configurations and motion of an incompressible mass, one of the main difficulties was found to lie in the determination of the gravitational potential. In Roche's model no such difficulty occurs; the whole mass is collected at a point and the gravitational potential is simply M/r. Thus the quantity $V + \frac{1}{2}\omega^2(x^2 + y^2)$, which has been denoted by Ω, assumes the simple form

$$\Omega = \frac{\gamma M}{r} + V_T + \frac{1}{2}\omega^2(x^2 + y^2) \quad \ldots\ldots\ldots\ldots\ldots(229\cdot1).$$

We have seen (§ 224) that the boundary of the compressible mass must be one of the equipotential surfaces $\Omega = $ constant.

For given values of V_T and ω, the surfaces $\Omega = $ constant will be a system of equipotentials of the usual type, and as Ω is uniquely determined as function of x, y and z, two different equipotentials can never intersect. Of the whole system of equipotentials only one is a possible boundary for the gravitating mass, this being picked out by the condition that the volume enclosed by it shall be just adequate to contain the whole amount of the compressible matter.

Or allowing either V_T or ω^2 to vary, we obtain a linear series of configurations by picking out the appropriate equipotential surface from each set. When V_T and ω^2 both vanish, the equipotentials are spheres and the boundary is spherical; as we pass along the linear series, the boundary departs more and more from the spherical shape.

With Roche's model there can be no points of bifurcation or turning points on any linear series. For when V_T and ω are given, the value of Ω is uniquely determined by equation (229·1), and hence the boundary is

---

* "Essai sur la Constitution et l'Origine du Système solaire" (1873). *Acad. de Montpellier, Sciences*, VIII. p. 235. See also Poincaré, *Leçons sur les Hypothèses Cosmogoniques*, Chap. III.

uniquely determined. But the condition for a point of bifurcation or a turning point is that there shall be two adjacent configurations of equilibrium, and hence (by § 226) two different boundaries, possible for the same value of V_T and ω.

Thus all possible configurations for a Roche's model must lie on one linear series, and this may in every case be supposed to start from the spherical configuration for which V_T and ω both vanish. As we proceed along this series, the different boundaries are equipotentials which differ more and more from spheres, until finally it may happen that the equipotential which forms the boundary coincides with one which marks a transition from closed to open equipotentials.

Such a transition must necessarily be through an equipotential which intersects itself, and therefore through an equipotential on which a point of equilibrium occurs. Such a point is determined by the equations

$$\frac{\partial \Omega}{\partial x} = \frac{\partial \Omega}{\partial y} = \frac{\partial \Omega}{\partial z} = 0.$$

Since Ω is necessarily constant over the surface of every equipotential, including the boundary, this condition may be put in the alternative form

$$\frac{\partial \Omega}{\partial n} = 0.$$

On moving still further along the linear series, we shall find that no closed equipotential is capable of containing the whole mass, so that there is no equilibrium configuration consistent with values of ω^2 and V_T beyond a certain limit.

Freely Rotating Mass.

230. To discuss the particular problem of a single mass rotating freely in space, we take $V_T = 0$, so that

$$\Omega = \frac{\gamma M}{r} + \tfrac{1}{2}\omega^2 (x^2 + y^2) \quad \dots\dots\dots\dots\dots(230 \cdot 1).$$

On sketching out the forms of the equipotential surfaces $\Omega = $ constant (cf. fig. 41), we see that as Ω decreases, the condition for a point of equilibrium will first be satisfied in the plane of xy. In this plane the condition becomes

$$\frac{\partial \Omega}{\partial x} = \frac{\partial \Omega}{\partial y} = 0,$$

which is satisfied if

$$-\frac{\gamma M}{r^3} + \omega^2 = 0,$$

or

$$\gamma M = \omega^2 \varpi_0^3 \quad \dots\dots\dots\dots\dots\dots(230 \cdot 2),$$

where ϖ_0 is the radius of the cross-section in the plane of xy.

This equation admits of a very simple interpretation. At a distance ϖ_0 from the nucleus, the gravitation force is $\gamma M / \varpi_0^2$ while centrifugal force is

$\omega^2 \varpi_0$. Equation (230·2) expresses that these forces are equal. Thus a point of equilibrium occurs when centrifugal force just balances gravity.

The same result follows from equation (224·4). The condition

$$\partial\Omega/\partial x = \partial\Omega/\partial y = \partial\Omega/\partial z = 0$$

which defines a point in equilibrium is seen to be the same as

$$\partial p/\partial x = \partial p/\partial y = \partial p/\partial z = 0,$$

so that gas-pressure exerts no force on the matter at the point of equilibrium, and its particles constitute small satellites which revolve about the central nucleus in orbits described under gravitation alone.

231. The particular equipotential on which the point of equilibrium occurs is readily found to be

$$\Omega \equiv \frac{\gamma M}{r} + \tfrac{1}{2}\omega^2 (x^2 + y^2) = \tfrac{3}{2} (\gamma M \omega)^{\frac{2}{3}},$$

or, using the relation $\varpi^3 = \gamma M/\omega^2$,

$$\frac{1}{r} + \frac{1}{2}\frac{\varpi^2}{\varpi_0^3} = \frac{3}{2}\frac{1}{\varpi_0} \qquad \ldots\ldots\ldots\ldots\ldots\ldots(231\cdot1),$$

where ϖ^2 stands for $x^2 + y^2$.

The surfaces $\Omega = $ constant are found to lie as in fig. 41, the critical equipotential defined by equation (231·1) being drawn thick.

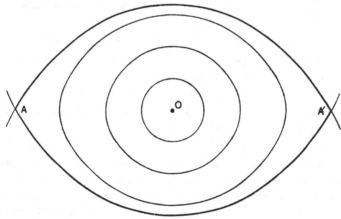

Fig. 41.

Putting $r^2 = z^2 + \varpi^2$, the equation of the critical equipotential (231·1) assumes the form

$$z = \frac{\varpi_0^2 - \varpi^2}{3\varpi_0^2 - \varpi^2} (4\varpi_0^2 - \varpi^2)^{\frac{1}{2}},$$

and its volume can be calculated to be *

$$4\pi \int_{\varpi=0}^{\varpi=\varpi_0} z\varpi\, d\varpi = 4\pi\varpi_0^3 \times 0\cdot180373.$$

* *Problems of Cosmogony and Stellar Dynamics*, p. 150.

If $\bar{\rho}$ denotes the mean density of all the matter inside this critical equi-potential, this volume is equal to the mass M divided by $\bar{\rho}$. Hence

$$\frac{\omega^2}{2\pi\gamma\bar{\rho}} = \frac{M}{2\pi\bar{\rho}\varpi_0^3} = \frac{\text{volume}}{2\pi\varpi_0^3} = 0.36075 \quad \ldots\ldots\ldots\ldots(231.2).$$

232. Let the value of $\omega^2/2\pi\gamma\bar{\rho}$ increase continuously in a mass of com-pressible matter in which the distribution of density is approximately that represented in Roche's model. The spherical configuration corresponds to $\omega^2/2\pi\gamma\bar{\rho} = 0$, and as $\omega^2/2\pi\gamma\bar{\rho}$ increases the boundary assumes in turn the shape of the different equipotentials shewn in fig. 41, until it reaches the value $\omega^2/2\pi\gamma\bar{\rho} = 0.36075$ at which the series of configurations comes abruptly to an end, through there being no closed equipotentials corresponding to higher values of $\omega^2/2\pi\gamma\bar{\rho}$.

The configurations remain symmetrical about the axis of rotation through-out, so that the series is clearly the series of pseudo-spheroids whose existence we discovered in § 227. Indeed, for small velocities of rotation, the configu-rations are strictly spheroidal and in every way analogous to the series of Maclaurin spheroids for a mass of incompressible matter.

When the surface density is different from zero, we have seen (§ 227) that the series of pseudo-spheroids necessarily has on it points of bifurcation corresponding to harmonic deformations of various orders, and that instability first enters this series at its intersection with a series of pseudo-ellipsoids analogous to the Jacobian ellipsoids.

In the present problem, in which the surface density is zero, we have now seen that no such point of bifurcation exists; the series of pseudo-spheroids continues stable up to its abrupt ending at the value

$$\omega^2/2\pi\gamma\bar{\rho} = 0.36075.$$

The question naturally arises as to the course of events when ω^2 first exceeds the critical value $0.36075 \times 2\pi\gamma\bar{\rho}$. When ω^2 has this value, ϖ_0 is given by equation (230.2), namely $\gamma M = \omega^2\varpi_0^3$. If ω increases further, ϖ_0 decreases, since M, the mass of the central nucleus, remains the same. Thus there is a new critical equipotential, of smaller radius, and so of higher density, for which both ω^2 and $\bar{\rho}$ are increased, but $\omega^2/2\pi\gamma\bar{\rho}$ retains its original value of 0.36075. An increase in ω can be met by the mass shrinking to this new configuration. But we have already seen that the particles which formed the sharp edge of the original configuration were in pure orbital motion under the gravitational attraction of the central nucleus alone. When the mass shrinks it can exercise no grip on these particles, so that they are left revolving in their original orbits with their original angular velocity.

Thus as ω^2 steadily increases, a rotating mass, formed after Roche's model, will pass through a series of pseudo-spheroidal configurations, rotating as a rigid body, until ω^2 reaches the critical value $0.36075 \times 2\pi\gamma\bar{\rho}$. At this stage the shape of the mass is that of a lenticular figure with a sharp edge. Beyond this, rotation as a rigid body is impossible. As ω^2 increases further, the central mass shrinks, $\bar{\rho}$ increasing so that $\omega^2/2\pi\gamma\bar{\rho}$ remains constantly equal to 0·36075. It retains its original lenticular configuration, but as it shrinks it leaves behind it successive rings of particles rotating in its equatorial plane. Thus the complete mass at any instant consists of a central lenticular mass rotating as a rigid body with an angular velocity given always by $\omega^2/2\pi\gamma\bar{\rho} = 0\cdot36075$, together with rings of particles occupying the equatorial plane, and rotating at slower speeds.

Laplace's Nebular Hypothesis.

233. This mathematical concept formed the basis of Laplace's famous nebular hypothesis *. Laplace believed that the normal astronomical mass shrunk continually as a result of the emission of radiation from its surface. If so, its density would continually increase, ω^2 also increasing so as to satisfy the conservation of angular momentum, with such a rate of increase that $\omega^2/2\pi\gamma\bar{\rho}$ also increased. Laplace consequently supposed that the normal astronomical mass passed through the sequence of configurations just described. He quoted Saturn surrounded by his rings as an example of the final configuration, and suggested that the satellites of Saturn had been formed by the condensation of similar rings, that the present rings also would in time condense into satellites, and that all the planets and satellites of the solar system had been produced by the condensation of rings of this type left behind in the shrinkage of a central cooling mass.

Strong reasons now compel the abandonment of this view of the origin of the solar system (cf. Chapter XVI below), but instances of the critical formation, the lens-shaped mass surrounded by rings of matter in the equatorial plane of the lens, are provided in abundance by the extra-galactic nebulae (cf. Chapter XIII, and particularly Plates II and XIII).

THE GENERALISED ROCHE'S MODEL.

234. The two models we have so far studied in detail, namely, Roche's model and the mass of incompressible matter, have provided examples of two entirely distinct methods of breaking up with increasing rotation, the incompressible mass breaking by fission into two parts, and Roche's model breaking up through the shedding of successive rings of matter from its equator. It is obviously very desirable to bridge over, if possible, the wide gap between these two extreme cases. To some extent a bridge is formed by the consideration of a third model, which combines some of the properties of both of the two models so far discussed.

Roche's model consisted of a nucleus of finite mass but infinitesimal volume, surrounded by an atmosphere of zero mass but finite volume; the density of the nucleus was infinite while that of the atmosphere was zero.

---

* "Exposition du Système du Monde, Note VII," in *Œuvres Complètes de Laplace*, published by the Acad. des Sciences (Paris, 1835), Vol. VI. p. 498.

In the new model the nucleus is supposed to be of finite extent, and therefore of finite density, while the atmosphere remains of finite extent but of infinitesimal density. Thus the potential of the mass V_M is no longer M/r but becomes identical with the potential of the nucleus. The nucleus will be supposed to be incompressible and of uniform density ρ_0, and the atmosphere being of infinitesimal density will be supposed to exert no appreciable pressure on the nucleus.

Let v_N denote the volume of the nucleus and v_A that of the atmosphere. The mass M is equal to $\rho_0 v_N$, so that the mean density is

$$\bar{\rho} = \frac{v_N}{v_A + v_N} \rho_0.$$

When the system rotates with uniform angular velocity each particle of the nucleus is subjected to exactly the same forces as though the nucleus were alone in space rotating with this angular velocity, the atmosphere being entirely non-existent. This completely determines the configurations of the nucleus; they consist of Maclaurin spheroids, Jacobian ellipsoids, etc.

The boundary of the atmosphere must be one of the equipotentials $\Omega = \text{constant}$; it must further be an equipotential of total volume $v_A + v_N$. Thus to obtain the complete figure of equilibrium corresponding to a given rotation, we must first draw a figure of equilibrium appropriate to this rotation for an incompressible mass of density ρ_0 and volume v_N. The boundary of this will be an equipotential $\Omega = \text{constant}$ of volume v_N. We must then draw successive outer equipotentials until a further volume v_A has been enclosed. The equipotential which just includes a further volume v_A will be the required boundary.

In drawing the equipotentials for a given speed of rotation, we may possibly find that closed equipotentials give place to open ones before a volume v_A has been enclosed. If so, there can be no figure of equilibrium corresponding to the given rotation. If v_A' is the volume enclosed by the last closed equipotential, the greatest atmosphere which can be retained at the given rotation will be one of volume v_A', and of the atmosphere of the original model, a volume $v_A - v_A'$ must already have been thrown off at the equator.

Let us examine the value of the critical volume v_A'. Let the surface of the nucleus be supposed to be the standard ellipsoid

$$\frac{x^2}{a^2} + \frac{y^2}{b^2} + \frac{z^2}{c^2} = 1,$$

the axis of z being the axis of rotation. The gravitational attraction at a point $x, 0, 0$ on the prolongation of the major axis is

$$X = 2\pi\gamma\rho_0 abc\, x \int_{x^2 - a^2}^{\infty} \frac{\lambda}{(a^2 + \lambda)^{\frac{3}{2}} (b^2 + \lambda)^{\frac{1}{2}} (c^2 + \lambda)^{\frac{1}{2}}},$$

which reduces, for a Maclaurin spheroid, in which $a = b$, to

$$X = 2\pi\gamma\rho_0 abc x \left[\frac{1}{\alpha^3} \sin^{-1}\frac{\alpha}{x} - \frac{(x^2 - \alpha^2)^{\frac{1}{2}}}{\alpha^2 x^2} \right] \quad \ldots\ldots\ldots(234\cdot1),$$

where $\alpha = (a^2 - c^2)^{\frac{1}{2}}$. The ratio of centrifugal force to gravity at any point on the x axis is $\omega^2 x/X$. This ratio is necessarily less than unity at a point on the boundary of the nucleus, but it increases as we pass outwards, and the point at which it attains the value unity is the critical point at which $\partial\Omega/\partial x = 0$. Hence we obtain this critical point by equating the right-hand member of the equation (234·1) to $\omega^2 x$; the resulting equation is

$$\frac{\omega^2}{2\pi\gamma\rho_0} = abc \left[\frac{1}{\alpha^3} \sin^{-1}\frac{\alpha}{x} - \frac{(x^2 - \alpha^2)^{\frac{1}{2}}}{\alpha^2 x^2} \right] \quad \ldots\ldots\ldots(234\cdot2).$$

The value of x which satisfies this equation determines the radius of the equator of the limiting equipotential.

In the special case in which the nucleus is a Maclaurin spheroid at its ellipsoidal point of bifurcation, the value of $\omega^2/2\pi\gamma\rho_0$ is 0·18712, and the root of equation (234·2) is found to be

$$x = 1\cdot6436\alpha = 1\cdot5990 \, (abc)^{\frac{1}{3}}.$$

The critical equipotential is drawn in fig. 42; it is clear that the value of v_A' here is quite small, being in point of fact rather less than one-third of v_N.

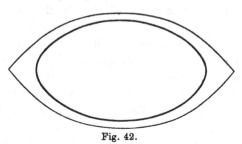

Fig. 42.

In the spherical configuration which corresponds to no rotation, there is no limit to the value of v_A', so that the value of the ratio v_A'/v_N steadily decreases from ∞ to about $\frac{1}{3}$ as we pass along the Maclaurin series to the point of bifurcation, and it is readily calculated that it decreases still further as we pass along the Jacobian series. At the pear-shaped point of bifurcation its value is about $\frac{1}{5}$.

We can now describe the series of equilibrium configurations assumed by this model as its angular momentum continually increases.

Suppose first that the ratio v_A/v_N is greater than $\frac{1}{3}$. For small values of ω, the boundary of the nucleus and the atmosphere are both spheroids of small eccentricity. For larger values of ω the boundary of the nucleus remains spheroidal, while that of the atmosphere is a pseudo-spheroid coinciding with one of the external equipotentials. As ω increases further this pseudo-spheroid develops a sharp edge, this occurring when the critical volume v_A' is equal to v_A. After this, matter is shed from the sharp edge on the equatorial plane of the mass. By the time the rotation is given by $\omega^2/2\pi\gamma\rho_0 = 0\cdot18712$, the atmosphere is reduced to about $\frac{1}{3}v_N$ in volume, so that $\bar{\rho} = \frac{3}{4}\rho_0$, and $\omega^2/2\pi\gamma\bar{\rho} = 0\cdot2496$. At this stage the figure loses its symmetry, no longer remaining a figure of revolution. The nucleus becomes ellipsoidal, while the boundary becomes a pseudo-ellipsoidal figure having two sharp pointed ends, and when the rotation increases further, two streams of matter will be ejected from these ends. As the nucleus gradually becomes more elongated, the atmosphere diminishes more and more, until the pear-shaped point of bifurcation is reached. After this the nucleus will divide into detached masses, each surrounded by a thin atmosphere.

If the original atmosphere were of volume less than $\frac{1}{3}v_N$, the course of events would be the same except that the atmosphere would not begin to be thrown off until after the symmetry of revolution had been lost. In this case the sequence of figures would be—spheroids of small eccentricity, pseudo-spheroids, pseudo-ellipsoids, pseudo-ellipsoids with pointed ends and a stream of matter emerging from each, finally ending in detached masses surrounded by thin atmospheres.

The Adiabatic Model.

235. The two models first discussed, namely, the mass of incompressible matter and Roche's model, may be regarded as representing the extreme limits of homogeneity and non-homogeneity. The composite model just discussed, formed by combining an incompressible nucleus of volume v_N with an extremely tenuous atmosphere of volume v_A, has provided a continuous transition between these two extremes; the limiting value $v_N/v_A = \infty$ gives the incompressible model while the limiting value $v_N/v_A = 0$ gives Roche's model.

The two original models may also be regarded as representing the extreme limits of compressibility, but when they are regarded in this light, the composite model fails to provide a transition from one to the other through varying degrees of compressibility.

Such a transition is provided by a model in which the pressure and density are supposed connected by a relation of the type*

$$p = K\rho^\kappa - p_0 \quad\quad\quad\quad\quad\quad (235\cdot1).$$

The value $\kappa = \infty$ provides an entirely incompressible mass, while we have already seen (§ 228) that the value $\kappa = 1\frac{1}{5}$ with $p_0 = 0$ provides approximately the distribution of density assumed in Roche's model.

Thus a general study of figures obeying the law (235·1) for values of κ from $1\frac{1}{5}$ to ∞ will provide a continuous transition from Roche's model to

---

* The term $-p_0$ is inserted so as to give a finite density $\rho = \sigma$ at the boundary, $p = 0$.

the incompressible model, through a series of figures of continually varying compressibility.

It has also the further advantage that on putting $p_0 = 0$ we can obtain the law

$$p = K\rho^\kappa,$$

which has been found (§ 85) to determine the approximate distribution of density in ideal wholly gaseous stars, at least so long as they remain in a spherical configuration. We accordingly proceed to study the configurations of rotating masses in which the relation between the pressure and density is that specified by equation (235·1).

236. When the pressure is given by

$$p = K\rho^\kappa - p_0 \quad\dots\dots\dots\dots\dots\dots(236\cdot1),$$

the value of the quantity $\int \dfrac{dp}{\rho}$ which we have denoted by $\phi(\rho)$ is

$$\phi(\rho) = \int \frac{dp}{\rho} = \frac{K\kappa}{\kappa - 1}\rho^{\kappa-1},$$

and the general equation of equilibrium (225·2) becomes

$$\frac{K\kappa}{\kappa - 1}\rho^{\kappa-1} = \Omega + C \dots\dots\dots\dots\dots\dots(236\cdot2),$$

in which, as before,

$$\Omega = V_n + \tfrac{1}{2}\omega^2(x^2 + y^2) \quad\dots\dots\dots\dots\dots(236\cdot3).$$

Operating with ∇^2 we obtain at once as the differential equation which must be satisfied by ρ,

$$\frac{K\kappa}{\kappa - 1}\nabla^2\rho^{\kappa-1} = -4\pi\rho + 2\omega^2 \quad\dots\dots\dots\dots(236\cdot4).$$

This of course reduces, in the case of spherical symmetry with $\omega^2 = 0$, to the equation (62·2) by which we discussed the equilibrium of a non-rotating star in Chapter III.

Let ρ_0 be the maximum density of the mass, and σ the density at its surface, so that $(\rho_0 - \sigma)$ is necessarily positive. Taking the point of maximum density as origin, it will be possible to expand ρ in the form

$$\rho = \rho_0 - \rho_2 - \rho_3 - \rho_4 - \dots \quad\dots\dots\dots\dots(236\cdot5),$$

where $\rho_2, \rho_3, \rho_4, \dots$ are functions of x, y, z, of degrees 2, 3, 4 ..., respectively. The value of ρ_2 is

$$\rho_2 = -\frac{1}{2}\left(x^2\frac{\partial^2\rho}{\partial x^2} + 2xy\frac{\partial^2\rho}{\partial x\partial y} + \dots\right)$$

in which the differential coefficients are evaluated at the origin, and, since the origin is supposed to be the point of maximum density, ρ_2 must be negative for all values of x, y and z. By a change of axes, it must be possible to express ρ_2 as a sum of squares, so that we may take the value of ρ_2 to be

$$\rho_2 = -(\rho_0 - \sigma)\left(\frac{x^2}{a^2} + \frac{y^2}{b^2} + \frac{z^2}{c^2}\right).$$

If we further put

$$\rho_3 + \rho_4 + \ldots = - e (\rho_0 - \sigma) P_0,$$

then equation (236·5) gives as the value of the density

$$\rho = \rho_0 - (\rho_0 - \sigma) \left(\frac{x^2}{a^2} + \frac{y^2}{b^2} + \frac{z^2}{c^2} + eP_0 \right) \quad \ldots\ldots\ldots\ldots(236\cdot6),$$

and the boundary, defined by the condition $\rho = \sigma$, has for its equation

$$\frac{x^2}{a^2} + \frac{y^2}{b^2} + \frac{z^2}{c^2} - 1 + eP_0 = 0 \quad \ldots\ldots\ldots\ldots\ldots(236\cdot7).$$

If eP_0 is small, this represents a distorted ellipsoid. Now the boundaries of all stable configurations for a perfectly incompressible mass have been seen to be spheroids and ellipsoids, and so are all included in equation (236·7) with $P_0 = 0$. The general argument of § 227 has further shewn that the corresponding configurations of compressible masses can be derived from these spheroidal or ellipsoidal configurations by continuous distortion. Thus it appears that the boundaries of compressible rotating masses may be supposed given by an equation of the form of (236·7), in which eP_0 will be small if the matter is only slightly compressible, but may become comparable with the other terms of the equation for highly compressible matter.

237. Let q be a function of the density ρ, defined by

$$q^2 = \frac{\rho_0 - \rho}{\rho_0 - \sigma} \quad \ldots\ldots\ldots\ldots\ldots\ldots(237\cdot1).$$

As we pass from the centre to the boundary, ρ varies continuously from ρ_0 to σ, so that q varies continuously from 0 to 1. The equation of the surface of constant density ρ is

$$\frac{x^2}{a^2} + \frac{y^2}{b^2} + \frac{z^2}{c^2} + eP_0 = q^2 \quad \ldots\ldots\ldots\ldots\ldots(237\cdot2),$$

so that this surface may be regarded as arrived at by distortion from an ellipsoid of semi-axes qa, qb, qc. Equation (236·7) is a special case of this equation, arrived at by taking $q = 1$.

Let the potential of a homogeneous mass of unit density bounded by the surface (237·2) be denoted by $V_0 (q)$ when evaluated at a point outside the surface, and by $V_i (q)$ when evaluated at a point inside the surface.

Then it is readily seen that the potential of the whole heterogeneous mass whose density is given by equation (236·6), and whose boundary is given by $\rho = \sigma$, must be

$$V_0 = \sigma V_0(1) + \int_\sigma^{\rho_0} V_0(q)\, d\rho \ldots\ldots\ldots\ldots\ldots\ldots(237\cdot31),$$

$$V_i = \sigma V_i(1) + \int_\sigma^{\rho'} V_i(q)\, d\rho + \int_{\rho'}^{\rho_0} V_0(q)\, d\rho \ldots\ldots\ldots(237\cdot32),$$

the first formula giving the potential at a point outside the mass, and the second formula giving the potential at an internal point at which the density is ρ'.

From these formulae I have proved* that the value of V_i is

$$V_i = \rho_0 V_i(1) - \pi\gamma abc(\rho_0 - \sigma) \int_0^\infty \left[\int_q^1 q^2 \left(1 - e\frac{d\phi(q)}{dq^2}\right) dq^2 \right] \frac{d\lambda}{\Delta} \quad \dots(237\cdot4),$$

where $\qquad \phi(q) = P - \tfrac{1}{4}fDP + \frac{1}{2^2}(\tfrac{1}{4}f)^2 D^2P - \dots - \tfrac{1}{8}e[\dots] + \dots \quad \dots(237\cdot5),$

being formally similar to the ϕ defined by our previous equation (99·7) except that f and D are now defined by

$$q^2 f = \frac{x^2}{a^2+\lambda} + \frac{y^2}{b^2+\lambda} + \frac{z^2}{c^2+\lambda} - q^2,$$

$$D = q^2 \left[\left(\frac{1}{a^2} - \frac{1}{a^2+\lambda}\right)\frac{\partial^2}{\partial\xi^2} + \left(\frac{1}{b^2} - \frac{1}{b^2+\lambda}\right)\frac{\partial^2}{\partial\eta^2} + \left(\frac{1}{c^2} - \frac{1}{c^2+\lambda}\right)\frac{\partial^2}{\partial\zeta^2} \right];$$

the lower limit in the integration with respect to q^2 is determined from the equation

$$\frac{x^2}{a^2+\lambda} + \frac{y^2}{b^2+\lambda} + \frac{z^2}{c^2+\lambda} + e\phi(q) - q^2 = 0.$$

We obtain the various equilibrium configurations by inserting the value for V_i just obtained, and the value for ρ given by equation (236·6) into the single equation of equilibrium, namely,

$$V_i + \tfrac{1}{2}\omega^2(x^2 + y^2) = \text{cons.} \dots\dots\dots\dots\dots\dots(237\cdot6).$$

Since this equation must be satisfied throughout the mass, we may equate the coefficients of all powers and products of x, y, z to zero separately, thus obtaining equations giving the coefficients in P_0.

The values of V_i and ρ are each equal to their values for the configurations of rotating incompressible masses plus terms in $(\rho_0 - \sigma)$ and powers of $(\rho_0 - \sigma)$. As a consequence the equations of the boundaries of the configurations of equilibrium for the rotating compressible mass are obtained in the form

$$F_0(x, y, z) + (\rho_0 - \sigma)F_1(x, y, z) + (\rho_0 - \sigma)^2 F_2(x, y, z) + \dots = 0 \quad \dots(237\cdot7),$$

where $F_0(x, y, z) = 0$ is the equation of a configuration of equilibrium for a mass of uniform density ρ_0, and so of a Maclaurin spheroid or a Jacobian ellipsoid. The function $F_1(x, y, z)$ is found to consist of terms of degrees 4, 2 and 0 in x, y, z, the function $F_2(x, y, z)$ consists of terms of degrees 6, 4, 2 and 0, and so on.

238. A configuration of special interest is that at the point of bifurcation between the pseudo-spheroidal series and the pseudo-ellipsoidal series. This is derived by distortion from the point of bifurcation at which the Maclaurin spheroids join the Jacobian ellipsoids, so that we take $F_0(x, y, z) = 0$ to be the equation of the boundary in this latter configuration.

---

* *Phil. Trans.* 218 (1919), p. 157, and *Problems of Cosmogony and Stellar Dynamics*, p. 166.

I have calculated the coefficients in this configuration as far as terms in $(\rho_0 - \sigma)^2$. The value of ω^2 is found to be given by

$$\frac{\omega^2}{2\pi\gamma\bar{\rho}} = 0{\cdot}18712 + 0{\cdot}06827\left(\frac{\rho_0 - \sigma}{\rho_0}\right) + [0{\cdot}01602 - 0{\cdot}07098\,(\kappa - 2)]\left(\frac{\rho_0 - \sigma}{\rho_0}\right)^2 + \dots$$

$$\dots\dots(238{\cdot}1),$$

so that the first effect of compressibility is to increase the value of $\omega^2/2\pi\gamma\bar{\rho}$. The intercept on the x-axis (the radius of the new equator) is found to be given by *

$$\frac{x^2}{a^2} = 1 + \left(\frac{\rho_0 - \sigma}{\rho_0}\right)[0{\cdot}54102 - 0{\cdot}49950\,(\kappa - 2)]$$

$$+ \left(\frac{\rho_0 - \sigma}{\rho_0}\right)^2[0{\cdot}74761 - 1{\cdot}13574\,(\kappa - 2) + 0{\cdot}83190\,(\kappa - 2)^2] + \dots$$

$$\dots\dots(238{\cdot}2),$$

so that for some values of κ at least the effect of compressibility will be to draw the equator of the figure further away from the axis of rotation.

This extension of the equator combined with the increase in the value of $\omega^2/2\pi\gamma\bar{\rho}$ increases the ratio of centrifugal force to gravity.

The question arises whether centrifugal force can ever equal or outweigh gravity at the equator of the pseudo-spheroid. Calculation shews that the two forces become precisely equal when

$$1 + \left(\frac{\rho_0 - \sigma}{\rho_0}\right)[0{\cdot}9990\,(\kappa - 2) - 1{\cdot}0500]$$

$$+ \left(\frac{\rho_0 - \sigma}{\rho_0}\right)^2[0{\cdot}4997\,(\kappa - 2)^2 + 0{\cdot}07140\,(\kappa - 2) - 0{\cdot}07998] + \dots = 0$$

$$\dots\dots(238{\cdot}3).$$

When κ is given, this equation determines a value of $(\rho_0 - \sigma)/\rho_0$, such that centrifugal force just balances gravity at the instant at which the pseudo-spheroidal form is giving place to the pseudo-ellipsoidal.

Alternatively the equation determines a critical value of κ when $(\rho_0 - \sigma)/\rho_0$ is assigned. It is this latter use of the equation which is of primary interest to us, the important case being $(\rho_0 - \sigma)/\rho_0 = 1$, which represents a mass such as an actual star in which the density falls to zero at the surface. Putting $(\rho_0 - \sigma)/\rho_0 = 1$ in equation (238·3) and ignoring terms in $[(\rho_0 - \sigma)/\rho_0]^2$ and beyond, the solution of the equation is found to be

$$\kappa = 2{\cdot}0501 \quad\dots\dots\dots\dots\dots\dots(238{\cdot}4),$$

while if we include the terms in $[(\rho_0 - \sigma)/\rho_0]^2$ and neglect those beyond, the solution is

$$\kappa = 2{\cdot}1252 \quad\dots\dots\dots\dots\dots\dots(238{\cdot}5).$$

We cannot state with great accuracy the value of κ to which these values are converging, but there is not likely to be any very great error in taking it to be $\kappa = 2{\cdot}2$. Assuming this value, it appears that a mass of gas for which

---

* See *Problems of Cosmogony and Stellar Dynamics*, equation (495).

$\kappa = 2\cdot2$ will begin to lose matter equatorially at precisely the moment at which the pseudo-spheroidal form becomes unstable and gives place to the pseudo-ellipsoidal form.

I have calculated the coefficients which occur in the equation of the boundary for the critical case of $\kappa = 2\cdot2$, and find the shape of boundary to be that of the outermost curve in fig. 43, but unfortunately it is not possible to draw the figure with much accuracy in the neighbourhood of the sharp edge. The interior curves are equipotentials and so are also surfaces of constant density and temperature.

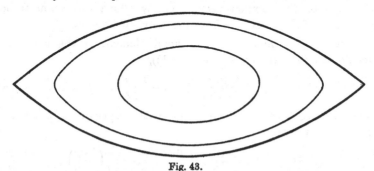

Fig. 43.

239. For the special value $\kappa = 2\cdot2$, equation (238·1) becomes

$$\frac{\omega^2}{2\pi\gamma\bar{\rho}} = 0\cdot18712 + 0\cdot06827 \left(\frac{\rho_0 - \sigma}{\rho_0}\right) + 0\cdot03022 \left(\frac{\rho_0 - \sigma}{\rho_0}\right)^2 + \dots \quad \dots(239\cdot1).$$

The general series of which the first three terms are here written down is probably convergent up to the value $(\rho_0 - \sigma)/\rho_0 = 1$, but it is not easy to determine the value to which it converges. At a guess the value of $\omega^2/2\pi\gamma\bar{\rho}$ appears to converge to about 0·31.

In discussing the incompressible mass, the Maclaurin spheroid was found to become unstable when the rotation was given by

$$\frac{\omega^2}{2\pi\gamma\bar{\rho}} = 0\cdot18712.$$

In Roche's model, which, in a sense, represents the extreme limit of compressibility, the rotation at which the mass began to break up was given by

$$\frac{\omega^2}{2\pi\gamma\bar{\rho}} = 0\cdot36075.$$

In the present model, we have modified the physical conditions until the two processes occur simultaneously for the same value of $\omega^2/2\pi\gamma\bar{\rho}$; it is then natural that this value of $\omega^2/2\pi\gamma\bar{\rho}$ should be intermediate between the values 0·18712 and 0·36075.

Equation (239·1) and the more general equation (238·1) shew that the first effect of compressibility is to increase the value of $\omega^2/2\pi\gamma\bar{\rho}$ at which the pseudo-spheroidal form first becomes unstable. Or, to put it in another way, out of a series of stars of varying degrees of compressibility, the wholly incompressible mass is the first to become unstable. The physical meaning of this result can be seen from the general considerations advanced in § 197.

A GENERAL THEOREM ON ROTATING MASSES.

240. This part of our discussion may terminate with a general theorem which forms a simple extension of one originally given by Poincaré[*]

Let the motion of a mass which is rotating approximately as a rigid body with angular velocity ω be referred to axes rotating with angular velocity ω, and let u, v, w be the components of velocity, which we assume to be small, of any element of the mass relative to these rotating axes.

The equations of motion of any small element of the mass are three equations of the form (cf. equations (224·1)),

$$\frac{du}{dt} = \frac{\partial V}{\partial x} + \omega^2 x - \frac{1}{\rho}\frac{\partial p}{\partial x} \quad\quad\quad\dots\dots\dots\dots\dots(240\cdot1).$$

Differentiating these three equations with respect to x, y, z and adding corresponding sides we obtain

$$\frac{d}{dt}\left(\frac{\partial u}{\partial x} + \frac{\partial v}{\partial y} + \frac{\partial w}{\partial z}\right) = -4\pi\gamma\rho + 2\omega^2 - \left[\frac{\partial}{\partial x}\left(\frac{1}{\rho}\frac{\partial p}{\partial x}\right) + \frac{\partial}{\partial y}\left(\frac{1}{\rho}\frac{\partial p}{\partial y}\right) + \frac{\partial}{\partial z}\left(\frac{1}{\rho}\frac{\partial p}{\partial z}\right)\right].$$

Let us multiply by the element of volume $dx\,dy\,dz$, and integrate throughout the whole of the rotating mass. On transforming the first and last integrals by Green's Theorem, we obtain

$$\frac{d}{dt}\iint (lu + mv + nw)\, dS = \iiint [2\omega^2 - 4\pi\gamma\rho]\, dx\, dy\, dz - \iint \frac{1}{\rho}\frac{\partial p}{\partial \nu}\, dS,$$

where the surface integrals extend over the whole surface of the mass, l, m, n are the direction-cosines of the outward normal to this surface at any point, and $\partial/\partial\nu$ denotes differentiation along this normal.

If A is the whole volume of the mass, the integral on the left measures the rate of increase of A, and the equation may be written in the form

$$\frac{d^2 A}{dt^2} = (2\omega^2 - 4\pi\gamma\bar{\rho})\, A - \iint \frac{1}{\rho}\frac{\partial p}{\partial\nu}\, dS \quad\quad\dots\dots\dots(240\cdot2),$$

where $\bar{\rho}$ is the mean density of the whole mass. Since p vanishes at the boundary of the mass and must be positive at all interior points, $\partial p/\partial\nu$ is necessarily negative, so that the last term is necessarily positive.

For the mass to be in a state of steady rotation, the left-hand member of the equation must vanish, so that we must have

$$\omega^2 < 2\pi\gamma\bar{\rho} \quad\quad\quad\dots\dots\dots\dots\dots(240\cdot3).$$

This is Poincaré's original theorem; whatever the arrangement of the mass, a rotation of speed greater than that given by equation (240·3) is inconsistent with a steady rotation.

---

[*] *Leçons sur les Hypothèses Cosmogoniques*, p. 22.

If inequality (240·3) is not satisfied, d^2A/dt^2 must be positive, so that the mass must continually increase its rate of expansion, or, if it is contracting, the contraction will be checked and ultimately replaced by an expansion.

According to the ideas of Laplace and Roche the ring of matter which was thrown off from the sun, and ultimately formed the planets, was rotating at one time as a closed ring with approximately the same angular velocity as the main mass of the sun. If $\bar{\rho}_s$ was the mean density of the sun, ω^2 would be given by

$$\omega^2 = 0·36075 \times 2\pi\gamma\bar{\rho}_s$$

whence, from inequality (240·3)

$$\bar{\rho} > 0·36075\,\bar{\rho}_s \quad\dots\dots\dots\dots\dots\dots\dots(240·4).$$

This shews that unless the ring condensed at once so as to have a density of at least a third of the mean density of the main mass, it could not rotate steadily but would continually expand under the centrifugal forces arising from its own rotation.

SUMMARY.

241. This and the preceding chapter have been occupied with an investigation into the configurations assumed by masses rotating freely in space under their own gravitational forces. Before leaving the theoretical discussion, and turning our attention to the actual problems of astronomy, it may be profitable to summarise the main theoretical results which have been obtained. Some of these results have been quite general, but we have also investigated in detail the behaviour of certain simplified model masses. These models have been four in number:

(*A*) The incompressible model, consisting of a mass of homogeneous incompressible matter of uniform density.

(*B*) Roche's model, consisting of a point nucleus of very great density, surrounded by an atmosphere of negligible density.

(*C*) The generalised Roche's model, consisting of a homogeneous incompressible mass of finite size and of finite density, surrounded by an atmosphere of negligible density.

(*D*) The adiabatic model, consisting of a mass of gas in adiabatic equilibrium, so that the pressure and density are connected at every point by the relation $p = K\rho^\kappa$, where K and κ retain the same values throughout the mass.

The two models A and B are limiting cases of the more general models C and D. If s denote the ratio of the volume of the atmosphere to that of the nucleus in the generalised Roche's model C, then model C degenerates into model A when $s = 0$, and degenerates into model B when $s = \infty$. Similarly the adiabatic model D degenerates into model A when $\gamma = \infty$ and into model B when $\gamma = 1\frac{1}{5}$ (cf. § 228). The relation between the four models is represented diagrammatically in fig. 44.

Independently of the study of any particular model, we have seen that an increase of rotation beyond a certain amount must break up the original mass.

For the incompressible model A, the mechanism of breaking up is very fully known to us, thanks mainly to the investigations of Maclaurin, Jacobi, Kelvin, Darwin and Poincaré. For small rotations the mass will be spheroidal in shape, but as soon as the angular velocity reaches the value of ω given by $\omega^2/2\pi\gamma\rho = 0.18712$, the spheroidal form no longer remains stable. It shews a tendency to crumple up about one of its equatorial diameters, with the result that the spheroidal shape gives place to an ellipsoidal figure. This first occurs when the axes of the spheroid are approximately in the ratio of $12:12:7$. With further increase of rotation, the ellipsoid lengthens while its two larger axes become increasingly unequal. When the axes of this ellipsoid are in the ratio of approximately $23:10:8$, a new development sets in. The ellipsoid is already so elongated as to be approximately cigar-shaped, and a tendency appears for a furrow to form near, although not actually at, the middle of this cigar-shaped structure. The deepening of this furrow causes a waist to develop,

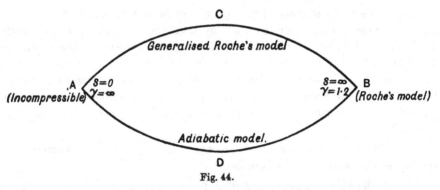

Fig. 44.

giving the figure a pear-shaped appearance. After this furrow has once started, the motion is cataclysmal until the mass divides into two detached parts.

For Roche's model B, the mechanism of break-up is also fully known. As the rotation gradually increases, the equator of the mass bulges more and more, until finally a sharp edge forms on the equator, so that the whole figure becomes lens-shaped (see fig. 41, p. 253). Any further increase of rotation now results in matter being shed from the equator in a continuous stream, centrifugal force outweighing gravity on the equator.

Models A and B both break up with increasing rotation, but they break up in very different ways. We have been able to shew quite generally that there are only these two distinct ways of breaking up; the method of breaking up of any other mass must be a variant of one or other of these two. It will be convenient to refer to the first method of break-up, that of the incompressible mass, as fissional break-up; and to the second method of break-up, that of Roche's model, as equatorial break-up.

As we pass along either of the chains of models C and D which connect A and B, or along any other chain of models connecting A and B, there must be some point on each at which fissional break-up gives place to rotational break-up. At such a point, the two methods of break-up must be about to begin simultaneously with the same rotation. Thus the condition determining such a point will be that centrifugal force is precisely equal to gravity on the equator of that particular configuration at which the rotation reaches such a value that a figure of revolution is no longer a stable form for the mass.

We have determined this critical point on each of the two chains of models C and D. Of these the adiabatic chain D is the more important. As we pass along this chain from A to B, the value of κ varies from ∞ to $1 \cdot 2$; the critical point is approximately given by $\kappa = 2 \cdot 2$. Thus a mass of gas or other compressible matter in adiabatic equilibrium will break up by fission if κ is greater than $2 \cdot 2$; it will break up equatorially if κ lies between $1 \cdot 2$ and $2 \cdot 2$. This latter range of course includes the values of κ for all gases whose density is so low that Boyle's law is approximately satisfied; for these κ is always less than $1 \cdot 66$.

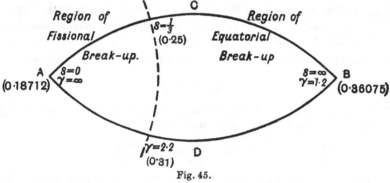

Fig. 45.

Thus a mass of gas rotating in adiabatic equilibrium, with Boyle's law satisfied, will necessarily break up by equatorial loss of matter, not by fission into two detached masses. The same is true of a mass of gas in radiative equilibrium, since we have that in these masses, provided the density is everywhere so low that Boyle's law is obeyed, the pressure and density are connected by a relation of the form $p = K\rho^\kappa$ where κ is always less than $1 \cdot 308$.

As we pass along the chain C of generalised Roche's models, the value of s, the ratio of the volume of the nucleus to that of the atmosphere, varies from ∞ to 0. The critical point is found to occur at about $s = \frac{1}{3}$. Thus when the atmosphere is less than a third of the volume of the nucleus, the mass will break up by fission; when the atmosphere is greater than this, the mass will break up equatorially.

These various results may be exhibited diagrammatically as in fig. 45 the numbers in brackets indicating the values of $\omega^2/2\pi\gamma\bar{\rho}$ at which the breaking-up process commences.

CHAPTER X

ROTATION AND FISSION OF STARS

242. THE rotating masses discussed in the last two chapters were supposed to rotate as rigid bodies, the angular velocity of rotation being the same throughout. Solid bodies such as the moon must necessarily rotate in this way, but no known reason compels semi-gaseous or semi-liquid bodies, such as the sun and stars, to rotate as rigid bodies. When Galileo first observed the rotation of the sun's surface disclosed by the motion of sunspots, he supposed that the whole of the sun would rotate with the same period as that of the sunspots, a period of 25 days. This assumption crept into astronomy and was tacitly accepted until 1926*, when I shewed that the inner parts of the sun and stars must rotate far more rapidly than their surfaces.

Thus before we can apply theoretical results, obtained on the supposition of uniform rotation, to real masses such as the sun and stars, we must examine to what degree the rotation of these latter bodies departs from uniformity and what is the effect of these departures from uniform rotation on their behaviour.

GENERAL EQUATIONS.

243. The motion of matter having a uniform coefficient of viscosity η is determined by the usual hydrodynamical equations† :

$$\rho \frac{Du}{Dt} = \rho X - \frac{\partial p}{\partial x} + \tfrac{1}{3}\eta \, \frac{\partial}{\partial x}\left(\frac{\partial u}{\partial x} + \frac{\partial v}{\partial y} + \frac{\partial w}{\partial z}\right) + \eta \nabla^2 u, \text{ etc. } \dots(243\cdot1),$$

where D/Dt denotes differentiation with respect to the time, following an element of the star in its motion, u, v, w are the components of velocity of the element, x, y, z are the components of force acting on the element, and p is the pressure at the element, including pressure of radiation.

Let us transform these equations to axes rotating with a uniform angular velocity ω about the axis of z. If the star is not rotating as a rigid body, the co-ordinates x, y, z of any element of the star referred to these axes will not remain constant. Let their rates of change be $\dot{x}, \dot{y}, \dot{z}$ and let us suppose these to be small, thus limiting our discussion to a star, or part of a star, which is rotating nearly, but not quite, with uniform angular velocity ω.

The velocities and accelerations in space are now given by

$$u = \dot{x} - \omega y, \quad v = \dot{y} + \omega x, \quad w = \dot{z},$$

$$\frac{Du}{Dt} = \ddot{x} - 2\omega\dot{y} - \omega^2 x, \quad \frac{Dv}{Dt} = \ddot{y} + 2\omega\dot{x} - \omega^2 y, \quad \frac{Dw}{Dt} = \ddot{z}.$$

---

* *Monthly Notices R.A.S.* LXXXVI. (1926), pp. 328 and 444.

† Lamb, *Hydrodynamics* (5th Edn.), p. 546.

Using these values, equations (243·1) transform into

$$\rho\ddot{x} = \rho\left(X + \omega^2 x\right) - \frac{\partial p}{\partial x} + \tfrac{1}{3}\eta\left(\frac{\partial \dot{x}}{\partial x} + \frac{\partial \dot{y}}{\partial y} + \frac{\partial \dot{z}}{\partial z}\right) + \eta\nabla^2\dot{x}, \text{ etc. } \dots(243\cdot2).$$

From these three equations we obtain the three equations

$$\frac{\partial \rho}{\partial y}\left(\ddot{x} - X - \omega^2 x\right) - \frac{\partial \rho}{\partial x}\left(\ddot{y} - Y - \omega^2 y\right) = \left(\rho\frac{d}{dt} - \eta\nabla^2\right)\left(\frac{\partial \dot{y}}{\partial x} - \frac{\partial \dot{x}}{\partial y}\right), \text{ etc.}$$

$$\dots\dots(243\cdot3).$$

Let us put
$$\xi_1 = \frac{\partial \dot{z}}{\partial y} - \frac{\partial \dot{y}}{\partial z},$$

and give corresponding meanings to ξ_2, ξ_3, so that ξ_1, ξ_2, ξ_3 are the three components of the vector which is generally called the "vorticity." To understand the meaning of this vector, let us suppose that any small element of the star has a rotation ω' about the axis of z, in addition to the rotation of the axes, so that its total angular velocity about the axis of z is $\omega + \omega'$. The values of $\dot{x}$, $\dot{y}$, $\dot{z}$ will be

$$\dot{x} = \dot{x}_0 - \omega'y, \quad \dot{y} = \dot{y}_0 + \omega'x, \quad \dot{z} = \dot{z}_0,$$

where $\dot{x}_0$, $\dot{y}_0$, $\dot{z}_0$ are the components of the velocity of translation of the element as a whole apart from its rotation. From these values for $\dot{x}$, $\dot{y}$, $\dot{z}$ we obtain

$$\xi_3 = \frac{\partial \dot{y}}{\partial x} - \frac{\partial \dot{x}}{\partial y} = 2\omega'.$$

Thus the vorticity ξ_3 is twice the rotation relative to the moving axes.

Multiplying the three equations (243·3) throughout by $\partial\rho/\partial x$, $\partial\rho/\partial y$, $\partial\rho/\partial z$, and adding corresponding sides, we obtain

$$\sum_{x, y, z} \frac{\partial \rho}{\partial z}\left(\rho\frac{d\xi_3}{dt} - \eta\nabla^2\xi_3\right) = 0 \dots\dots\dots(243\cdot4).$$

The simplest case occurs when the whole of the vorticity is about the axis of z, so that $\xi_1 = \xi_2 = 0$; the equation then reduces to

$$\frac{d\xi_3}{dt} = \frac{\eta}{\rho}\nabla^2\xi_3 \dots\dots\dots\dots(243\cdot5).$$

When η/ρ is constant, this is identical with the well-known Fourier equation of conduction of heat, shewing that when ξ_3 is not initially uniform throughout the mass, it will tend to equalise itself according to the same laws, and at the same rate, as inequalities of temperature are equalised in a medium whose coefficient of conduction of heat is η/ρ.

In an actual star the values of η and ρ will of course be very far from constant, but in a general way the rate of equalising inequalities of rotation (measured by $\tfrac{1}{2}\xi_1$, $\tfrac{1}{2}\xi_2$, $\tfrac{1}{2}\xi_3$) will be comparable with that given by equation (243·5). Thus we may say that the inequality of angular velocity between two points in a star at a distance r apart will be reduced to half in a time comparable with

$$\frac{\rho r^2}{\eta} \dots\dots\dots\dots\dots(243\cdot6).$$

RADIATIVE VISCOSITY.

244. Before using this result, we must evaluate the coefficient of viscosity η. In a gas in which all the molecules moved with the same velocity c and had the same free path l, this would be given by the usual kinetic theory of formula[*]

$$\eta = \tfrac{1}{3}\rho c l \quad\dots\dots\dots\dots\dots\dots\dots(244 \cdot 1).$$

This formula is obtained by treating the molecules of the gas as carriers of momentum. Our view of the constitution of stellar interiors has led us to recognise the existence of three types of carriers of energy inside a star, namely, ionised atoms, free electrons and radiation. These three types must also carry momentum. Radiation has been seen entirely to outstrip the two others in the transport of energy, and I have recently shewn[†] that the same is true, although in a lesser degree, of the transport of momentum.

To transform formula (244·1) so as to apply to the transport of momentum by radiation, we replace ρ by aT^4/C^2, the mass of radiant energy per unit volume; we replace c by C, the velocity of light; and, as in § 72, we replace l, the free path, by $1/k\rho$ where k is the coefficient of opacity. The value of η_R, the coefficient of viscosity arising from the transport of energy by radiation, is now found to be

$$\eta_R = \frac{aT^4}{3Ck\rho} \quad\dots\dots\dots\dots\dots\dots(244 \cdot 2).$$

The value of η_R calculated by this very simple method can hardly be expected to be exact; indeed, the original formula $\eta = \tfrac{1}{3}\rho c l$ which we borrowed from the theory of Gases, is itself far from exact. By a much more elaborate calculation than that just given I have found[‡] that the exact value of η_R is

$$\eta_R = \frac{2aT^4}{15Ck\rho} \quad\dots\dots\dots\dots\dots\dots(244 \cdot 3),$$

which is only 40 per cent. of the approximate value given above.

Inserting the value of k obtained from Kramers' theory (§ 77), namely,

$$k = \frac{cF\rho}{\mu T^{3 \cdot 5}},$$

this becomes

$$\eta_R = \frac{2a\mu T^{7 \cdot 5}}{15Cc\rho^2 F} \quad\dots\dots\dots\dots\dots\dots(244 \cdot 4).$$

We see that, except for small variations in μ and F, η_R would be constant throughout a star arranged so that $\rho \propto T^{3\cdot 75}$. We have seen that stars in general are arranged roughly according to a law of the form $\rho \propto T^n$, where n is about 3·25 for a star of very small mass, and runs up to about 5 for a star of very great mass. Thus in a star of small mass η_R reaches its largest value at the centre of the star, but in a star of great mass it reaches its greatest

---

[*] Jeans, *Dynamical Theory of Gases*, p. 273.
[†] *M.N.* LXXXVI. (1926), p. 329.   [‡] *Ibid.* p. 444.

value in the outer layers; it will be approximately constant ($n = 3.75$) for a star of mass about equal to that of Sirius.

245. The absolute value of η_R is most simply determined in terms of the star's radiation. At a distance r from the centre of the star the flux of radiation H is (§ 72)

$$H = -\frac{4aT^3C}{3k\rho}\frac{dT}{dr} \qquad \ldots\ldots\ldots\ldots\ldots(245\cdot1),$$

in ergs per sq. cm. per second, so that relation (244·3) can be put in the form

$$\eta_R = -\frac{HT}{10C^2}\bigg/\left(\frac{dT}{dr}\right) \qquad \ldots\ldots\ldots\ldots\ldots(245\cdot2).$$

If T fell uniformly along the radius of the star from its central value T_c to its surface value of nearly zero, the value of dT/dr would be equal to $-T_c/R$ at every point, where R is the radius of the star. A study of the graphs of T given by Emden* shews that, whatever the build of star, dT/dr never varies by a factor of more than about 2 from its mean value $-T_c/R$, except in the regions very near to the centre of the star, and in the outermost regions of very massive stars; in these special regions dT/dr is only a small fraction of T_c/R. Excluding these regions, we see that, to within a factor about 2,

$$\eta_R = \frac{HRT}{10C^2T_c} \qquad \ldots\ldots\ldots\ldots\ldots\ldots(245\cdot3).$$

For the sun, $R = 7 \times 10^{10}$ cms. and at the sun's surface $H = 6.25 \times 10^{10}$. The greater part of the sun's mass is enclosed within a sphere of radius $\frac{1}{4}R$, so that approximately the same stream of energy must cross this sphere as crosses the sun's outer surface. Thus the value of H at this sphere must be about 16 times its value at the surface, or about 10^{12} ergs per sq. cm. per second. If for the moment we assume that the gaseous model gives a sufficiently good picture of the sun's interior, the temperature at this distance from the sun's centre is about $\frac{1}{2}T_c$. Inserting these values in equation (245·3) we find that at the point just discussed the value of η_R is about 4.

At this point the density is rather over $\frac{1}{4}\rho_c$. With the values for T_c and ρ_c already calculated in § 91, we obtain a temperature of about 29 million degrees and a density of about 35.

The coefficient of ordinary gaseous viscosity which is of the order of 10^{-4} at ordinary temperatures, probably rises to about unity at 29 million degrees. Persico† has specially studied the viscosity of a mixture of free electrons and atomic nuclei, and finds that at a temperature of 33 million degrees and a density of 36 (which he takes to be the central temperature and density of the sun) the coefficient of material viscosity would be about 4·2. For η_R we have obtained the value 4 at $T = 29$ million degrees and $\rho = 40$. Our formula (244·4) shews that the value of η_R at $T = 33$ million degrees and $\rho = 36$ would

---

* *Gaskugeln*, p. 86.   † *M.N.* LXXXVI. (1926), p. 93.

be about 3·24 times as great, and so equal to about 13. This is roughly three times Persico's value for the material viscosity at the same temperature.

Still supposing the gaseous model to give a sufficiently good approximation, we have seen that ρ varies approximately as $T^{3\cdot3}$ in the sun, whence equation (244·4) shews that η_R varies about as $T^{0\cdot9}$. This is about comparable with the rate of variation of material viscosity with the temperature, so that the rates of η_R to the material coefficient of viscosity is not likely to vary enormously throughout the sun's central regions.

When we discard the gaseous model, calculation becomes far more difficult. The simplest procedure is probably to pass from the ideal gaseous configuration just discussed to the actual liquid configuration by the method described in § 129, namely, imagining the density to remain unaltered while the temperature is everywhere reduced to about three-quarters of its value. Formula (245·3) shews that when this is done the value of η_R remains unchanged, and so equal to its value just calculated for the gaseous model, while the 25 per cent. (or so) reduction of temperature, ρ remaining unchanged, reduces the coefficient of material viscosity by anything from $12\frac{1}{2}$ to $22\frac{1}{2}$ per cent. The predominance of radiative over material viscosity is accordingly slightly increased.

Thus, although accurate calculations are difficult, it seems likely that the coefficient of radiative viscosity will substantially exceed that of ordinary material viscosity, both at the centre of the sun and in its outer layers, although in neither case is there any excessive preponderance, such as was found in discussing the coefficients of conduction of heat.

As between one star and another, equation (245·3) shews that the values of η_R at corresponding points in different stars will be proportionate to HR, where R is the radius of the star and H is the flux of energy per sq. cm. at its surface ($H = \sigma T_e^4$). In giant main sequence stars both H and R are greater than for the sun, while ordinary material viscosity remains about the same. In stars on the giant branch H is somewhat less than for the sun, but R is far greater, and the product HR is greater, while ordinary material viscosity is less as a consequence of low internal temperatures. Thus the preponderance of radiative over material viscosity, which is only moderate for the sun, becomes very marked in giant stars. In these practically the whole viscous transfer of momentum must be by radiation. On the other hand, it seems quite likely that ordinary viscosity may preponderate over radiative viscosity in stars of quite small mass such as the two components of Kruger 60.

Rate of Equalisation of Angular Velocity.

246. In general we have seen that inequalities of angular velocity over a distance r will fall to half-value in a time of the order of

$$\frac{\rho r^2}{\eta} \text{ seconds} \quad \dots\dots\dots\dots\dots\dots(246\cdot1),$$

where η is the total coefficient of viscosity, material plus radiative. In a star of large mass we may neglect material viscosity in comparison with radiative, and assign to η the value already obtained for η_R in equation (245·3). The time given by formula (246·1) now becomes

$$\frac{10C^2r^2\rho}{HR}\left(\frac{T_c}{T}\right) \text{ seconds } \quad \dots\dots\dots\dots\dots(246\cdot2).$$

For a star of mass even as small as the sun, the error in giving this value to η will be negligible to the degree of accuracy to which we are now working.

Let us first apply this formula to the sun. If we give to η a value of about 20 in formula (246·1), or if we insert the numerical data already given into formula (246·2), we find that inequalities in the angular velocity of the sun between its centre and points a quarter-radius distant will be reduced to half values in times of the order of 3×10^{21} seconds, or 10^{14} years. This is so large in comparison with the age of the sun that there can have been but little equalisation in the values of ω in the sun's interior since it assumed its present state.

As between one star and another, formula (246·2) shews that the times of equalisation, over corresponding fractions of the star's radii, are proportional to

$$\frac{\rho R}{H} \quad \dots\dots\dots\dots\dots\dots\dots\dots\dots(246\cdot3).$$

A star's mass being proportional to ρR^3, and its emission of energy to HR^2, its mean rate of generation of energy per unit mass, $\bar{G}$, is proportional to $H/\rho R$. Hence formula (246·3) shews that the times of equalisation in different stars vary inversely as $\bar{G}$. Thus the times of equalisation are far less for more massive stars than for the sun. V Puppis generates energy some 580 times as rapidly per unit mass as the sun, so that its time of equalisation is only about $\frac{1}{580}$ times that of the sun, or, say, 2×10^{11} years.

A star loses mass owing to its emission of radiation at a rate given by

$$-C^2\frac{dM}{dt} = M\bar{G},$$

so that the time required for a star's mass to change by any specified fraction of the whole is proportional to $1/\bar{G}$, and so is proportional to the time required for a specified equalisation of ω. In other words, the time required for a given equalisation of ω is the same as the time required for the loss of a certain fraction of a star's mass, this fraction being the same for all stars. We have found that the loss of almost the sun's whole mass would produce only an insignificant equalisation of angular velocity in its inner regions. The same must consequently be true of all stars, independently of their radii, masses and ages, and we conclude in general that *throughout the whole life of any star whatever, the equalisation of angular velocity produced by viscosity is negligible in the central regions of the star.*

The foregoing discussion has been concerned only with the central regions of a star and regions in which the values of T and ρ are comparable with their values at the centre. When we proceed to regions nearer the surface, the value of ρ in formula (246·1) falls much more rapidly than η, and we soon come to regions in which the time of equalisation is first comparable with, and then less than, the age of the star. We now proceed to examine the rotations of the outer layers of stars, considering not only viscosity but also a second phenomenon connected with the transport of momentum to which I first drew attention in 1926*.

The braking action of Radiation.

247. This phenomenon is as follows. If any small element of a star, at a distance ϖ from the axis of rotation, rotates with angular velocity ω, the velocity of the element in space will be $\varpi\omega$. Relative to the rotating star, the stream of radiation through this element will on the average move along the direction of the temperature gradient in the star, in a direction which intersects the axis of rotation. Hence the velocity of this stream in space will be compounded of two velocities, a tangential velocity $\varpi\omega$ in a direction at right angles to the axis of rotation, and a radial velocity C in a direction intersecting the axis of rotation.

Let us fix our attention on any small amount of radiation of energy MC^2, and so of mass M, threading its way through the star. As this traverses the element in question, its moment of momentum about the axis of rotation is $M\varpi^2\omega$. As the radiation passes outwards through the star, with continual absorptions and re-emissions, ϖ continually increases. Thus if ω had a uniform value throughout the star, $M\varpi^2\omega$ would continually increase; in other words, the radiation, as it passed further from the axis of rotation, would continually gain in moment of momentum about the axis of rotation. From the principle of the conservation of angular momentum, this gain must involve a corresponding loss to the layers of the star through which the radiation is passing when the gain occurs. Thus the passage of radiation through a rotating star produces a braking effect on the rotation of the star, and as this effect is different in different layers of the star, it must tend to set up inequalities in the angular velocities of different parts of the star. Let us attempt to express these ideas in mathematical language.

In fig. 46 let AB, CD represent the areas cut off by a small cone of angle $\sin\theta\,d\theta\,d\phi$ from spheres of radii r and $r+dr$ in a star. Let H be the flux of radiant energy per unit area crossing the area AB, and let $\dot\phi$ denote the angular velocity of the element AB of the star about the axis of rotation. As the area of AB is $r^2\sin\theta\,d\theta\,d\phi$ the energy of the radiation which crosses it per unit time is $Hr^2\sin\theta\,d\theta\,d\phi$. This radiation has mass equal to

$$\frac{H}{C^2}r^2\sin\theta\,d\theta\,d\phi,$$

* *M.N.* LXXXVI. (1926), p. 444.

and as it has an average tangential velocity in space equal to $r \sin \theta \dot{\phi}$, the moment of momentum of the radiation crossing AB per unit time is

$$\frac{H}{C^2} r^4 \sin^3 \theta \, d\theta \, d\phi \, \dot{\phi} \quad \dots\dots\dots\dots\dots\dots(247\cdot1).$$

Each second angular momentum of this amount enters the volume $ABCD$ of the star, while a corresponding amount, obtained by evaluating expression $(247\cdot1)$ at a radial distance $r + dr$, flows out of this volume. Hence the volume $ABCD$ experiences a net loss of angular momentum per second equal to

$$\frac{d}{dr}\left(\frac{H}{C^2} r^4 \sin^3 \theta \, d\theta \, d\phi \, \dot{\phi}\right) dr \quad \dots\dots\dots\dots(247\cdot2).$$

The mass of this element of the star is $\rho r^2 \sin \theta \, d\theta \, d\phi \, dr$, so that its angular momentum about the axis of rotation is

$$\rho r^4 \sin^3 \theta \, d\theta \, d\phi \, \dot{\phi} \, dr \quad \dots\dots\dots\dots\dots\dots(247\cdot3).$$

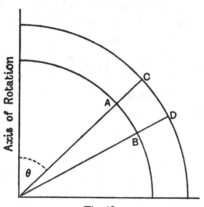

Fig. 46.

We have found that the passage of radiation through this element causes its angular momentum $(247\cdot3)$ to decrease at the rate given by expression $(247\cdot2)$. The flow of momentum caused by ordinary viscosity produces a further change. The viscous flow of momentum across the area AB is $\eta r \sin \theta \dfrac{d\dot{\phi}}{dr}$ per unit area, so that the total flow is $\eta r^3 \sin^2 \theta \, d\theta \, d\phi \dfrac{d\dot{\phi}}{dr}$, and its angular momentum about the axis of rotation is $r \sin \theta$ times this, or

$$\eta r^4 \sin^3 \theta \, d\theta \, d\phi \frac{d\dot{\phi}}{dr} \quad \dots\dots\dots\dots\dots\dots(247\cdot4).$$

The direction of this flow is inward along the radius. There is a similar inward flow across CD, the amount of which is obtained by evaluating expression $(247\cdot4)$ at a distance $r + dr$ from the star's centre. The net result is a gain to the angular momentum of the element of amount

$$\frac{d}{dr}\left(\eta r^4 \sin^3 \theta \, d\theta \, d\phi \frac{d\dot{\phi}}{dr}\right) dr \quad \dots\dots\dots\dots(247\cdot5).$$

Equating the change in momentum to the gain produced by viscosity and the loss produced by radiation, we obtain the equation

$$\rho r^2 \frac{D}{Dt}(r^2 \phi) = \frac{d}{dr}\left[\eta r^4 \frac{d\phi}{dr} - \frac{H}{C^2} r^4 \phi\right] \quad \dots\dots\dots(247 \cdot 6).$$

In obtaining this equation we have divided throughout by $\sin^3 \theta\, d\theta\, d\phi$, and find that θ has disappeared from the equation entirely. Thus the changes in the angular velocity are independent of the colatitude θ.

This equation, like the earlier equation for radiative viscosity, is only approximate. As the same elements of radiation have been supposed to account for both the viscous change of momentum and the changes caused by the passage of radiation, it is inevitable that some of their effects should have been counted twice over.

Somewhat arduous analysis is necessary to obtain the exact equation which must replace equation (247·6) when the problem is treated with full mathematical rigour. Fortunately the final result is simple. I have shewn* that equation (247·6) must be replaced by

$$\rho r^2 \frac{D}{Dt}(r^2 \phi) = \frac{\partial}{\partial r}\left[\eta r^4 \frac{d\phi}{dr} - \frac{3}{5}\frac{H}{C^2} r^4 \phi\right] \quad \dots\dots\dots(247 \cdot 7),$$

which only differs from the approximate equation by the presence of the factor $\frac{3}{5}$ in the last term.

Rotational Steady Motion.

248. If we confine our attention to the first term on the right of equation (247·7) we again find that viscosity tends to equalise inequalities in $\dot\phi$, the time necessary to reduce inequalities at a distance r apart to half value being of the order of $\rho r^2/\eta$. If we similarly confine our attention to the second term on the right of equation (247·7), we find that the flow of radiation tends to set up inequalities in $\dot\phi$, and that the inequalities at a distance r become appreciable after a time comparable with $rC^2\rho/H$, which in turn is comparable with $\rho r^2/\eta_R$. Thus the two times are of the same order of magnitude.

In the dense central regions of the star we have found that the first, and therefore also the second, of these times is large in comparison with the whole age of the star. Thus neither viscosity nor the flow of radiation produces any appreciable effect on the star's rotation in these regions, and to a good enough approximation equation (247·7) may be replaced by

$$\frac{D}{Dt}(r^2 \phi) = 0 \quad \dots\dots\dots\dots\dots\dots(248 \cdot 1).$$

In the outer regions of the star conditions are different on account of the smallness of ρ. Equation (247·7) shews that $\rho r^4 \dot\phi$ must change at a rate

---

* *M.N.* lxxxvi. (1926), p. 453. In this paper the left-hand of our equation (247·7) was written as $\frac{D}{Dt}(\rho r^4 \dot\phi)$ because ρ and r were not there supposed capable of change.

approximately equal to its right-hand member, so that if ρ is small, ϕ must change very rapidly, and this succession of changes in ϕ will be towards a steady state determined by the vanishing of the right-hand member of equation (247·7). When this steady state is attained,

$$\frac{\partial}{\partial r}\left[\eta r^4 \frac{d\phi}{dr} - \frac{3}{5}\frac{H}{C^2} r^4 \phi\right] = 0,$$

so that

$$\eta r^4 \frac{d\phi}{dr} - \frac{3}{5}\frac{H}{C^2} r^4 \dot{\phi} = f(\theta)\dots\dots\dots\dots\dots(248\cdot2),$$

where $f(\theta)$ denotes any function of θ.

It is easy to interpret the equation. Each layer of a rotating star possesses angular momentum and so may be thought of as a flywheel, whose rotation is checked or accelerated by viscosity, and also may be checked by the necessity of imparting increased angular momentum to the radiation passing through it, or may be accelerated if this radiation reaches the layer with more angular momentum than it takes away with it. The outer layers of the stars, being of low density, possess so little angular momentum that their flywheel action may be disregarded. Having no capacity for storing angular momentum they can only transmit it, and the amount any layer receives in any interval must be precisely equal to the amount it passes on to the next layer. This is the physical meaning of equation (248·2).

249. Let us suppose that a sufficiently good approximation is obtained by treating the whole viscosity as radiative, so that, from equation (245·2),

$$\eta = -\frac{HT}{10C^2}\bigg/\left(\frac{dT}{dr}\right)$$

and equation (248·2) assumes the form

$$\frac{r^4 H}{C^2}\left(-\frac{1}{10}\frac{T}{\dfrac{dT}{dr}}\frac{d\phi}{dr} - \frac{3}{5}\phi\right) = f(\theta) \dots\dots\dots(249\cdot1).$$

In the outer layers of the star the rate of generation of energy per unit volume, $G\rho$, may be disregarded on account of the smallness of ρ.

Thus if E is the total emission of the star in ergs per second, we may put $4\pi r^2 H = E$ and so write equation (249·1) in the form

$$\dot{\phi} - \frac{A}{r^2} = -\frac{1}{6}\frac{T}{\dfrac{dT}{dr}}\frac{d\phi}{dr} \dots\dots\dots\dots(249\cdot2)$$

where A is a function of θ only.

To a first approximation, the solution of this equation may be taken to be

$$\dot{\phi} = \frac{A}{r^2} \dots\dots\dots\dots\dots(249\cdot3),$$

for with this value of $\dot{\phi}$, the value of the right-hand member of equation (249·2) is found to be

$$-\frac{1}{6}\frac{T}{\dfrac{dT}{dr}}\frac{d\phi}{dr} = \frac{T\phi}{3r\dfrac{dT}{dr}} \qquad \dots\dots\dots\dots\dots(249\cdot4).$$

Near the surface of a star, dT/dr is comparable with $-T_c/r$, so that this last expression is of the order of $-\dfrac{T}{3T_c}\phi$, and on account of the smallness of T/T_c this is small in comparison with ϕ.

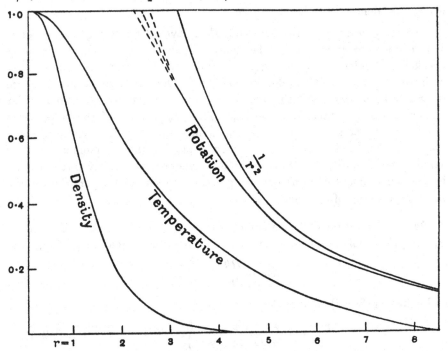

Fig. 47. Variation of angular velocity of rotation, density and temperature in a gaseous star.

We obtain a better approximation by utilising equation (249·4) and writing equation (249·2) in the form

$$\phi\left(1 - \frac{T}{3r\dfrac{dT}{dr}}\right) = \frac{A}{r^2} \qquad \dots\dots\dots\dots\dots(249\cdot5).$$

Since dT/dr is everywhere negative, we see that ϕ has the value A/r^2 at the surface of the star, and gradually falls below this value as we pass inwards into the star.

In fig. 47 the curve marked rotation shews the values of ϕ that I have calculated for a star which is assumed to be arranged, in its outer layers, like a mass of gas in adiabatic equilibrium with $\kappa = 1\cdot300$, the value we obtained for a hypothetical gaseous sun in § 91.

The dotted part of the rotation curve refers to regions in which the agencies just considered fail to reduce the rotation to law and order throughout the whole life of the star, so that the angular velocity is indeterminate. Such regions may cover about a third of the star's radius, the $1/r^2$ law being obeyed outside. Thus the central core of a star should rotate with about nine times the angular velocity of its surface. Its outer layers have been reduced nearly to rest by the braking action of radiation, and may be regarded as slowly rotating veils drawn round the stars concealing the more energetic rotations which are taking place within.

The Configurations of a Mass in Steady Rotation.

250. The motion of any element inside a star is determined by three dynamical equations, corresponding to the three directions in space. Equation (247·7), which has just been discussed, fixes the changes in $\dot{\phi}$, and so determines the motion of the element in the direction in which it moves as a consequence of the star's rotation. If the star has no motion except that of the rotations of its different parts, the two remaining equations must express that there shall be no motion in the two other directions in space.

In cylindrical co-ordinates ϖ, ϕ, z, where

$$\varpi = r \sin \theta \text{ and } z = r \cos \theta,$$

these equations may be obtained as equations of relative equilibrium for motion in the directions of ϖ and z increasing. They are found to be

$$\frac{1}{\rho}\frac{\partial p}{\partial \varpi} - \frac{\partial V}{\partial \varpi} = \varpi \dot{\phi}^2 \quad \text{............................(250·1),}$$

$$\frac{1}{\rho}\frac{\partial p}{\partial z} - \frac{\partial V}{\partial z} = 0 \quad \text{..............................(250·2).}$$

In the special case in which p is a function of ρ only, we differentiate the two equations with respect to z, ϖ and subtract. This gives $\partial\dot{\phi}/\partial z = 0$, so that $\dot{\phi}$ is a function of ϖ only. Thus the rotation is the same at every point of a cylinder drawn about the axis of rotation, a result due to Poincaré.

When a star is in radiative equilibrium, p does not depend solely on ρ. In this case we eliminate p from the two above equations and obtain

$$\varpi \frac{\partial}{\partial z}(\rho \dot{\phi}^2) = \frac{\partial \rho}{\partial \varpi}\frac{\partial V}{\partial z} - \frac{\partial V}{\partial \varpi}\frac{\partial \rho}{\partial z} \quad \text{.................(250·3),}$$

or, in r, θ co-ordinates

$$r^2 \sin \theta \cos \theta \frac{\partial}{\partial r}(\rho \dot{\phi}^2) - r \sin^2 \theta \frac{\partial}{\partial \theta}(\rho \dot{\phi}^2) = \frac{\partial \rho}{\partial \theta}\frac{\partial V}{\partial r} - \frac{\partial \rho}{\partial r}\frac{\partial V}{\partial \theta} \quad \text{...(250·4).}$$

Since the star is supposed symmetrical about its axis of rotation, the gravitational potential V can be expanded in the form

$$V = V_0 + V_2 P_2(\cos \theta) + V_4 P_4(\cos \theta) + \dots \quad \text{.........(250·5),}$$

$P_2, P_4, \dots$ representing the usual zonal harmonics of even orders. Poisson's relation $V^2 V = -4\pi\rho$ at once gives the corresponding density ρ in the form

$$\rho = \rho_0 + \rho_2 P_2 (\cos \theta) + \rho_4 P_4 (\cos \theta) + \dots \quad \dots\dots\dots(250\text{·}6),$$

where

$$-4\pi\rho_n = \frac{1}{r^2} \frac{\partial}{\partial r} \left(r^2 \frac{\partial V_n}{\partial r} \right) - n(n+1) V_n \quad \dots\dots\dots(250\text{·}7).$$

A star devoid of rotation has been found to arrange itself in spherical layers with a high central condensation of mass. When the star is in rotation, the layers will no longer be spherical, but the high central condensation of mass will persist. In the extreme case in which the mass is entirely concentrated at the centre, as in Roche's model, V reduces to M/r, so that $V_2, V_4 \dots$ all vanish, as also of course do $\rho_2, \rho_4, \dots$. In less extreme cases, in which the mass is not wholly concentrated at the centre, $V_2, V_4 \dots$ will not vanish, but V_2 will be small in comparison with V_0 in the star's outer layers of low density; V_4 will be small in comparison with V_2, and so on. It follows from equations (250·7) that ρ_2 will be small compared with ρ_0, ρ_4 will be small compared with ρ_2, etc.

We can now re-write equation (250·4) in the form

$$r^2 \sin \theta \cos \theta \frac{\partial}{\partial r} (\rho \dot\phi^2) - r \sin^2 \theta \frac{\partial}{\partial \theta} (\rho \dot\phi^2) = \left(\rho_2 \frac{\partial V_0}{\partial r} - V_2 \frac{\partial \rho_0}{\partial r} \right) \frac{\partial}{\partial \theta} P_2 (\cos \theta)$$

$$+ \left(\rho_4 \frac{\partial V_0}{\partial r} - V_4 \frac{\partial \rho_0}{\partial r} \right) \frac{\partial}{\partial \theta} P_4 (\cos \theta) + \left(\rho_2 \frac{\partial V_2}{\partial r} - V_2 \frac{\partial \rho_2}{\partial r} \right) P_2 (\cos \theta) \frac{\partial}{\partial \theta} P_2 (\cos \theta)$$

$$+ \dots \quad \dots\dots(250\text{·}8),$$

in which the term which occupies the top line on the right-hand side is large compared with those in the second line, while these are large compared with all the remaining terms which are not written down.

Confining ourselves for the moment to the first term on the right of this equation, the simplest solution is found to be given by

$$r^2 \frac{\partial}{\partial r} (\rho \dot\phi^2) = -\frac{1}{3} \left(\rho_2 \frac{\partial V_0}{\partial r} - V_2 \frac{\partial \rho_0}{\partial r} \right) \quad \dots\dots\dots\dots(250\text{·}9),$$

so that $\rho\dot\phi^2$ is a function of r only. We obtain the complete solution by adding to this the most general solution of equation (250·8) with the right-hand member put equal to zero. As a glance at equation (250·3) shews, this is merely $f(\varpi)$, the most general function of ϖ. Thus the value of $\rho\dot\phi^2$ is

$$\rho\dot\phi^2 = -\int \frac{1}{3r^2} \left(\rho_2 \frac{\partial V_0}{\partial r} - V_2 \frac{\partial \rho_0}{\partial r} \right) dr + f(\varpi).$$

The boundary of the star is given by $\rho = 0$, so that the equation of this boundary is obtained by equating the right-hand member of the equation to zero. We see that the additional solution represented by $f(\varpi)$ is necessary in order that the boundary may have shapes other than spherical.

In a star such as the sun, the speed of rotation of the outer layers is so slight that the star's surface may, to a good approximation, be treated as spherical. For such a star $f(\varpi)$ must be zero, and the most general solution possible is that given by equation (250·9). To a first approximation ρ may be replaced by ρ_0, and so is a function of r only, so that $\dot{\phi}^2$ is a function of r only. The circumstance that θ did not enter into the equation of motion (247·6) has already shewn that if $\dot{\phi}$ is initially a function of r only, neither viscosity nor the transmission of radiation can destroy this property. Our present discussion has further shewn that for equilibrium $\dot{\phi}$ must be a function of r only, at least so long as terms after the first on the right-hand of equation (250·8) are neglected. To this approximation, the rotation is by spherical shells, each spherical shell rotating as a rigid body with its own angular velocity, which in the outer layers of the star varies inversely as the square of the radius of the shell.

251. To pass to a second approximation, we must take into account the second term on the right of equation (250·8). The additional terms required in the solution are seen to be proportional to $\sin^2 \theta$ and $\cos^2 \theta$, multiplied by functions of r only. Thus to a second approximation, the rotation in any sphere of given radius r is of the form

$$\phi = A + BP_2(\cos \theta) \quad \ldots\ldots\ldots\ldots\ldots(251\cdot1),$$

where A and B are constants, or, in terms of the latitude λ,

$$\phi = a - b \sin^2\lambda \quad \ldots\ldots\ldots\ldots\ldots(251\cdot2),$$

where b is of the same sign as B. To a still further approximation, we must introduce into ϕ a small term proportional to the fourth zonal harmonic $P_4(\cos \theta)$ giving a law of rotation

$$\phi = a - b \sin^2\lambda - c \sin^4\lambda \ldots\ldots\ldots\ldots\ldots(251\cdot3).$$

The physical explanation of these laws is that the more rapidly rotating layers in the inner regions of a star are of distinctly spheroidal shape, whereas the less rapidly rotating layers outside are nearly spherical. The inner spheroidal layers consequently approach nearest to a star's surface at its equator, and their drag results in an equatorial acceleration, particles near the star's equator revolving more rapidly than those near its poles.

The mechanism can be simply illustrated by considering a star with a liquid centre of uniform density, built according to the extended Roche's model. Let the liquid core rotate with a constant angular velocity ω, and let its surface be

$$\frac{x^2 + y^2}{a^2} + \frac{z^2}{b^2} = 1, \text{ or } \frac{\cos^2 \lambda}{a^2} + \frac{\sin^2\lambda}{b^2} = \frac{1}{r^2}.$$

If the angular velocity varies as $1/r^2$ throughout the atmosphere which surrounds this core, the speed of rotation at the point R, λ on the boundary of the star is

$$\phi = \frac{\omega r^2}{R^2} = \frac{\omega a^2}{R^2}\left[1 - \frac{e^2}{1 - e^2}\sin^2\lambda + \frac{e^4}{(1 - e^2)^2}\sin^4\lambda - \ldots\right]\ldots(251\cdot4),$$

where e^2 stands for $(a^2 - b^2)/a^2$, so that e is the eccentricity of the surface of the core.

Solar Rotation.

252. Observations on the rates at which faculae, sunspots and calcium flocculi travel round the sun shew that the periods of rotation are different in different latitudes, and formulae of the general type of (251·2) and (251·3) are found to fit the observations best. In fig. 48* the dots shew the period of

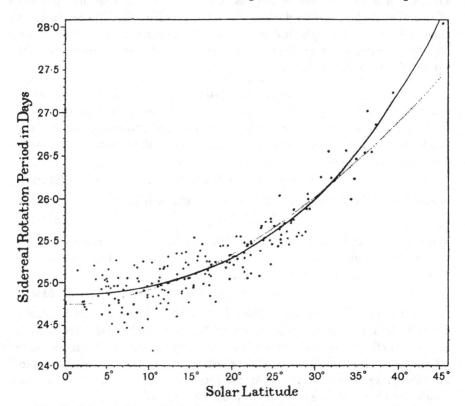

Fig. 48. Solar rotation periods in different latitudes derived from measures of faculae
(Greenwich Observatory).

Formula $\dot{\phi} = 14°\cdot49 - 1°\cdot78 \sin^2 \lambda - 3°\cdot16 \sin^4 \lambda$, thus ——
„ $\dot{\phi} = 14°\cdot54 - 2°\cdot81 \sin^2 \lambda$ „   ......

rotation of various faculae in different latitudes which had been observed at Greenwich from 1888 to 1923. The two curves across the diagram shew the best fits that can be obtained from formulae of the type of (251·2) and (251·3) respectively. Other determinations of the rotations in different latitudes, as fitted to formulae of this type, are shewn in fig. 49.

* *M.N.* LXXXIV. (1924), Plate 10. I am indebted to Sir F. W. Dyson, Astronomer Royal, for permission to reproduce this figure and also fig. 49.

If we attempt to interpret these formulae in terms of the extended Roche's model, a comparison with formula (251·4) shews that the observed equatorial acceleration of the sun's rotation would require the inner core to have an eccentricity e such that $e^2/(1-e^2) = 0·15$ or $e = 0·36$. The ratio of its axes is given by $b = 0·93\,a$. To produce this degree of ellipticity in a core of uniform

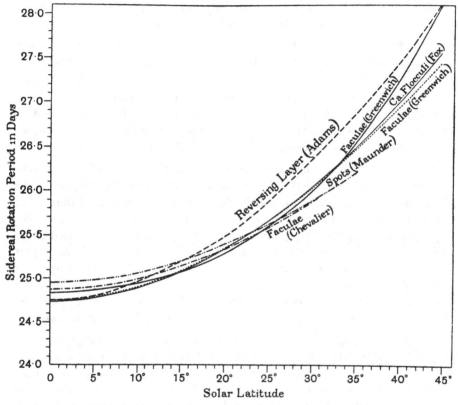

Fig. 49. Solar rotation period determined by various methods.

Reversing Layer (Adams)	$\dot{\phi} = 14°·54 - 3°·50 \sin^2\lambda$	– – –
Calcium Flocculi (Fox)	$\dot{\phi} = 14°·56 - 2°·98 \sin^2\lambda$	———
Faculae (Greenwich)	$\begin{cases} \dot{\phi} = 14°·49 - 1°·78 \sin^2\lambda - 3°·16 \sin^4\lambda \\ \dot{\phi} = 14°·54 - 2°·81 \sin^2\lambda \end{cases}$	▬▬▬ ············
Faculae (Chevalier)	$\dot{\phi} = 14°·47 - 2°·27 \sin^2\lambda$	– · – · –
Spots (Maunder)	$\dot{\phi} = 14°·43 - 2°·13 \sin^2\lambda$	– · · – · · –

density ρ, $\omega^2/2\pi\gamma\rho$ must be equal to 0·035. The value of $\omega^2/2\pi\gamma\rho$ for the sun's outer surface is only 0·000014, so that if the sun's equatorial acceleration is to be explained in terms of this model, the angular velocity of the core must be very much greater than that of the surface. If the law $\dot{\phi} \propto 1/r^2$ is obeyed, the core must be of small radius or, alternatively, $\dot{\phi}$ may increase much more rapidly than $1/r^2$ in the central regions of the sun.

The Internal Condition of Rotating Stars.

253. The variation of the rotation of the sun's atmosphere with latitude is of interest as providing direct observational evidence of the variations of rotation which theory shews must inevitably occur in every rotating star. From the point of view of cosmogony in general, however, variations in the speeds of rotation of stellar surfaces are only unimportant by-products of the far more extensive variations of rotation in the star's interior.

The knowledge of these rotations which we have gained from theory puts us in a position to form a fairly definite picture of the internal condition of a rotating star.

In Chapter III we studied the configurations of a non-rotating star in some detail. The conditions which were found to obtain in a star devoid of rotation can best be described by imagining the star divided into two distinct regions which we may describe as the core and the envelope.

In fig. 47 (p. 278) the innermost curve gives the distribution of density as we pass along the radius of a non-rotating gaseous star of mass about equal to that of our sun. This shews that the density falls to about a twentieth of the central density at a point only a third of the radius out from the centre of the star. We may describe as the core that part of the star which lies within a sphere whose radius is about a third of that of the star, this region containing the main part of the star's mass, and being the only region in which the density is comparable with that at the star's centre. In the outer region, which we describe as the envelope, the density is only a small fraction of the central density, and the mass of the envelope is only a small fraction of the total mass of the star.

The curves in fig. 47 are drawn on the assumption that the gas-laws are obeyed throughout the star, but we have seen that considerations of stability demand deviations from the gas-laws. A star in which the gas-laws were obeyed would slowly shrink or contract until it had reached a configuration in which the gas-laws were not obeyed in the central regions of the star; stability can only be maintained when this central region offers a greater resistance to compression than a purely gaseous structure can provide. Thus fig. 47 cannot represent an actual stable star; to represent this, substantial deviations from the gas-laws must be introduced, with the result (§ 132) that both the density and temperature curves become very much flatter in this central region than the curves shewn in fig. 47. We may legitimately suppose that this modification of the density and temperature curves is confined to the core, the envelope continuing to obey the gas-laws.

Thus we think of the star as consisting of a gaseous atmosphere and of a core whose structure may lie almost anywhere between the two extremes provided by a structure obeying the gas-laws as represented in fig. 47, and an almost incompressible mass of approximately uniform density.

The investigations of the present chapter have provided a picture of the same star when in rotation. We have found that viscosity has so slight an effect in the core that the whole life of the star is inadequate for it to establish uniformity of angular velocity. The different layers of the core continue to rotate as though viscosity were non-existent, with angular velocities determined mainly by the past history of the star. On the other hand, viscosity and the braking effect of the transmission of radiation are extremely potent in the envelope, and we have seen that in a comparatively short time the angular velocity ω in the envelope will conform to a law which, to a sufficient approximation for our present purpose, may be expressed in the form $\omega \propto 1/r^2$.

In Chapters VIII and IX we studied the configurations assumed by masses rotating with angular velocities which were supposed uniform throughout each mass. For a fairly incompressible mass we found series of configurations which, under slow rotation, approximated to spheroids, and with higher velocities of rotation were pseudo-spheroids, pseudo-ellipsoids, or pear-shaped figures, according to the amount of angular momentum with which the system was endowed, the series finally ending, for high angular momentum, in two detached and separate masses revolving about one another. If, however, the compressibility approximated to that of a perfect gas, the configurations were pseudo-spheroidal throughout, the series of configurations coming to an abrupt end through the pseudo-spheroid forming a sharp edge along its equator, so that its form became that of a double convex lens. Beyond this no configurations existed in which the mass rotated with uniform velocity throughout; if the mass acquired further angular momentum beyond that corresponding to the lenticular figure, it shed matter from its equator, and this shed matter rotated at rates different from that of the main mass.

We have seen that the angular velocity cannot be uniform in an actual star. If the star is spherical, or approximately spherical, in shape we have seen that the angular velocity in its outermost layers will fall off approximately as $1/r^2$. When the star is not of spherical shape this law is no longer generally true, but the same physical principles from which we deduced it shew that in the equatorial plane of the star the angular velocity must fall off as $1/\varpi^2$, where ϖ is the distance from the axis of rotation. When once this law is established, the lens-shaped figure with a sharp edge can never be reached. The lens-shaped figure is determined by the condition that centrifugal force exactly balances gravity on its equator. Now if ω falls off as $1/\varpi^2$, centrifugal force $\omega^2\varpi$ must fall off as $1/\varpi^3$, and so falls off more rapidly than gravity, which only falls off about as rapidly as $1/\varpi^2$. In an actual rotating star the law $\omega \propto 1/\varpi^2$ is reached with infinite rapidity in the outermost layers of all, so that the lenticular figure can never be reached, and the shedding of matter from the equator of a lenticular figure (which, incidentally, formed the basic conception of Laplace's nebular hypothesis) cannot occur in an actual star.

Thus if a star acquires an ever-increasing amount of rotation, its outer surface must always remain smooth. It can never acquire a sharp edge, but may acquire a more and more drawn out equatorial cross-section. Apart from this, the star is in most respects similar to the modified Roche's model discussed in § 234, which consisted of an approximately incompressible core surrounded by a tenuous envelope. A star with a sufficiently incompressible core may behave in the way in which we found the modified Roche's model to behave. With increasing angular momentum the core may depart from the spheroidal configuration; it may become first ellipsoidal, then pear-shaped, and may end by fission into two detached masses rotating about one another.

During this process the envelope, owing to its small mass, exerts only a negligible influence on the dynamics of the core. The rôle of the envelope is merely that of adjusting itself to the changes going on in the core; it will arrange itself so that its outer surface always forms an equipotential in the field formed jointly by its own rotation and the ever-changing gravitational field of the core. When the core finally breaks into two detached masses, the envelope may either form a continuous surface surrounding both masses, or may itself divide into two envelopes, one surrounding each of the two masses. Or, to put the matter in a slightly different form, the final result will be two detached masses surrounded by two envelopes which may or may not overlap.

The Sequence of Configurations of a Shrinking Star.

254. Our discussion has so far supposed that the rotating mass continually acquires more and more angular momentum. The separate stars in the sky are so widely spaced that their mutual gravitational influence on one another is quite negligible under normal circumstances, and no agency is known by which the angular momentum of a star can increase under normal circumstances. The emission of radiation from a rotating star, by carrying away angular momentum, will cause a decrease in the total angular momentum residing in the star, but apart from this the angular momentum must remain constant.

On the other hand, our discussion has also supposed that the density of the rotating mass remains constant, and this supposition also is equally un-fulfilled in nature. In discussing the evolution of the stars in Chapter VI, we saw how a star on any one series (*M*-ring, *L*-ring, or *K*-ring) might at any instant reach the limit of stable configurations, and start to contract with comparative rapidity until it reached the next lower series as a star of far higher density.

During such a contraction the angular momentum of the whole star will remain approximately constant, so that its average angular velocity must increase. Since the effect of viscosity is negligible in the core, the motion of the core will conform to the well-known hydrodynamical equation*

$$\frac{D}{Dt}\int_A^B (u\,dx + v\,dy + w\,dz) = \left[-\int \frac{dp}{\rho} - \Omega + \tfrac{1}{2}(u^2 + v^2 + w^2) \right]_A^B.$$

* Lamb, *Hydrodynamics*, Equation (2), p. 34.

On taking the integration round a circle of constant angular velocity $\dot{\phi}$, this takes the form

$$\frac{D}{Dt}\left(\varpi^2\dot{\phi}\right) = 0,$$

so that $\varpi^2\dot{\phi}$ remains constant as the star shrinks. If the shrinkage is uniform, $\dot{\phi}$ changes uniformly, and is always proportional to ϖ^{-2} or to $\rho^{\frac{2}{3}}$. Thus the average rotation ω of the core will remain proportional to $\rho^{\frac{2}{3}}$ throughout the shrinkage of the star, and the value of $\omega^2/2\pi\gamma\rho$ will be proportional to $\rho^{\frac{1}{3}}$. Although it has been convenient to regard our different configurations as belonging to masses of a given density ρ, we have seen that each separate shape of configuration corresponds to a given value of $\omega^2/2\pi\gamma\rho$, so that there are an infinite set of configurations of any specified shape corresponding to the different values of ω and ρ which are consistent with $\omega^2/2\pi\gamma\rho$ having an assigned value. The angular momenta of the various configurations of any given shape are proportional to $Mr^2\omega$, where r is any linear dimension of the mass, or, since M is proportional to ρr^3 and $\omega^2/2\pi\gamma\rho$ is assigned, the angular momenta are proportional to $M^{\frac{5}{3}}\rho^{\frac{1}{6}}$. Thus if M is kept constant and ρ is allowed to increase, while the angular momentum remains constant, the mass will pass through the same sequence of configurations as though M and ρ were both kept constant while the angular momentum increased. This is precisely the series of configurations we have had under review.

255. We now see that the shrinkage which accompanies the passage of an actual star from one region of stable configurations to another—or if we so prefer to put it, the shrinkage which accompanies the ionisation of one ring of electrons—may result in fission of the star into two distinct masses.

As we have seen, the final stages of the process of fission are probably cataclysmic, with the result that it is difficult to follow them out dynamically. Had the pear-shaped figure proved to be stable, the final stages of the fissional process would have consisted merely in a gradual deepening of the furrow round the pear until we were left with two detached masses in contact rotating as one single rigid body, the periods of rotation of the two masses and their period of revolution about one another all three coinciding. There is no longer the same justification for this supposition when it is recognised that fission occurs only after cataclysmic motion.

We may, however, notice that only one vibration is unstable at the point of bifurcation at which cataclysmic motion begins, this vibration being one in which neither half of the mass gains upon the other, either in rotation or revolution. When the elongation of the pear-shaped figure first takes place, the pointed end of the pear must lag somewhat behind the rotation of the blunter end, as a consequence of conservation of angular momentum, but any such difference of rotation produces a distortion which corresponds to a stable vibration: forces of restitution at once come into play and equalise the angular velocities. Similar forces of restitution will be in operation right up to the instant of fission, so that in the final system the rotations may be expected to agree with the revolution, both in period and in phase. The two constituents will accordingly rotate about one another in a motion which must at least approximate to a rigid body motion, the periods of rotation and revolution being approximately identical and the eccentricity of the orbits being very small.

These are precisely the conditions found in binary stars which have, according to all available criteria, just broken up by fission. The primary criterion for recent fission is the nearness of the two components, since we shall find that as a binary ages its two components move further apart. A class of stars exists, commonly described as stars of the β Lyrae type, in which the light curve varies continuously as the components undergo successive eclipses, this indicating that the two constituent masses must be either in actual contact or very close to actual contact.

This class of star seems to fulfil all the conditions which our theoretical investigation has shewn ought to be expected in stars which have just broken up by fission. The periods of rotation and revolution appear to be identical; any difference in these periods would shew itself in non-periodicity of the light curve; of this there is no evidence whatever. The eccentricity of orbit is invariably small, being almost zero for X Carinae and RR Centauri, in which the separation, calculated from the light curve, is zero or negative (corresponding to imperfect fission), and ranging from 0 to 0·02 for stars such as β Lyrae H.D. 1337 and W Crucis in which the separation is excessively small.

CHAPTER XI

THE EVOLUTION OF BINARY SYSTEMS

256. In the last chapter we found that binary systems of the β Lyrae type shew all the characteristics to be expected in stars which have just broken up by fission.

These stars form one end of a continuous chain of binary systems. As we proceed along this chain the physical characteristics of the systems vary widely, substantial departures occurring from the characteristics shewn by systems of the β Lyrae type. In the present chapter we shall investigate what changes are likely to be produced in binary systems by the passage of time and the play of natural forces, with a view to examining to what extent the observed chain of binary formations can be interpreted as an evolutionary chain.

OBSERVATIONAL MATERIAL.

257. The following table gives particulars of a few typical binary systems which shew the small separation, short period, low eccentricity of orbit and other features which must be regarded as the primary indication that a system has recently been formed by fission.

The last column but one gives the radii of the two stars in terms of the radius of their relative orbit, and the final column gives the sums of these radii. If this sum were equal to unity the stars would be in contact.

Table XIX. *Newly-formed Binaries.*

Star	Spectral Type	Mass	Eccentricity of Orbit	Period in Days	Radii in terms of Orbit	Sum of Radii
H.D. 1337	$O\,8\frac{1}{2}$	36·3, 33·8	0	3·52	0·59, 0·39	0·98
V Puppis	$B\,1$	19·2, 17·9	0·08	1·45	0·46, 0·42	0·88
u Herculis	$B\,3$	7·66, 2·93	0·05	2·05	0·31, 0·37	0·68
TX Cassiop.	$B\,3,\ B\,5$	—	—	2·93	0·57, .0·30	0·87
β Lyrae	$B\,5$	—	0·018	12·92	0·68, 0·27	0·95
RZ Centauri	A	—	0	1·88	0·49, 0·24	0·73
WZ Cygni	A	—	0	0·58	0·46, 0·38	0·84
S Antliae	$F\,0$	—	—	0·65	0·50, 0·39	0·89
RR Centauri	F	—	0	0·30*	0·50, 0·50	1·00
W Ursae Maj.	$F\,8$	0·74, 0·52	0	0·33	0·37, 0·37	0·74
SW Lacertae	$G\,2$	—	—	0·32	0·42, 0·46	0·88
W Crucis	$G\,0$	—	0	198·5	0·61, 0·34	0·95

* J. Voute, *Annalen v. d. Bosscha-Sterrenwacht*, Lemberg (Java), II. (1927), 2e gedeelte. The elements of orbit are those calculated by Shapley from an earlier light curve by Roberts, but in any case the light curve shews that the two components must be nearly or quite in contact.

On plotting spectral type against absolute magnitude, practically all of these and similar stars are found to lie on the main sequence. *W* Crucis forms an exception; the mean densities of its two components are given by Shapley[*] as $1·3 \times 10^{-6}$ and $3·1 \times 10^{-6}$, and both components are of giant type. Its origin may well be of the kind suggested at the end of § 165.

258. Aitken[†] has analysed the orbits of 119 spectroscopic binaries, classified by period and eccentricity, with results shewn in the following table:

Table XX. *Spectroscopic Binaries classified by Period and Eccentricity* (Aitken).

	Period of Orbit in Days						Total Number
	0–5	5–10	10–20	20–50	50–150	Over 150	
Eccentricity							
0 to 0·1	40	9	6	3	3	1	62
0·1 to 0·2	5	4	1	0	2	4	16
0·2 to 0·3	1	5	1	1	2	2	12
0·3 to 0·4	0	0	2	1	2	1	6
0·4 to 0·5	0	1	0	2	1	3	7
0·5 to 0·6	0	0	1	3	3	2	9
0·6 to 0·7	0	0	1	0	0	1	2
0·7 to 0·8	0	0	0	3	1	0	4
0·8 to 0·9	0	0	0	0	1	0	1
0·9 to 1·0	0	0	0	0	0	0	0
Total number	46	19	12	13	15	14	119
Average period	2·75	7·80	15·17	30·24	106·4	1035	
Average eccentricity	0·047	0·147	0·202	0·437	0·371	0·328	

He has analysed the orbits of 68 visual binaries in the same way, obtaining the result shewn in Table XXI opposite, in which the periods are now measured in years instead of days.

Both tables shew a marked increase of eccentricity with period, and the phenomenon runs on from one table to the other, the eccentricities of the visual binaries which have long periods being far higher than those of the spectroscopic binaries whose periods are far shorter. The general uniformity of this progression is clearly shewn in Table XXII, in which the whole 187 orbits are divided, according to their periods, into seven groups of approximately equal size by adding together suitable groups of columns of the two preceding tables.

[*] *Contributions from the Princeton University Observatory*, No. 3 (1915).
[†] *The Binary Stars*, New York, 1918.

Table **XXI.** *Visual Binaries classified by Period and Eccentricity* (Aitken).

	Period of Orbit in Years				Total Number
	0–50	50–100	100–150	Over 150	
Eccentricity					
0 to 0·1	0	0	0	0	0
0·1 to 0·2	5	1	0	1	7
0·2 to 0·3	4	0	0	0	4
0·3 to 0·4	7	4	2	1	14
0·4 to 0·5	5	6	1	2	14
0·5 to 0·6	4	4	1	5	14
0·6 to 0·7	1	1	1	1	4
0·7 to 0·8	3	3	0	1	7
0·8 to 0·9	1	1	0	1	3
0·9 to 1·0	0	0	1	0	1
Total number	30	20	6	12	68
Average period	31·3	74·4	124·5	243	
Average eccentricity	0·423	0·514	0·558	0·529	

Table **XXII.** *Eccentricities of Binaries of various Periods* (Aitken).

Type of Star	Number	Mean Period	Mean Eccentricity
Spectroscopic Binary	46	2·75 days	0·047
	19	7·80 ,,	0·147
	25	23·00 ,,	0·324
	29	1·5 years	0·350
Visual Binary	30	31·3 years	0·423
	20	74·4 ,,	0·514
	18	170·0 ,,	0·539

This makes it abundantly clear that statistically period and eccentricity progress together, although of course this is not necessarily true for individual stars. This line of progress characterises the chain of binary configurations already mentioned, binaries which have been newly formed by fission occupying the extreme end of the chain at which the periods and eccentricities are both very small.

259. Aitken has also studied the distribution of spectral type in binary systems. His results are shewn in the following table, which gives the average periods and the average eccentricities of binaries, both spectroscopic and visual, of different spectral classes:

Table XXIII. *Binary Systems classified by Spectral Type* (Aitken).

Type of Star	Spectral Class	Number	Mean Period	Mean Eccentricity
Spectroscopic Binary	O to $B\,8$	48	26·76 days	0·189
	$B\,9$ to $A\,3$	36	13·35 ,,	0·187
	$A\,5$ to $F\,2$	10	32·76 ,,	0·252
	$F\,5$ to $G\,0$	13	267·4 ,,	0·129
	$G\,5$ to $M\,4$	8	152·9 ,,	0·196
Visual Binary	O to $B\,8$	0	—	—
	$B\,9$ to $A\,3$	13	98·9 years	0·529
	$A\,5$ to $F\,2$	9	100·6 ,,	0·512
	$F\,5$ to $G\,0$	30	78·7 ,,	0·478
	$G\,5$ to $K\,2$	12	86·0 ,,	0·432
	$K\,5$ to $M\,4$	4	126·7 ,,	0·402

In the spectroscopic binaries it is impossible to say whether the eccentricity increases or decreases with advancing spectral type; it is at any rate safe to conclude that there is no marked correlation between spectral type and eccentricity. As regards periods, the later type spectroscopic binaries have periods which are distinctly longer than those of earlier type, but there is no very marked correlation extending throughout the table.

Visual binaries shew a complete absence of correlation between period and spectral type, but an appreciable correlation between eccentricity and spectral type in the sense of a decrease in eccentricity accompanying an advance in spectral type.

Aitken's classifications, given above, include only those binary orbits for which the data are thoroughly well determined. We must be very cautious in assuming that these orbits are representative of binary orbits in general, or in supposing that the characteristics we have noticed would prevail in a random selection of binaries.

The spectral lines of early type stars are singularly ill-defined, so that the velocities in early type spectroscopic binaries can only be measured accurately when they are very large, the Doppler effect now becoming so marked that the breadth and fuzziness of the lines is of little account. As a consequence, early type binaries in which the periods are long, and the velocities consequently small, are apt to be excluded from the table. This results in a tendency for the periods of early type binaries entered in the table to be short in comparison with the periods of late type binaries which are unaffected by this restriction. This consideration may account for some or all of the correlation between period and spectral type shewn in the table for spectroscopic binaries.

Visual binary systems of classes O, B and A are in general bright but remote from us, so that the two components must be remote from one another also if the limited power of our telescopes is to disclose their binary nature. There is thus a tendency for the table to include no binaries of early type except those whose orbits are large and so of long period. In view of the correlation we have already discovered between period and eccentricity, this also gives a preference to orbits of high eccentricity. This consideration may account for some or all of the correlation between eccentricity and spectral type shewn in the second half of the table.

Aitken's own conclusion is that, even after allowing for these factors, "the evidence is definitely in favour of an increase of the period of binary stars with advancing spectral class." He reaches this conclusion, however, by taking into account the difference in spectral class between spectroscopic and visual binaries, the spectroscopic binaries being mainly of early class, and the visual binaries of late class. When we regard these two sets of binaries as distinct types of object, as we shall find reason to do, the correlation between spectral type and period is seen to amount to very little. As regards the correlations, if any, between eccentricity and spectral type in the two classes of binaries separately, Aitken expresses the opinion that "it is doubtful whether any significance attaches to either."

Thus the only correlation which is firmly established is that of advancing period with advancing eccentricity of orbit, as shewn in Table XXII.

TIDAL FRICTION.

260. We turn now to a theoretical discussion of the changes which are likely to occur in the orbit of a binary star after fission has occurred.

In Chapter VIII (§ 218), we saw that two stars rotating about one another, with the three periods of rotation and revolution all equal, are secularly unstable until their distance apart reaches a certain limit, and that even after this critical distance is exceeded, secular stability exists only in a partial sense. Our investigation, however, dealt only with incompressible masses. In an actual star, there may be no secular stability until the two components into which the core has divided are at a considerable distance apart, and yet the atmospheric envelopes surrounding these two components may form a continuous surface, or may so increase the radii of the complete stars as to cause them to appear to be almost in contact.

After this point has been passed, and the configurations possess partial secular stability, it becomes important to investigate in what sense they remain secularly unstable.

Fission has been supposed to occur as a consequence of shrinkage of the star, and there is no reason why this shrinkage should suddenly cease at the moment at which fission takes place. If the two components continue to shrink further after fission, the rate of rotation of each will increase in accordance with the conservation of angular momentum, so that the rotations of the

separate masses will gain on their revolution about one another, with the
result that after a time the arrangement of the masses will be of the general
nature shewn in fig. 50. In a configuration such as this, each mass exerts a
couple on the other in such a direction as to augment the orbital motion of
revolution already taking place. These couples are the direct successors of the
forces of restitution, mentioned in § 255, which tend to equalise the periods of
rotation and of revolution. The effect of these couples on the orbits of the
masses has been very fully investigated by Darwin under the designation of
" Tidal Friction."

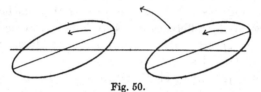

Fig. 50.

261. Suppose that a star originally of mass $M + M'$ has divided into two
components of masses M, M', each of which describes an approximately
elliptic orbit about the centre of gravity of the two. Let e be the eccentricity and a the semi-major-axis of the orbit described by either mass
relative to the other.

If the tidal friction couples were non-existent, there would be the usual
two first integrals of the motion,

$$\frac{MM'}{M + M'}\, r^2 \dot{\theta} = h \qquad \qquad \dots \dots \dots \dots \dots \dots (261 \cdot 1),$$

where
$$h^2 = \gamma^2 \frac{M^2 M'^2}{M + M'}\, a\, (1 - e^2) \dots \dots \dots \dots \dots (261 \cdot 2),$$

$$\text{Energy} = E, \quad \text{where} \quad E = - \gamma MM'/2a \dots \dots \dots \dots (261 \cdot 3).$$

Let the couples produced by tidal friction be supposed to act for a short
interval dt, each component of the system exerting a couple G on the other
in the direction of the orbital motion. As a consequence, the orbit will be
disturbed and at the end of the interval dt a new orbit will be described.
The eccentricity and semi-major-axis of this may be denoted by $e + \dot{e}\, dt$,
$a + \dot{a}\, dt$, in which $\dot{e}$ and $\dot{a}$ are regarded as rates of increase during the action
of the couple G. These rates of change are readily found. Differentiating
equation (261·3), we find

$$\frac{1}{a^2}\frac{da}{dt} = \frac{2}{\gamma MM'}\frac{dE}{dt} = \frac{2}{\gamma MM'}\, G\dot{\theta} \qquad \dots \dots \dots \dots (261 \cdot 4).$$

Since G acts in the direction of θ increasing, $G\dot{\theta}$ must be positive. Thus
da/dt is positive, so that tidal friction increases a.

By logarithmic differentiation of equation (261·2),

$$\frac{1}{1 - e^2}\frac{d}{dt}(1 - e^2) = \frac{2}{h}\frac{dh}{dt} - \frac{1}{a}\frac{da}{dt} \qquad \dots \dots \dots \dots (261 \cdot 5)$$

The angular momentum h increases at a rate G, so that $dh/dt = G$. On further using the value for da/dt provided by equation (261·4), and substituting for $\dot{\theta}$ from equation (261·1), this becomes

$$\frac{1}{1-e^2}\frac{d}{dt}(1-e^2) = \frac{2G}{h}\left[1 - \frac{a^2}{r^2}(1-e^2)\right],$$

whence, using the polar equation of the orbit in the form

$$a(1-e^2) = (1 + e\cos\theta)\, r,$$

$$\frac{de}{dt} = \frac{G}{h}[2\cos\theta + e(1 + \cos^2\theta)] \quad\ldots\ldots\ldots\ldots(261\cdot6).$$

Since G acts in the direction of θ increasing, G/h is positive, while $e(1 + \cos^2\theta)$ is necessarily positive. Tidal friction acts mainly when the masses are closest together—i.e. when $\cos\theta$ is nearest to $+1$. Hence it is readily found that $G\cos\theta/h$ is preponderatingly positive and de/dt integrated through a whole orbit must be positive.

Thus tidal friction increases both a and e, and as the evolution of a binary star progresses, tidal friction must produce—(i) increasing separation, (ii) increasing period, (iii) increasing eccentricity.

262. So far, then, tidal friction has been seen to be capable of producing, at least in a general qualitative way, exactly those phenomena which appear in actual fact to occur as the evolution of a binary system progresses. But tidal friction falls off very rapidly as the distance between the components increases. When the two components are at a distance r apart, the tides raised by either on the other are proportional to $1/r^3$, and the tidal couple, which arises from the gravitational pull of tides on tides, falls off approximately as $1/r^6$. Equation (261·4) shews that the change in $1/a$ which occurs in a whole revolution is proportional to G and so to $1/a^6$, so that $1/a^5$ will decrease by approximately the same amount at each revolution of the star. Clearly the changes become inappreciable when a is at all large.

263. A still more clearly-defined limitation to the scope of tidal friction has been indicated by Russell[*] and others. Any increase in the separation of the two components of a binary star, whether under tidal friction or otherwise, involves an increase in the orbital momentum of the system. So long as the system remains free from external disturbance, the total angular momentum of the system remains constant, so that the increase in orbital momentum is necessarily gained at the expense of the rotational momenta of the constituent stars. When the masses of the two components are very unequal, there is a large store of angular momentum in the rotation of the more massive one, and separation can proceed very far before this has all been transferred to angular momentum. But conditions are very different

[*] *Astrophys. Journal*, xxxi. (1910), p. 185.

when the components are of approximately equal mass, as is the case with the majority of binary stars.

264. In § 217 we obtained the total angular momentum **M** of a newly-formed binary system in which the components revolved in a circular orbit of radius R, in the form

$$\mathbf{M} = \left[Mk^2 + M'k'^2 + \frac{MM'}{(M + M')} R^2 \right] \gamma^{\frac{1}{2}} (1 + \zeta)^{\frac{1}{2}} (M + M')^{\frac{1}{2}} R^{-\frac{3}{2}} \dots (264\cdot1).$$

Consider first the extreme case in which the masses are supposed homogeneous and incompressible. With a view to obtaining some idea of the ratio of division of **M** into its rotational and orbital parts, I have calculated the ratios of the separate terms in **M** for Darwin's figures of closest approach from the data already tabulated in § 218, with the following results:

$\dfrac{M'}{M} =$	0	0·4	0·5	1·0
Rotational momentum of M'	0	·039	·046	·077
„ „ M	1	·160	·135	·077
Orbital momentum	0	·801	·819	·846
Total	1·000	1·000	1·000	1·000

With very few exceptions all known binary stars have values of M'/M which lie between 0·4 and 1·0. Excluding the few systems for which M'/M is less than 0·4, it appears that the orbital momentum must initially be at least 80 per cent. of the whole if the components move in circular orbits; if they moved in stable elliptical orbits, it would of course be greater still.

Thus no matter for how long tidal friction or other similar tendencies act, the orbital momentum cannot increase to more than 1·25 times its initial value throughout the whole course of a binary star's history. In the subsequent motion of the star, the orbital momentum is

$$\frac{MM'}{(M + M')^{\frac{1}{2}}} \gamma^{\frac{1}{2}} (1 + \zeta)^{\frac{1}{2}} l^{\frac{1}{2}} \dots\dots\dots\dots(264\cdot2),$$

where l is the semi-latus-rectum of the elliptical orbit. It follows that, throughout the whole life of the star, $(1 + \zeta) l$ cannot increase to more than $(1\cdot25)^2$ or 1·56 times its initial value. For bodies at a considerable distance apart $\zeta = 0$; for two similar ellipsoids in contact $\zeta = 0\cdot22$, which is the maximum value of ζ. Thus in the whole course of evolution the value of $1 + \zeta$ cannot decrease more than in the ratio $1\cdot22 : 1$. It follows that l cannot at the very most increase in a ratio greater than $1\cdot56 \times 1\cdot22$ or 1·90.

These calculations have referred to a perfectly homogeneous mass. To study the effect of heterogeneity, let us pass to the extreme case of matter so compressible that Roche's model (§ 228) may be supposed to give an approximation to the arrangement of density. We may now put $k^2 = k'^2 = 0$

and $\zeta = 0$. The whole momentum is orbital, and in this extreme case the constancy of **M** requires that l shall remain absolutely constant. We may reasonably suppose that all compressibility lessens the possible range of increase in l, so that the increase of 90 per cent. just calculated for an incompressible mass is the maximum possible, always provided the mass ratio does not exceed $2\frac{1}{2} : 1$, and that the system remains free from external disturbance.

265. Under these same assumptions there is a simple relation between the dimensions of the present orbit of a binary star generated by fission and those of the parent star out of which the system originated.

Let r and $\bar{\rho}$ denote the mean radius and mean density of the parent star at the instant at which the pseudo-spheroidal form first becomes unstable, so that $M + M' = \frac{4}{3}\pi\bar{\rho}r^3$, and let θ denote the value of $\omega^2/2\pi\gamma\bar{\rho}$ at the same instant. Then the angular momentum at this instant is

$$\mathbf{M} = (M + M')\,k^2\omega = \gamma^{\frac{1}{2}}(M + M')^{\frac{3}{2}}\frac{k^2}{r^2}\,r^{\frac{1}{2}}(\tfrac{3}{2}\theta)^{\frac{1}{2}}.$$

When fission has taken place and the components have become thoroughly separated, the orbital momentum is

$$\frac{MM'}{(M + M')^{\frac{1}{2}}}\gamma^{\frac{1}{2}}(1 + \zeta)^{\frac{1}{2}}\,l^{\frac{1}{2}}.$$

Since this must always be less than **M**, it follows that at any stage of the star's orbital motion

$$\frac{l}{r} < \frac{(M + M')^4}{M^2M'^2}\frac{k^4}{r^4} \times \tfrac{3}{2}\theta.$$

If the parent star was wholly incompressible, the critical figure is a Maclaurin spheroid at its point of bifurcation, for which $k^2/r^2 = 0.3838$ and $\theta = 0.18712$, so that the inequality becomes

$$\frac{l}{r} < 0.04135\,\frac{(M + M')^4}{M^2M'^2} \quad\quad\quad\ldots\ldots\ldots\ldots\ldots\ldots(265.1).$$

For a compressible mass the value of k^2/r^2 is less, but θ is greater. Calculation makes it clear that compressibility decreases k^2/r^2 much more rapidly than it increases θ, so that compressibility lessens the numerical factor 0.04135—for instance, for Roche's model it reduces to zero. Thus the inequality (265.1) remains true independently of the compressibility of the mass.

If the two components are of very unequal mass

$$(M + M')^4/M^2M'^2$$

is very large, so that l may be very large compared with r—the components can separate to a distance large compared with the dimensions of the parent star. But for the normal binary in which $M/M' < 2\frac{1}{2}$, the value of $(M + M')^4/M^2M'^2$ is less than 24.01 and inequality (265.1) takes the form

$$l < 0.9928r,$$

so that under no circumstances can the semi-latus-rectum of a binary system in which $M/M' < 2\frac{1}{2}$ exceed the mean radius of the parent star at the instant at which the spheroidal or pseudo-spheroidal form first became unstable.

266. For α Centauri the present value of l is $2\cdot5 \times 10^{14}$ cms., subtending an angle of $12\cdot75''$. If the system was generated by rotational fission and has been under no influence beyond that of tidal friction, the mean radius of the pseudo-spheroid just before elongation commenced must have been at least $2\cdot5 \times 10^{14}$ cms., so that the major-axis must have been at least 6×10^{14} cms., subtending (at its present distance) an angle of at least $30''$. The mass of the system being $3\cdot8 \times 10^{33}$ grms., the mean density must have been less than 6×10^{-11}.

These figures are so remote from possibility as to dispel at once the supposition that the present long periods and large orbits of binaries can be ascribed to the workings of tidal friction on systems evolved by rotational fission, and we are compelled to turn elsewhere for a solution of the problem.

SECULAR DECREASE OF MASS.

267. The foregoing discussion has supposed the masses of the two components of the binary to remain constant; we have seen that in actual fact they must undergo a slow secular decrease as a consequence of the emission of radiation from the star. In 1924* I investigated whether this secular change could produce the lengthening of the period and increase in the eccentricity which are observed to occur in binary systems.

When a force F acts upon a body of mass m which is losing mass by radiation at a rate $-dm/dt$ per unit time, the momentum of the mass mv experiences an increase at a rate F from the action of the force, and a loss at a rate $\left(-\dfrac{dm}{dt}v\right)$ from the momentum carried away by the emitted radiation. Thus the change of momentum is governed by the equation

$$\frac{d}{dt}(mv) = F + \frac{dm}{dt}v,$$

or, simplifying,
$$m\frac{dv}{dt} = F \quad\dots\dots\dots\dots\dots\dots\dots\dots\dots(267\cdot1).$$

The ordinary Newtonian equations are accordingly applicable without modification to a body whose mass is changing as a result of the emission of radiation.

Motion of Particle about a Body of Decreasing Mass.

268. Consider first the abstract problem of the motion of a particle of small mass about a gravitating body whose mass is decreasing as a result of the emission of radiation.

---

* *M.N.* LXXXIV. (1924), pp. 2 and 912.

Taking polar co-ordinates with the centre of force as origin, it follows from equation (267·1) that the motion of the particle will be governed by the usual equations

$$r^2 \dot{\theta} = h \quad \dots\dots\dots\dots\dots\dots(268\cdot1),$$

$$\ddot{r} - r\dot{\theta}^2 = -\gamma m / r^2 \quad \dots\dots\dots\dots\dots(268\cdot2),$$

in which h remains constant throughout the motion, but m varies with the time. From these we readily deduce

$$\frac{d}{dt}\left[\tfrac{1}{2}(\dot{r}^2 + r^2 \dot{\theta}^2) - \frac{\gamma m}{r} \right] = -\frac{\gamma}{r}\frac{dm}{dt} \quad \dots\dots\dots(268\cdot3).$$

The quantity in square brackets on the left is the total energy of the particle in its orbit. This is equal to $-\gamma m / 2a$, where a is the semi-major-axis of the elliptic orbit being described at the instant, so that equation (268·3) may be put in the form

$$\frac{d}{dt}\left(\frac{m}{2a}\right) = \frac{1}{r}\frac{dm}{dt} \quad \dots\dots\dots\dots\dots(268\cdot4).$$

The average value of $1/r$ taken over a complete revolution is $1/a$, so that on averaging equation (268·4) over a complete revolution we obtain

$$\frac{d}{dt}\left(\frac{m}{2a}\right) = \frac{1}{a}\frac{dm}{dt} \quad \dots\dots\dots\dots\dots(268\cdot5),$$

in which d/dt of course only refers to secular changes. Integrating we obtain

$$ma = \text{constant.} \quad \dots\dots\dots\dots\dots(268\cdot6).$$

The eccentricity of the orbit described at any instant is given by

$$1 - e^2 = \frac{l}{a} = \frac{h^2}{\gamma m a},$$

and since ma remains constant, it follows that e must remain constant.

The period P is given by

$$P = \frac{2\pi}{n} = 2\pi\left(\frac{a^3}{\gamma m}\right)^{\tfrac{1}{2}},$$

and since ma remains constant, P varies as $1/m^2$.

Motion of a Binary System, in which both Components decrease in Mass.

269. In an actual astronomical binary system, both components are losing mass at the same time, and usually at different rates. This case is best treated in terms of Cartesian co-ordinates.

Let x, y, z and $\dot{x}$, $\dot{y}$, $\dot{z}$ denote the positions of the two components of a binary system at any instant, their masses being M and M' respectively.

As in equation (267·1), the equations of motion of the components will be

$$\ddot{x} = \frac{\gamma M'(x' - x)}{r^3}, \text{ etc. } \dots\dots\dots\dots(269\cdot1).$$

From these equations we readily deduce that

$$\frac{d}{dt}\left[\tfrac{1}{2}M\left(\dot{x}^2+\dot{y}^2+\dot{z}^2\right)+\tfrac{1}{2}M'\left(\dot{x}'^2+\dot{y}'^2+\dot{z}'^2\right)-\frac{\gamma MM'}{r}\right]$$

$$=\frac{dM}{dt}\left[\tfrac{1}{2}\left(\dot{x}^2+\dot{y}^2+\dot{z}^2\right)-\frac{\gamma M'}{r}\right]+\frac{dM'}{dt}\left[\tfrac{1}{2}\left(\dot{x}'^2+\dot{y}'^2+\dot{z}'^2\right)-\frac{\gamma M}{r}\right]$$

$$\ldots\ldots\ldots\ldots(269\cdot2).$$

The truth of this equation can be seen without detailed analysis, for the terms other than those in dM/dt and dM'/dt must necessarily vanish in order that the conservation of energy may be satisfied when M and M' do not vary with the time.

The motion of the mass M is that of a particle describing a gravitational orbit under a force

$$\frac{\gamma M'^3}{(M+M')^2}\frac{1}{S^2}$$

to the centre of gravity of the two masses, where S is the distance to this centre of gravity, so that

$$S=\frac{M'}{M+M'}\,r.$$

If a is the semi-major-axis of the relative orbit of the two particles, the semi-major-axis of the orbit of M is similarly

$$\frac{M'}{M+M'}\,a.$$

Thus the equation expressing that the energy of an elliptic orbit is $-\gamma M/2a$ assumes the form

$$\tfrac{1}{2}\left(\dot{x}^2+\dot{y}^2+\dot{z}^2\right)-\frac{\gamma M'^2}{(M+M')r}=-\frac{\gamma M'^2}{(M+M')2a}\quad\ldots\ldots(269\cdot3),$$

and from this and the corresponding equation for the orbit of the mass M',

$$\tfrac{1}{2}M\left(\dot{x}^2+\dot{y}^2+\dot{z}^2\right)+\tfrac{1}{2}M'\left(\dot{x}'^2+\dot{y}'^2+\dot{z}'^2\right)-\frac{\gamma MM'}{r}=-\frac{\gamma MM'}{2a}$$

$$\ldots\ldots\ldots\ldots(269\cdot4).$$

Since the average value of $1/r$ throughout the description of an elliptic orbit is $1/a$, we obtain, on averaging equation (269·3),

$$\text{average value of }\tfrac{1}{2}\left(\dot{x}^2+\dot{y}^2+\dot{z}^2\right)=\frac{\gamma M'^2}{(M+M')2a}.$$

On averaging equation (269·2) through a complete revolution of the orbit, we now obtain

$$\frac{d}{dt}\left(-\frac{MM'}{2a}\right)=\frac{dM}{dt}\left[\frac{M'^2}{(M+M')2a}-\frac{M'}{a}\right]+\frac{dM'}{dt}\left[\frac{M^2}{(M+M')2a}-\frac{M}{a}\right],$$

which, after some simplification, becomes

$$\frac{d}{dt}\left(\frac{1}{2a}\right)=\frac{1}{(M+M')2a}\frac{d}{dt}(M+M').$$

This has the integral
$$a(M + M') = \text{constant}. \quad\ldots\ldots\ldots\ldots\ldots\ldots(269\text{·}5).$$

From equations (269·1) we readily obtain
$$(\ddot{x}' - \ddot{x})(y' - y) - (\ddot{y}' - \ddot{y})(x' - x) = 0,$$
whence, on integration,
$$(\dot{x}' - \dot{x})(y' - y) - (\dot{y}' - \dot{y})(x' - x) = h, \text{ a constant.}$$

The eccentricity of orbit being described at any instant is given by
$$1 - e^2 = \frac{l}{a} = \frac{h^2}{\gamma a(M + M')} \quad\ldots\ldots\ldots\ldots\ldots(269\text{·}6),$$
where l is the semi-latus rectum of the relative orbit. Since both h and $a(M + M')$ have been seen to remain constant throughout the whole motion, it follows that e also remains constant. The period P is given by
$$P = 2\pi \left(\frac{a^3}{\gamma(M + M')}\right)^{\frac{1}{2}},$$
so that, by equation (269·5), P varies as $(M + M')^{-2}$.

Summing up, we see that as a binary system of total mass $M + M'$ dissolves into radiation:

The eccentricity of orbit remains constant.

The linear dimensions of the orbit increase, varying as $(M + M')^{-1}$.

The period of the orbit increases, varying as $(M + M')^{-2}$.

The velocity of the orbit decreases, varying as $(M + M')$.

270. If a binary system of mass $M + M'$ emits E ergs of radiation per second, the resulting loss of mass is determined by the equation
$$E = -C^2 \frac{d}{dt}(M + M') \quad\ldots\ldots\ldots\ldots\ldots(270\text{·}1).$$

Differentiation of equation (269·5) with respect to the time now gives
$$\frac{1}{a}\frac{da}{dt} = -\frac{1}{(M + M')}\frac{d}{dt}(M + M') = \frac{E}{C^2(M + M')} = \frac{\bar{G}}{C^2} \quad\ldots\ldots(270\text{·}2),$$
where $\bar{G}$ is the average rate of generation of energy in ergs per gramme per second of the combined mass. For instance, regarding sun and earth as a binary system, $\bar{G} = 1\text{·}90$, and we readily calculate that the radius of the earth's orbit is increasing at the rate of $0\text{·}31 \times 10^{-7}$ centimetres a second, or a metre per century. The orbits of the other planets are of course expanding at proportionate rates. The year, defined as the periodic time of the earth's orbit round the sun, is increasing at the rate of $0\text{·}00042$ seconds a century, while the periodic times of the other planets are increasing at proportionate rates.

It is, however, clear that changes of the type just considered cannot change the orbits of binaries which have been formed by fission into the orbits of binaries as actually observed. Throughout such changes the eccentricity would retain always the exceedingly small value it had immediately after fission, while the expansion of orbit determined by equation (269·5) is far too slow to reconcile with the observed orbits of binary systems in general. We must travel still further to understand the evolution of binary systems.

DISTURBANCES FROM PASSING STARS.

271. Having found that the forces acting from within the system itself cannot explain the observed evolution of binary systems, it is natural to proceed to a study of the forces originating from outside; in other words, we proceed to examine the effect of other stars.

The problem before us has a close analogy with the problems of the Kinetic Theory of Gases. The stars may be compared to the molecules of a gas, single stars being represented by monatomic molecules, binary systems by diatomic molecules, and so on. So far we have merely been considering the dynamics of molecules in a gas in which no collisions occur; we must now, so to speak, consider the dynamics of collisions. It will, however, be best to state and discuss our problem in a form which does not borrow anything from, or presuppose any analogy with, the Kinetic Theory of Gases.

272. With the notation already in use, let the velocity of the component M' of a binary system relative to M be ξ, η, ζ, so that

$$\xi = \dot{x}' - \dot{x}, \text{ etc.}$$

The energy of the orbit is readily found to be

$$E = \frac{1}{2} \frac{MM'}{M + M'} (\xi^2 + \eta^2 + \zeta^2) - \frac{\gamma MM'}{r} \quad \ldots\ldots\ldots\ldots(272·1).$$

The forces acting on the components M, M' consist of those arising from inside the system itself, and those arising from other stars. In each case the equations of motion are (cf. § 267) of strictly Newtonian form, whence it follows, from the general principles of statistical mechanics, that the co-ordinates and components of velocity in a great number of stars, which have attained a final steady state through being subjected to one another's gravitational force for a sufficiently great time, will approximate to the statistical distribution specified by Maxwell's law. In the special case of binary systems this law takes the form

$$A e^{-2HE} d\xi\, d\eta\, d\zeta\, 4\pi r^2 dr \quad \ldots\ldots\ldots\ldots\ldots(272·2),$$

where E is the energy of the system, as given by equation (272·1), A is a constant which is proportional to the total number of systems under discussion, and H is another constant, determined by the total energy of all the systems, or, what is the same thing, by their average energy.

In the final state specified by Maxwell's law (272·2), equipartition of energy obtains, so that the average value of each squared term $\left(\dfrac{MM'}{M + M'} \xi^2 \text{ etc.} \right)$

in expression (272·1) for the energy is the same. Its actual value is readily found to be $1/2H$.

The components of velocity u, v, w of the centres of gravity of the binary systems in space will also, in this final state, conform statistically to a law of distribution of Maxwell's type, the actual law being

$$Be^{-H(M+M')(u^2+v^2+w^2)}\, du\, dv\, dw,$$

where H is the same as in the law of distribution (272·2). The average values of $(M + M')\, u^2$, etc., are again each equal to $1/2H$, so that for systems of a given total mass $(M + M')$ the average value of $(u^2 + v^2 + w^2)$, which we shall denote by C^2, will be given by

$$C^2 = \frac{3}{2H\,(M + M')} \quad\dots\dots\dots\dots\dots\dots(272\cdot3).$$

Halm*, Seares† and others have found that to a rough approximation, the value of $C^2(M + M')$ is the same for binary stars of different masses. We shall discuss the full significance of this fact later; for the present we merely note that formula (272·3) provides a means of determining H by direct observational methods.

When H is known, formula (272·2) gives the law of distribution of the velocity components ξ, η, ζ, and of the separation in space r, which would prevail in a system of binary orbits after a sufficiently great time had elapsed. But the observational material on the orbits of binaries is not in a form convenient for comparison with this theoretical statistical distribution of ξ, η, ζ and r, orbits of binary stars being generally tabulated in terms of their periods and eccentricities. We accordingly proceed to transform the theoretical law of statistical distribution (272·2) into a form which will express the statistical distribution of periods and eccentricities‡.

273. The law of distribution (272·2) depends only on $\xi^2 + \eta^2 + \zeta^2$ and r. Since the values of $\xi^2 + \eta^2 + \zeta^2$ and r, and hence also the whole law of distribution, are invariant as regards different directions of the axes of reference in space, we may select any directions we please for these axes. Let us choose the x-axis in the direction of r increasing, so that ξ becomes identical with $\dot{r}$, the radial velocity. When this is done $\eta^2 + \zeta^2$ must be the square of the tangential velocity τ. Let ϕ be the azimuth of the direction of τ in any plane we please perpendicular to the axis of x. Then we may replace $\eta^2 + \zeta^2$ by τ^2, and $d\eta\, d\zeta$ by $\tau\, d\tau\, d\phi$. Making these substitutions, and integrating with respect to ϕ, the law of distribution (272·2) transforms into

$$8\pi^2 A e^{-2HE'} d\dot{r}\, \tau\, d\tau\, r^2\, dr \quad\dots\dots\dots\dots\dots(273\cdot1),$$

where
$$E' = \frac{1}{2} \frac{MM'}{M + M'} (\dot{r}^2 + \tau^2) - \frac{\gamma MM'}{r} \quad\dots\dots\dots\dots(273\cdot2).$$

---

* *M.N.* LXXI. (1911), p. 634.

† *Astrophys. Journ.* LV. (1922), p. 165, or *Mount Wilson Contributions*, No. 226. See § 276 below.

‡ A somewhat fuller discussion of this problem will be found in *M.N.* LXXIX. (1919), p. 408.

The relative orbit is the same, and has the same total energy E', as that of a particle of mass $\dfrac{MM'}{M+M'}$, moving about a centre of force $\dfrac{\mu}{r^2}$, where $\mu = \gamma(M+M')$. Let us put

$$h = r\tau; \quad \sigma = r - \frac{\mu}{r\tau}; \quad E'' = \tfrac{1}{2}(\dot{r}^2 + \tau^2) - \frac{\gamma(M+M')}{r} \quad \dots(273\cdot3),$$

so that h is the rate of description of double areas in this *relative* orbit, and E'' is the energy for a particle of unit mass.

Solving these equations for r, T and $\dot{r}$, we find

$$\tau = \sigma + \frac{\mu}{h} \quad \dots\dots\dots\dots\dots\dots\dots\dots\dots\dots(273\cdot4),$$

$$r = \frac{h}{\sigma + \mu/h} \quad \dots\dots\dots\dots\dots\dots\dots\dots(273\cdot5),$$

$$\dot{r}^2 = 2E'' - \sigma^2 + \mu^2/h^2 \quad \dots\dots\dots\dots\dots(273\cdot6).$$

The first two equations shew that the range of values from $r = 0$ to $r = \infty$ and from $\tau = 0$ to $\tau = \infty$ is exactly covered by letting both $\sigma + \mu/h$ and h vary from 0 to ∞. In permanent binary systems, E'' is always negative, and since $\dot{r}^2$ is necessarily positive, this limits σ and h to values such that $-\sigma^2 + \mu^2 h^2$ is positive. If a definite negative value is assigned to E'' and a definite positive value to h, the value of σ can range within the limits $\pm[2E'' + \mu^2/h^2]^{\frac{1}{2}}$.

We first transform variables in formula (273·1) from $\dot{r}$, T, r to E'', h, σ. The Jacobian of transformation is found, after slight simplification, to be

$$\frac{\partial(E'', h, \sigma)}{\partial(\dot{r}, \tau, r)} = r\dot{r},$$

so that

$$d\dot{r}\,\tau\,d\tau\,r^2\,dr = \frac{r\tau}{\dot{r}}\,dE''dh\,d\sigma.$$

Let us next write

$$2E'' + \mu^2/h^2 = p^2 \quad \text{and} \quad \sigma = p\cos\theta,$$

so that the limits for σ are from $\theta = 0$ to $\theta = \pi$. Transforming to the variables h, E'' and θ, the law of distribution (273·1) becomes

$$8\pi^2 A e^{-\frac{2HMM'}{M+M'}E''} \frac{h^2 dh\,dE''d\theta}{k^4(p\cos\theta + \mu/h)^2}.$$

On integrating from $\theta = 0$ to $\theta = \pi$ this becomes

$$8\pi^3 A\gamma(M+M')\,e^{-\frac{2HMM'}{M+M'}E''}\frac{dE''h\,dh}{(-2E'')^{\frac{3}{2}}} \quad \dots\dots\dots(273\cdot7).$$

This is easily transformed so as to give the law of distribution of values of the eccentricity and the period. Temporarily denoting the eccentricity by ϵ (to avoid confusion with $e = 2\cdot7182$), we have

$$1 - \epsilon^2 = -\frac{2E''h^2}{\mu^2},$$

so that, keeping E'' constant,

$$\epsilon\,d\epsilon = \frac{2E''h\,dh}{\mu^2}.$$

Using this and the further relation

$$-2E'' = \left[\frac{2\pi\mu}{P}\right]^{\frac{2}{3}},$$

the law of distribution (273·7) takes the form

$$\tfrac{2}{3}\pi^2 A e^{\frac{HMM'}{(M+M')^{\frac{1}{3}}}\left(\frac{2\pi\gamma}{P}\right)^{\frac{2}{3}}}\,dP2\epsilon\,d\epsilon \dots\dots\dots\dots\dots(273\cdot8),$$

which gives the law of distribution of orbits classified according to periods and eccentricities.

274. This law of distribution falls naturally into two parts, one of which, $2\epsilon\,d\epsilon$, involves only the eccentricity e, while the other involves only the period P. This shews that in the final steady state which is obtained after a suffi- cient amount of interaction between different systems, there will be no correlation between the periods and eccentricities of binary systems.

A very different state of things is disclosed by the tables given in § 258, which reveal a pronounced correlation between these two quantities, the periods and eccentricities of orbits increasing together in a very marked way. We must conclude that the orbits of binary stars are still far from having at- tained a statistical steady state.

This result is not altogether disadvantageous to the progress of cosmogony. If the theoretical law of steady-state distributions had been exactly obeyed by actual stellar orbits, we should have obtained a result which, while strik- ing and concise, would have closed the door against all further progress in the direction in which we are now working. As it proves, the law is far from being obeyed, and the deviations from the law, which come from the stars not having interacted for a sufficient time, must represent surviving vestiges of the initial conditions of the stars : they provide material, then, for discussing the origins and early histories of binary systems. If the present distribution of eccentricities and periods had agreed with the theoretical steady-state law, the problem of discussing these origins and early histories would have been comparable only to that of trying to decipher the writing on a slate after a wet sponge had been rubbed over it. Our discussion has shewn that the sponge was not thoroughly wet; the writing, although doubt- less smudged, is not wholly obliterated. The eccentricities and periods have not yet been levelled to the extent required by the theoretical steady-state law, and from the outstanding irregularities we may hope to read something of the course of events which has led to the present state of things.

Distribution of Eccentricities.

275. In the final steady-state law, the eccentricities are distributed according to the law $2e\,de$, so that all values of e^2 are equally probable. Observation shews that actually small values of e^2 predominate to an enormous extent.

Taking spectroscopic binaries first, we may compare the eccentricity of orbits of the 119 spectroscopic binaries given in Table XX (p. 290) with those required by the steady-state law. The second column of the following table gives the observed eccentricities, while the third column shews the distribution to be expected if the steady-state law $2e\,de$ were in operation:

	Observed	Theoretical
e from 0·0 to 0·2	78	5
e „ 0·2 to 0·4	18	14
e „ 0·4 to 0·6	16	24
e „ 0·6 to 0·8	6	33
e „ 0·8 to 1·0	1	43
	119	119

The theoretical law is nothing like obeyed, so that clearly spectroscopic binaries have not existed long enough for even an approximation to a final steady state to have become established. The predominance of low values of e suggests that spectroscopic binaries as a class started with low values of e, which interaction with other stars has nothing like raised to the steady-state values.

The eccentricities of the 68 visual binaries tabulated on p. 291 may be treated in a similar way and give the results shewn in the following table :

	Observed	Theoretical
$e = 0·0$ to 0·2	7	6
$e = 0·2$ to 0·4	18	18
$e = 0·4$ to 0·6	28	30
$e = 0·6$ to 0·8	11	42
$e = 0·8$ to 1·0	4	54
	68	—

The agreement between the observed and theoretical number of stars is seen to be good up to $e = 0·6$ but fails entirely beyond. The 68 binaries in question seem likely to provide a fair sample of all visual binaries as regards distribution of eccentricities, since there is no reason which makes binaries of abnormally high or low eccentricity specially liable to discovery. Of course the 68 binaries do not form anything like a fair sample as regards distribution of periods, since special reasons make binaries both of long and of short periods difficult of detection, but the correlation between eccentricity and period, which is very marked in the spectroscopic binaries, is almost entirely absent in the visual binaries (see Table XXII on p. 291).

The observed agreement up to about $e = 0\cdot6$, and deficiency for higher values of e, is naturally explained by the supposition that visual binaries as a class start with low values of e; that encounters with other stars tend to adjust these to the law $2e\,de$; and that there has not yet been time for complete adjustment, but only for adjustment as far as to about $e = 0\cdot6$.

We can perhaps best survey the whole situation by thinking in terms of average eccentricities. In the final steady-state law $2e\,de$, the average value of e is $0\cdot667$. If any class of binaries starts life with nearly circular orbits ($e = 0$), the effect of encounters with other stars must be to increase the average value of e progressively until it approximates asymptotically to $0\cdot667$, so that the average value of e observed in any class gives a rough measure of the age of the class as binaries. The tables at the beginning of the present chapter now become full of meaning. Table XXIII suggests that spectroscopic binaries of types O, B and A are not substantially younger than those of the so-called later types, F, G, K, M, while Table XX suggests that spectroscopic binaries of short period are younger than those of long period, and Table XXII suggests that spectroscopic binaries as a class are younger than visual binaries, but these conclusions would need some modification if the process of adjustment were quicker in some classes of stars than in others, a possibility to which we shall return later.

Distribution of Periods.

276. The law of distribution of periods in the theoretical steady state is given by (cf. formula (273·8))

$$De^{\dfrac{HMM'}{(M+M')^{\frac{3}{2}}}\left(\dfrac{2\pi\gamma}{P}\right)^{\frac{3}{2}}}\,dP \qquad\qquad\ldots\ldots\ldots\ldots\ldots\ldots(276\cdot1),$$

where D is a constant which depends on how many stars are under discussion.

Formula (272·3) provides a means of determining the constant H from the observed velocities of the stars in space. From a very thorough study of the question, Seares* has concluded that stars of different masses all give approximately the same value for H, thus shewing that the translational motions of the stars very nearly conform to the steady-state law.

In the following table, taken from his paper†, the second column gives the values of the mean masses of stars of different spectral types, the mass of the sun being taken as unity, while the second column gives the mean values of C^2, the square of the velocity in space, the unit being one kilometre a second.

---

* *Astrophys. Journ.* LV. (1922), p. 165.   † *l.c.* p. 190.

The third and fourth columns give values of MC^2, which would be precisely constant if the steady-state law were exactly obeyed.

TABLE XXIV. *Stellar Equipartition of Energy* (Seares).

Spectral Type	log M	log C^2	log MC^2	MC^2
B 3	0·95	2·34	(3·29)	(1950)
B 8·5	0·81	2·40	(3·21)	(1620)
A 0	0·78	2·78	3·56	3630
A 2	0·70	2·87	3·57	3720
A 5	0·60	2·95	3·55	3550
F 0	0·40	3·11	3·51	3240
F 5	0·19	3·36	3·55	3550
G 0	$\bar{1}$·99	3·62	3·61	4070
G 5	$\bar{1}$·88	3·78	3·66	4570
K 0	$\bar{1}$·83	3·80	3·63	4270
K 5	$\bar{1}$·79	3·74	3·53	3390
$M\ a$	$\bar{1}$·77	3·78	3·55	3550
Means, excluding B type stars ...			3·57	3754

Excluding B-type stars, which are not included in the mean, we see that MC^2 is fairly uniform for all types of stars, its average value being 3754 in the units used by Seares, or $7·50 \times 10^{46}$ ergs. Putting the mean value of MC^2 equal to $3/2H$ as in formula (272·3) we find

$$H = 2 \times 10^{-47} \quad \text{..........................(276·2).}$$

If P is measured in years, and M, M' in terms of the sun's mass as unity, the law of distribution (276·1) becomes

$$De^{0·28 \frac{MM'}{(M+M')^{\frac{1}{3}}} P^{-\frac{2}{3}}} dP \quad \text{....................(276·3).}$$

The exponential factor becomes very large when P is very small. For instance, if M, M', the masses of the two components, are each equal to that of the sun, the exponential factor is found to have the following values:

When $P = 1$ year   the factor $= 1·25$,

,, $P = 1$ month   ,,   ,, $= 3·17$,

,, $P = 4$ days   ,,   ,, $= 86$.

For large values of P the factor approximates to unity, so that the law approximates to DdP, shewing that in binaries of long period the steady-state distribution is one in which the periods are evenly distributed over all values up to $P = \infty$. In binaries of short period the exponential factor gives an enormous preponderance of orbits having the shortest periods of all.

Binaries which have been newly formed by fission have periods of only a few days (cf. Table XIX, p. 289). It now appears that the ultimate effect of encounters with other systems must be to spread these periods out over all possible lengths of period up to $P = \infty$, the periods being fairly uniformly distributed except that a certain preponderance of short periods remains.

277. The whole kinetic energy of a binary system can be expressed in the form

$$\tfrac{1}{2}(M + M')(u^2 + v^2 + w^2) + \tfrac{1}{2}\frac{MM'}{M + M'}(\dot{r}^2 + r^2\dot{\theta}^2 + r^2\sin^2\theta\dot{\phi}^2) \quad (277\cdot1)$$

or
$$\tfrac{1}{2}(M + M')c^2 + \tfrac{1}{2}\frac{MM'}{M + M'}(\dot{r}^2 + \tau^2).$$

In the final steady state, the average value of each of the squared terms in formula (277·1) must be the same, namely, $1/2H$, so that the average value of τ^2 will be

$$\bar{\tau}^2 = \frac{M + M'}{HMM'} \quad\dotfill(277\cdot2).$$

Since $\tau = r\dot{\theta}$, the mean value of τ^2 taken through the whole description of a single orbit is

$$\bar{\tau}^2 = \frac{1}{P}\int_0^P \tau^2\,dt = \frac{1}{P}\int_0^{2\pi} r^2\dot{\theta}\,d\theta = \frac{2\pi h}{P} = \left(\frac{2\pi\mu}{P}\right)^{\frac{2}{3}}(1 - e^2)^{\frac{1}{2}}.$$

For any value of P, formula (273·6) has shewn that the law of distribution of values of e is $2e\,de$. On averaging over all values of e, the mean value of τ^2 for all orbits of given period P is

$$\bar{\tau}^2 = \frac{2}{3}\left(\frac{2\pi\mu}{P}\right)^{\frac{2}{3}}.$$

Corresponding to the value of $\bar{\tau}^2$ given by equation (277·2), the period is found to be

$$P = \frac{2\pi\gamma(HMM')^{\frac{3}{2}}}{(M + M')^{\frac{1}{2}}} \quad\dotfill(277\cdot3).$$

When a binary system has this period, the tangential velocity τ has precisely the value appropriate to the final steady state, so that on the average encounters with other stars tend neither to increase nor to decrease this tangential velocity, and so tend on the average to leave the period as it is.

Suppose, however, that a binary system has a period far greater than this. The tangential velocity τ then has a value far below the average steady-state value given by equation (277·2), so that encounters with other stars are likely on the whole to increase this value and so to lessen the period. Similarly if a binary system has a period less than that given by equation (277·3), encounters with other systems are likely to increase the period. The law of distribution (276·3) may in a sense be regarded as a distribution ranged about the period given by equation (277·3) as median, but with infinite dispersion.

On inserting the value $H = 2 \times 10^{-47}$ already found, and measuring M

and M' in terms of the sun's mass, the critical period P_0 given by formula (277·3) takes the form

$$P_0 = 0.21 \frac{(MM')^{\frac{4}{3}}}{(M + M')^{\frac{1}{3}}} \text{ years} \quad \dots\dots\dots\dots(277\cdot4),$$

and in terms of this critical period the law of distribution (276·3) becomes

$$De^{0.79\,(P/P_0)^{-\frac{2}{3}}} dP \quad \dots\dots\dots\dots\dots(277\cdot5).$$

When each constituent is of mass equal to the sun, the critical period is found to be about 55 days. For more massive stars it is longer in proportion to the mass.

278. Known binary stars fall into the almost distinct classes of spectroscopic and visual binaries. The majority of spectroscopic binaries have periods of less than the critical period, while the visual binaries, without exception, have periods greater than the critical period.

The known binaries cannot, however, be assumed to form a fair sample of binaries as a whole. They represent the fruits of two distinct methods of search, and the apparent division between the visual and spectroscopic binaries may merely represent a no-man's-land in which neither method of search is effective, rather than a region in which no binaries exist.

The best way of obtaining a fair sample of binaries as a whole is by fixing our attention on regions of space so near to the sun that few binaries are likely to have remained undiscovered.

We have seen that six binaries are known to exist within four parsecs of the sun; four of their periods are known with tolerable accuracy and are approximately 39, 50, 55 and 80 years, while the other two have much longer periods. Every one of these six is a visual binary, and it is reasonably certain that there is not a spectroscopic binary within four parsecs of the sun.

Table III (p. 22) contains Hertzsprung's list of 21 binaries of known periods lying within ten parsecs of the sun. Here we found two spectroscopic binaries with periods of 9 and 22 months, the remainder being visual binaries with periods ranging from about 25 to 10,000 years. Even the spectroscopic binaries had periods well above the critical period.

Now if the binaries in the sky had reached a final steady state, or even approximated to such a state, there ought to be about as many binaries whose periods are being lengthened by encounters, and so are now less than the critical period, as there are binaries whose periods were being shortened by encounters, and so are over the critical period. Actual observation shews an enormous predominance of binaries of long periods. Every binary within four parsecs of the sun, and probably every binary within ten parsecs of the sun, has a period greater than the critical period of about 55 days, so that, speaking statistically, every binary within these regions is at present having its period reduced by encounters with other systems. Since the periods are decreasing they must originally (still speaking statistically) have been greater than they now are.

The average periods of binaries are so large in comparison with the critical period of 55 days, that the argument can be put in a very simple form. In a binary of period 100 years, which is about the average period of observed binaries, the velocity of each component relative to the centre of gravity of the system is about four kilometres a second. This system is disturbed by stars and other binary systems moving with velocities relative to it of the order of forty kilometres a second, and it is excessively improbable that such disturbance can either leave the relative velocity at the low figure of four kilometres a second or decrease it still further. The normal event is for the relative velocity to be increased, and this may either break up the system by causing its components to describe hyperbolic orbits or may lessen the period if they continue to describe elliptic orbits. Either event results in a decrease of the average period of the whole system of binaries.

279. We accordingly conclude that the average period of the binaries in the sky is at present being reduced by their encounters with other stars. Hence the average of these periods in the past must have been greater than the present average of about 100 years, and we must suppose that they had some other origin than fission of the kind we have had under consideration.

The most probable account of the long-period binaries would seem to be that they represent the remains of independent condensations in the parent nebula which failed to get clear of one another's gravitational fields and have been describing orbits about one another ever since, their periods being on the average continually lessened by encounters with other stars.

On this view binary systems fall into two distinct classes—

(i) Systems formed by rotational fission, whose periods are short but lengthening.

(ii) Systems formed out of independent condensations in the parent nebula, whose periods are long but shortening.

The former class have, on the average, periods of less than 55 days, and the latter class periods of more than 55 days. The two classes correspond broadly to the two observational classes of spectroscopic and visual binaries.

The Genesis of Triple and Multiple Systems.

280. After a binary system has been formed by fission, each of its two components may undergo a further shrinkage, under the same conditions of approximate constancy of angular momentum as produced fission in the parent star, and these conditions may produce fission in the components, thus generating triple or multiple systems. Actually the angular momenta of the two components will not remain absolutely constant, but will be diminished to a certain extent by tidal friction, and this diminution will certainly delay, and may entirely prevent, the fission of the two constituents separately.

Let us first disregard the action of tidal friction and treat the angular momentum of each component after fission has once occurred as a constant quantity. Let us further treat the masses as incompressible or nearly so.

Let M be the mass of the parent star and $\mathbf{M}$ its angular momentum at the moment of fission, and after fission let one component have mass $f_m M$ and angular momentum $f_a \mathbf{M}$. When the rotations of two masses result in their having configurations of similar shape, although of different dimensions, $\omega^2/2\pi\gamma\rho$ is the same for both, so that their angular momenta are proportional to $M^{\frac{5}{3}}\rho^{-\frac{1}{6}}$. Let the angular momentum $\mathbf{M}$ be supposed to be

$$\mathbf{M} = \alpha M^{\frac{5}{3}} \rho^{-\frac{1}{6}},$$

where α depends only on the shape of the mass.

Apply this equation first to the parent star at the moment of fission, and next to the constituent star when, if ever, fission overtakes it. If ρ_P, ρ_C are the densities of the parent and constituent at their moments of respective fission, we find

$$\mathbf{M} = \alpha M^{\frac{5}{3}} \rho_P^{-\frac{1}{6}},$$

$$f_a \mathbf{M} = \alpha \left(f_m M \right)^{\frac{5}{3}} \rho_C^{-\frac{1}{6}}.$$

From these two equations

$$\frac{\rho_C}{\rho_P} = \frac{f_m^{10}}{f_a^{6}}.$$

Values of f_a corresponding to different values of f_m have already been calculated in § 264. From the table there given we deduce

Value of f_m	0	0·29	0·33	0·50	0·66	0·71	1
„ f_a	0	0·039	0·046	0·077	0·135	0·160	1
„ ρ_C/ρ_P		1040	1740	4760	2590	2460	1
„ $\left(\dfrac{\rho_C}{f_m \rho_P}\right)^{\frac{1}{3}}$	∞	15	17	21	16	15	1

We see that a second fission cannot take place until the density has increased to some thousands of times the density at the first fission. The last line gives the values of $(\rho_C/f_m\rho_P)^{\frac{1}{3}}$, this being the ratio of the dimensions of the orbit produced by the first fission to that produced by the second fission.

The last line shews that when a triple system is formed in the way we have imagined, the separation of the two components of the main system will be from 10 to 20 times that of the components of the subsidiary system, and if still further subdivision takes place, a similar ratio must prevail.

The relation
$$P = 2\pi \frac{a^{\frac{3}{2}}}{\gamma^{\frac{1}{2}}(M + M')^{\frac{1}{2}}}$$

now shews that the period as the main system must be at least about

30 times that of the first subsidiary system, while the period in this must be about 30 times that of the second subsidiary systems if such exist, and so on.

This general result was first obtained by Russell[*], but his method was different from the foregoing, and the detailed figures he gave were somewhat different from those just obtained.

281. A rather extreme example of the type of multiple system predicted by the foregoing theory is to be found in Polaris. This shews spectroscopically periods of 4 days and about 20 years, while Courvoisier finds that the spectroscopic triple system is in orbital motion with a fourth visible star, the period being 20,000 years.

A typical visual system of the kind predicted by theory is illustrated in fig. 51, this being the star 1502 in Jonckheere's *Catalogue*[†]. The figure is drawn to scale to represent the projection of the system on the celestial sphere, except that the distance Cc has been somewhat increased. The actual separations (epoch 1908·9) are

$$Cc = 3{\cdot}10'', \; CD = 22{\cdot}67'', \; AB = 24{\cdot}17'', \; AC = 235{\cdot}72''.$$

282. Generally speaking, all that can be observed of a multiple system is its projection on the celestial sphere at a single instant of time. Even if the orbital elements of the close pair can be determined, it is generally impossible to determine those of the wide pair. Thus effects of foreshortening and ellipticity of orbit make it impossible to decide whether any observed individual system conforms to the demands of theory or not.

Fig. 51.

If a large number of orbits are discussed statistically, allowance can of course be made for foreshortening and ellipticity. A group of triple systems having the same ratio of their semi-parameters l_2/l_1, and oriented at random in space, would shew projections on the celestial sphere such that the ratio s_2/s_1 of their observed separations obeyed a definite statistical law of distribution. The summarised results of a statistical discussion by Russell[‡] are shewn in Table XXV (p. 314).

The material for discussion consisted of 74 triple or multiple systems given in Burnham's *Catalogue*; since multiple systems appear two or even three times in the list, the total number of entries is 83. These systems are divided into two classes according to the ratio of the separation of the wide pair to the annual motion of the system in the sky; Class I consists of 64 systems in which the separation of the wide pair is less than 1000 years' proper motion, while Class II consists of 19 systems in which the separation of the wide pair is

---

[*] *L.c.* p. 196.   [†] *Memoirs R.A.S.* LXI. (1917).
[‡] *Astrophys. Journ.* XXXI. (1910), p. 200.

greater than this. This gives a rough classification according to the actual dimensions of the system. The last column gives the theoretical distribution s_2/s_1 to be expected for 45 systems for which l_2/l_1 has the uniform value 0·09.

TABLE XXV. *Separations in Triple Systems* (Russell).

Observed ratio of separations (s_2/s_1)	Number of systems		
	Class I	Class II	Theory ($l_2/l_1 = 0·09$)
>0·40	1	5	1
0·40 to 0·30	2	—	2
0·30 to 0·20	3	—	3
0·20 to 0·15	5	—	$4\frac{1}{2}$
0·15 to 0·10	8	3	$9\frac{1}{2}$
0·10 to 0·05	$17\frac{1}{2}$	2	16
0·05 to 0·025	$14\frac{1}{2}$	2	8
<0·025	13	7	1

The table shews that the systems in Class II do not conform at all to the theoretical law of distribution. Down to a separation ratio of about 0·05, the systems in Class I conform closely to the distribution to be expected in systems with an actual separation ratio 0·09 oriented at random in space. Below an apparent separation ratio of 0·05 there are too many systems to be accounted for in this way, but the excess could be attributed to a further group of systems having a separation ratio of less than 0·09—somewhere about 0·04.

Apart from these precise figures, it is clear that, so far as Class I is concerned, the law of distribution of s_2/s_1 is just such as might statistically be expected for a number of systems in which l_2/l_1 had values ranging from about 0·09 downwards. This distribution fully conforms to the theoretical requirements for systems generated by successive processes of fission.

The systems in Class II fall into two sharply defined groups. A group of 14 for which s_2/s_1 is less than 0·15 may very possibly be interpreted as systems with a separation ratio ranging from about 0·06 downwards arranged at random in space, and so may possibly have been formed by fission; but a group of 5 with a separation ratio greater than 0·40, cannot possibly be so explained. Russell, following an earlier suggestion of Moulton's, supposes that these may have evolved out of separate condensations in the nebula from which the stars were originally formed.

283. Russell interprets his investigation as shewing that the majority of binary systems whose separation is less than 1000 years' proper motion, have been evolved by fission. The investigation shews that they may have been so evolved, but not that they must; it shews that a group of systems which had evolved by fission would shew the observed arrangement, but the argument is not reversible and the investigation does not prove that no other origin could result in close and wide pairs.

Considerations of another kind make it unlikely that the normal triple or multiple system can have been formed by repeated fission of the type we have had under consideration. The table on p. 312 shews that with the mass ratios normally observed in binaries, there would have to be a thousand-fold increase in density between the first and second fission, and if a third fission should occur, a further thousand-fold increase between the second and third fission, making a million-fold increase in all. The observed range of stellar densities, wide though it may be, does not encourage us to postulate increases of this order in the densities of ordinary stars. Further, we have seen that rotational fission can only occur in masses which approximate to the liquid state; in a gaseous mass in equilibrium the central condensation of density is too great for fission to occur at all, and it is difficult to imagine two liquid stars with their liquid densities in the ratio of a million to one.

Let us imagine two adjacent condensations in a nebula, which are destined ultimately to form separate stars, failing to get clear of one another's gravitational fields, and so describing elliptic orbits about one another. After a time it seems likely that their periods of rotation and revolution will approximately coincide. As soon as this happens, the system becomes dynamically indistinguishable from one which has originated by fission. The dynamical investigation just given may be supposed to begin at this stage, and we see that one further fission will result in the production of a normal triple system.

Even this fails to explain the existence of systems in which three separations have occurred in succession, or of triple systems in which the close pair has too long a period for it to have originated by fission.

To explain these we probably have to consider the shrinkage of a condensation which has been formed out of a tenuous nebula and is contracting until its density is great enough to ensure stability. Although the details require working out, it may be that during the process of shrinkage the density remains far more uniform than would be the case if the mass were in equilibrium throughout, and the central condensation of mass may be so slight that fission can occur in the manner in which it occurs in a liquid of uniform density. A succession of fissions of this kind, all occurring during the actual act of shrinkage, before equilibrium had been attained, would account for all observed varieties of triple and multiple systems.

CHAPTER XII

THE AGES OF THE STARS

284. THE discussion of the last chapter shewed that the orbits of binary stars, both spectroscopic and visual, are still far from conforming to the statistical laws which must finally prevail after the stars have interacted with one another for an unlimited length of time. The same is true of the components of the velocities of the stars in space. After a sufficiently long time of interaction between stars, these ought to conform to the well-known Maxwell law of distribution of velocities. The investigation of Seares already given has shewn that the resultant velocities conform well enough, at least to the extent of obeying the law of equipartition of energy, but the distribution of their directions is far from conforming to this law. After a sufficiently long time of interaction, stellar velocities must be distributed in all directions equally, their motion not favouring any one special direction. As Kapteyn shewed in 1904, the actual velocities of the stars shew a very marked favouritism for a definite direction in space, so that the law of distribution appropriate to the final state is far from being obeyed.

If the statistical laws which specify the final steady state had proved to be exactly obeyed, we could have concluded that the stars had been interacting with one another for a very long time, but we could not have estimated the length of this time except possibly in terms of a lower limit. As we find that these laws are only partially obeyed, we conclude that the stars have not been interacting with one another for an indefinitely long time, and the extent to which the laws are obeyed makes it possible to form estimates, although necessarily vague ones, of the actual ages of different types of stars.

We shall find (§ 291 below) that a star's motion is determined mainly by the gravitational forces from distant stars, and only to a very minor degree by the forces from near stars, so that to a first approximation each star describes an orbit under the gravitational field of the universe as a whole. Although the amount and direction of the star's velocity continually change, the description of this orbit does nothing towards bringing about the final statistical steady state. For the stars in any region have velocities whose amounts are determined by their equations of energy, and whose directions are determined by their orbits as a whole, and there is no reason why these should conform to the final steady state law.

The tendency to obey this law is produced by the smaller forces which arise out of the interactions with the nearer stars, and we can estimate the ages of the stars by calculating the length of time necessary for these forces to produce the observed approximation to the steady state laws.

In forming such estimates our unit of time is virtually the interval between one stellar encounter and the next, so that we begin our investigation by considering the frequency of stellar encounters.

THE DYNAMICS OF STELLAR ENCOUNTERS.

285. When two stars come so close as to exert appreciable forces on one another each describes a hyperbolic orbit about the centre of gravity of the two. In fig. 52 let S be the centre of gravity of two stars of masses m, m', which are pulling each other appreciably out of their courses, let V_0 be the velocity of m' before the encounter began, and let V be its velocity at the moment of closest approach, both velocities being measured relative to the centre of gravity S. Let p be the perpendicular distance of the undeflected path from S, and let a be the distance at the instant of closest approach.

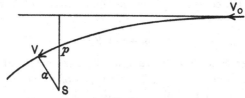

Fig. 52.

The orbit described by m' is that which would be described under a gravitational force $\gamma m^3/(m + m')^2 r^2$ directed towards S. Thus the principles of conservation of energy and momentum supply the relations

$$V^2 - V_0^2 = \frac{2\gamma m^3}{(m + m')^2 a} \quad \dots\dots\dots\dots\dots\dots(285\cdot1),$$

$$pV_0 = aV \quad \dots\dots\dots\dots\dots\dots\dots\dots(285\cdot2).$$

The elimination of V between these equations gives

$$p^2 = a^2 + \frac{2\gamma m^3 a}{(m + m')^2 V_0^2} \quad \dots\dots\dots\dots\dots(285\cdot3).$$

The eccentricity of the orbit, e, is given by

$$\frac{e+1}{e-1} = \frac{p^2}{a^2} = 1 + \frac{2\gamma m^3}{(m + m')^2 V_0^2 a},$$

and as total angle of deflection ψ of either orbit is equal to $2 \operatorname{cosec}^{-1} e$, we obtain

$$\frac{\sin \frac{1}{2}\psi}{1 - \sin \frac{1}{2}\psi} = \frac{\gamma m^3}{(m + m')^2 V_0^2 a} \quad \dots\dots\dots\dots(285\cdot4).$$

This gives the relation between a and ψ; to find the relation between p and ψ we eliminate a between this and equation (285·3) and obtain

$$\tan \frac{1}{2}\psi = \frac{\gamma m^3}{(m + m')^2 V_0^2 p} \quad \dots\dots\dots\dots\dots(285\cdot5).$$

286. Consider a group of stars all moving parallel to one another with the same velocity V_0 relative to S. In unit time the number of stars which approach the star of mass m in orbits such that the perpendicular distance is less than a given length p, is

$$\pi \nu V_0 p^2,$$

where ν is the density of these stars in space. Using the relation between p and ψ, as given by equation (285·5), we find that the number of encounters per unit time which result in a deflection greater than ψ is

$$\frac{\pi \nu \gamma^2 m^6}{(m+m')^4 V_0^3} \cot^2 \tfrac{1}{2} \psi \quad\dots\dots\dots\dots\dots(286\text{·}1).$$

The direction of a star's motion may be said to be completely changed when ψ is equal to a right angle or more, so that $\cot \tfrac{1}{2}\psi$ is less than unity. Taking $\cot \tfrac{1}{2}\psi = 1$ in formula (286·1), we find that the number of encounters per unit time which completely change the direction of a star's motion is

$$\frac{\pi \nu \gamma^2 m^6}{(m+m')^4 V_0^3} \quad\dots\dots\dots\dots\dots\dots(286\text{·}2).$$

Since the star-density ν occurs linearly in this formula, we can add together the effects produced by different types of stars, or, more briefly still, we can take ν to be the whole star-density in the sky and assign average values to m, m' and V_0.

Inverting, we find as the average interval of time between two such encounters,

$$\frac{(m+m')^4 V_0^3}{\pi \nu \gamma^2 m^6} \text{ seconds } \quad\dots\dots\dots\dots(286\text{·}3).$$

To represent present conditions in the neighbourhood of the sun, let us take $\nu = 4 \times 10^{-57}$ and $m = m' = 2 \times 10^{33}$. The effect of an encounter depends on V_0 through the factor V_0^{-3}, so that encounters with small values of V_0 are far more effective than those for which V_0 is large. For this reason, on averaging we must take a rather small value for V_0, and shall select $V_0 = 10$ kms. a second $= 10^6$. With these values, the interval of time given by formula (286·3) is found to be 7×10^{22} seconds or about 2×10^{15} years.

The distance of closest approach in an encounter in which the paths of the two stars are each turned through a right angle is obtained on putting $\psi = 90°$ in formula (285·4). This gives for a, the closest approach to the common centre of gravity,

$$a = \frac{(\sqrt{2}-1)\,\gamma m^3}{(m+m')^2 V_0^2},$$

and on inserting the numerical values just used, a is found to be $1·4 \times 10^{13}$ cms., which is approximately the radius of the earth's orbit. As this is far larger than the radius of the normal star, we see that in general a star's direction of motion can be entirely changed long before an actual collision occurs.

To study the frequency of actual collisions, let us examine how often stars of radius equal to the sun's radius 7×10^{10} cms. will approach so near as to touch or to collide. Putting $a = 7 \times 10^{10}$ in formula (285·3), and using the values for m, m' and V_0 already mentioned, we find that the value of p below which collisions occur is $2\cdot2 \times 10^{12}$ cms., so that the time-interval between collisions is

$$\frac{1}{\pi \nu V_0 p^2} = 6 \times 10^{17} \text{ years.}$$

Thus collisions between stars are so rare that they may be disregarded.

287. A star's course may not only be turned by violent encounters of the kind we have been considering, but also by a succession of feeble encounters, none of which is of much effect by itself but which have a cumulative effect equal to one big encounter.

By differentiation of formula (286·1), we find that there are

$$\frac{\pi \nu \gamma^2 m^6}{(m+m')^4 V_0^3} \frac{\cos \frac{1}{2} \psi}{\sin^3 \frac{1}{2} \psi} d\psi$$

encounters in unit time which produce a deflection of path between ψ and $\psi + d\psi$. For small deflections, this may be put in the form

$$\frac{8\pi \nu \gamma^2 m^6}{(m+m')^4 V_0^3} \frac{d\psi}{\psi^3} \quad \ldots\ldots\ldots\ldots\ldots\ldots(287\cdot1).$$

The cumulative effect of encounters which produce small deflections $\psi_1, \psi_2, \ldots$ is to produce a deflection of which the expectation Ψ is given by

$$\Psi^2 = \psi_1^2 + \psi_2^2 + \cdots \quad \ldots\ldots\ldots\ldots\ldots\ldots(287\cdot2).$$

Let $\psi_1, \psi_2, \ldots$ be all the deflections of amount between two limits α and β which occur within a time t. Then, from formula (287·1), we find that

$$\Psi^2 = t \int_{\alpha}^{\beta} \frac{8\pi \nu \gamma^2 m^6}{(m+m')^4 V_0^3} \frac{d\psi}{\psi}$$

$$= \frac{8\pi \nu \gamma^2 m^6}{(m+m')^4 V_0^3} t \log_e \left(\frac{\beta}{\alpha}\right) \quad \ldots\ldots\ldots\ldots(287\cdot3).$$

Let us take the upper limit of deflection to be $\beta = \frac{1}{2}\pi$, thus considering the cumulative effect of deflections less than those considered in § 286. It might at first be thought that to take account of all deflections of amount less than $\frac{1}{2}\pi$, we ought to take $\alpha = 0$, but such a procedure would be erroneous for the following reason.

Formula (287·2) is only accurate if the deflections $\psi_1, \psi_2, \ldots$ are independent, and this requires that they should originate in distinct encounters. If ψ is allowed to become very small, the corresponding distance a of closest approach, as given by equation (285·4), becomes very large, so that there are several stars within a distance a at the same instant, and their effects tend to neutralise one another. To obtain correct results we must stop off the integration before it brings us to values of a as large as this.

We must accordingly choose the lower limit α so as to correspond to a distance of closest approach which is about equal to the distance between adjacent stars, and so to $\nu^{-\frac{1}{3}}$. By equation (285·4), this value of α is given by

$$\alpha = \frac{2\gamma m^3 \nu^{\frac{1}{3}}}{(m+m')^2 V_0^2}.$$

Assigning this value to α and putting $\beta = \frac{1}{2}\pi$, equation (287·3) becomes

$$\Psi^2 = \frac{8\pi\nu\gamma^2 m^6}{(m+m')^4 V_0^3} \log_e \left(\frac{\pi (m+m')^2 V_0^2}{4\gamma m^3 \nu^{\frac{1}{3}}} \right) t \quad \dots\dots(287\cdot4),$$

or, inserting the numerical values already mentioned,

$$\Psi^2 = \frac{8\pi\nu\gamma^2 m^6}{(m+m')^4 V_0^3} \times 11\cdot9t.$$

The time necessary for deflections less than a right angle to produce a resultant deflection equal to a right angle is obtained by putting $\Psi = \frac{1}{2}\pi$, and is found to be

$$t = 0\cdot026 \frac{(m+m')^4 V_0^3}{\pi\nu\gamma^2 m^6} \text{ seconds} \quad \dots\dots\dots(287\cdot5).$$

Comparing with formula (286·3) we see that this time is only about one-fortieth of that needed for a single encounter to deflect the path by a right angle. With the values already used it is equal to about 5×10^{13} years.

Estimate of Stellar Ages.

288. It is now clear that changes in the direction of a star's path, and hence also exchanges of energy and momentum, are produced far more through the cumulative effect of a large number of small encounters, than by the occurrence of single big effects. With conditions such as now prevail in the neighbourhood of the sun, it appears that after 5×10^{13} years only about one star in eighty will have had its course turned through a right angle by a single close encounter with another star, but one star in every two will have had its course turned through a right angle by an accumulation of small effects.

We are not free to imagine a period of 5×10^{13} years with present conditions prevailing uniformly throughout, for we have already seen that in a period of this length, the stars would have lost the greater part of their masses by radiation. If the observed approach to equipartition of energy is to be explained as the effect of stellar encounters in past ages, we have to go back to times when stellar masses were greater than now and other conditions were consequently different from now.

The whole galactic system must be experiencing a decrease of mass at a rate comparable with that of the average star. If, as was at one time suggested by Poincaré, the Milky Way may be treated as an encircling band of stars held in relative equilibrium by slow rotation, then the radius of this belt must increase as the mass of the encircled matter decreases, the relation between the radius a and the mass M of the galaxy being the relation $Ma = constant$ found in § 268. If, as is more probable, the galaxy must be treated as a system of independently-moving stars, Ma must still remain constant, where a is the radius of the orbit of a single star surrounding the galaxy. Whichever way we regard the matter, the radius of the galaxy must be expanding approximately in accordance with the relation $Ma = constant$.

Thus when, if ever, the galaxy had four times its present mass, its radius must have been only a quarter of its present radius, the star density must have been 64 times as great as now, and the density of matter 256 times as great as now. Formula (287·5) shews that if stellar velocities were the same as now, progress towards equipartition would have proceeded at 1024 times the present rate, and division by a factor of 1024 reduces the time-interval just calculated from 5×10^{13} years to 5×10^{10} years.

289. We are probably not entitled to assume that in these earlier epochs stellar velocities were the same as now, for the investigation of § 269 has suggested that loss of mass by radiation must in general be accompanied by a slowing down of stellar motion.

Poincaré's Theorem (§ 62) shews that, when a system of stars is in steady motion, the mean square of their velocity is proportional to the mean gravitational potential throughout the mass, and therefore approximately to M/a. If a varies inversely as M, the average stellar velocity must be proportional to the average stellar mass, a general result of which § 269 has already provided an illustration. Thus if the galaxy was then, and is now, in a state of steady motion, the velocities when the masses were four times as great as now must have been four times as great as the present velocities.

Let M denote the present mass of a single star which we take to be typical of the galactic system as a whole. From formula (118·4) we may suppose that at a time t back, its mass M' was given by

$$M'^2 = \frac{M^2}{1 - 2\alpha t M^2} \qquad \dots\dots\dots\dots\dots\dots(289\cdot1),$$

where $\alpha = 5\cdot2 \times 10^{-33}$. If, as we have supposed, V_0 varies as M', while ν varies as M'^3, equation (287·4) shews that the rate of change of Ψ^2 varies approximately as M'^2, the logarithmic function varying by so little that its changes may be disregarded.

Thus if C denotes the present rate of change of Ψ^2, the rate of change at a time t back is given by

$$\frac{d}{dt}(\Psi^2) = \frac{C}{1 - 2\alpha t M^2},$$

and the change $\delta\Psi^2$ produced in Ψ^2 in the whole time t is

$$\delta\Psi^2 = \int_0^t \frac{C}{1 - 2\alpha t M^2}\, dt = \frac{C}{2\alpha M^2} \log(1 - 2\alpha t M^2).$$

The time necessary to produce any specified change $\delta\Psi^2$ in Ψ^2 is accordingly

$$t = \frac{1}{2\alpha M^2}(1 - e^{-2\alpha M^2 \delta \Psi^2/C}) \quad\ldots\ldots\ldots\ldots\ldots(289\cdot 2).$$

No matter how great $\delta\Psi^2$ may be, this time can never exceed $1/2\alpha M^2$, which, as formula (289·1) shews, is the time since the typical star was of infinite mass.

If we take $M = 2 \times 10^{33}$, the value of α already given makes this time equal to 8×10^{12} years, while the value of C is found to be 2×10^{13} years, from formula (287·4). With these numerical values, formula (289·2) becomes

$$t = (1 - e^{-0\cdot 45\Psi^2}) \times 8 \times 10^{12} \text{ years},$$

from which we may calculate the following values for t (in years):

$\Psi = 30°$	$60°$	$90°$	$180°$,
$t = 2 \times 10^{12}$	3×10^{12}	5×10^{12}	8×10^{12}.

290. It is by no means clear what value of Ψ will best represent the degree of approach to the final steady state which is shewn by the velocities of the stars, the more so as the actual velocities seem to conform much better to the steady state law than the distribution of their directions. The final steady state is of course only given by $\Psi = \infty$, but $\Psi = 180°$ ought to give a very good approximation to it, and possibly something of the order of $\Psi = 90°$ would represent the observed degree of approach. Without specifying the actual angle, we may say that the calculations just given indicate that a time of the general order of 5×10^{12} years would suffice to bring about the observed degree of approach. This time must probably be extended substantially to allow for the fact that the stars spend part of their lives in regions in which the star-density is far less than that we have assumed.

Such calculations as the foregoing can lay but little claim to accuracy, but are important as providing positive information as to the actual ages of the stars. In § 118 we calculated the time needed for a star to radiate away a specified amount of its mass, and this gave the age of the star if we assumed that it had originally been far more massive than now. But we are now in possession of a means of estimating the time, at least as regards order of magnitude, throughout which the stars have actually existed.

For, unless the observed approach to equipartition of energy is a pure accident, which is almost incredible, it can only have been produced by gravitational encounters between the stars themselves. If the kinetic energy of the stars is interpreted as a physical temperature, the value of H already calculated shews that this temperature must be of the order of $1\cdot 8 \times 10^{62}$ degrees centigrade, and this figure amply rules out all possibility of the approach to equipartition having resulted from the action of physical agencies.

For equipartition of energy in the stars to be produced by any physical agency whatever, pressure of radiation, high speed electrons, molecular bombardment, or other physical agency of any kind whatever, the agency in question must have been in thermodynamical equilibrium with matter at a temperature of 1.8×10^{62} degrees. No such physical temperature is known, or can be imagined, in the universe, so that we must conclude that the observed equipartition arises from the gravitational interaction of the stars themselves, and this inevitably leads to ages of the order of those just calculated.

THE DYNAMICS OF BINARY SYSTEMS.

291. In terms of the analogy with the Kinetic Theory of Gases (§ 271) we have treated the stars as molecules of a gas, and have investigated the time necessary for their velocities to approximate to the distribution specified by Maxwell's law; we have in fact calculated what Maxwell describes as the "Time of Relaxation."

In terms of the same analogy, binary systems may be treated as diatomic molecules. We have already found the distribution of orbits in the final steady state (§ 273), and have examined to what extent observed binary systems conform to this distribution. We can form a second estimate of stellar ages by calculating the time necessary to establish this approximation to the final steady state law of distribution.

Let us first examine the effect of the forces from passing or distant stars on the eccentricity and period of a single binary system.

The gravitational forces which an outside star of mass M_0 at x_0, y_0, z_0 exerts at a point x, y, z of a binary system can be derived from a potential V, or $\gamma M_0/r$, which can be expanded in the form

$$V = \frac{\gamma M_0}{[(x_0 - x)^2 + (y_0 - y)^2 + (z_0 - z)^2]^{\frac{1}{2}}}$$

$$= \frac{\gamma M_0}{R} + \frac{\gamma M}{R^3}(xx_0 + yy_0 + zz_0)$$

$$+ \frac{\gamma M}{2R^5}[3(xx_0 + yy_0 + zz_0)^2 - (x^2 + y^2 + z^2)(x_0^2 + y_0^2 + z_0^2)] + \dots (291.1),$$

where R stands for $(x_0^2 + y_0^2 + z_0^2)^{\frac{1}{2}}$. The forces are of the type

$$X = \frac{\partial V}{\partial x} = \frac{\gamma M_0}{R^3}x_0 + \frac{\gamma M_0}{R^5}[x(2x_0^2 - y_0^2 - z_0^2) + 3yx_0y_0 + 3zx_0z_0] + \dots (291.2),$$

and the total force at x, y, z will be the sum of a number of such forces, one from every star whose gravitational field of force is perceptible.

The first term $\Sigma \gamma M_0 x_0/R^3$ in the total force is constant over the whole system, and so merely gives rise to an acceleration of the system as a whole, without affecting the orbit of its components. The remaining terms represent forces which vary from point to point of the system, and so change the period and eccentricity of its orbit.

The number of distant stars which lie within a small cone of solid angle $d\omega$ and at a distance between R and $R + dR$ is $\nu R^2 dR d\omega$, where ν is the star-density in this element of volume. If these stars have an average mass $\bar{M}$, their contribution to the force X is

$$\gamma \bar{M} \nu d\omega dR \left[\frac{x_0}{R} + \frac{1}{R^3} [x(2x_0^2 - y_0^2 - z_0^2) + 3y x_0 y_0 + 3z x_0 z_0] + \dots \right] \quad (291 \cdot 3).$$

In this expression x_0, y_0, z_0 are each comparable with R. In summing over stars at all distances, the various contributions to the first term have the same relative importance as in the integral $\int dR$, so that the important contributions come from distant stars. Thus the motion in space of the binary system, as of course of any other star or system, is determined mainly by the distant stars.

In the next term, different distances have the same relative importance as the corresponding terms in the integral $\int dR/R$. The important contributions to this integral come from large and from small values of R, the intermediate values being relatively unimportant.

Although the contributions from large and from small values of R appear to be of approximately equal importance, they are effective to very different degrees in altering the period and eccentricity of the orbit of the binary system. The forces arising from very distant stars alter but little during the description of a complete orbit by the binary system, whence it is readily seen that the effects they produce at different points of the orbit tend to neutralise one another, and their total effect is very slight. The forces from near stars do not neutralise themselves in any such way, with the result that near stars are mainly effective in altering the orbit of the binary.

The same remains true when we pass to the consideration of terms beyond those written down in formula (291·3), and we reach the general conclusions that

(1) Distant stars are mainly effective in determining the motion in space of a binary system or of a single star.

(2) Near stars are mainly effective in producing changes in the period and eccentricity of a binary system.

292. The period P and angular momentum h of an elliptic orbit are given in terms of the energy E of the orbit by the equations

$$P = \frac{2\pi\gamma}{(m+m')^{\frac{1}{2}}} \left(\frac{mm'}{-2E} \right)^{\frac{3}{2}}; \quad h = \frac{\gamma mm'}{(m+m')^{\frac{1}{2}}} \left(\frac{mm'}{-2E} \right)^{\frac{1}{2}} (1 - e^2)^{\frac{1}{2}}.$$

Differentiating with respect to the time, keeping m and m' constant, we obtain

$$\frac{1}{P}\frac{dP}{dt} = \frac{3}{2E}\frac{dE}{dt},$$

$$\frac{1}{h}\frac{dh}{dt} = -\frac{e}{(1-e^2)}\frac{de}{dt} + \frac{1}{2E}\frac{dE}{dt}.$$

Thus to obtain the rates at which the period and eccentricity change during an encounter with another star we must calculate dE/dt and dh/dt.

We may calculate dE/dt as the rate at which the forces already evaluated do work on the two components of the binary system. Thus

$$dE = \mathop{\Sigma}_{\substack{m\\1,\,2}}\left(X\frac{dx}{dt} + Y\frac{dy}{dt} + Z\frac{dz}{dt}\right)dt,$$

where X, Y, Z are the components of the force already evaluated, and the summation is with respect to the two components of the binary system. Inserting the value of X from equation (291·2) and integrating (by parts) throughout a complete encounter, we find that the increment of energy dE resulting from the encounter is

$$dE = -\mathop{\Sigma}_{\substack{m\\1,\,2}}\left[\tfrac{1}{2}x^2\frac{d}{dt}\left\{\frac{\gamma M}{R^5}(2x_0^2 - y_0^2 - z_0^2)\right\} + \tfrac{1}{2}y^2\frac{d}{dt}\{\ldots\} + \tfrac{1}{2}z^2\frac{d}{dt}\{\ldots\}\right.$$

$$\left. + 3yz\frac{d}{dt}\left\{\frac{\gamma M}{R^5}y_0z_0\right\} + 3zx\frac{d}{dt}\{\ldots\} + 3xy\frac{d}{dt}\{\ldots\}\right]\ldots\quad(292\text{·}1).$$

In this formula, all terms of the type $\dfrac{d}{dt}\{\ldots\}$ depend solely on the motion of the passing star; the orbit of the binary enters only through the quadratic terms in $\tfrac{1}{2}x^2$, $3yz$, etc. Thus if all other factors were the same, the change of energy dE produced by an encounter would be proportional to the square of the linear dimensions of the orbit of the binary. It is readily seen that the same is true as regards dh, and so also as regards dP and de. These various formulae are only valid if the passing star keeps outside the orbit of the binary.

293. The orbits of the spectroscopic binaries have dimensions of the order of 10^{11} or 10^{12} cms., whereas the analysis of § 286 shews that during a star's whole life it is unlikely that there will be a single approach of another star to within a distance of 10^{14} cms. Thus formula (292·1) is entirely appropriate to encounters with spectroscopic binaries.

On the other hand the orbit of the average visual binary has linear dimensions of the order of 10^{15} cms. (representing a period of 390 years when each component is of mass equal to the sun). Thus within a lifetime of the order of 5×10^{12} years, a number of stars are likely to pass so near to either component of a binary as to be momentarily nearer than the other component. In such a case formula (292·1) does not apply, but the two components of the binary may momentarily be considered as separate systems, and the interaction between the passing star and the near component will be approximately the same as though the other component were non-existent. It follows that the time necessary to change completely the orbit of a visual binary is about the same as that necessary to set up an approximation to equipartition in the motions of translation of the stars. Thus the observed law of distribution of eccentricities in visual binaries suggests that their age, as binary systems, is about equal to that we have estimated for the stars in general, namely an age of millions of millions of years.

For spectroscopic binaries formula (292.1) shews that the effect of the encounters which occur within a specified time must be reduced at least by a factor of the order of r^2/R^2, where r is the linear dimension of the binary orbit, of the order of 10^{11} or 10^{12} cms., and R is the distance of effective encounters, of the order of 10^{13} or 10^{14} cms. The factor r^2/R^2 is roughly of the order of 10^{-4}, and we see that, within the ages we have found it necessary to allot to the stars, the effect of encounters on the periods and eccentricities of binary systems must be almost negligible. Thus the eccentricities and periods of spectroscopic binaries can be increased for a short time after their birth by tidal friction; subsequent diminution of mass will increase their periods, but not their eccentricities, while encounters with other stars produce effects which are so small as to be almost negligible.

These conclusions are entirely in accord with the results of observation as shewn in Tables XIX—XXIII of the last chapter. Tables XX and XXIII shew that the spectroscopic binaries, regardless of their periods and spectral class, and so presumably of their ages, reach a limiting average eccentricity of 0.2 or 0.3; this must be the average eccentricity attained when tidal friction ceases to have any appreciable further action. The periods of spectroscopic binaries increase with advancing spectral type to just about the extent that might be expected as a result of decrease of mass.

On the other hand, visual binaries shew a partial approach to the final steady-state law in their distribution of eccentricities and, in so far as it is possible to form a judgment, in their periods also. This can be attributed to the effect of encounters with other stars, and enables us to assign to them an age of the order of millions of millions of years.

MASS-RATIO IN BINARY SYSTEMS.

294. Another quantity associated with a binary system has been reserved for separate discussion, namely the ratio of the masses of the two components. Aitken * has tabulated the mass-ratio of 67 binary systems, taken from the Lick "Third Catalogue of Spectroscopic Binary Stars†," with results shewn in the following table:

---

* *Lick Obs. Bull.* No. 365 (1925), p. 46.   † *Ibid.* No. 141 (1924).

TABLE XXVI. *Mass-ratios in Spectroscopic Binaries* (Aitken).

Number of Stars	Spectral Class	Average Period (days)	Average m/m'	Range of m/m'
16	B	5·16	0·73	0·39 to 1·00
29	A	9·58	0·69	0·17 to 0·99
11	F	8·73	0·93	0·51 to 1·00
10	G	9·06	0·89	0·71 to 1·00
1	$G5$	5·4	0·88	0·88

Owing to the emission of radiation, the masses m, m' of the components of a binary system will change in accordance with the equation

$$\frac{dm}{dt} = - C^2 E.$$

Simple differentiation gives

$$\frac{d}{dt} \log \left(\frac{m}{m'}\right) = \frac{1}{m} \frac{dm}{dt} - \frac{1}{m'} \frac{dm'}{dt} = - C^2 \left(\frac{E}{m} - \frac{E'}{m'}\right) = - C^2 (\bar{G} - \bar{G}') \dots (294\cdot1).$$

Here $\bar{G}$ and $\bar{G}'$ are the rates of generation of energy, in ergs per gramme, of the two components. In spectroscopic binaries the more massive component has almost invariably the larger value of G, so that equation (294·1) shews that, as a spectroscopic binary ages, the ratio m/m' of its masses ought continually to approximate to unity. Table XXVI reveals a definite, although not very marked, tendency in this direction in actual binaries.

If we like to introduce definite assumptions, we can calculate the ages of stars of different spectral classes from the extent to which their two components have approached equality of mass. Let us, for instance, assume that the emission of radiation is given by our approximate law $L = M^3$, so that, as in § 118, the mass M at any time t is given in terms of the mass M_0 at time $t = 0$, by the relation

$$M = M_0 (1 + 2\alpha t M_0^2)^{-\frac{1}{2}}.$$

The ratio of the masses of the two constituents of a binary is now given by

$$\left(\frac{m}{m'}\right)^2 = \left(\frac{m_0}{m_0'}\right)^2 \frac{1 + 2\alpha t m_0'^2}{1 + 2\alpha t m_0^2}.$$

Solving for t we find that the time-interval between mass-ratios m_0/m_0' and m/m' is given by

$$2\alpha t = \frac{1}{m_0^2} \left[\left(\frac{m}{m'}\right)^2 - \left(\frac{m_0}{m_0'}\right)^2 \right] \Big/ \left[1 - \left(\frac{m}{m'}\right)^2 \right] \quad \dots\dots(294\cdot2).$$

As an illustration let us calculate the time necessary for an A or B type binary of mass ratio 0·70, whose smaller component has five times the mass of the sun, to change into an F or G type binary with a mass ratio 0·90. Putting

$m = 10^{34}$, $m_0/m_0' = 0.70$ and $m/m' = 0.90$, formula (294.2) gives $2at = 1.7 \times 10^{-68}$, whence, inserting the value $a = 5.2 \times 10^{-88}$, we find

$$t = 1.7 \times 10^{20} \text{ secs.} = 5.4 \times 10^{12} \text{ years.}$$

This gives us an entirely independent estimate of the time·interval between types B and F or G. It is of the same order of magnitude as the previous estimates, and to this extent is entirely satisfactory, but it must be admitted that it is based on rather slender evidence, and for this reason cannot claim great precision.

More detailed calculations have been given by Vogt[*], Shajn[†] and Smart[‡], who have discussed the question from different aspects. Vogt and Smart have reached conclusions agreeing in the main, although different in detail, from those we have just reached.

The Birth of the Stars.

295. If its exact age could be assigned to every star, the present distribution of stars would enable us to study the rates at which stars had come into being at different epochs in the past. We have nothing like sufficient information for a detailed study of this question to be profitable, but we may examine the broader question of whether the galactic system of stars was born in a continuous steady stream or by an approximately instantaneous creation.

Let $f(m)$ denote the luminosity-function in any system of stars, so that $f(m)\,dm$ is the number of stars of absolute magnitudes between m and $m + dm$. Consider the hypothetical case in which exactly similar stars are born in a continuous steady stream so that the distribution of stars by luminosity does not change, at least in respect of its brighter members. The analytical condition that the distribution by luminosity shall not change is readily found to be

$$f(m)\frac{dm}{dt} = \text{constant} \quad \text{......................(295.1)},$$

where dm/dt denotes the rate at which m changes for a star of absolute magnitude m.

Let a star's luminosity L be supposed connected with its mass M by the relation $L = M^3$ of § 118 (p. 130). Combining this with the fundamental equation for the loss of mass by radiation

$$L = - C^2 \frac{dM}{dt},$$

we readily find that $\dfrac{dm}{dt}$ varies as M^2, and so as $10^{-0.267\,m}$. From equation (295.1) it follows that the luminosity-function in a steady distribution must be of the form

$$f(m) = C\,10^{0.267\,m} \quad \text{........................(295.2)},$$

* *Zeitschrift für Physik*, 2 (1924), p. 142.   † *M.N.* LXXXV. (1925), p. 245.   ‡ *Ibid.* p. 423.

where C is a constant; thus the luminosity-function changes by a factor of 4·64 for each change of 2·5 magnitudes.

This theoretical relation is comparatively insensitive to the exact relation which is assumed to hold between L and M. If this is taken to be $L = M^{3\cdot4}$ instead of $L = M^3$ (cf. § 118) the factor 4·64 is replaced by 5·08.

The luminosity-function which expresses the observed distribution of luminosities in the galactic system has already been given in Table V (p. 34). The entries for intervals of 2·5 magnitudes are repeated in the table below, together with values calculated from the theoretical law (295·2). In calculating these values the constant has been selected so as to make the theoretical and observed numbers agree for stars of the luminosity of the sun ($m = 5$).

Abs. Mag.	Observed no. of stars per mag.	Calculated no. in steady state
− 5	1	430
− 2·5	90	2,000
0	3,300	9,290
2·5	42,000	43,000
5	200,000	200,000
7·5	350,000	929,000
10	500,000	4,300,000
12·5	600,000	20,000,000

By comparison with the theoretical steady-state distribution, observation reveals a great deficiency of both very bright and very faint stars, which we may, for the moment, identify with very old and very young stars. This indicates that the galactic system is not the outcome of a continuous steady creation of stars. The process must have started long enough back for the faintest stars in the system to have attained their present small masses, but seems to have reached a very marked maximum of intensity at the time of birth of the stars which are now of absolute magnitudes from 2 to 5, subsequently declining so as to become almost negligible. If this maximum rate of creation had been maintained there would be 430 times as many very young stars of absolute magnitude − 5 as actually exist.

Recapitulation.

296. Let us attempt to sum up the results obtained in this and the preceding chapter.

We have examined the various agencies which can produce changes in the periods and eccentricities of binary systems, and have found that only one, namely encounters with other stars, is capable of producing effects comparable with those put in evidence by the observed periods and eccentricities.

It is possible to calculate the rate at which the periods and eccentricities of binary systems are being changed by encounters, and hence we can calculate the time during which encounters must have acted to produce the observed distribution of periods and eccentricities.

In the case of visual binaries we have found that both periods and eccentricities shew a certain degree of conformity to the final steady state law which would be attained after encounters had been in progress for an eternity of time. The observed partial conformity to this law is found to indicate an age of the order of 5 million million years.

In the case of spectroscopic binaries, the orbits are so small that encounters with other stars can get almost no grip on the system, and so leave its period and eccentricity almost unchanged, at any rate through periods of millions of millions of years. This is in agreement with the observed fact that spectroscopic binaries shew very little progressive increase, either of period or of eccentricity, with advancing age, and what progressive changes are observed can properly be attributed to tidal friction and loss of mass by radiation. Thus we cannot estimate the ages of spectroscopic binaries from their orbits.

The ratio of the masses of the two components of a binary system ought to tend towards unity with the passage of time, as a consequence of the more massive star changing its mass more rapidly than the less massive. Observation reveals such a tendency, and its amount provides a means of estimating the ages of binaries. The ages of spectroscopic binaries are in this way found to be millions of millions of years, and of the same order as those of visual binaries.

Observation has disclosed a very marked tendency towards equipartition of energy in the translational velocities of the stars. This provides material for an alternative estimate of the ages of the stars, which is again found to indicate ages of the order of millions of millions of years.

In Chapter XIV below (§ 349) we shall obtain yet another estimate of the ages of the stars by considering the time necessary to break up and scatter moving star clusters, and this will be found to confirm the estimates of stellar ages made in the present chapter.

The ages suggested by these various modes of investigation are all in substantial agreement, and are such as to indicate that a star lives long enough to lose the main part of its original mass by transformation into radiation, as has been suggested by various other observational facts. This rules out as inadequate all sources of energy except the complete annihilation of matter or some subatomic equivalent; nothing else can provide sufficient total radiation for the calculated lives of the stars.

A detailed discussion of the stars of the galactic system has suggested that the majority of these were born many millions of millions of years ago, that the process of creation was specially active at the time of birth of the stars which are now of absolute magnitudes 2 to 5, a period from 2 to 8 million million years ago, and that since then it has almost ceased.

PLATE VII

Part of the Andromeda Nebula *M* 31 shewing resolution into stars.

CHAPTER XIII

THE GREAT NEBULAE

297. It has already been noticed how the "great nebulae" form what Herschel described as a system of "island-universes," distinct and detached both from one another and from the galactic system of stars. Hubble has found that these nebulae are all of comparable size, being, as fig. 2 (p. 15) has shewn, of size comparable with, although smaller than, the galactic system.

This of itself would encourage the conjecture that the great nebulae may be star-clouds, of the same general nature as the cloud of stars surrounding the sun. This view of the nature of the great nebulae has been very prevalent since the time of the Herschels, and various items of recently gained knowledge appear to give it support rather than the reverse.

Viewed from a fairly remote nebula, our galactic system of stars would appear as a cloud of faint light, which telescopes of terrestrial power would be unable to resolve into separate stars. Since the average light from these stars gives a spectrum of F or G type, the composite spectrum of this cloud of stars would closely resemble a stellar spectrum of F or G type, and this is precisely the type of spectrum shewn by the great nebulae, their spectra even being crossed by dark lines of the same general character as the Fraunhofer lines in the solar spectrum.

Viewed from a near nebula, through a telescope of terrestrial power, the galactic system would be fairly easily resolved into separate stars, so that if the near nebulae are clouds of stars similar to the galactic system, they ought to admit of resolution in our telescopes. In actual fact some of them have been so resolved, at least in their outermost regions. Plate VII shews a small area of the outer region of the Andromeda nebula M 31, photographed with the 100-inch telescope, and the resolution into distinct stars can be easily seen *. In M 33 (Plate XI) the resolution into separate stars is even easier. By resolving such regions into distinct stars, and detecting Cepheid variables in them, Hubble has been able to estimate the distances of these nebulae. He has further found the nearer nebulae to be of the same general size and luminosity as the two Magellanic clouds, and as these latter are quite obviously and unmistakably clouds of stars, it would appear reasonable to conjecture that the nebulae also may be. Finally, those who maintain that the nebulae are merely remote clouds of stars, island universes like our own, can point to the fact that they are of the same general shape and build as the galactic system, namely, flattened discs with high central condensation.

Against this, all spiral nebulae, so far as can be judged from those we see

---

* I am indebted to Dr Hubble for preparing and sending me this and many other photographs.

edge on, shew a far greater degree of flattening than the galactic system. Moreover, although the outer regions of the nearer spiral nebulae can be resolved into stars, the central regions, although of course at the same distance and submitted to the same telescopic power, obstinately defy resolution, as do. all spherical and elliptical nebulae. Plate VIII shews the central regions of the Andromeda nebula, and there is no suspicion of resolution into separate stars here. Finally, although a study of the superficial appearance of the great nebulae may suggest that they are star-clouds, and nothing but star-clouds, we shall find that a theoretical study gives no support to this view.

The Classification of the Great Nebulae.

298. The great nebulae exhibit an enormous difference of structural detail, but Hubble, who has devoted much skill and care to their classification*, finds that most of the observed forms can be reduced to law and order. They fall primarily into the two great classes of nebulae of regular shape which exhibit no spiral arms, and true "spirals," consisting of a rather vaguely-defined central region from which two spiral arms emerge. Hubble finds that the true spiral forms, as regards their main features at least, can be specified in terms of three distinct quantities, which he describes as (1) the relative size of the unresolved nuclear region, (2) the extent to which the arms are un-wound (the openness or angle of the spiral), (3) the degree of condensation in the arms. When these quantities are estimated or evaluated observationally for different nebulae, they are found to vary very approximately in unison, with the result that practically all observed nebular configurations can be arranged to form linear series.

Hubble finds that it is not possible to place all observed nebulae in one continuous sequence; their proper representation demands a **Y**-shaped diagram such as is shewn in Fig. 53.

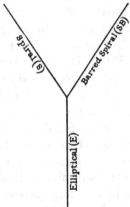

The lower half of the **Y** is formed by nebulae of approximately elliptical or circular shape. These are subdivided into eight classes, designated $E\,0$, $E\,1$, $E\,2, \ldots E\,7$, the numerical index being the integer nearest to $10\left(\dfrac{a-b}{a}\right)$, where a and b are the greatest and least diameters of the nebulae as projected on the sky. Thus $E\,0$ consists of nearly circular nebulae, $(b > 0{\cdot}95a)$, while $E\,7$ consists of nebulae for which b is about $0{\cdot}3a$, this being the greatest inequality of axes observed in the "elliptical" nebulae. Examples of typical nebulae are shewn in the upper half of Plate IX (p. 334).

Fig. 53. Diagrammatic representation of the observed configurations of the Great Nebulae.

* *Astrophys. Journ.* LXIV. (1926), p. 321 and *The Observatory*, L. (1927), p. 276.

PLATE VIII

Mt Wilson Observatory

The Central region of the Andromeda Nebula *M* 31. (Cf. Plate I)

PLATE VII.

The upper half of the Y-shaped diagram consists of two distinct branches, one of which is found to contain a far larger number of nebulae than the other. The principal branch contains the normal "spiral" nebulae, which are characterised by a circular nucleus from which emerge two (or occasionally more) arms of approximately spiral shape. Typical examples are shewn in the left-hand half of Plate X. These nebulae are subdivided into three classes, designated *Sa, Sb, Sc*, class *Sa* fitting almost continuously on to class *E*7.

The minor branch contains a special class of spirals characterised by the circumstance that the spiral arms appear to emerge from the two ends of a straight bar-shaped or spindle-shaped mass. These are sometimes called ϕ-nebulae, on account of their characteristic shape, although θ-nebulae would appear to be a better designation. Typical examples are shewn on the right-hand half of Plate X. Both here and in Plate IX, both of which are taken from Hubble's original paper, the examples shewn are the actual nebulae which Hubble selects as the milestones to mark the various sub-classes.

About 97 per cent. of known extra-galactic nebulae are found to fit into this Y-shaped classification. The remaining 3 per cent. are of irregular shape, and refuse to fit into the classification at all. These include the two Magellanic clouds and other star-clouds which Hubble treats as nebulae for purposes of classification. The irregular nebulae are distinguished by a complete absence of symmetry of figure and also by the absence of any central nucleus. Typical examples of "irregular nebulae" are shewn at the bottom of Plate IX.

Apart from the irregular nebulae, Hubble states that, out of more than a thousand nebulae examined, less than a dozen refused to fit into the Y-shaped diagram at all, while in less than 10 per cent. of the cases was there any considerable doubt as to the proper position of a nebula in the diagram. Clearly then, the Y-shaped diagram provides a highly satisfactory working classification.

Physical Interpretation.

299. Obviously the proper physical interpretation of the classification just described is of the utmost importance to cosmogony.

A first and most important clue is provided by the fact that numbers of the great nebulae are known to be in rotation. In 1914 Wolf[*] detected rotation in the spiral *M* 81 (see Plate XII, p. 351), and in the same year Slipher[†] discovered it in the type nebulae N.G.C. 4594 of class *Sa* (see Plate IX). Pease measured spectroscopically the velocity of rotation along the major axis of this latter nebulae in 1916[‡] and along the major axis of the Andromeda nebula *M* 31 in 1918[§].

The symmetry of figure shewn by nebulae of the *E* and *Sa* types is precisely such as rotation might be expected to produce, and thus suggests an inquiry

[*] *Vierteljahrsschrift der Astron. Gesell.* LXIX. (1914), p. 162.
[†] Lowell, *Obs. Bull.* No. 60 (1914).   [‡] *Proc. Nat. Acad. Sci.* II. (1916), p. 517.
[§] *Ibid.* IV. (1918), p. 31.

as to how far the observed figures of the regular nebulae can be explained as the figures assumed by masses rotating under their own gravitation. The theoretical material necessary for such an inquiry has already been assembled in Chapters VIII and IX.

In Chapter IX we examined the configurations assumed under rotation by a mass of matter of a type which we described as the extended Roche's model, this consisting of a central nucleus of homogeneous incompressible matter surrounded by a light atmosphere of negligible density. The possible configurations for a given mass rotating with a given uniform angular velocity are of the general type of those shewn in fig. 54. Here the shaded part represents the central mass, which assumes the shape of the Maclaurin spheroid appropriate to the given rotation, while the outer curves are the external closed

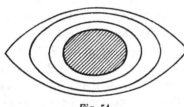

Fig. 54.

equipotentials of this mass in rotation. The rotating mass may have any one of these external equipotentials as its boundary, and selects that particular one whose volume is just adequate to contain its own atmosphere. If even the outermost of the closed equipotentials is inadequate to contain the whole atmosphere, the mass fills the outermost lenticular shaped equipotential, and the remainder spills over into the equatorial plane.

We notice at once that this series of equipotentials have very much the shape of the "elliptic" nebulae which occupy the lower half of the Y-diagram. A mass rotating with given angular velocity assumes these various forms according to the extent of the atmosphere which surrounds it. We add a bit more atmosphere to an $E\,3$ figure and it becomes $E\,4$; subtract a bit and it becomes $E\,2$. But the sequence of figures possible for a given velocity of rotation is limited at both ends; it is limited at one end by the bare Maclaurin spheroid and at the other by the last closed (sharp-edged) equipotential of the Maclaurin spheroid.

The limits in both directions can be extended by varying the angular velocity of rotation; an increase of speed changes an $E\,3$ mass into $E\,4$, while a decrease of speed changes an $E\,3$ mass into $E\,2$. With zero rotation every mass, no matter how great or how small its atmosphere, becomes $E\,0$, so that this fixes the limit in one direction. The limit in the other direction is that of the sharp-edged equipotential surrounding the critical Maclaurin spheroid which is just about to elongate itself into a Jacobian ellipsoid. This is not, however, of shape $E\,7$ but of shape $E\,5\cdot3$.

The same general sequence of configurations is exhibited by almost any model in which the mass is well concentrated towards the centre, as for instance the adiabatic model discussed in § 235. The limits here are again $E\,0$ at one end, and at the other end the sharp-edged equipotential of the

PLATE IX

ELLIPTICAL NEBULAE

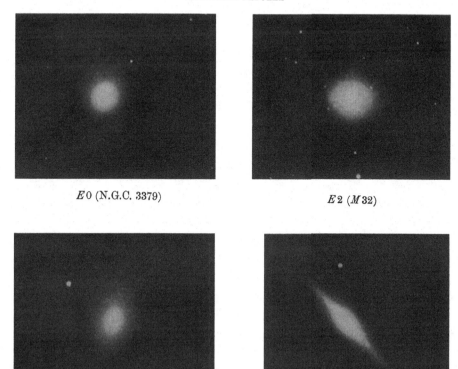

E 0 (N.G.C. 3379)   *E* 2 (*M* 32)

E 5 (*M* 59)   *E* 7 (N.G.C. 3115)

IRREGULAR NEBULAE

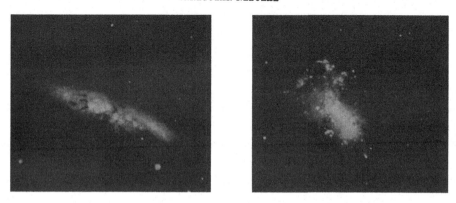

M 82   N.G.C. 4449

Mt Wilson Observatory (Dr E. Hubble)

Elliptical and Irregular Nebulae

critical figure ($\kappa = 2\cdot2$) at which the pseudo-spheroid is just about to elongate into a pseudo-ellipsoid. This critical figure is that shewn in fig. 43 (p. 263), and its shape is about $E\,6\cdot1$, but it is extraordinarily difficult to calculate and draw the figure with any accuracy.

When $\kappa = \infty$, the corresponding figure is the Maclaurin spheroid of shape $E\,4\cdot2$, while when $\kappa = 1\cdot2$, the figure is the limiting Roche's equipotential shewn in fig. 41 (p. 253); the shape of this is $E\,3\cdot3$.

300. The S and SB branches, as well as the E branch, of the Y-shaped diagram, permit of interpretation as the configurations of rotating masses.

Imagine that we gradually increase the rotation of a mass arranged according to the simple Roche's model in which the dense central nucleus is so small that it may be treated as a point. Its configuration passes through a sequence of figures of increasing ellipticity until it reaches the lenticular figure of shape $E\,3\cdot3$ shewn in fig. 41 (p. 253); at this $\omega^2/2\pi\gamma\bar{\rho} = 0\cdot36075$. Further rotation does not increase the ellipticity of figure beyond this, but, as we have seen, causes an ejection of matter from the sharp edge which forms the equator of the lens. The matter so ejected describes orbits in the equatorial plane and as there are no forces to move it out of this plane, it forms a thin annular layer of matter in this plane. While this ejection of matter is in progress, the remaining central mass retains the shape of the critical lens-shaped figure, while $\omega^2/2\pi\gamma\bar{\rho}$ retains the constant value $0\cdot36075$, the increase in ω^2 being met by a proportionate increase in $\bar{\rho}$, caused by light matter being removed from the atmosphere while the whole of the dense central mass remains.

In its essentials, the series of configurations so obtained consists of a central lenticular figure of shape $E\,3\cdot3$, extended in its equatorial plane by an annular layer which revolves around it much as Saturn's rings revolve around Saturn. Apart from details of structure, this gives the general characteristics of the nebular configurations which Hubble classifies as S, the normal spirals.

The generalised Roche's model gives a similar series of configurations, as has already been indicated in fig. 54. An additional series is however possible for this model, since the central mass may possess so much angular momentum that the Maclaurin spheroids give place to Jacobian ellipsoids. The figure then becomes a pseudo-ellipsoid with three unequal axes, and may, if its atmosphere is of sufficient extent, proceed to shed matter in two streams from the ends of its major axis. The general type of configuration obtained in this way is shewn in fig. 55 (p. 336), which represents a cross-section *in the equatorial plane.* These configurations reproduce the general features of the nebular configurations which Hubble classifies as SB (barred spirals). In the limiting case in which there is no atmosphere at all, these configurations reduce to ordinary Jacobian ellipsoids.

301. We have already noticed (§ 235) that the gap between the simple Roche's model and the incompressible model can be bridged in other ways than through the generalised Roche's model. An alternative way, which lends itself to convenient discussion is through the series of adiabatic models discussed in § 235. Roche's model is represented by taking $\kappa = 1\cdot2$, the incompressible model is obtained by taking $\kappa = \infty$, and the bridge is formed by allowing κ to vary continuously from $1\cdot2$ to ∞.

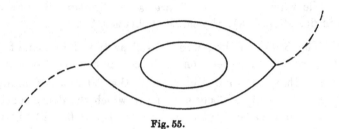

Fig. 55.

Again we obtain a series of elliptical figures, which gives place when $\kappa < 2\cdot2$ to a series of pseudo-spheroids with rings of matter in the equatorial plane, and gives place when $\kappa > 2\cdot2$ to a series of pseudo-ellipsoids, with two streams of matter shed from the ends of the longest diameter.

The general series of configurations which have been obtained, as the fruit of the theoretical research summarised in Chapter IX, may be represented

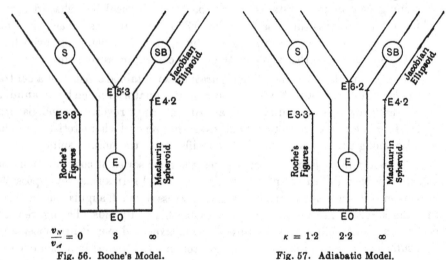

Fig. 56. Roche's Model.　　Fig. 57. Adiabatic Model.

Figs. 56 and 57. Diagrams representing theoretical configurations of Rotating Masses.

diagrammatically as in figs. 56 and 57. Fig. 56 refers to the generalised Roche's model, v_N/v_A denoting the ratio of the volume of the nucleus to that of the atmosphere. Fig. 57 refers to the adiabatic model, κ denoting the index which occurs in the law $p \propto \rho^\kappa$.

The general arrangement of series of configurations presented by these two models is seen to be very similar. Perhaps, however, this is hardly surprising

PLATE X

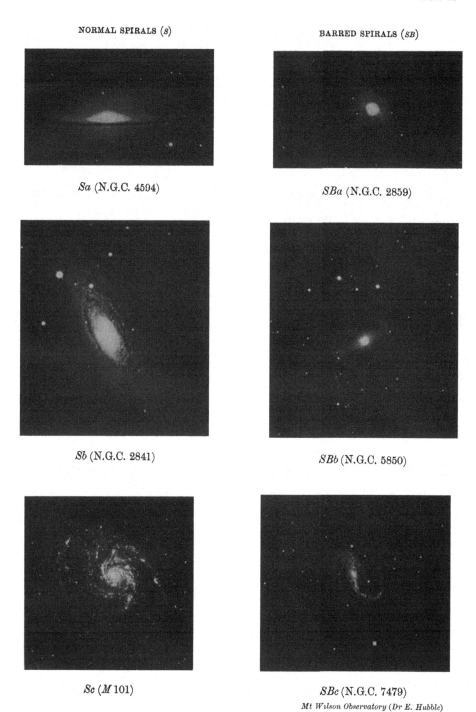

NORMAL SPIRALS (*s*)

BARRED SPIRALS (*SB*)

Sa (N.G.C. 4594)

SBa (N.G.C. 2859)

Sb (N.G.C. 2841)

SBb (N.G.C. 5850)

Sc (*M* 101)

SBc (N.G.C. 7479)

Mt Wilson Observatory (Dr E. Hubble)

Normal and Barred Spiral Nebulae

in view of the fact that both the outermost series are identical in the two diagrams. Both models are seen to be equally capable of explaining the observed sequences of nebular configurations.

302. Remembering that rotation has actually been observed in a number of nebulae, there seem to be strong reasons for conjecturing that the observed configurations of the nebulae may be explained in general terms as the configurations of rotating masses. There is of course no need to tie our explanation down to either of the two models discussed. If either of these can provide an explanation, no doubt hundreds of other models could do the same.

General dynamical theory has shewn that the boundary of any uniformly rotating mass which is acted on solely by its own gravitational field must have an equation of the form

$$V + \tfrac{1}{2}\omega^2 (x^2 + y^2) = \text{constant} \quad \dots\dots\dots\dots\dots(302\cdot 1)$$

where V is the gravitational potential of the mass. The only supposition necessary to arrive at this equation is that there must be a definite pressure p, of the usual hydrostatic type, at each point of the rotating mass. When such a pressure exists, the equations of equilibrium relative to the rotating axes are of the type

$$\frac{\partial p}{\partial x} = \rho \, \frac{\partial V}{\partial x} + \omega^2 \rho x, \text{ etc.,}$$

and it follows at once that any surface over which p is constant, including the boundary over which $p = 0$, must have an equation of the form of (302·1). We have discussed the surfaces specified by this equation for certain definite models, but any other model of the same general nature would have given similar results. Indeed the general argument of § 226 has shewn that the only possible types of configuration, for any rotating mass whatever, are those we have designated as pseudo-spheroids and pseudo-ellipsoids. The pseudo-spheroids give Hubble's sequence E of configurations before equatorial ejection begins and the sequence S afterwards. The pseudo-ellipsoids give the sequence SB.

Theory has, however, shewn throughout that pseudo-ellipsoids are possible figures of equilibrium only for rotating bodies in which there is no great central condensation of mass. It is extremely difficult to see how this condition could be satisfied in a nebula rotating in equilibrium, and the proposed interpretation of the SB nebulae, the barred spirals, must be correspondingly under suspicion. The remainder of our discussion accordingly deals almost exclusively with E and S—elliptical and ordinary spiral—nebulae.

303. To a quite rough approximation we may suppose that the potential V at the surface of any gravitating mass is equal to $\gamma M/r$, where γ is the gravitation constant, M is the total mass, and r is the distance of the point from the centre of the mass. Equation (302·1) shews that the values of

$$V + \tfrac{1}{2}\omega^2 (x^2 + y^2)$$

at the ends of the major and minor axes of any figure of equilibrium must be the same, so that, if $2a$ and $2b$ are the major and minor axes respectively of the rotating mass,

$$\frac{\gamma M}{a} + \tfrac{1}{3}\omega^2 a^2 = \frac{\gamma M}{b}.$$

Putting $(a-b)/a = e$ as before, this equation becomes

$$\tfrac{1}{3}\omega^2 a^2 b = \gamma M\left(1 - \frac{b}{a}\right) = \gamma Me \quad \dots\dots\dots\dots(303\cdot1).$$

If the figure is symmetrical about its axis of rotation, we may, again to a quite rough approximation, put

$$M = \tfrac{4}{3}\pi\bar{\rho}a^2 b \quad \dots\dots\dots\dots\dots\dots(303\cdot2)$$

where $\bar{\rho}$ is the mean density of the mass, and equation $(303\cdot1)$ assumes the form

$$e = \frac{3}{4}\frac{\omega^2}{2\pi\gamma\bar{\rho}} \quad \dots\dots\dots\dots\dots\dots(303\cdot3).$$

Thus for any rotating mass whatever, the shape of figure is determined approximately by the value of $\omega^2/2\pi\gamma\bar{\rho}$, and depends on nothing else.

It follows that if the nebulae are gravitating masses in rotation, we can determine the values of $\omega^2/2\pi\gamma\bar{\rho}$ from their observed ellipticities of figure. When the figure is of type S, relation $(303\cdot2)$ is only true as regards the central mass, the equatorially ejected matter being disregarded. For this central mass, e is generally observed to be about $0\cdot5$, so that, as regards order of magnitude at least, we must have

$$\frac{\omega^2}{2\pi\gamma\bar{\rho}} = \frac{2}{3} \quad \dots\dots\dots\dots\dots\dots(303\cdot4).$$

304. According to Pease*, the line of sight (spectroscopic) velocity of the Andromeda nebula $M\,31$ at a distance of x seconds of arc from the centre is given by

$$v = -0\cdot48x - 316 \text{ kms. a sec.} \dots\dots\dots\dots(304\cdot1).$$

This, in conjunction with Hubble's estimate of 285,000 parsecs for the distance of the nebula, fixes its period of rotation at about 16 million years. This represents an angular velocity $\omega = 1\cdot2 \times 10^{-14}$, and the linearity of formula $(304\cdot1)$ shews that the angular velocity has this uniform value throughout the range within which this formula is valid. Equation $(303\cdot4)$ now shews that the mean density $\bar{\rho}$ of the lenticular-shaped central mass is

$$\bar{\rho} = 5 \times 10^{-22}.$$

Pease† similarly found the spectroscopic velocity of the nebula N.G.C. 4594 in Virgo (Plate X) to be given by the formula

$$v = -2\cdot78x + 1180 \text{ kms. a sec.} \dots\dots\dots\dots(304\cdot2).$$

The distance of this nebula is not known from direct observation. The average absolute magnitude of nebulae whose distances are known is however

---

* *Proc. Nat. Acad. Sci.* IV. (1918), p. 21.   † *Ibid.* II. (1916), p. 517.

– 15·2, and if we suppose the absolute magnitude of this nebula to be – 15·2, its observed apparent magnitude of 9·1 would assign to it a distance of 700,000 parsecs.

Further, Hubble's statistical investigation shews that nebulae of the Sa class, to which this nebula belongs, have an average diameter of 1450 parsecs. The apparent diameter of this nebula is 7·1 minutes of arc, and a comparison of these figures would again assign to the nebula a distance of precisely the 700,000 parsecs previously calculated.

Adopting 700,000 parsecs as the distance of the nebula, one second of arc represents 3·5 parsecs, and as a distance of 3·5 parsecs represents a velocity difference of 2·78 kilometres a second, the period of rotation is found to be $2·52 \times 10^{14}$ secs. or about 8 million years. This gives as the mean density of the lenticular-shaped central mass,

$$\bar{\rho} = 2 \times 10^{-21}.$$

Generalising from these two instances, we may suppose the mean density of the lenticular centres of nebulae of this type to be of the order of 10^{-21}.

Statistical Study of Nebulae.

Surface brightness.

305. Hubble * has divided four hundred nebulae into the various classes just described and tabulated m_T, the integrated total apparent magnitude, and d, the diameter in minutes of arc, using observational data provided by Holetschek †, Hardcastle ‡, Hopman§ and Curtis‖, as well as about 300 original photographs taken at Mount Wilson.

In any one class, there is found to be a marked correlation between m_T and $\log d$, as might of course be expected; it merely means that the nebulae which look largest in the sky, and so are presumably the nearest, send us most light. On attempting to give mathematical expression to this correlation, it is found that it can be represented very closely by an equation of the form

$$m_T = C - K \log d \qquad\qquad\qquad (305\cdot1)$$

where C is different for each type of nebula, but K is the same for all, being very approximately equal to 5. Thus $m_T + 5 \log d$ is approximately the same for all nebulae of any given class, but varies from class to class.

If a nebula or other astronomical body has area a^2 and emission E per unit area, its apparent luminosity when at a distance R is proportional to Ea^2/R^2. Since a/R is proportional to d, the angular diameter, this is proportional to Ed^2. Hence the apparent magnitude is given by

$$-0\cdot4\,m = \log E + 2 \log d + \text{a constant},$$

whence $\qquad\qquad m + 5 \log d = \text{constant} - 2\cdot5 \log E.$

---

* *Astrophys. Journ.* LXIV. (1926), p. 328.   † *Ann. d. Wienen Sternwarte*, xx. (1907).

‡ *M.N.* LXXIV. (1914), p. 699.   § *Ast. Nach.* CCXIV. (1921), p. 425.

‖ *Publications of Lick Obs.* XIII. (1918).

Thus Hubble's investigation proves that all nebulae of a given class have approximately the same value of E, and consequently the same surface brightness, independently of their sizes. The nebulae are, however, so nearly transparent that the surface brightness at any point represents the integrated light from the whole thickness of the nebula. Thus all nebulae of the same class have approximately the same value for this integrated light, and as they are all of the same shape, it is probable, although not proved, that they are all similar structures.

306. The mean values of $m + 5 \log d$ vary substantially from one class of nebula to another, the average values being shewn in the following table :

TABLE XXVII. *Relative Surface Brightness of Nebulae* (Hubble).

Type	$m + 5 \log d$	$m + 2\cdot5 \log (ab)$	Type	$m + 5 \log d$
$E\,0$	10·38	10·38	Sa	13·35
$E\,1$	10·54	10·43	Sb	13·90
$E\,2$	11·08	10·84	Sc	14·44
$E\,3$	11·33	10·94		
$E\,4$	11·90	11·45	SBa	13·00
$E\,5$	11·42	10·67	SBb	13·16
$E\,6$	12·03	11·03	SBc	14·41
$E\,7$	12·82	11·51	Irreg.	13·68

The third column gives the values of $m + 2\cdot5 \log (ab)$ for elliptical nebulae, the term $\log (ab)$ giving a better approximation to the logarithm of the area than $2 \log d$.

It will be noticed that the surface brightness increases fairly steadily as we advance up the Y-shaped diagram, and this of itself would have suggested a Y-shaped classification.

307. There is of course considerable dispersion about the mean surface brightness in each class. The amount of this is shewn in fig. 58. The points represent nebulae of all classes, but their surface brightnesses are corrected for differences of type before plotting, by reducing them to a fictitious standard type $S\,0$ which is imagined to occur at the fork of the Y at which $E\,7$, Sa and SBa coalesce, and at which $m + 5 \log d$ is supposed to have the standard value 13·0. Thus the distance of any point from the median line represents the divergence of the value of $m + 5 \log d$ from the median value for the particular class to which the corresponding nebula belongs. The two top points represent the two Magellanic Clouds, which Hubble treats as irregular nebulae for purposes of classification, the next point ($m = 5\cdot0$) represents the Andromeda nebula $M\,31$, the next point ($m = 7\cdot0$) the nebula $M\,33$ in Triangulum, and so on.

In the elliptical nebulae, there is found to be no appreciable connection between the deviation from the mean and the orientation of the nebula, but

PLATE XI

The Spiral Nebula *M* 33 in Triangulum

Mt Wilson Observatory

The Greater Magellanic Cloud

Franklin Adams Chart

Nebulae of late type

there is a marked connection in the spiral nebulae, those which are seen nearly edge-on being appreciably less luminous than the average for their class. This may no doubt be attributed to the dark bands of absorbing matter which generally lie along the equators of spiral nebulae.

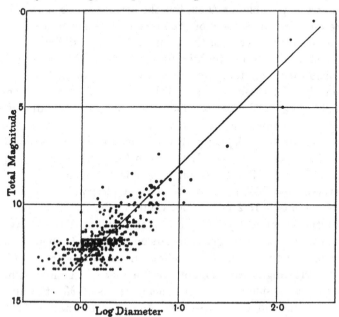

Fig. 58. The relation between Nebular Diameters and Total Magnitudes.
(The slant line represents constant surface brightness.)

Distances and Absolute Magnitudes.

308. There are seven nebulae and star-clouds whose distances are known and of which it is consequently possible to estimate the absolute total luminosity M_T. These are shewn in the first five columns of the following table:

TABLE XXVIII. *Systems whose distances are known* (Hubble).

System	Class	Distance (parsecs)	m_T	M_T	M_S	$M_S - M_T$
Galaxy	—	—	—	—	− 5·5	—
G. Magellanic Cloud	Irreg.	34,500	0·5	− 17·0	− 8·0	9·0
S. „ „	Irreg.	31,600	1·5	− 16·0	− 5·5	10·5
M 31 (Andromeda)	*Sb*	285,000	5·0	− 17·2	− 6·5	10·7
M 32 „	*E* 2	285,000	8·8	− 13·4	—	—
M 33 (Triangulum)	*Sc*	263,000	7·0	− 15·1	− 6·5	8·6
M 101	*Sc*	450,000	9·8	− 13·5	− 6·3	7·2
N.G.C. 6822 ...	Irreg.	300,000	8·5	− 13·7	− 5·8	7·9
Mean				− 15·1	− 6·3	9·0

The mean of the seven absolute magnitudes is $-15\cdot1$, and the dispersion about this mean is comparatively slight.

The last column but one gives the absolute magnitude M_S of the brightest known star in the system. No stars can be detected in elliptical nebulae, so that no entry is possible for $M\,32$, but the corresponding entry for the galaxy may take its place. The mean of the seven entries so obtained is $-6\cdot3$, which is $8\cdot8$ magnitudes fainter than the mean of M_T. The difference in magnitude between a star system and its brightest star can be estimated without knowing the distance of the star system, and so can be estimated for a large number of systems beyond those shewn in Table XXVIII. From a comprehensive study of 32 systems, Hubble concludes that the average value of $M_S - M_T$ is $9\cdot0$, which, as it happens, is also the average for the six systems in the table for which values of $M_S - M_T$ can be given. This would give a hypothetical value of $M_T = -14\cdot5$ for the galaxy. If this is included in averaging M_T instead of the value of M_T for $M\,32$, the average M_T is found to be $-15\cdot3$, this referring only to systems in which stars can be actually observed.

Hubble adopts $-15\cdot2$ as a mean value for M_T for all systems. It will be clear that the average M_T for spirals, star-clouds and the galaxy is near to $-15\cdot2$, and that there is no great dispersion (about ±1) about this mean, but the evidence that the mean M_T for elliptic nebulae is near to $-15\cdot2$ is rather less strong. It might be thought that evidence on this point could be obtained from a consideration of the mean values of M_T, the apparent total magnitudes of the different classes of systems, since distance effects would be likely to affect all classes of systems equally. This would be true if we had at our disposal a test of all the nebulae in a given region of space, but the tests used by Hubble are limited not to a given region of space but to a given range of apparent magnitudes, namely up to about $12\cdot5$. The faint nebulae in each class naturally outnumber the bright to a very great extent, so that the average apparent magnitude in each case is just below $12\cdot5$, ranging from $10\cdot93$ for $E\,7$ to $11\cdot99$ for $E\,3$, and the approximate equality of the averages for the different classes proves nothing very definitely, except that there is no enormously great difference in the mean values of M_T for the various classes.

Absolute Dimensions.

309. On the supposition that the total absolute magnitude M_T is equal to $-15\cdot2$ for all classes of nebulae, it is possible to calculate the actual dimensions of a typical nebula of each class from the observed values of $M_T + 5\log d$. Table XXIX gives the greatest diameters, as calculated by Hubble, after allowing statistically for foreshortening effects in the elliptical nebulae on the supposition that their directions are oriented at random in space.

The minor axes are calculated from these values of the greatest diameter, by use of the relation that in class Ex,

$$\frac{a-b}{a} = \frac{x}{10}.$$

TABLE XXIX. *Dimensions of systems of different types* (Hubble).

Type	Greatest diameter (parsecs)	Least diameter (parsecs)	Type	Diameter (parsecs)
E 0	360	360	Sa	1450
E 1	430	390	Sb	1900
E 2	500	400	Sc	2500
E 3	590	410	SBa	1280
E 4	700	420	SBb	1320
E 5	810	400	SBc	2250
E 6	960	380		
E 7	1130	340	Irreg.	1500

It will be noticed that the minor axes are all approximately equal, so that the sequence of elliptical configurations can be obtained by a process of equatorial expansion of a sphere of about 360 parsecs diameter, the polar axis remaining unaltered.

THE PHYSICAL STRUCTURE OF THE NEBULAE.

310. We have seen that the density of matter in the central lenticular masses of nebulae is of the order of 10^{-21} grammes per cubic centimetre. The free path in a gas at this density is of the order of 10^{14} centimetres. The central mass of the Andromeda nebula, with an apparent diameter of about 12 minutes of arc must have an actual diameter of about $1·6 \times 10^{21}$ cms. or 500 parsecs. Table XXIX shews that the elliptical nebulae have diameters which are at least of the same order of magnitude. Thus if nebular matter is gaseous its free path is quite insignificant in comparison with the dimensions of the nebula, with the result that the concept of gas-pressure may be legitimately introduced into the discussion of nebular dynamics. It follows that the various nebular configurations may legitimately be interpreted as configurations of masses of rotating gas.

If matter of the same density exists in the form of particles larger than molecules the free path is correspondingly longer; if each particle has n times the dimensions of a molecule, the free path is n times as long as in a gas. If we make n equal to 10^7, so that the matter exists in the form of particles about a millimetre in diameter, the free path becomes of the order of 10^{21} cms. and so is comparable with the dimensions of the nebula itself. When this condition is attained, there can be no definite hydrostatic pressure at each point, so that the configurations of a mass in this state will differ from those of a gas or a liquid such as we have so far had under discussion. It follows that the various nebular configurations cannot legitimately be interpreted as those of rotating masses formed of particles of one millimetre or more in diameter.

This result would dispel all possibility of interpreting the observed nebular configurations as those of clouds of stars, were it not for the circumstance, already noticed in § 285, that the gravitational field of a star endows it, as regards its meeting with other stars, with an effective radius which is far greater than the radius of its material structure.

311. To examine whether the elliptical nebulae and the lenticular centres of the spiral nebulae can be interpreted as clouds of stars, let us consider an elliptical or lenticular nebula of mean density $\rho = 10^{-21}$, supposed composed of stars each of mass equal to the sun's mass, 2×10^{33} grammes.

The number of stars per cubic centimetre is found to be 5×10^{-55}. The average velocity of these stars can be found from Poincaré's Theorem (§ 62) that $2T + W = 0$, or $\Sigma m(v^2 - \frac{1}{2}V) = 0$. This shews that the mean value of v^2 is $\frac{1}{2}V$, and this is of the order of $\frac{1}{2}\gamma M/R$, where M is the mass and R the radius of the whole system. For an average nebula this is of the order of 10^{14}, so that v is of the order of 10^7, or a hundred kilometres a second.

In § 287, taking full account of the gravitational field of the stars, we found the time necessary for a star's path to be deflected through a right angle by encounters with neighbouring stars to be

$$t = 0.026 \frac{(m + m')^4 V^3}{\pi v \gamma^2 m^6} \text{ seconds.}$$

Inserting the numerical values $m = m' = 2 \times 10^{33}$, $V = 10^7$ and $v = 5 \times 10^{-55}$ we find a time of the order of 10^{22} seconds. Thus the "free path" of a star is the distance travelled in 10^{22} seconds with a mean velocity of 10^7 cms. a second. This is of the order of 10^{29} cms. or about fifty million diameters of the nebula. It is at once clear that the conception of gas-pressure cannot legitimately be used in connection with nebular dynamics when the nebula is supposed to be a cloud of stars.

No reason remains why a cloud of stars should assume the special shapes of observed nebulae. We shall discuss the actual shapes which can be assumed by a cloud of stars in the next chapter, and shall find that, while the shapes of actual nebulae are possible shapes for clouds of stars, they are only a few possible shapes out of an almost infinite variety. The forces between stars could not restrict the cloud to these special shapes, and even if the star-clouds were initially forced to assume these shapes, they could not be expected to retain them for any length of time. When we discuss the problem in detail, we shall find that if the nebulae were pure clouds of stars, they would have no clearly defined structure or features, but would extend indefinitely into space in the way in which the globular clusters are observed to do and the galactic system is believed to do. To restrain the nebulae to their observed definite shapes, the concept of a gaseous free path is necessary.

312. The foregoing considerations make it extremely improbable that elliptical nebulae and the central lenticular masses of other nebulae can be formed of stars or of any other type of particle averaging more than about an inch in diameter. The calculation has of course no application with respect to the outer regions of spiral nebulae. It is still open for these to be of stellar structure; indeed we have seen that the outer regions of many of these have actually been resolved into stellar points which to all appearances are normal ordinary stars.

GRAVITATIONAL INSTABILITY.

313. Imagine for a moment that the outer regions of a spiral nebula (the equatorial extensions outside the lenticular central mass) were composed of small particles similar to those which constitute the central mass; to make the picture perfectly definite let us imagine that they are purely gaseous. So long as we disregard the gravitational attraction of the gas-particles on one another, a small disturbance set up at any point will travel through the gas with the velocity of sound until it becomes dissipated by viscosity. Any slight condensation of the gas at a point P, accompanied by a compensating rarefaction at a near point Q, results in a gain in the total energy of the medium, and this additional energy travels about until it is dissipated by viscosity, when it remains scattered through the medium in the form of heat energy.

The matter assumes a different aspect when the mutual gravitational attractions of the gas-particles are taken into account. A condensation at P with an equivalent rarefaction at Q causes, as before, an increase in the thermodynamical energy of the gas, but it also causes a decrease in the gravitational energy. And if P is sufficiently distant from Q, the decrease of gravitational energy is numerically greater than the increase of thermodynamical energy, so that the total energy is decreased. In this case there is no excess of energy to travel in the form of wave-motion through the gas until it becomes dissipated by viscosity; instead, the medium has found a way of acquiring kinetic energy of unlimited amount at no expense to itself, and so will go on acquiring energy in this way indefinitely. In other words it has become unstable, the instability entering through displacements in which condensations and rarefactions occur in pairs at sufficiently distant points.

314. We shall refer to the type of instability just explained as " gravitational instability." We have no laboratory experience of it, because it can only become operative in masses of matter far larger than any at our disposal in terrestrial laboratories. On the other hand it proves to be of the utmost importance to cosmogony, for it provides simple, adequate and natural explanations of the creation of four generations of astronomical bodies, namely the great nebulae, the stars, the planets and the satellites of the planets.

The simplest picture of the mode of operation of gravitational instability is provided by the following analysis, which is an extension and modification of analysis I originally gave in 1902 *.

* *Phil. Trans.* 199 A (1902), p. 49.

Consider any motion of a continuous mass of gas or other compressible matter, this being determined by the usual hydrodynamical equations

$$f_x = X - \frac{1}{\rho}\frac{\partial p}{\partial x}, \text{ etc. } \quad\dots\dots\dots\dots\dots(314\cdot1)$$

where f_x, f_y, f_z are the components of acceleration of the particle which is momentarily at x, y, z. Let us assume that the pressure is a function of the density only, and put

$$\int \frac{dp}{\rho} = \phi(\rho),$$

so that the equations of motion become

$$f_x = X - \frac{\partial \phi(\rho)}{\partial x}, \text{ etc. } \quad\dots\dots\dots\dots\dots(314\cdot2).$$

Let us compare the motion with a slightly varied motion such that the particle which is at x, y, z at the time t in the original motion is at $x + \xi,\ y + \eta,\ z + \zeta$ at time t in the varied motion. The acceleration of this particle in the varied motion has components

$$f_x + \frac{d^2\xi}{dt^2}, \text{ etc.}$$

so that the particle which is at x, y, z at time t in the varied motion has components of acceleration

$$f_x + \frac{d^2\xi}{dt^2} - \xi\frac{\partial f_x}{\partial x} - \eta\frac{\partial f_x}{\partial y} - \zeta\frac{\partial f_x}{\partial z}, \text{ etc.}$$

As the result of varying the motion let the density and the components of force at x, y, z be changed to

$$\rho + \delta\rho,\ X + \delta X,\ Y + \delta Y,\ Z + \delta Z.$$

Then the equations which govern the varied motion are

$$f_x + \frac{d^2\xi}{dt^2} - \xi\frac{\partial f_x}{\partial x} - \eta\frac{\partial f_x}{\partial y} - \zeta\frac{\partial f_x}{\partial z} = X + \delta X - \frac{\partial}{\partial x}\phi(\rho + \delta\rho) \ \dots(314\cdot3).$$

Subtracting corresponding sides of this and equation (309·2), we obtain

$$\frac{d^2\xi}{dt^2} - \xi\frac{\partial f_x}{\partial x} - \eta\frac{\partial f_y}{\partial y} - \zeta\frac{\partial f_z}{\partial z} = \delta X - \frac{\partial}{\partial x}\left(\frac{\partial\phi}{\partial\rho}\delta\rho\right) \quad\dots\dots(314\cdot4)$$

where

$$\frac{\partial\phi}{\partial\rho}\delta\rho = \frac{1}{\rho}\frac{\partial p}{\partial\rho}\delta\rho = -\left(\frac{\partial\xi}{\partial x} + \frac{\partial\eta}{\partial y} + \frac{\partial\zeta}{\partial z}\right)\frac{\partial p}{\partial\rho}.$$

Equation (314·4) and its two companions form a set of three equations which are linear in ξ, η, ζ. The condition that these equations shall have a solution, other than $\xi = \eta = \zeta = 0$, is expressed by the vanishing of a quantity which involves only the coefficients of ξ, η, ζ and their differentials in equations (314·4), etc.

When conditions are such that this quantity vanishes, we are at what may be described as a dynamical point of bifurcation. Two alternative motions are open, both of which satisfy the dynamical equations of motion. In one the particle which was at $x - dx$, $y - dy$, $z - dz$ at time $t - dt$ moves to x, y, z at time t; in the other it moves to $x + \xi$, $y + \eta$, $z + \zeta$.

As in the corresponding statical problem, there may or may not be a transfer of stabilities at a point of bifurcation. To discuss the question of stability we merely have to examine whether small displacements ξ, η, ζ determined by equations (314·4) will increase beyond limit or not. If they are found to increase beyond limit, the original unvaried motion was unstable, and the system changes over to the varied motion at the point of bifurcation; in the reverse case the original motion was stable and the displacement ξ, η, ζ, if set up, merely behaves as a small oscillation about a stable state of motion.

Equations (314·4) are too complex to be solved in the most general case, but we can obtain a knowledge of the general nature of the solution by considering the simple case in which f_x, f_y, f_z are approximately constant throughout a large extent of the medium, this including the case of a medium at rest.

In this case equations (314·4) assume the form

$$\frac{d^2\xi}{dt^2} = \delta X - \frac{\partial}{\partial x}\left(s\frac{dp}{d\rho}\right) \quad\ldots\ldots\ldots\ldots\ldots\ldots\ldots(314\text{·}5)$$

where s is written for the "condensation" $\delta\rho/\rho$.

Differentiating this and the two companion equations with respect to x, y, z and adding, and using the relations

$$\frac{\partial}{\partial x}\delta X + \frac{\partial}{\partial y}\delta Y + \frac{\partial}{\partial z}\delta Z = -4\pi\gamma\delta\rho; \quad s = \frac{\delta\rho}{\rho} = -\left(\frac{\partial\xi}{\partial x} + \frac{\partial\eta}{\partial y} + \frac{\partial\zeta}{\partial z}\right),$$

we obtain
$$\frac{d^2 s}{dt^2} = 4\pi\gamma\rho s + \nabla^2\left(s\frac{dp}{d\rho}\right) \quad\ldots\ldots\ldots\ldots\ldots(314\text{·}6).$$

This equation involves only the one variable s and so determines the way in which s changes throughout the motion.

315. On omitting the first term on the right, we obtain the equation of propagation of rarefactions and condensations of the medium, when the gravitational attraction of the medium on itself is neglected. In this case the equation reduces to Laplace's equation, indicating propagation in the form of waves of sound, with the usual velocity $\sqrt{\left(\dfrac{dp}{d\rho}\right)}$.

To discuss the more general problem in its simplest form, let us confine our attention to a region of space within which $dp/d\rho$ may be treated as constant, and consider pure wave-motion along the axis of x, the value of s being supposed proportional to

$$\cos\left(\frac{2\pi x}{\lambda}\right)$$

so that λ is the wave-length. Equation (314·6) becomes

$$\frac{d^2 s}{dt^2} = \left[4\pi\gamma\rho - \left(\frac{2\pi}{\lambda}\right)^2\frac{dp}{d\rho}\right]s \quad\ldots\ldots\ldots\ldots(315\text{·}1).$$

There is a solution in which s is proportional to $e^{\pm iqt}$ where

$$q^2 = \left(\frac{2\pi}{\lambda}\right)^2 \frac{dp}{d\rho} - 4\pi\gamma\rho \quad\ldots\ldots\ldots\ldots\ldots\ldots(315\cdot2),$$

and it is easily seen that this represents wave-motion along the axis of x, with a velocity of propagation

$$\left[\frac{dp}{d\rho} - 4\pi\gamma\rho\left(\frac{\lambda}{2\pi}\right)^2\right]^{\frac{1}{2}} \quad\ldots\ldots\ldots\ldots\ldots\ldots(315\cdot3).$$

If we again omit the gravitational term $-4\pi\gamma\rho$, we have a wave-motion which travels with a uniform velocity $(dp/d\rho)^{\frac{1}{2}}$ independently of the wavelength. The restoration of the gravitational term invariably lessens the velocity of propagation, but since the term in question is multiplied by λ^2, we see that the effect of gravitation is inappreciable for waves of short wavelength. For waves of longer wave-length the gravitational term becomes more important. Finally a value of λ is reached at which the velocity of propagation, as given by formula (315·3), disappears altogether and subsequently becomes imaginary. For such values of λ there can be no proper propagation of waves; the value of q^2, as given by equation (315·2), becomes negative, so that the time factors $e^{\pm iqt}$ assume the form $e^{\pm\theta t}$, where θ is real. This represents unstable motion, the initial condensations and rarefactions increasing exponentially with the time.

316. We have seen that equation (314·6) determines possible distributions of condensation (s) and rarefaction ($-s$) which may be superposed on to the original motion. It now appears that all distributions which vary harmonically are unstable if their wave-length is greater than a critical wave-length λ_0 defined by

$$\lambda_0^2 = \frac{\pi}{\gamma\rho}\left(\frac{dp}{d\rho}\right) \quad\ldots\ldots\ldots\ldots\ldots\ldots(316\cdot1).$$

If p and ρ are connected by a relation of the adiabatic type $p \propto \rho^\kappa$, then

$$\frac{dp}{d\rho} = \kappa\frac{p}{\rho}.$$

If we further put $p = \frac{1}{3}\rho c^2$, where c^2 is the mean of the squared velocity of the particles of which the medium is formed, we may express λ_0^2 in the equivalent forms

$$\lambda_0^2 = \frac{\pi\kappa p}{\gamma\rho^2} = \frac{\pi\kappa c^2}{3\gamma\rho} \quad\ldots\ldots\ldots\ldots\ldots\ldots(316\cdot2).$$

A medium of dimensions much greater than λ_0 would tend to form condensations whose mean distance apart would be comparable with λ_0.

317. It might at first appear that any mass of sufficiently great extent must break up into condensations, but this is obviated by the circumstance that λ_0 tends to increase *pari passu* with the size of the mass.

Consider the most general case of a crowd of particles, either gas-molecules or otherwise, which form a single mass in equilibrium. Poincaré's theorem

(§ 62) shews that the average value of c^2 throughout the system is equal to the average value of $\tfrac{1}{2} V$, and this is of the order of magnitude of $\gamma M/R$, where M is the total mass and R the radius of the system.

Using this value for c^2, equation (316·2) shews that, as regards order of magnitude, λ_0^2 is given by

$$\lambda_0^{\,2} = \frac{\pi \kappa}{3\gamma \rho}\left(\frac{\gamma M}{R}\right) = 4\kappa \left(\frac{\pi}{3}\right)^2 R^2$$

so that λ_0 is about equal to $2R$, the diameter of the mass.

Thus a single mass in a state of equilibrium has no tendency to break up into condensations at distances apart less than its diameter, and this does not leave room for any subordinate condensations: the mass itself constitutes the one and only condensation possible.

In general, λ_0^2 is proportional to c^2, which in turn is proportional to the temperature, so that a sudden cooling reduces the value of λ_0^2. Cooling will ultimately result in contraction and by the time the mass has so far contracted as to be again in equilibrium, the value of $2R$, the diameter, will again be equal to λ_0, and no subsidiary condensations can be formed. But if a mass is cooled so rapidly that its linear dimensions cannot keep pace, or for any other reason do not keep pace, with its fall of temperature, then λ_0^2 becomes less than the dimensions of the mass, and subsidiary condensations will form at distance apart of the order of λ_0.

318. This suggests that condensations cannot form in elliptical nebulae which rotate as systems in a state of steady motion, or in the central masses of spiral nebulae, but that they must inevitably form in the equatorial extensions of the spiral nebulae. As the central mass shrinks, these equatorial extensions are left behind, endowed with an angular momentum which precludes the possibility of their shrinking, while their temperature must fall to a quite low value as the result of their being practically unprotected against loss of heat by radiation into space.

It is difficult to estimate the final temperature likely to be attained ultimately by these equatorial extensions, and it is almost as difficult to estimate their density. Let us, as a very rough estimate, put $c = 10^4$ which is the velocity of atoms or molecules of atomic or molecular weight 600 at $-30°$ C., or of weight 200 at $450°$ C., and $\rho = 10^{-22}$ which is a tenth of the mean density of each of the two nebulae whose density it has been possible to calculate. Using these values for c and ρ and putting $\kappa = 1\tfrac{2}{3}$, equation (316·2) gives $\lambda_0 = 5 \times 10^{18}$ cms. $= 1·6$ parsecs.

This is quite minute in comparison with the linear dimensions of the nebula, so that a large number of condensations ought to form throughout the matter which has been shed equatorially, and now lies in the equatorial plane of the nebula, the mean distance of adjacent condensations being about 1·6 parsecs. This is at least of the order of magnitude of the distance between actual adjacent condensations in the nebula. In the Andromeda nebula, chains of condensations can be observed at about one second of arc apart, and at a distance of 285,000 parsecs one second of arc represents an actual distance of 1·4 parsecs, or somewhat more if we allow for foreshortening.

The mass of matter surrounding each of these condensations is of course $\lambda_0^3\rho$, and with the figures we have already selected, this comes out at $12\cdot5 \times 10^{33}$ grammes or about six times the mass of the sun.

THE BIRTH OF STARS.

319. The last calculation would seem to give the clue to the physical meaning of the process we have been considering: the nebular matter is condensing into stars; and we have been contemplating the birth of the stars. Our calculations have, it is true, been based on rather arbitrarily assumed values of c and ρ, so it is well to examine what difference would have resulted from a different choice of values of c and ρ. The following table gives values of λ_0 and M (equal to $\lambda_0^3\rho$) calculated from formula (316·2) for a variety of values of c and ρ. The value of κ is kept equal to $1\frac{2}{3}$ throughout, because no possible change in the value of κ could appreciably alter the result.

TABLE XXX. *Distance apart and Masses of Condensations formed in nebular matter by Gravitational Instability.*

ρ	$c = 10^4$ cms. a sec.		$c = 10^5$ cms. a sec.	
	λ_0	M	λ_0	M
	(parsecs)	(grammes)	(parsecs)	(grammes)
10^{-32}	160,000	$1\cdot3 \times 10^{39}$	1,600,000	$1\cdot3 \times 10^{42}$
10^{-30}	16,000	$1\cdot3 \times 10^{38}$	160,000	$1\cdot3 \times 10^{41}$
10^{-22}	1·6	$1\cdot3 \times 10^{34}$	16	$1\cdot3 \times 10^{37}$
10^{-21}	0·5	4×10^{33}	5	4×10^{36}
10^{-20}	0·16	$1\cdot3 \times 10^{33}$	1·6	$1\cdot3 \times 10^{36}$
	(cms.)		(cms.)	
10^{-6}	5×10^{10}	$1\cdot3 \times 10^{26}$	5×10^{11}	$1\cdot3 \times 10^{29}$
10^{-4}	5×10^9	$1\cdot3 \times 10^{25}$	5×10^{10}	$1\cdot3 \times 10^{28}$
10^{-2}	5×10^8	$1\cdot3 \times 10^{24}$	5×10^9	$1\cdot3 \times 10^{27}$
1	5×10^7	$1\cdot3 \times 10^{23}$	5×10^8	$1\cdot3 \times 10^{26}$

We see that condensations of stellar mass are only formed in media whose density is of the general order of magnitude of that of the great nebulae.

The observed condensations in the outer regions of such nebulae as M 31 and M 33 are almost certainly stars. A number of them have been found to be Cepheids of entirely normal type. Moreover, Hubble's sequence of nebulae passes through nebulae such as M 31 and M 33 directly into pure star-clouds M 33 being in many respects similar, as Hubble has shewn *, to the Magellanic

* *Astrophys. Journ.* LXIII (1926), p. 236.

PLATE XII

Mt Wilson Observatory

The Spiral Nebula *M* 81

Clouds (see Plate XI). The masses of the condensations are probably some-what greater than that of the sun, for if they are stars they must be at the beginnings of their lives and so in their most massive states. The observed periods of the Cepheid variables indicate high luminosities and high masses.

In many nebulae such as M 81 (Plate XII) the outer regions appear to be resolved into stars, while regions which extend well out beyond the nucleus do not. It may be that the stars have not yet formed here, although there is the alternative possibility that they are merely obscured by surviving wisps of nebular matter.

320. We have seen that the elliptical nebulae and the central masses of the spiral nebulae cannot be clouds of stars, since they could not retain their observed shapes if they were; the observed shapes of the elliptical nebulae and the central masses of the spiral nebulae can only be explained by the supposition that they consist of quite small particles, which it is natural to think of as molecules of a gas. On the other hand the outer regions of the spiral nebulae almost certainly are clouds of stars. We have further found that the whole sequence of observed nebular configurations can be interpreted as the sequence assumed by a single rotating mass whose physical conditions are changing. This might in any case have led us to suspect that this change was accompanied by the transformation of nebular matter into stars. But, to clinch the matter, we have found that such a transformation is demanded by the simplest of dynamical principles; when a nebula shrinks and its rotation increases, matter is inevitably ejected from the nebular equator and spread over the equatorial plane. The temperature of this matter falls, but its angular momentum prevents it shrinking to keep pace with its falling temperature, and as a consequence the dynamical principle we have described as "gravitational instability" compels it to condense into distinct drops, just as cooled steam condenses into drops of water, although for a very different physical reason. The dynamical principle which compels the matter to con-dense into drops enables us to calculate the masses and distances apart of the separate drops. We have calculated that if the cooled nebular matter was in the gaseous state with a molecular velocity of the order of 10^4 cms. a second, the condensations would form at distances apart just about equal to those of the condensations observed in actual nebulae, while the masses of these con-densations would be about equal to those of the stars.

These results would seem to leave little room for doubt that the main process which occurs in the spiral nebulae is the condensation of nebular gas into stars, and that in these nebulae we have found the birthplaces of the stars.

While the mathematical analysis and the detailed calculations relevant to the process of "gravitational instability" were first given by myself in 1902 [*] and 1918 [†], the general conception of the stars having been formed out of a

---

[*] *Phil. Trans.* 199 A (1902), p. 1.   [†] *Problems of Cosmogony and Stellar Dynamics*, Ch. VIII.

homogeneous medium by a process of condensation under gravity is of course very old, being indeed almost as old as the law of gravitation itself. We find Sir Isaac Newton in his first letter to Dr Bentley (Dec. 10, 1692) writing as follows:

"It seems to me, that if the matter of our sun and planets, and all the matter of the universe, were evenly scattered throughout all the heavens, and every particle had an innate gravity towards all the rest, and the whole space throughout which this matter was scattered, was finite, the matter on the outside of this space would by its gravity tend towards all the matter on the inside, and by consequence fall down into the middle of the whole space, and there compose one great spherical mass. But if the matter were evenly disposed throughout an infinite space, it could never convene into one mass; but some of it would convene into one mass and some into another, so as to make an infinite number of great masses, scattered great distances from one to another throughout all that infinite space. And thus might the sun and fixed stars be formed, supposing the matter were of a lucid nature."

Nebular Evolution.

321. As regards their general order of magnitude, the masses of the nebulae can be estimated from the densities and linear dimensions already given. We can also calculate the mass of a lenticular nebulae, or of the central lenticular figure of a spiral nebula, from the circumstance that the particles which form the sharp equatorial edge are in orbital motion under the gravitational attraction of the mass. Thus if ω is the angular velocity and a the radius of the equatorial sharp edge,

$$\gamma M = \omega^2 a^3.$$

The only two nebulae whose angular velocities of rotation are known are M 31 (Andromeda) and N.G.C. 4594 (Virgo). By a method similar to that just explained, Hubble estimates the masses of these to be

M 31, Mass $= 7 \times 10^{42}$ grammes $= 3.5 \times 10^9 \times$ mass of sun.

N.G.C. 4594, Mass $= 4 \times 10^{42}$ grammes $= 2 \times 10^9 \times$ mass of sun.

The masses are of the same order of magnitude. As M 31 is abnormally bright it is likely to be abnormally massive, whereas N.G.C. 4594 which is normal at least to the extent that $M_T + 5 \log d$ represents exactly the mean surface brightness of its class, is more likely to be of normal mass. Thus we may suppose the normal nebula to have a mass of 4×10^{42} grammes or two thousand million suns, which is significantly near to the probable total mass of the cluster of stars surrounding our sun.

Hubble estimates the average distance apart of the great nebulae to be 570,000 parsecs, this representing a density of distribution of 9×10^{-18} nebulae per cubic parsec, or 3×10^{-73} nebulae per cubic centimetre. Assuming the average mass of a nebula to be 4×10^{42}, the average density of matter in space is 1.2×10^{-30} grammes per cubic centimetre. Hubble, on grounds which seem to me unconvincing, takes the mean mass of a nebula to be only 2.6×10^8 times that of the sun, and so obtains a mean density of only 1.5×10^{-31}, stating, however, that both estimates must be regarded as lower limits.

322. Let us imagine the matter in each nebula to be spread out until it uniformly fills a cube of such size that the cubes of the different nebulae just touch. Since the nebulae are all of about equal mass and are approximately scattered uniformly in space, these cubes will be all about equal and the density of matter in each will be about the same, namely $1·2 \times 10^{-30}$ grammes per cubic centimetre. We accordingly have space filled with matter of this uniform density.

This brings us to the picture which the majority of cosmogonists have adopted as their conception of the primaeval universe. Kant started from it, but made the initial mistake of supposing that such a distribution of matter would acquire angular momentum with the mere passage of time. Laplace, avoiding Kant's error, postulated that the initial nebulous mass should be in rotation and argued that its inevitable shrinkage would produce solar systems. As we have just noticed, Newton thought that "some of it would convene into one mass and some into another,...and thus might the sun and fixed stars be formed."

We have seen that the principle of gravitational instability requires that such a continuous distribution of matter should convene into distinct masses, but Table XXX (p. 350) suggests that with any reasonable molecular velocity, a medium of density $1·2 \times 10^{-30}$ would form condensations whose masses would be those of spiral nebulae rather than stars, and the case is still stronger if we accept Hubble's lower estimate of density. With the higher density of $1·2 \times 10^{-30}$, and with a molecular velocity of 3 kms. a second, the condensations would form at average distances of 520,000 parsecs apart, and have average masses of $3·6 \times 10^{42}$ grammes.

It is difficult to imagine that the spiral nebulae were created precisely as they now are, and if we try to peer back one stage further into the history of the universe, the picture to which we are naturally led is that of matter scattered uniformly through space. It is satisfactory to find that if this matter were in the gaseous state, its molecules moving with reasonable velocities of thermal agitation, then the next stage in the evolution of the universe would be the formation of distinct aggregations having masses comparable with those of the spiral nebulae.

323. Imagine that a condensation of mass M initially forms a sphere of radius r, and that it begins to fall in under its own gravitational attraction, each particle falling towards the centre under gravity unchecked by any forces whatever. The time until the radius of the sphere is reduced to r' is readily found to be

$$\frac{r^{\frac{3}{2}}}{(2\gamma M)^{\frac{1}{2}}}(y + \sin y \cos y),$$

where $r' = r \cos^2 y$. If r' is only a small fraction of r, y will be very nearly equal to $\frac{1}{2}\pi$, so that the time necessary for a sphere to shrink under gravity to any small fraction of its original radius r is approximately

$$\frac{\pi}{(\gamma M)^{\frac{1}{2}}}\left(\frac{r}{2}\right)^{\frac{3}{2}}.$$

If we put $M = 4 \times 10^{42}$ and $r = 285,000$ parsecs $= 9 \times 10^{23}$ cms. in this formula, we find that the time necessary for a condensation in the primaeval medium to shrink to the dimensions of a spiral nebula is of the order of 6×10^{10} years. This supposes that each particle falls solely under the attraction of the condensation to which it belongs. Actually each particle would be under the attractions of all the masses in the neighbourhood, and these various attractions would to a large extent neutralise one another, while as soon as the motion had proceeded to any extent, it would be further retarded by molecular collisions. Thus the actual time would be many times longer than that just calculated.

324. Any currents or motion in the original medium would contribute angular momentum to the nascent nebulae, and as these shrank to nebular dimensions, the constancy of angular momentum would result in fairly rapid rotations of the shrunken masses; currents of well below a kilometre a second in the primaeval gas would be adequate to produce the observed rotations of the two nebulae whose rotations have been measured.

The question arises as to what length of time must elapse before a mass whose angular momentum was initially scattered in the form of random currents can assume a state of approximately uniform rotation about a definite axis. We saw in § 243 that inequalities of rotation would be reduced to half-value across a distance r in a time of the order of magnitude of $\rho r^2/\eta$, where η is the coefficient of viscosity. Since a nebula is almost transparent, radiative viscosity is non-existent, so that η may be taken to be the coefficient of material viscosity $\frac{1}{3}\rho c l$, where c is the mean velocity and l is the mean free path, and the time in question becomes $3r^2/cl$.

For a final nebula, let us put $r = 4 \times 10^{20}$, $l = 10^{14}$, $c = 10^7$, and we find that inequalities of rotation are halved over a distance of about a quarter of the radius of an average nebula in a period of $1\cdot6 \times 10^{13}$ years.

In an earlier stage of the shrinkage the time for equalisation over a corresponding fraction of the radius is less. When the nebula has n times its final dimensions, equalisation takes place n times as rapidly. In view of the slowness of the earlier stages of nebular contraction, it seems likely that a nebula may acquire fairly uniform rotation in the process of shrinkage, but we have seen that if this does not happen, the whole of astronomical time will barely suffice to smooth out such inequalities of rotation as remain.

The calculation hardly suggests that a nebula will acquire absolutely uniform rotation, and neither does observation suggest that the rotations of the nebulae are uniform. Photographs of different lengths of exposure may be

PLATE XIII

I. N.G.C. 3115

II. N.G.C. 5866

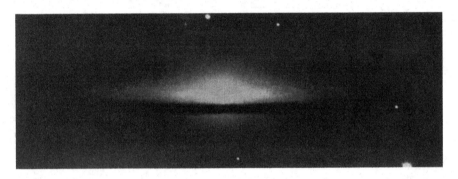

III. N.G.C. 4594

IV. N.G.C. 5746

V. N.G.C. 4565

Mt Wilson Observatory

The Sequence of Nebular Evolution

interpreted as revealing approximately the surfaces of constant density in the nebulae, and it is a remarkable fact that short exposures shew an area which exhibits the elongated shape of the main nebula but on a smaller scale. If the nebula were rotating with a uniform angular velocity, the inner surfaces would be less elongated than the outer, and the innermost surfaces of all would be nearly spherical. Exposures of different lengths shew that this is not the case, suggesting very forcibly that the inner layers are rotating far more rapidly than the outer. Incidentally this may explain the circumstance that nebulae of types $E\,6$ and $E\,7$ shew more elongation than is possible in a uniformly rotating mass.

325. In the final stages of its shrinkage, the nebular condensation approximates to the shrinking rotating nebula, endowed with a constant amount of angular momentum, which formed the starting point of Laplace's cosmogony, and also the subject of the theoretical researches collected in Chapter IX. As such a nebula shrinks, it must pass through the sequence of configurations already described. After being at first almost spherical, it will become spheroidal, then will develop a sharp edge in its equatorial plane. Matter will then be shed off from this sharp edge and left describing orbits in the equatorial plane. Individual nebulae may of course stop at any point in the sequence from want of angular momentum.

As the shrinkage proceeds, the central regions become continually more dense, and must in time assume a stellar condition in which the matter is opaque to radiation. As with the stars discussed in Chapter V, such a nebula cannot exist stably with density so low that the gas-laws are obeyed throughout, so that the shrinkage must proceed until substantial deviations from the gas-laws occur, at any rate in the central regions of the nebula.

So long as these deviations from the gas-laws are not too great, we can use Poincaré's theorem to calculate the mean molecular velocity in the nebula. The calculation is identical with that already given in § 311, and predicts a molecular velocity of 10^7 cms. a second if the average gravitational potential throughout the nebula is taken to be $\gamma M/R$. In view of the high central condensation of mass, the average must be far higher than this, so that 10^8 cms. a second is probably an underestimate for the mean molecular velocity.

Velocities of this order of magnitude are impossible for complete atoms or molecules, since they correspond to temperatures far above those at which atoms or molecules can exist without electronic dissociation. To attain consistency we must suppose the molecules to be completely broken up into their constituent electrons and nuclei; a velocity of 10^8 cms. a second with an effective molecular weight of 2·5 corresponds to a mean temperature of 125 million degrees, with a central temperature of perhaps double this.

Thus the physical conditions at the centre of a nebula must be precisely the same as those we have found to prevail at the centres of the white dwarfs, and, apart from their rotation and outer structure, the extra-galactic nebulae may be regarded as being merely white dwarfs of colossal mass.

326. If the foregoing calculation had been presented in another order, we could have traced the changing physical conditions which succeed one another as the nebula shrinks, and would have found that the final " white dwarf" state was reached when the nebula had shrunk to a radius of the order of magnitude of 400 parsecs, this being the radius that corresponds to a " white dwarf" temperature. On the hypothesis we are now considering, it is this that determines the dimensions of the great nebulae, and accounts for their being all of approximately equal size, as shewn in Table XXIX.

The equilibrium of the nebula along its polar axis is determined by the usual hydrostatic equation

$$\frac{dp}{dr} = -g\rho,$$

into which the rotation does not enter at all. Thus when a nebula shrinks in the way just imagined, the final length of its polar axis ought to be approximately independent of its velocity of rotation. This latter quantity determines the ratio of the two axes, but the length of the minor axis must be determined almost entirely by the amount of matter in the mass.

Thus we can imagine the original nebular medium to condense, under gravitational instability, into nebulae of approximately equal mass M, rotating with different angular velocities. The minor axes of these masses will be all equal, but their major axes will depend on the varying degrees of rotation. Or, more precisely, the minor axes will be ranged about a mean with the same dispersion no matter what their rotation, while the equatorial extension is determined by the rotation.

This is exactly what Hubble's measurements reveal in actual nebulae.

327. If the rotation of the nebulae represents angular momentum arising from random currents in the primaeval nebular medium, then the angular momenta of the different nebulae about any one axis in space ought to shew a Gaussian distribution. The total angular momenta **M** ought accordingly to shew a Maxwellian distribution of the form

$$\mathbf{M}^2 e^{-\kappa \mathbf{M}} \, d\mathbf{M} \quad \dots\dots\dots\dots\dots\dots\dots(327\cdot1),$$

the axes of rotation being oriented at random. If the nebulae are treated as being all of the same mass and dimensions, **M** is proportional to the angular velocity, and so by equation (303·3) to $\epsilon^{\frac{1}{2}}$, the square root of the ellipticity, so that the law of distribution of ellipticities ought to be of the form

$$\epsilon^{\frac{1}{2}} e^{-k\epsilon} d\epsilon \quad \dots\dots\dots\dots\dots\dots\dots(327\cdot2).$$

PLATE XIV

N.G.C. 7217

M 64

Mt Wilson Observatory

Spiral Nebulae with nearly circular arms

Actually this formula can only be valid up to about $\epsilon = 0.6$, all higher values being represented by the spiral nebulae. The fact that the spirals far out-number the elliptical nebulae shews that k must be quite small, so that up to $\epsilon = 0.6$ the effect of the factor $e^{-k\epsilon}$ is probably inappreciable. If we neglect the factor $e^{-k\epsilon}$ altogether, the law of distribution reduces to $\epsilon^{\frac{1}{2}} d\epsilon$. The following table shews the numbers of actually observed nebulae as given by Hubble after allowing for foreshortening on the supposition that the nebulae are oriented at random in space, together with a calculated distribution which obeys the law $\epsilon^{\frac{1}{2}} d\epsilon$.

Ellipticity	Observed (Hubble)	Calc.
0 to 0·05	9·9	1·27
0·05 to 0·15	5·6	5·3
0·15 to 0·25	7·6	7·6
0·25 to 0·35	9·3	9·2
0·35 to 0·45	10·0	10·7
0·45 to 0·55	12·1	12·1
0·55 to 0·65	10·8	13·2
0·65 to 0·75	21·7	14·2

There is good enough agreement except at the two ends of the curve. The large observed preponderance of nearly spherical nebulae may arise from the outer layers rotating less rapidly than the inner layers, and so not having the elongation appropriate to the angular momentum of the nebula, while the less marked excess of nebulae of high ellipticity could be easily explained in the not improbable event of a few Sa nebulae having been classified as $E\,7$.

THE SPIRAL ARMS.

328. Laplace, discussing the sequence of configurations assumed by his rotating nebula, supposed that it had a universe to itself, and so was acted on by no forces beyond its own gravitation. In this case symmetry required that the matter shed in the equatorial plane should form perfect circles.

An actual nebula must be acted on by tidal forces arising from the general gravitational field of the universe as well as from other nebulae. When first it assumes the lenticular form, the forces of gravitation and centrifugal force do not suddenly become equal at all points of the equator simultaneously. Two opposite points of the equator will be distinguished from the others as the points at which the tidal forces give the greatest height of tide, and the ejection of matter will first commence at these two points. The two points are points fixed in space, determined by the tidal field in space, and not points rotating with the nebula, although of course their position may gradually change as the gravitational forces of the ejected matter modify the general tidal field.

For these reasons ejection of matter ought to occur symmetrically at two antipodal points on the equator of the nebula, and the ejected matter ought to form two symmetrical streamers or arms in the equatorial plane of the nebula, branching out from two opposite points of the equator*.

The typical nebula shews precisely such arms. Plate XIV shews two nebulae in which the arms are closely coiled around the central mass, while Plate XV shews two others in which the arms are much more open.

329. Van Pahlen†, Groot‡ and Reynolds§ have measured the curves formed by these spiral arms and find that in normal cases they approximate to the equiangular spirals

$$r = Ae^{a\theta} \quad \dots\dots\dots\dots\dots\dots\dots\dots\dots\dots(329\cdot1),$$

where A and α are constants, the value of α determining the degree of openness or closeness of the spiral.

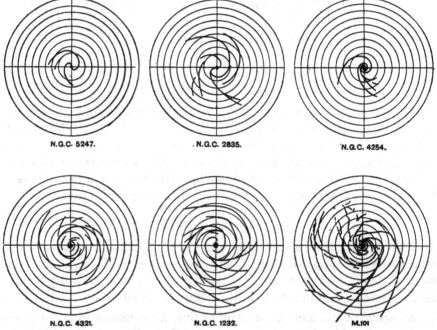

Fig. 59. The curves formed by the spiral arms of six nebulae (Reynolds).

The agreement is not at all close. Van Pahlen measured the three spirals M 33 (Plate XI), M 51 (Plate XV) and M 74, and found substantial departures from the strictly spiral shape in each. Groot found that in eight out of nine spirals considered, the curve was tolerably close to a true equiangular spiral. Reynolds measured six other spirals which he regarded as typical and found substantial deviations from the spiral form in each. Fig. 59 shews the

* For a fuller and more precise discussion, see *M.N.* LXXXIII. (1923), p. 453.
† *Ast. Nach.* CLXXXVIII. (1911), No. 4503.   ‡ *M.N.* LXXXV. (1925), p. 535.
§ *Ibid.* p. 1014.

PLATE XV

M 101

M 51

Mt Wilson Observatory

Spiral Nebulae with open arms

curves as measured by Reynolds after allowing for the foreshortening arising from the inclination of the nebular planes to the line of sight. If the curves were true equiangular spirals they would cut the background of circles always at precisely the same angle.

No satisfactory explanation has so far been advanced as to why the spiral arms should have these particular shapes.

The most obvious conjecture to make would be that the arms are orbits described by the ejected matter, but this does not survive examination. The orbits under gravitation ought to be very nearly circular or elliptical; even if open orbits, such as equiangular spirals, could be obtained, they ought continually to increase in length with increasing age, and nebulae of average age ought to shew many thousands of convolutions, whereas in fact almost all nebulae shew just about two convolutions and no more.

Van Maanen* and Lundmark† have attempted to measure the motion in the spiral arms by direct comparison of photographs taken at intervals of several years, but the results they obtain are not consistent either with one another or with Hubble's determination of nebular distances.

Even the geometry of the spiral arms raises great difficulties. Some nine-tenths at least of the spiral nebulae shew the characteristic equiangular spiral shape of arms, so that any motion occurring in these arms must be such as to transform one equiangular spiral into another. The most general motion which does this is compounded of a motion along the arms and a tangential motion proportional to $r(a + b \log r)$‡. Any motion along the arms seems to be precluded by the circumstance that the length of the arms remains permanently equal to about two convolutions; it does not appear to be a case of stars fading into invisibility after describing two convolutions, for in a number of nebulae (as, for example, M 51, Plate XV) the arms terminate in definite secondary nebulae. The coefficient b in the tangential motion represents secular changes in the angle of the nebula and so must be very small. Thus the only remaining possible motion is one of pure rotation.

It seems almost impossible to explain pure rotation dynamically in terms of known forces, and we are led to the disconcerting, but almost inevitable conjecture, that the motions in the spiral nebulae must be governed by forces unknown to us§. E. W. Brown‖ has attempted to escape this conclusion by regarding the spiral arms not as orbits but as envelopes of orbits described under gravitational forces, the nebular matter being supposed only to shew where these orbits are greatly concentrated, as at their points of contact with an envelope. The law of force necessary to produce orbits whose envelope is an equiangular spiral is, however, found to be very complicated, demanding a highly artificial distribution of matter to produce it.

* See a series of papers in the *Astrophysical Journal*, xliv.–lvii. (1916–1923), or *Mount Wilson Contributions*, Nos. 118–260.

† *M.N.* lxxxv. (1925), p. 865.   ‡ *M.N.* lxxxiii. (1923), p. 61.

§ *Ibid.* p. 73.   ‖ *Astrophys. Journ.* lxi. (1925), p. 97.

The only result that seems to emerge with some clearness is that the spiral arms are permanent features of the nebulae. They appear to have been formed in the process of shrinkage, two convolutions or thereabouts being formed by each nebula, and to have been perpetuated in static form ever since. Their further interpretation forms one of the most puzzling, as well as disconcerting, problems of cosmogony.

Not only so, but until the spiral arms have been satisfactorily explained, it is impossible to feel confidence in any conjectures or hypotheses in connection with other features of the nebulae which seem more amenable to treatment. Each failure to explain the spiral arms makes it more and more difficult to resist a suspicion that the spiral nebulae are the seat of types of forces entirely unknown to us, forces which may possibly express novel and unsuspected metric properties of space. The type of conjecture which presents itself, somewhat insistently, is that the centres of the nebulae are of the nature of " singular points," at which matter is poured into our universe from some other, and entirely extraneous, spatial dimension, so that, to a denizen of our universe, they appear as points at which matter is being continually created.

CHAPTER XIV

THE GALACTIC SYSTEM OF STARS

330. As we have seen (§ 13), our sun is a member of a huge system of stars whose number must be counted in thousands of millions. In general shape this system may be compared to an oblate spheroid with very unequal axes, or, less mathematically, to a coin or round biscuit. The stars are not uniformly distributed throughout this system, being much more thickly scattered in its central parts than in its outer regions. Probably there is no clearly defined boundary, the star-density diminishing indefinitely as we recede from the centre, but never becoming quite zero. The sun lies almost exactly in the central plane of the system, although not precisely at the centre. Those stars which lie near the edge of the coin or biscuit are so remote as to appear very faint to us and constitute the Milky Way. The system of stars bounded by the Milky Way is commonly called the Galactic System.

The stars shew so little motion that for a long time astronomers failed to detect any motion at all, and they became known as "fixed stars" to distinguish them from the planets or "wandering stars" whose motion was obvious to everyone. But modern astronomy finds it possible to measure the motions of a great number of stars. By measuring a star's apparent displacement in the sky over a number of years, it is possible to determine its angular motion round the sun, and if the star's distance is known this can immediately be translated into an actual velocity of kilometres per second. By measuring the displacement of lines in the star's spectrum, it is often possible to determine the speed with which the star is approaching, or receding from, the sun. The combination of both methods makes it possible to determine the star's motion completely, and to announce that it is moving at so many kilometres a second in such or such a direction.

There are two ways of studying the motion of a large collection of independently moving bodies, such as stars, the molecules of a gas, or the asteroids in the solar system. It may be possible to measure the motion of each individual and so acquire a complete knowledge of the motion in question. If this is impossible it may still be possible, by the use of statistical methods, to discover certain general characteristics of the motion. In the kinetic theory of gases, for instance, no attempt is made to follow the motions of individual molecules, but a general statistical treatment shews that they must necessarily move equally in all directions, no preference being shewn for any one direction over others, and Maxwell's well-known law gives the statistical distribution between different velocities. By contrast, the motion of the asteroids shews very marked preferential motion, being confined very closely to a single plane, namely the plane of the orbits of the outer planets, while in the neighbourhood of any one point, the motion of the asteroids is nearly confined to one direction in that plane, namely the direction at right angles to the direction of the sun.

The question arises as to whether the motion of the stars is in any sense an ordered motion like that of the asteroids, or whether it is purely random like the motion of the molecules of a gas. Until 1904 most astronomers would probably have conjectured that the motion was of the latter type, but solely on the grounds that the motion had not so far been found to be anything but characterless. In 1904 the situation was abruptly changed by Kapteyn's discovery of the phenomenon of "star-streaming."

Kapteyn found that the motion of the stars in the neighbourhood of the sun shewed a distinct preference for one direction in space. The preference was not entirely like the preferential motion of the asteroids, since some stars moved forwards and some backwards along the same direction, but this distinction is one which can be removed by altering the frame of reference to which the motions are referred. The stars do not all move along this particular direction; indeed, the preference for this direction is not very highly marked, many stars moving absolutely athwart it.

The direction of the preferential motion is found by Kapteyn and many others to be exactly in the galactic plane. Kapteyn attempted a dynamical explanation on the general lines that the direction of preferential motion was at right angles to the direction pointing towards the centre of the galactic system, the motion of the stars about the centre of the galactic system thus being supposed similar to that of the asteroids about the sun, except that Kapteyn imagined two swarms of "asteroids" intermingled with one another and moving in opposite directions.

An alternative mathematical expression of the observed fact was suggested by Schwarzschild. The molecules of a gas obey Maxwell's law of distribution, according to which the number whose components of velocity u, v, w lie within a small range $du\,dv\,dw$ is of the form

$$A e^{-h(u^2+v^2+w^2)}\,du\,dv\,dw.$$

Schwarzschild proposed that the motions of the stars conformed to an "ellipsoidal" law of distribution of the type

$$A e^{-h(u^2+v^2)-kw^2}\,du\,dv\,dw.$$

If h is made equal to k in this formula, the motion reduces to the random motion specified by Maxwell's law. If h is infinite in comparison with k, u and v must be zero for every star, so that the motion reduces to a pure to-and-fro motion along the axis of z. If h is larger than k, the motion shews a preference for the two directions along the axis of z, the amount of this preference being determined by the value of the ratio h/k. The observed degree of preferential motion is represented by assigning to h/k a value of about 2.

Steady motion of a System of Stars.

331. With a view to searching for an explanation of this and other observed phenomena in the galactic system, we proceed to a dynamical discussion of the motions of a system of stars, which move under one another's gravitational attractions.

Consider first a spherical space in which stars are scattered with fairly uniform density, the mean density of matter being ρ. The gravitational force at a distance r from the centre is a force $\frac{4}{3}\pi\gamma\rho r$ towards the centre, and under this force a star will describe an elliptic orbit with a period equal to

$$\sqrt{\frac{3\pi}{\gamma\rho}} \text{ seconds } \dots\dots\dots\dots\dots\dots\dots(331\cdot1).$$

It is noteworthy that this period depends only on the density and not on the radius r of the system.

In the neighbourhood of the sun, we have supposed there to be a star density of one star per ten cubic parsecs, and if the average star is supposed to be of mass equal to the sun, the average density of matter in space is found to be 7×10^{-24} grammes per cubic cm. The period of an orbit described in a field of matter of this density is found, from formula (331·1), to be about 150 million years.

Thus a star moving in the neighbourhood of the sun, or in regions of space of equal star density, will have its path turned through an angle of 360° in about 150 million years by the gravitational forces arising from the stars as a whole. Our process of averaging the density through space has not allowed for the special deviations of path produced by near encounters with individual stars, but formula (287·4) shews that in 150 million years the expectation of the total deflection produced by such near encounters is only a few minutes of arc.

Apart from the precise figures we have used, this makes it clear that the motion of a star is determined almost exclusively by the main gravitational field of the system to which it belongs, so that, apart from exceptional cases, the influence of near stars may be neglected.

332. This at once explains the continued existence of moving star clusters (§ 26). The members of these clusters are so far apart that the encounters with near stars must be different for each. These encounters produce only very slight effect on the motion, so that the stars continue to describe approximately parallel courses under the uniform gravitational field of the stars as a whole. Detailed calculations are given in § 349 below.

333. The problem of tracing the paths of stars now reduces to the problem of motion in a steady gravitational field, the field being produced by the stars themselves. In brief, the statistical problem of stellar motion is that of the kinetic theory of gases with the collisions left out and a gravitational field thrown in. Just as, in the kinetic theory, the gas may be imagined divided up into a system of showers of parallel-moving molecules, so in stellar dynamics the stars may be imagined divided up into a system of parallel-moving clusters. But there is the essential difference that in stellar dynamics these clusters retain their identity through long periods of time, whereas in gas-theory they do not.

Confining our attention to a small region $dx\,dy\,dz$ of space, let us suppose that the number of stars within this region, whose velocity-components u, v, w lie within prescribed limits $du\,dv\,dw$, is

$$f(u, v, w, x, y, z, t)\,du\,dv\,dw\,dx\,dy\,dz\ldots\ldots\ldots\ldots(333\cdot1).$$

If V is the gravitational potential of the whole system of stars, the motion of each of these stars will be determined by the equations

$$\frac{du}{dt} = \frac{\partial V}{\partial x}, \quad \frac{dv}{dt} = \frac{\partial V}{\partial y}, \quad \frac{dw}{dt} = \frac{\partial V}{\partial z}.$$

After a time dt the parallel motion of these stars will take them to a position $x + u\,dt$, $y + v\,dt$, $w + z\,dt$, while their gravitational accelerations will have increased their velocity components to $u + \frac{\partial V}{\partial x}\,dt$, etc. Hence expression $(333\cdot1)$ must be equal to

$$f\left(u + \frac{\partial V}{\partial x}\,dt, \quad v + \frac{\partial V}{\partial y}\,dt, \quad w + \frac{\partial V}{\partial z}\,dt, \quad x + u\,dt, \quad y + v\,dt, \quad z + w\,dt, \quad t + dt\right),$$

since the stars specified in both groups are identical. We must accordingly have

$$\frac{df}{dt} + \frac{\partial V}{\partial x}\frac{\partial f}{\partial u} + \frac{\partial V}{\partial y}\frac{\partial f}{\partial v} + \frac{\partial V}{\partial z}\frac{\partial f}{\partial w} + u\frac{\partial f}{\partial x} + v\frac{\partial f}{\partial y} + w\frac{\partial f}{\partial z} = 0 \ldots(333\cdot2).$$

This is the differential equation which must be satisfied by the distribution-function f throughout any motion whatever of a system of stars. It will be seen to be identical with the corresponding equation in the kinetic theory of gases, except that the terms arising from collisions are left out.

334. If the stars are in a state of steady motion, f does not change with the time, so that the first term df/dt in equation $(333\cdot2)$ must be omitted.

To solve the resulting equation for f, Lagrange's rule directs us to find all possible integrals $E_1 = $ cons., $E_2 = $ cons., etc., of the system of equations

$$\frac{du}{\dfrac{\partial V}{\partial x}} = \frac{dv}{\dfrac{\partial V}{\partial y}} = \frac{dw}{\dfrac{\partial V}{\partial z}} = \frac{dx}{u} = \frac{dy}{v} = \frac{dz}{w} \ldots\ldots\ldots\ldots\ldots(334\cdot1).$$

The solution of the equation is then simply

$$f = \phi(E_1, E_2, \ldots) \ldots\ldots\ldots\ldots\ldots\ldots(334\cdot2)$$

where ϕ is any arbitrary function.

Clearly, however, equations (334·1) are merely the equations of motion of the star in the general gravitational field of the whole system, so that E_1, E_2, ... are the first integrals of the equations of motion.

One such integral can be written down at once, namely the integral of energy

$$E_1 \equiv \tfrac{1}{2}(u^2 + v^2 + w^2) - V = \text{constant} \quad \dots\dots\dots\dots(334·3),$$

and it is only in special cases that other integrals exist.

Systems with Spherical Symmetry.

335. If the gravitational field of the cluster is spherically symmetrical, so that V is a function only of r, the distance from the centre, equations (334·1) assume the form

$$\frac{du}{\dfrac{x}{r}\dfrac{\partial V}{\partial r}} = \frac{dv}{\dfrac{y}{r}\dfrac{\partial V}{\partial r}} = \frac{dw}{\dfrac{z}{r}\dfrac{\partial V}{\partial r}} = \frac{dx}{u} = \frac{dy}{v} = \frac{dz}{w} \quad \dots\dots\dots\dots(335·1),$$

and there are three integrals

$$\varpi_1 \equiv yw - zv = \text{cons.},$$
$$\varpi_2 \equiv zu - xw = \text{cons.},$$
$$\varpi_3 \equiv xv - yu = \text{cons.},$$

expressing that the moments of momentum ϖ_1, ϖ_2, ϖ_3 per unit mass about the axes of co-ordinates remain constant.

There are no other integrals except for special values of V. For instance, if $V = \alpha r^2$, where α is constant, there are additional integrals of the type $u^2 - 2\alpha x^2 = \text{cons.}$, etc., the corresponding motion being one in which each particle describes a continually repeated elliptic orbit about the centre.

Apart from special artificial cases such as this, there can be no integrals beyond those already mentioned, so that the distribution-function f must be of the form

$$f(E_1, \varpi_1, \varpi_2, \varpi_3)\, du\,dv\,dw\,dx\,dy\,dz \dots\dots\dots\dots(335·2).$$

Since $\nabla^2 V = -4\pi\rho$, it follows that if the gravitational field is spherically symmetrical, ρ can depend only on r, so that the density is arranged in spherical shells. The density is obtained by integrating the distribution law (335·2) with respect to all values of u, v and w. For the resulting density to depend only on r, it is necessary (as is most easily shewn by rotation of axes) that this law should be of the form

$$f(E_1, \varpi_1^2 + \varpi_2^2 + \varpi_3^2)\, du\,dv\,dw\,dx\,dy\,dz \dots\dots\dots\dots(335·3).$$

If c^2 is written for $u^2 + v^2 + w^2$, the law of distribution can be written in the form

$$f(\tfrac{1}{2}c^2 - V, \; r^2 c^2 \sin^2\alpha) \quad \dots\dots\dots\dots\dots(335·4),$$

where α is the angle between the directions of r and c. At any single point in space the law of distribution f depends on c and on α. With this law of distribution, the velocities of the stars are not uniformly distributed over all directions in space, for this would require that f should depend on c only. The velocity-diagram for the motions of the stars near a given point will not be spherically symmetrical, but will be a figure of revolution, having the radius through the point to the centre of the system as origin.

H. H. Turner at one time suggested[*] that the observed star-streaming in our own universe might be explained in this manner, supposing it to orginate in the backward and forward motions of stars describing orbits of high eccentricity about the centre of the whole system. Eddington subsequently shewed[†] that steady states of the type included in formula (335·4) were possible, but failed to notice that (as we shall soon see) they were only possible in a strictly spherical universe. It is fairly certain that our system of stars is nothing like spherical, being a lenticular or biscuit-shaped structure, and the observed star-streaming is almost certainly not along radii but nearly at right angles to radii. Thus a formula of the type of (335·4) cannot account for the observed stellar motions in the galactic system.

Finally, we may notice that the total angular momentum of a system of stars whose steady motion is determined by this formula is zero, so that this state cannot be assumed by a system of stars which has originated out of a rotating nebula or other rotating body.

Systems with Axial Symmetry.

336. After the system whose gravitational field is spherically symmetrical, the next system in order of simplicity is one in which there is symmetry about an axis, so that the surfaces, both of equal potential and of equal density, are surfaces of revolution

Taking the axis of symmetry to be the axis of z, the only integrals of equations (334·1) are seen to be the energy-integral, and the integral $\varpi_3 = \text{cons.}$, which expresses that the moment of momentum of a star about the z-axis remains constant. Hence the only possible state of steady motion is one in which the law of distribution is of the form

$$f(E_1, \varpi_3) \qquad \qquad \qquad (336\cdot1).$$

Systems with no symmetry.

337. The only remaining type of system is that in which there is no symmetry at all. Here the only integral is the energy integral, so that the law of distribution must be of the form

$$f(E_1) \equiv f\left[\tfrac{1}{2}(u^2 + v^2 + w^2) - V\right] \qquad \qquad (337\cdot1).$$

Integrating over all values of u, v and w, we find that the density at any point is a function of V only, so that there must be a relation of the type

$$V = \phi(\rho).$$

---

[*] *M.N.* LXXII. (1912), pp. 387 and 474.

[†] *Ibid.* LXXIV. (1914), p. 5; LXXV. (1915), p. 366; and LXXVI. (1916), p. 37.

Now this is precisely the type of relation we obtained in Chapter VIII in discussing the configurations of equilibrium of compressible masses. But when such a system is acted on by no forces except its own gravitation, the only solution of the equation represents a spherically symmetrical configuration in which both V and ρ depend only on r. This brings us back to the spherically symmetrical configurations discussed in § 335, of which the systems now under discussion are seen to form a special class. Thus the search for systems in steady motion with no symmetry at all has failed; no such motion is possible.

Stability.

338. We are left with the result that, apart from spherical systems in which the spherical symmetry is complete, the only possible states of steady motion are those included in formula (336·1), which represents motions and configurations possessing symmetry about an axis. The total angular momenta of these systems is not necessarily zero, and all systems formed out of rotating masses must be of this type.

Formula (336·1) includes the spherically symmetrical systems discussed in § 337 as a particular case. It does not include the spherical systems with radial star-streaming discussed in § 335, and the mere fact that it does not at once suggests that these latter systems are of a special artificial kind. They have no angular momentum themselves and possess no counterparts which have angular momentum. Thus the slightest amount of angular momentum imparted to such a system destroys the state of steady motion. It must then pass through a series of states of unsteady motion until it ends up in one of the states specified by the law $f(E_1, \varpi_3)$. These spherical systems with radial star-streaming must accordingly be regarded as unstable, and the only states of stable steady motion are included under the law $f(E_1, \varpi_3)$; their configurations possess symmetry about one axis, which may be thought of as a sort of axis of rotation.

Stable States of Steady Motion.

339. To consider these states of steady motion more fully, let us transform to cylindrical co-ordinates ϖ, θ, z and let the components of the velocity c in these three directions be denoted by Π, Θ, Z. Then $\varpi_3 = \varpi \Theta$, and the law of distribution $f(E_1, \varpi_3)$ assumes the form

$$f[\tfrac{1}{2}(\Pi^2 + \Theta^2 + Z^2) - V, \varpi \Theta] \dots\dots\dots\dots\dots(339\cdot1).$$

The velocities are not distributed uniformly in space but shew preferential motion in the directions $\pm \Theta$, i.e. in directions parallel to the central plane $z = 0$, and at right angles to the radius to the centre of the system.

Thus the observed star-streaming in the galactic system exhibits the qualities predicted by formula (339·1) and, so far as these qualities go, the star-streaming is capable of interpretation as resulting from steady motion of the only kind which is dynamically stable.

STAR-STREAMING IN THE GALACTIC SYSTEM.

340. The hypothesis that the observed star-streaming arises from a state of steady-motion admits of a quantitative test which is more exacting than the qualitative test so far considered.

As we have already seen (§ 15), Kapteyn has given a first approximation to the density of star-distribution in the regions surrounding the sun, according to which the surfaces of equal density are similar ellipsoids. It is easy to calculate the gravitational potential arising from such a distribution of stars, so that V may be regarded as known at every point. Different forms of the function f in formula (339·1) will of course give different distribution of star-density, and on excluding all those which do not give the observed distribution of stars, we are left with all the states of steady-motion which are possible for the observed distribution of stars in the sky. It is of interest to discuss how far the star-streaming of such steady motions agrees with the star-streaming which is actually observed.

341. The first attack on the problem was made by Kapteyn in 1922[*]. It was based on the assumption that each star belonged to one or other of two streams of stars which were revolving about the galactic axis, and that, relative to the general motion of the streams to which they belonged, the motions of the individual stars obeyed a Maxwellian law of distribution, the mean velocity being the same throughout the galactic system. Kapteyn was of opinion that this last assumption might be "considered doubtful" in its application to the stellar system. From a study of radial velocities he took the mean value of a single component of velocity to be 10·3 kilometres a second. As we have already seen from Poincaré's theorem, the mean velocity in a system of stars moving in steady motion is determined by the mean gravitational potential throughout the system. Kapteyn having estimated the star-density throughout the system, could estimate the gravitational potential as soon as he assigned a definite mass to the average star. He found that this gravitational potential would give his assumed stellar velocities if each star had an average mass equal to about 1·7 times that of the sun.

This calculation was based on Kapteyn's estimate that the distribution of stars in the central regions of the galactic system was at the rate of 0·0451 per cubic parsec. This figure is too small, since more stars than this are already known in the neighbourhood of the sun, and the average mass of a star must be correspondingly reduced. Kapteyn's result shewed, in effect, that to produce the observed velocities, the average density of matter in the neighbourhood of the sun must be equal to 0·0451 × 1·7 times the mass of the sun per cubic parsec. If we take the density of stars to be one per ten cubic parsecs (§ 8), we get the same density of matter by supposing the mass of the average star to be 0·77 times the mass of the sun.

[*] *Astrophys. Journ.* LV. (1922), p. 302.

Kapteyn found that, for his assumed distribution of density to be possible for stars moving in a state of steady motion with star-streaming of the kind already described, the velocity of star-streaming had to have a certain velocity at each point in the system, a result which is of course also clear from the general discussion of § 339. He calculated that in the galactic plane, the necessary velocity of star-streaming would be about 13·0 kilometres a second at 1010 parsecs from the centre, and that it would be fairly uniformly equal to 19·5 kilometres a second at all distances greater than 2000 parsecs. This would give a relative velocity for the two streams of 39 kilometres a second.

Kapteyn estimated the relative velocity of the observed star-streaming near the sun to be about 40 kilometres a second, so that his investigation would seem, at first glance, to shew that star-streaming could only be explained if the sun was something over 2000 parsecs from the centre of the system. This would however be antagonistic to the hypothesis from which the investigation started, that the sun is near the centre of the system. Correcting for this, Kapteyn found that the relative velocity of star-streaming would be 39 kilometres a second at a distance of rather over 1000 parsecs, and would be 33 kilometres a second at 500 parsecs. Taking 35 kilometres a second to be the minimum velocity consistent with observation, Kapteyn concluded that the distance of the sun from the centre of the system must be greater than 600 parsecs.

As Kapteyn believed that the observed symmetry of brightness of the sky precluded a distance greater than 700 parsecs, he adopted 650 parsecs as the distance of the sun from the centre of the galactic system.

342. An investigation of the problem of star-streaming which I published in the same year shewed[*] that Kapteyn's assumption as to the distribution of velocities being Maxwellian was not only unjustified, but was also unnecessary. As the discussion of § 339 has shewn, the problem is fully determinate when the gravitational field of the stars is given, so that any extraneous assumption is superfluous, and can only lead to erroneous results.

My own investigation amounted in effect to examining what form of the distribution function f in § 339 would give rise to the field of star-densities estimated by Kapteyn. The data introduced were the axes of the Schwarzschild ellipsoid, at a point which was ultimately to be identified with the position of the sun. The corresponding mean velocity of peculiar motion is considerably greater than the 10·3 kilometres a second estimated by Kapteyn. As a consequence the density of matter necessary to produce these velocities was found to be greater than Kapteyn's value. My ultimate figure was $0·0451 \times 4·8 \times 10^{33}$ grammes per cubic parsec, which with ten stars per cubic parsec gives an average mass of $2·16 \times 10^{33}$ grammes per star, or 1·08 times the mass of the sun.

* *M.N.* LXXXII. (1922), p. 122.

Kapteyn's adjusted figure of 0·77 times the mass of the sun is probably nearer to the mass of the average star than my figure of 1·08 times this mass. The majority of stars are either dwarfs of types K and M, or else perhaps white dwarfs, and an average of 1·08 times the sun's mass seems impossibly high. From a statistical study of their orbits, Jackson and Furner* estimated the mean total mass of the two components of visual binaries to be 1·60 times the mass of the sun, giving 0·80 times the sun's mass for each component. We have seen that visual binaries as a class have not been formed by fission, so that 0·80 times the sun's mass would represent the average mass of a single star. This is very near to Kapteyn's value. On the other hand Kapteyn's value leaves no margin for the gravitational field of dark stars, of stars still undiscovered in excess of ten to the cubic parsec, of diffuse or nebular matter in space, or of aggregations or star-clouds such as are known to exist within distances less than those with which we are now concerned (§ 15). When these factors are taken into account a distribution of matter equal to 1·08 times the sun's mass for each visible star is probably fairly near to the truth.

From either investigation, and even more so from the two jointly, we seem entitled to conclude that the gravitational field of known stars is just about adequate to account for the observed velocities of stellar motions and of star-streaming. It is, perhaps, rather remarkable that if there had been neither planets nor binary systems to reveal the masses of the stars, the phenomena of stellar motions would have enabled us to estimate the average stellar mass to within a few per cent. of the truth.

My investigation gave 1090 parsecs as the distance of the sun from the centre of the system. The observations summarised in § 15 indicate that the system has no true centre, the stars surrounding the sun belonging in part to a small local system and in part to the much larger general galactic system. When Seares and van Rhijn† treated these stars as all belonging to a single system, they estimated its centre to be 1200 parsecs distant from the sun.

Final State.

343. We have seen that a law of distribution of velocities of the type

$$f(E_1, \varpi_3) \quad \dots\dots\dots\dots\dots\dots\dots\dots\dots(343\cdot1)$$

will give steady motion except for the disturbing effects of encounters of near stars. Further, this formula has been found to include all possible cases of stable steady motion.

The effect of near encounters will be slowly to change the character of the motion, and after a sufficiently long time, of the order of magnitude of the times considered in Chapter XII, the system of stars will tend to a steady state in which even close encounters do not disturb the statistical specification of motion.

* *M.N.* LXXXI (1920), p. 4. † *Astrophys. Journ.* LXII. (1925), p. 320.

During this process the form of the function f must change, and when the final steady state is attained, the general principles of statistical mechanics indicate[*] that the form of the function f must be

$$f(E_1, \varpi_3) = Ae^{-2hm(E_1 + \omega\varpi_3)} \quad\quad\quad\quad\dots\dots\dots\dots(343\cdot2)$$

where A, h and ω are constants. Inserting their values for E_1 and ϖ_3, the law of distribution becomes

$$f(E_1, \varpi_3) = Ae^{-hm[(u^2 + v^2 + w^2) - 2V + 2\omega(xv - yu)]}$$
$$= Ae^{-hm[(u - \omega y)^2 + (v + \omega x)^2 + w^2] + 2h[V + \frac{1}{2}\omega^2(x^2 + y^2)]} \dots(343\cdot3).$$

Integrating over all values of u, v, w from $-\infty$ to $+\infty$, we find that the density at x, y, z must be of the form

$$\rho = Ce^{2h[V + \frac{1}{2}\omega^2(x^2 + y^2)]} \quad\quad\quad\quad\dots\dots\dots\dots\dots(343\cdot4)$$

where C is a constant. Thus, just as in a rotating mass of gas, the surfaces of equal density have equations of the form

$$V + \tfrac{1}{2}\omega^2(x^2 + y^2) = \text{cons.} \quad\quad\dots\dots\dots\dots\dots(343\cdot5).$$

These surfaces have already been discussed theoretically in Chapters VII and VIII.

344. Formula (343·3) shews that, when the steady state is attained, the stars at any point have a mass-velocity of components ωy, ωx, 0, which is the velocity of a pure rotation with angular velocity ω about the axis of z. Superposed on to this mass-velocity are velocities of individual stars distributed according to a Maxwellian distribution. In this motion there is no star-streaming.

Thus the fact of star-streaming being observed is evidence that the stars are not yet in the final steady state now under consideration. As we have seen, star-streaming can be explained on the supposition that they are in a state of steady motion which is permanent except for the effects of near encounters, but they must still be far from the later state in which even near encounters do not disturb their motion. The discussions of § 272 and § 276 have nevertheless suggested that they may be a good distance on towards this final state.

345. The law of density (343·4) which must obtain in the final state gives infinite density at an infinite distance from the centre except when $\omega = 0$. Even when $\omega = 0$, it gives a finite density at all distances from the centre, so that the system of stars is of infinite extent in space; it is in fact arranged like a mass of gas in isothermal equilibrium without rotation.

When ω is different from zero, the formula shews that there can be no steady state until all the stars have been scattered to infinity. Actually, as we have seen (§ 229), the surfaces of equal density (343·5) consist of some closed surfaces and some open surfaces. If the density at the last of the

---

[*] *Dynamical Theory of Gases* (4th ed.), §§ 107, 113.

closed surfaces is quite small, then the stars inside it form an *approximately* permanent system, although there is a continual slow loss of stars across this surface. If however the density at the last closed surface, and so also at the first open surface, is not quite small, there will be a rapid loss across these surfaces, the stars streaming off in all directions in their efforts to establish the law of density (343·4), and as the velocity of many of these is, by formula (343·3), greater than the velocities of escape $\sqrt{(2V)}$, a great part of the loss is permanent.

The escaping stars are of high velocity and so take away more than their due share of angular momentum, so that in time the value of ω may become so small that the system assumes a nearly spherical shape with only slight escape of stars. It seems probable that the globular star clusters may be formations of this kind.

The galactic system may possibly be going through a process of the kind just described, but if so it cannot be anywhere near its final state.

The question naturally suggests itself as to whether the spiral nebulae admit of interpretation, in terms of the foregoing conceptions, as clouds of stars on their way to a steady state. The elliptical nebulae have, as we have seen, surfaces whose equations are of the form of (343·5), so that to this extent they admit of interpretation as rotating clouds of stars. But the spiral nebulae do not, since such an interpretation fails to explain the concentration of stars in the equatorial plane of the nebula. If the lenticular central part of a spiral nebula consists of stars, then by formula (343·4) the star-density will be uniform over the whole surface of this lens, and since the velocities are also fairly uniform, stars will stream away in approximately equal numbers from all parts of the surface. The stars can in no way be restricted either to escaping only in the equatorial plane or to moving only in that plane after they have escaped. For this reason we cannot interpret the inner parts of the spiral nebulae as clouds of stars, and must fall back on the explanation already given in Chapter XIII. As the spiral nebulae appear from all available evidence to be merely an extension of the series of elliptical nebulae, this indicates that the latter also must be interpreted as clouds of gas.

THE DYNAMICS OF MOVING CLUSTERS.

346. We have seen that the general gravitational field of the galactic system as a whole can adequately explain both the general random motion of the stars and their ordered motion of star-streaming. Apart from these motions, groups of stars, the "moving clusters," are seen pursuing a steady motion through the random motion of the surrounding stars, all the stars of a cluster moving with approximately parallel and equal velocities so that the cluster retains its identity. This phenomenon of course admits of dynamical discussion.

Any one star of a moving cluster will be acted on by forces of three kinds:

(a) The forces arising from the general gravitational field of the galactic system as a whole.

(b) The forces arising from other stars of the same cluster.

(c) The forces from near stars, not belonging to the cluster, which are undergoing encounter with the star in question in the sense that the forces between the two stars are of appreciable amount.

We shall only consider motion in the central dense parts of the galactic system, and here the star-density may be supposed to be approximately uniform, giving rise to a uniform mean density ρ of matter. If, following Kapteyn, the surfaces of equal density in the galactic system as a whole are supposed to be similar spheroids, then the gravitational forces in the central region of the galactic system will be derived from a potential

$$V = K - A(x^2 + y^2) - Cz^2 \quad \dots\dots\dots\dots(346 \cdot 1),$$

where A, C, K are constants, the galactic plane being the plane of xy. The ratio of C to A depends only on the shape of the spheroids. With Kapteyn's ratio 5·102 for their semi-axes, we find that $C = 6 \cdot 31A$. Poisson's relation $\nabla^2 V = -4\pi\gamma\rho$ gives the further equation $2A + C = 2\pi\gamma\rho$, so that

$$A = 0 \cdot 755\gamma\rho; \quad C = 4 \cdot 78\gamma\rho \quad \dots\dots\dots\dots(346 \cdot 2).$$

Let the centre of gravity of the moving cluster be supposed to be at the point x, y, z. The components of force here are $-2Ax, -2Ay, -2Cz$. At an adjacent point $x + \xi, y + \eta, z + \zeta$ the components of force are $-2A(x + \xi)$, $-2A(y + \eta), -2C(z + \zeta)$. Thus a star whose co-ordinates relative to the centre of the cluster are ξ, η, ζ, will experience an acceleration relative to the centre of the cluster, of components $-2A\xi, -2A\eta, -2C\zeta$. Thus if forces (a) alone were operative, the equations of motion of a star in the cluster, relative to the cluster as a whole, would be

$$\frac{d^2\xi}{dt^2} = -2A\xi, \quad \frac{d^2\eta}{dt^2} = -2A\eta, \quad \frac{d^2\zeta}{dt^2} = -2C\zeta \quad \dots\dots\dots(346 \cdot 3).$$

Since A and B are both positive, the resulting motion of the star in the cluster will be compounded of three harmonic oscillations, two parallel to the galactic plane, each of period $2\pi/(2A)^{\frac{1}{2}}$, and one perpendicular to this plane of period $2\pi/(2C)^{\frac{1}{2}}$. With an average of one star of mass 2×10^{33} per ten cubic parsecs, the actual period of the oscillation parallel to the galaxy is found to be 236,000,000 years, and that of the oscillation perpendicular to the galaxy is 91,000,000 years.

If we wish to include the forces (b) arising from the other cluster stars, we may assume the cluster to be of uniform density and of ellipsoidal shape. Its gravitational potential at a point whose co-ordinates relative to its centre are ξ, η, ζ, is then of the form

$$V = k - a\xi^2 - b\eta^2 - c\zeta^2,$$

and we can take account of the resulting forces by replacing equations (346·3) by

$$\frac{d^2\xi}{dt^2} = -2(A+a)\,\xi, \text{ etc.} \dots\dots\dots\dots\dots(346\cdot4).$$

Let us at the same time include the forces (c) which arise from chance encounters with passing stars. Suppose that at any instant, as the result of such an encounter, a star experiences a force per unit mass of components F_x, F_y, F_z. During such an encounter equations (346·4) must be replaced by

$$\frac{d^2\xi}{dt^2} = -2(A+a)\,\xi + F_x, \text{ etc.} \dots\dots\dots\dots\dots(346\cdot5).$$

The general solution of this equation is

$$\xi = a\cos pt + \beta\sin pt + \frac{1}{p}\int_0^t F_x\sin p\,(t-t')\,dt' \dots\dots\dots(346\cdot6)$$

where $\alpha\cos pt + \beta\sin pt$ represents the oscillation which is being executed by the co-ordinate ξ at the instant $t=0$, and $p^2 = 2A$.

The period of the term $\sin p\,(t-t')$ has been seen to be millions of years, and this is so long in comparison with the time of a single encounter that this term may be removed outside the sign of integration, and equation (346·6) may be rewritten in the form

$$\xi = a\cos pt + \beta\sin pt + \Sigma\frac{I_x}{p}\sin p\,(t-t') \dots\dots\dots(346\cdot7)$$

where $I_x = \int F_x\,dt$. The integral is taken through a single encounter occurring at time $t=t'$, and the summation refers to all encounters. Following the usual procedure of squaring and averaging, we find for the expectation of the value of ξ at time t,

$$\xi^2 = \tfrac{1}{2}(a^2 + \beta^2) + \Sigma\frac{I_x^2}{4(A+a)} \dots\dots\dots\dots\dots(346\cdot8)$$

where the summation is over all encounters which have occurred up to time t.

From its meaning, the last term in this equation necessarily increases steadily with the time, so that a cluster expands in all directions as it ages. This of course arises from the scattering effects of encounters with other stars. If the cluster was originally small in comparison with its present size, we may neglect the first term altogether, and rewrite the equations in the form

$$\xi^2 = \frac{\Sigma I_x^2}{4(A+a)}, \quad \eta^2 = \frac{\Sigma I_y^2}{4(A+b)}, \quad \zeta^2 = \frac{\Sigma I_z^2}{4(C+c)} \dots(346\cdot9).$$

347. These equations shew that in general a moving cluster will expand at different rates in different directions, so that even if it is spherical initially, it will not remain so.

Suppose first that a cluster approximately at rest is bombarded by stars reaching it equally from all directions. The values of ΣI_x^2, ΣI_y^2 and ΣI_z^2

will in this case be approximately equal, so that the value of ξ^2, η^2, ζ^2 in the expanded cluster will be inversely proportional to $A + a$, $A + b$, $C + c$.

If the star-density of the cluster is small in comparison with that of the field of stars in which it moves, a, b and c may be neglected in comparison with A and C. The semi-axes of the cluster are now proportional to $A^{-\frac{1}{2}}$, $A^{-\frac{1}{2}}$, $C^{-\frac{1}{2}}$, so that, as C is greater than A, the cluster flattens out in the galactic plane; its cross section in this plane remaining circular. With the relation $c = 6\cdot31\,A$ already used, the ratio of the axes of the cluster would be $2\cdot51 : 2\cdot51 : 1$.

If the system of the B-stars discussed by Charlier* is treated as a moving cluster, or rather as a cluster at rest, we see that it ought to form a spheroidal system, flattened in the galactic plane, the ratios of its axes being as just stated. This is approximately what Charlier finds, except that he gives the ratio of the axes as $2\cdot8 : 2\cdot8 : 1$.

348. In only one known cluster, the Taurus cluster, is the star-density of the cluster itself other than small in comparison with that of the field of stars. Rasmuson† estimates the star-density in this cluster to be about one star per 8 cubic parsecs, which is greater than the density of the galaxy itself. The shape of this cluster is accordingly determined more by the values of a, b and c arising from the field of the cluster itself, than from the values of A, A and C. If ΣI_x^2, ΣI_y^2, ΣI_z^2 were all equal, then the ratio of the axes would be $2\cdot51$ to unity if the ratio were determined solely by A, A and C, and would be one of equality if the ratio were determined by a, b and c. Detailed calculation shews that the actual star-densities require theoretically an intermediate ratio of flattening of about $1\cdot5$ to unity, and this is almost precisely the degree of flattening found by Rasmuson from observation.

In addition to this flattening in the plane of the galaxy, Rasmuson finds the axes in this plane to be slightly unequal; the velocity of the cluster in space is ample to account for this.

Apart from the Taurus cluster, Rasmuson has studied the space distribution of four other principal moving clusters. Three of these (Perseus, Scorpio-Centaurus and 61 Cygni) exhibit a distinct flattening parallel to the galactic plane, the degree of flattening being in each case less than the maximum permitted by theory, in which the ratio of the axes is $2\cdot51$ to 1. The greatest observed ratio is $2\cdot34$ to 1 in the Scorpio-Centaurus cluster.

The fourth cluster, the Ursa Major cluster, does not shew galactic flattening, but is flattened at right angles to its direction of motion, as also in varying degrees are the three clusters just mentioned. If a cluster is bombarded by stars moving relatively to it in directions mainly along the axis of x, it is readily found that ΣI_y^2 and ΣI_z^2 will be equal, while ΣI_x^2 will

---

* *Lund Meddelanden*, Series II, No. 14 (1916).   † *Ibid.* No. 26 (1921).

be small in comparison, and this will result in a flattening perpendicular to the direction of motion of the cluster. Thus clusters ought to shew flattening at right angles both to the galactic axis and to their own direction of motion. Rasmuson's investigation indicates that most clusters shew precisely the required flattening.

Expansion and Evolution of Clusters.

349. The calculations already given in Chapter XII enable us to estimate the rate at which a cluster expands. For simplicity let us consider a cluster bombarded in all directions equally, so that

$$\Sigma I_x^2 = \Sigma I_y^2 = \Sigma I_z^2 = \tfrac{1}{3}\Sigma I^2,$$

where I denotes the total impulsive force at an encounter.

From formula (285·5) we find that

$$I = \tfrac{1}{2} V_0 \psi = \frac{\gamma m}{4 V_0 p},$$

where the stars are all supposed to have equal mass m. On summing over all encounters which occur in unit time, after the manner adopted in § 287, we find

$$\frac{d}{dt}(\xi^2) = \frac{\Sigma I^2}{12A} = \frac{\pi \gamma^2 m^2 \nu}{48 A V_0} \log_e\left(\frac{p_2}{p_1}\right),$$

where $2p_2$ is the mean distance between stars, and $2p_1$ is the distance of approach for an encounter which deflects a star's orbit through a right angle. On inserting the numerical values (cf. § 287)

$$\log_e\left(\frac{p_2}{p_1}\right) = 11·9, \quad A = 0·755\gamma\rho, \quad V_0 = 10^6, \quad m = 2 \times 10^{33},$$

this becomes

$$\frac{d}{dt}(\xi^2) = 0·33\,\frac{\pi \gamma m}{V_0} = 1·4 \times 10^{20}.$$

For a point at which $\xi = 20$ parsecs, the expectation of rate of growth is only about a centimetre a second, and a growth of a further 20 parsecs would require 3×10^{12} years.

While the cluster is growing in size, the velocities of its members will be gradually deviating from their parallel paths, so that the cluster is losing its identity both by its members being scattered in space and by their motions becoming scattered in direction. The formulae of § 287 shew that with the numerical values already used ($m = m' = 2 \times 10^{33}$, $\nu = 4 \times 10^{-57}$, $V = 10$ kms. a second) a period of 20,000 million years will see about one star in 1000 knocked entirely out of the cluster by a violent encounter, while the average angle between the directions of motion of the surviving members will be about one degree. After 500,000 million years one-fortieth of the original members of the cluster will have been lost by violent encounters, while the mean angle of the velocities of the remainder is about 5 degrees. We see that the complete disintegration of the cluster takes about 10^{12} years, and results from gradual scattering rather than from single violent encounters.

This calculation has supposed the stars all to be of solar mass. The formulae of § 287 shew that the time needed to produce a specified deflection in the path of a star of mass m' is proportional to $(m + m')^4$, where m is the mass of the average star. Thus for a cluster of stars of five times the mass of the sun, the times of disintegration just calculated must be increased 81 times, so that complete disintegration takes 10^{14} years rather than 10^{12} years.

The foregoing rates of disintegration are the speediest possible, for they have been calculated for the maximum possible density of stars, namely that prevailing at the centre of the galactic system. A cluster can prolong its life almost indefinitely if it travels mainly in the outlying regions of the galaxy, where disturbing stars are sparsely scattered. Nevertheless, the foregoing figures give some indication of the ages of the star-clusters.

Stars of mass equal to that of the sun would be knocked out of moving clusters after a few million million years, so that, with the ages we have already calculated for the stars, only stars considerably more massive than the sun ought, as a general rule, to be left in the moving clusters.

In view of this result, it is significant that most recognised clusters consist mainly of stars of types B and A, which, as we have seen, are considerably more massive than the sun. The stars of a cluster do not range over all spectral types; there is generally a clearly defined limit* both in spectral type and absolute magnitude, and this limit fixes the age of the cluster. It is impossible to give very precise figures, but a limit corresponding to a mass double that of the sun would fix the age of the cluster at about five million million years.

The Origin of the Galactic System.

350. Within a few million million years the clusters we now observe in the sky will become inextricably mixed with the general mass of stars in the sky, and a few million million years ago a large number of stars which now appear to be moving at random must have been recognisable as members of clusters. It is interesting to consider whether our whole galactic system may have been formed simply out of a collection of moving clusters and the *débris* of disintegrated clusters. We can imagine a great number of such clusters thrown together, passing through one another in their motion and gradually becoming inextricably mingled. Occasionally a cluster would break free from the main mass and form an approximately spherical structure under its own gravitation; such clusters would constitute the "globular clusters" described in § 27.

* See, for instance, Curtis, *Publ. Astron. Soc. Pac.* 27, p. 248, Rasmuson, *Lund Medde-landen*, Series II, No. 26 (1921), and Trumpler, *Lick Obs. Bulletins*, 333 (1921) and 361 (1925).

When such a mingling of clusters takes place, the dimensions of the resulting system may be far greater than those of the component systems. Let T_1, W_1 denote the kinetic and potential energies of a single cluster before mingling, and let J_1 denote its energy of translation relative to the centre of gravity of all the clusters. Then if T_0 and W_0 denote the kinetic and potential energy of the resulting single cluster, the equation of energy is

$$T_0 + W_0 = \Sigma \, (T_1 + W_1 + J_1).$$

If the constituent clusters were each in a state of steady motion before the mingling took place, Poincaré's theorem gives

$$2T_1 + W_1 = 0$$

for each, while by the time the final cluster has attained to a state of steady motion we must have

$$2T_0 + W_0 = 0.$$

By simple algebra we obtain from these equations

$$W_0 = \Sigma \, W_1 + 2\Sigma J_1.$$

Thus in the act of combination an amount of energy $2\Sigma J_1$, which is double the total original energy of translation of all the clusters, is changed into potential energy, and so is spent in expanding the final cluster against its own gravitational attraction. With moderate initial velocities of translation, the ultimate expansion may be enormous. If we regard the galactic system as having been formed by the commingling of a number of clusters, or of the stellar products of a number of spiral nebulae, there is no difficulty in the circumstance that its dimensions are far greater than those of the constituent nebulae or clusters are likely to have been.

351. The galactic system is too symmetrical in shape and shews too clearly defined a structure to have been formed merely out of a random conglomeration of moving clusters; if it was formed by moving clusters, something must have guided their motion into an ordered shape. No hypothesis as to the origin of the galactic system can be accepted which does not account for the very clearly marked galactic plane.

This receives a simple and natural explanation in the hypothesis that the main part at least of the system represents the final stage of development of a single huge spiral nebula. Observation shews that the spiral nebulae retain their characteristic flattened shape throughout the greater part of their evolutionary history; the stars in their outer regions must move almost exactly in the plane of the nebula, or they would, after a comparatively short time, shew more scattering about this plane than is actually observed.

The galactic plane accordingly admits of simple explanation as the plane of the parent nebula. The persistence of the stellar velocities in this plane combined with the rotational velocity of the original nebula, provides a satisfactory explanation of the observed star-streaming in the galactic plane, this now being interpreted in terms of the ellipsoidal velocity distribution of Schwarzschild, and not in terms of the two intermingled star-streams of Kapteyn. On this view the velocities of the stars in any small region may be resolved into:

(1) a uniform motion of rotation about the axis of the galaxy,

(2) an individual motion, superposed on to the foregoing, with the law of distribution given by Schwarzschild's ellipsoidal law.

The combined motion gives a velocity distribution of the type included in formula (339·1), and so represents a possible state of steady motion.

The foregoing view of the origin of the galactic system accords with the fact that the planes of the orbits of binary stars shew a preference for parallelism to the plane of the galaxy. In the past different investigators have reached different conclusions on this question*, but a recent investigation by Kreiken† provides strong evidence that there is a real tendency to parallelism.

It is clear from photographs of spiral nebulae that, when condensation of their outer parts first sets in, these do not immediately break into a uniformly distributed cloud of stars. Most nebulae in the early stages of development, and some in late stages, exhibit condensations which are far too large to be single stars, and are probably clusters each containing a great number of stars. The theory of gravitational instability makes it easy to understand how these large clusters come to exist. For it shews that all displacements of a gaseous medium which have a wave-length above a certain critical length are unstable, but that those of greatest wave-length are most unstable. The result must be that the condensations form on the largest possible scale first, and then gradually smaller condensations form inside these until the shortest wave-length is reached which gives rise to instability, the final condensations being of stellar mass. Nebulae such as M 51 (Plate XV) and M 81 (Plate XII) shew this process going on.

These larger condensations may very possibly be precisely those groups of stars whose relics appear as moving clusters or open clusters in the galaxy, while the few which succeed in escaping altogether from the main mass, before the process of disintegration is far advanced, may be the observed globular clusters. This conception reconciles the strong general impression which the galaxy produces of being a chaos of moving clusters, with the equally strong impression it produces of being the last stage in the development of a spiral nebula.

* Aitken, *The Binary Stars*, p. 218.

† E. A. Kreiken, *M.N.* lxxxvii. (1927), p. 101.

352. Against this must be noticed the apparent difficulty that the stars which constitute the galactic system seem to be of very different ages; at first sight it hardly seems probable that the M-type dwarfs can have been born out of the same nebula as the huge and vigorous O and B-type stars. We have however seen that the atoms at the centres of the spiral nebulae (§ 325) and of the white dwarfs must be completely ionised, and that this renders them immune from annihilation (§ 120). No limit can be assigned to the ages of the atoms which form the giant M or the giant O and B type stars if only we can suppose that they have been shielded from decay in the hot central regions either of a spiral nebula or of a white dwarf throughout the main part of their lives.

In a spiral nebula, as in a star, the atoms of highest atomic weight must sink to the centre, so that it is these, broadly speaking, which are preserved from decay. As the development of the nebula proceeds, layer after layer is shed by the shrinking main mass and condenses, first into clusters and then into stars. The stars which are born first, coming from the outermost layers of the nebula, will have the lowest atomic weights, and as they contain the highest proportion of atoms of the "permanent" elements, will be least luminous per unit mass. Those which are born last will contain the atoms of highest atomic weight and so will have the greatest luminosity per unit mass.

At any instant the ages of the stars in existence, as measured from the time when their atoms first condensed into stars, will vary greatly, but the ages of the atoms of these various stars will all be the same, all dating back to the creation of the original parent nebula. We shall find the lowest atomic weights and the lowest luminosity in the stars which appear to be oldest, and high atomic weights and high luminosity in stars which appear to have been recently born. This is in general agreement with what is observed in the galactic system, but in addition certain specific facts of observation seem to have some bearing on the question.

Shapley[*] has found that in various globular clusters, the brightest stars are red, and the faintest are blue. If the stars of a globular cluster are arranged in a temperature-luminosity diagram (§ 57) the upper half of the main sequence is missing. In the open clusters precisely opposite conditions prevail. The Pleiades[†], Perseus, Ursa Major and other clusters[‡] do not possess a single red giant; all the giant stars are main-sequence stars, and the giant branch which leads up to the giant M type stars is absent. In clusters such as M 11, which are of intermediate type, both kinds of giant stars appear and the temperature-luminosity diagram is of the usual reversed-

[*] *Mount Wilson Contributions*, Nos. 115–117 (1915), 129 (1917), 133 (1917), 151–157 (1918), 160, 161 (1919).

[†] Trumpler, *Lick Obs. Bulletin*, No. 333 (1921), p. 114.

[‡] Lundmark, *Lick Obs. Bulletin*, No. 338 (1922), p. 151.

γ type*. Interpreted with the help of the information obtained in § 167, this means that the stars in globular clusters generate more energy per unit mass than stars of similar mass in the open clusters. On the view we are now considering this would mean that the former were born later than the latter, which accords with the formation of the globular clusters remaining intact while the open clusters are largely broken up.

This view of the origin of the stars demands a modification of the simple view of stellar evolution which was propounded in Chapter VI. For, on the view we are now considering, Kruger 60 can never have been similar to Betelgeux, and Betelgeux will probably never be similar to Kruger 60. The latter star was born of the feebly-luminous matter which floated to the outside of the parent nebula, while Betelgeux was born out of the richly-luminous matter which collected at the centre of the nebula. The former calculation of 10^{14} years as the age of Kruger 60 must be abandoned, since it was based on the assumption that the star had travelled the whole road from being a star similar to Betelgeux.

Only the dynamical evidence as to the ages of the stars now remains valid. The evidence collected in Chapter XII indicates ages of the general order of 10^{12} or 10^{13} years, and this is in very good agreement with the evidence obtained from the rates of expansion and disintegration of star-clusters given in § 349 of the present chapter. There is no longer any evidence that any star is more than about 10^{13} years old, and, indeed, a good many lines of evidence converge in indicating ages of the order of from five to ten million million years for the main mass of the stars, these ages being measured from the time at which the stars first condensed out of the parent nebula. The atoms of the parent nebula must have ages which are at least of the same order of magnitude, and as all the nebulae in the sky may be of the same age, there is no reason against supposing that the whole universe may have been created, or come into being, at the same instant.

* Trumpler, *Lick Obs. Bulletin*, No. 361 (1925), p. 15.

CHAPTER XV

VARIABLE STARS

353. Over 2000 stars are known to be variable, and of these about 1000 are definitely periodic. These periodic variables fall into the two main classes of Cepheid and long-period variables.

It is still uncertain whether Cepheid and long-period variables are essentially different objects or varieties of essentially similar objects. If the latter, the varieties are quite distinct. Long-period variables have periods ranging from about 60 to 500 days, whereas no Cepheid is known whose period exceeds 38·7 days (U Carinae), and most have periods substantially shorter than this. Apart from their different ranges of period, the two classes of variables have many features in common. The light curve of Cepheid variables does not shew a regular symmetrical rise and fall, but rather a fairly rapid rise to maximum brightness followed by a slow decline to minimum, and many long-period variables shew the same features, although generally to a less degree. The Cepheid variables shew a very marked correlation between period and spectral type, shorter periods accompanying the earlier spectral types. Adams and Joy* find a similar correlation in the long-period variables, and this proves to be a direct extension of that already established for Cepheids. In a diagram in which spectral type and period are taken as co-ordinates, they find that a single smooth curve runs through the positions occupied by the long-period variables, the normal Cepheid variables and the cluster variables which form a special short-period group of Cepheids.

Without deciding whether these variables are different or similar types of object, it will be convenient to discuss them together until we are compelled to differentiate between them.

354. Observationally the most marked characteristic of both classes is their extreme rarity. In part, this is a necessary consequence of the fact that they are extremely bright, since extremely bright stars are in any case rare. Cepheids have an average absolute magnitude of about −2, and we have seen (Table V) that for every million stars as bright as the sun there are only 450 stars of this absolute magnitude. But Cepheids are even more rare than this, only a small fraction of stars of the requisite degree of brightness being Cepheids. Within a distance of 100 parsecs of the sun, there must be about 400,000 stars. Only two of these (Polaris and β Cephei) shew Cepheid characteristics, and both are so non-typical that there is some doubt as to whether they are true Cepheids or not. Within a sphere of radius 1000 parsecs there

---

* *Proc. Nat. Acad. Sci.* xiii. (1927), p. 391.

are probably some 100 million stars. Shapley * finds that only about 50 known Cepheids lie within this distance of the sun. Thus probably only about one star in a million is a Cepheid variable.

The proportion of long-period variables is necessarily even smaller, since less than one star in a million has the luminosity of the long-period variables and of the stars of the requisite luminosity only a fraction are variables. At a guess perhaps about one star in ten millions is a long-period variable.

355. In searching for a clue to the physical interpretation of variability we naturally consider first whether it is the effect of some peculiar accident which happens only to a few stars, or whether it is a normal condition which affects all or many stars in the course of their natural development.

The only accidents which we can imagine are of the nature of collisions or of close approaches by other stars. Calculations have shewn (§ 286) that actual collisions must be excessively rare, while encounters at distances close enough to produce physical effects in the stars can be only one degree less so. It is difficult to estimate for how long the effects of a collision or encounter would continue to produce a variation in a star's light, but unless we allow a very long period indeed the number of variable stars in the sky would appear to be too great for such an origin to be assigned to their variability. Moreover, stars of all masses and of all luminosities would be equally liable to accidents of this type, so that, on this hypothesis, it would be hard to explain why, as a rule, only stars of high mass and high luminosity shew variability. On the other hand, it should not be overlooked that the cluster-variables are especially frequent in globular star-clusters (from which they take their name), and that in these the stars are packed so closely that near encounters must be far more frequent than in the galactic system as a whole.

Nevertheless, surveying the question as a whole, it seems very improbable that variability can be attributed to the occurrence of an accident, and we seem' forced to conclude that it is more of the nature of a passing phase in the normal development either of every star or at least of a considerable proportion of stars. In favour of this view is the circumstance that most, and possibly all, of the M-type giants of high luminosity are found to be long-period variables. This of itself would almost suffice to rule out the accident theory, at least in its application to long-period variables.

If variability is a phase in the development of the normal star, the figures given above shew that it must be of short duration. If a star's average life is 10^{13} years, and only one star in a million is variable at any given instant, then variability can only last for about 10^7 years. Hertzsprung † finds that the period of the typical Cepheid variable δ Cephei is decreasing at the rate of a tenth of a second per annum. This rate of decrease would reduce the present

---

* *Astrophys. Journ.* xlviii. (1928), p. 279, or *Mt Wilson Contribution*, No. 151.
† *Observatory*, xlii. (1919), p. 338.

period of 5·366 days to zero in about four and a half million years, which again would suggest a duration of the whole variability of the order of 10^6 or 10^7 years.

356. As we have already noticed (§ 48), the main physical feature of the variation is a fluctuation in the star's visible light rather than in its total emission of radiation, and this requires a change in the star's effective temperature which shews itself observationally as a fluctuation of spectral type. This fluctuation might either arise solely from surface causes or from deep-seated events affecting the whole star. A large mass of observational evidence favours the latter alternative. Interferometer measurements indicate that the angular diameter of Betelgeux changes with its light-variation, a range of over 25 per cent. in all having been already recorded, and what is true of one long-period variable is probably true of all. And the spectral lines of Cepheids shew periodic advances and recession which, if interpreted in the most obvious way, indicate periodic changes in the star's radius. If this is the correct interpretation, the radial velocity, integrated through a half period, must give the total change in the radius of the star. This change of radius is found to be as large as $8\frac{1}{2}$ million kilometres for l Carinae and over 6 million kilometres for X Cygni; for Cepheids in general it averages a million kilometres*, the average radii of the stars themselves, as determined from the luminosities and effective temperatures, being of the order of 20 million kilometres.

The radial velocity shewn by the spectral lines of Cepheids was at first supposed to arise from orbital motion, the Cepheid being regarded as a binary system of which only one component gave a visible spectrum. Many lines of evidence now make this interpretation untenable. In 1918 Shapley† adduced arguments to prove that the then prevalent view of Cepheids as binary systems must be discarded, and that they ought rather to be regarded as single spherical stars in a state of pulsation or oscillation. A similar suggestion had been put forward by Plummer‡ some years earlier in respect of the short period cluster-variables. Some of the consequences of this view of Cepheid variation are in good agreement with observation although, as we shall see, they could equally well be deduced from a somewhat wider view of the cause of the variation.

357. From Poincaré's Theorem, we have found (§ 62) that the mean velocity of thermal agitation inside a gaseous star of mass M and radius R must be of the order of $(\gamma M/R)^{\frac{1}{2}}$ or of $2R(\frac{1}{3}\pi\gamma\rho)^{\frac{1}{2}}$, since $M = \frac{4}{3}\pi\rho R^3$. The slowest dynamical oscillation of a spherical mass has a period approximately equal to the time needed for a wave of compression to travel the length $2R$ of a diameter. As the velocity of such waves is about equal to the velocity of thermal agitation, the slowest dynamical oscillation must have a period of the order of $(\frac{1}{3}\pi\gamma\rho)^{-\frac{1}{2}}$ or say $(\gamma\rho)^{-\frac{1}{2}}$. Thus the period P must be equal to $(\gamma\rho)^{-\frac{1}{2}}$

* Eddington, *The Internal Constitution of the Stars*, p. 182.
† *M.N.* LXXIX. (1918), p. 1.
‡ *Ibid.* LXXIII. (1913), p. 665, LXXIV. (1914), p. 662, and LXXV. (1915), p. 575.

multiplied by a numerical factor which will depend on the precise model on which the star is built, but will always be near to unity.

This approximate relation $P = (\gamma\rho)^{-\frac{1}{2}}$ may be put in the form

$$P^2\rho = 0.0020 \quad\quad\quad\quad\quad\quad (357.1),$$

where P is the period measured in days and ρ is the mean density of the star.

This enables us to calculate the density which corresponds to a given period of pulsation. Calculations for various types of stars are shewn in Table XXXI. The third column gives the observed period of the star, which is identical, on the pulsation-theory, with the period of the star's oscillation, while the last column gives the mean densities of stars of the type in question and of mass ten times that of the sun, as estimated by Seares*.

The calculated mean densities are seen to agree tolerably well with the estimated values.

TABLE XXXI. *Periods and Densities of Variable Stars.*

Star	Spectral Type	Period	ρ (Calc.)	ρ (Seares)
Long Period	*M*	300	0.00000002	0.0000006
Cepheid	*K*	18	0.000006	0.00001
,,	*G*	4	0.00012	0.00002
,,	*F*	0.9	0.0025	0.0004
Cluster	*A*	0.3	0.022	0.008

The Period-luminosity law for Cepheids.

358. In the case of Cepheids a more precise test of the relation $P \propto \rho^{-\frac{1}{2}}$ is provided by the period-luminosity law (§ 11).

This law is usually exhibited in the form of a curve in which $\log P$ is plotted against absolute visual magnitude m_{vis}. For periods greater than a day, the curve shews an approximately linear relation between $\log P$ and m_{vis}, which may be expressed by the equation (see Table XXXII below, col. 6)

$$\log P + 0.30\, m_{\text{vis}} = \text{constant} \quad\quad\quad\quad (358.1).$$

In a group of Cepheids all of which have the same mass and the same effective temperature, the visual luminosity is proportional to the square of the radius, and therefore to $\rho^{-\frac{2}{3}}$. In terms of visual magnitude this relation becomes

$$0.4\, m_{\text{vis}} = \tfrac{2}{3} \log \rho + \text{a constant} \quad\quad\quad\quad (358.2).$$

If P varies as $\rho^{-\frac{1}{2}}$, this equation is seen to be precisely identical with equation (358.1), which expresses the period-luminosity law. This law is

---

* *Astrophysical Journ.* LV. (1922), p. 165, or *Mount Wilson Contrib.* No. 226. Substantially better agreement can be obtained by allowing for the differences in mass of Cepheids of different types (cf. Table XVII of Seares' paper).

accordingly seen to represent the relation $P \propto \rho^{-\frac{1}{2}}$, the relation which we have just seen (§ 357) must connect the period P and the density ρ of a pulsating mass of gas.

359. Let us examine the form assumed by the law when differences in mass and effective temperature are taken into account.

A star's bolometric luminosity is proportional to $4\pi R^2 T_e^4$, so that, if m denotes the star's absolute bolometric magnitude,

$$- 0\!\cdot\!4m = 4 \log T_e + 2 \log R + \text{constant} \quad\ldots\ldots\ldots\ldots(359\!\cdot\!1),$$

and from the approximate relation (§ 118) that the bolometric luminosity is proportional to M^3 or to $(\frac{4}{3}\pi\rho R^3)^3$,

$$- 0\!\cdot\!4m = 3 \log \rho + 9 \log R + \text{constant} \quad\ldots\ldots\ldots\ldots(359\!\cdot\!2).$$

Eliminating R between these two equations and replacing ρ from the relation that $P \propto \rho^{-\frac{1}{2}}$, we obtain

$$\log P + 0\!\cdot\!23m + 3 \log T_e = \text{constant} \quad\ldots\ldots\ldots\ldots(359\!\cdot\!3).$$

The agreement of this formula with observation is almost uncanny. In the following table the first four columns express observational data collected and averaged by Shapley*, and the fifth column gives the absolute bolometric magnitude obtained by applying the bolometric correction from § 48. The sixth column gives the quantity which ought to be constant according to formula (358·2), while finally the last column gives the value of the left-hand member of equation (359·3).

TABLE XXXII. *Observed and Calculated Data for Cepheid Variables.*

Spectral Type	Effect. Temp.	$\log P$	m_{vis}	m_{bol}	$\log P + 0\!\cdot\!3\,m_{\text{vis}}$	$\log P + 0\!\cdot\!23\,m + 3 \log T_e$
$A\,0$	10000	$-0\!\cdot\!56$	$-0\!\cdot\!3$	$-0\!\cdot\!6$	$-0\!\cdot\!6$	$11\!\cdot\!3$
$A\,5$	8500	$-0\!\cdot\!31$	$-0\!\cdot\!3$	$-0\!\cdot\!4$	$-0\!\cdot\!4$	$11\!\cdot\!4$
$F\,0$	7400	$-0\!\cdot\!06$	$-0\!\cdot\!6$	$-0\!\cdot\!6$	$-0\!\cdot\!2$	$11\!\cdot\!4$
$F\,5$	6500	$+0\!\cdot\!23$	$-1\!\cdot\!0$	$-1\!\cdot\!0$	$-0\!\cdot\!1$	$11\!\cdot\!4$
$F\,7\!\cdot\!5$	6000	$+0\!\cdot\!40$	$-1\!\cdot\!4$	$-1\!\cdot\!4$	$0\!\cdot\!0$	$11\!\cdot\!4$
$G\,0$	5500	$+0\!\cdot\!59$	$-1\!\cdot\!8$	$-1\!\cdot\!9$	$0\!\cdot\!0$	$11\!\cdot\!4$
$G\,2\!\cdot\!5$	5050	$+0\!\cdot\!85$	$-2\!\cdot\!4$	$-2\!\cdot\!6$	$+0\!\cdot\!1$	$11\!\cdot\!4$
$G\,5$	4600	$+1\!\cdot\!22$	$-3\!\cdot\!9$	$-4\!\cdot\!2$	$0\!\cdot\!0$	$11\!\cdot\!3$
$G\,7\!\cdot\!5$	4300	$+1\!\cdot\!62$	$-5\!\cdot\!4$	$-5\!\cdot\!9$	$0\!\cdot\!0$	$11\!\cdot\!2$

360. The success of the law $P\rho^{\frac{1}{2}} = \text{constant}$ as shewn in this table is so striking that one is tempted at first sight to suppose that the Cepheid variables, at any rate within the range covered by the table, might unhesitatingly be treated as pulsating spheres, in accordance with Shapley's suggestion. When, however, the absolute values of the quantities are evaluated

* *Harvard Circular*, No. 314 (1927).

the agreement appears very much less good, as has already been indicated in Table XXXI.

The mathematical theory of pulsating spheres has been discussed very fully by Eddington *. Apart, however, from all detailed mathematical treatment, the pulsation theory of Cepheid variation appears to encounter a serious and probably fatal objection at the very outset, which makes all further mathematical treatment superfluous.

As in § 98, the total flow of radiation from the star is

$$4\pi r^2 H = - \frac{\alpha r^2 T^{6\cdot5}}{\rho^2}\frac{dT}{dr} \dots\dots\dots\dots(360\cdot1),$$

where α is a constant, and all the other quantities are evaluated near, but not quite at, the surface of the star. As the star pulsates, all the quantities on the right will vary harmonically, and I have shewn † that the variations of T and ρ will be in phase with one another, so that the variations in $4\pi r^2 H$, the total emission of radiation, will also be in phase with T and with R, the radius of the star. Since this total emission is also proportional to $R^2 T_e^4$, the variation in T_e must also be in phase with R.

Now the spectral lines of a Cepheid shew displacements which indicate a rhythmical advance and recession. On the pulsation theory this must be caused by the changes in the value of R. But the displacement of the spectral lines shews that the changes in T_e and in the emission of radiation are nothing like in phase with R; in general they are almost exactly a quarter period out, being in phase not with R but with dR/dt.

This objection, which I first pointed out in 1926‡, was also advanced independently by Reesinck§ and its validity was finally conceded by Eddington‖. The pulsation theory might possibly be saved, as Eddington has remarked, if the displacements of the spectral lines had been wrongly interpreted, but the following considerations suggest that the hope is a very slender one.

Since the bolometric luminosity at any instant is proportional to $R^2 T_e^4$, we have, as in equation (359·1),

$$m_{\mathrm{vis}} = \Delta m - 10 \log T_e - 5 \log R + \text{constant} \dots\dots(360\cdot2),$$

where Δm is the bolometric correction. This depends only on T_e, so that if m_{vis} were in phase with R, as the pulsation theory requires, both would be in phase with T_e and so with the spectral type of the star. Shapley¶ has examined the spectral changes of 20 Cepheids in detail and finds that they do not coincide with those of m_{vis}. I have found** that the light curves of most

---

* *M.N.* LXXIX. (1918), p. 2, LXXIX. (1919), p. 177 and *The Internal Constitution of the Stars*, Ch. VIII.

† *Ibid.* LXXXVI. (1926), pp. 86 and 574. ‡ *l.c.* p. 90.

§ *Onderzoekingen over δ-Cephei en overhet Cepheidenprobleem.* (Dissertation, Amsterdam, 1926), and *M.N.* LXXXVII. (1927), p. 414.

‖ *M.N.* LXXXVII. (1927), p. 539. ¶ *Astrophys. Journ.* XXIV. (1916), p. 273.

** *M.N.* LXXXV. (1925), p. 810.

Cepheids can be interpreted as the joint result of two distinct variations, a variation in T_e which follows the observed changes of spectral type, and a variation in R which is generally out of phase with the foregoing. The double maximum which occurs in the light curves of Cepheids of the S Sagittae type can be naturally interpreted as being caused by T_e and R attaining their maxima at different epochs. This argument shews that the pulsation theory is untenable, quite independently of the interpretation of the observed displacements of the spectral lines.

Thus all the evidence appears to combine to shew that the pulsation theory is inadequate to explain Cepheid variation.

The Fission Theory of Cepheid Variation.

361. An alternative theory which I put forward in 1925* escapes the particular difficulties which, to all appearances, prove fatal to the pulsation theory, although it is yet to be seen what new difficulties it may encounter in their place.

Let us, as in § 253, regard a star as consisting of a liquid or semi-liquid core surrounded by a gaseous atmosphere. With slow rotation the core will take the shape of a pseudo-spheroid, analogous to a Maclaurin spheroid, and the atmosphere will adjust itself to the gravitational field of this core. The rotation of the atmosphere is unimportant; the considerations of § 249 suggest that it will in all probability rotate more slowly than the core, in which case it must exhibit an equatorial acceleration like that of the sun.

If the star shrinks, the ellipticity of its core must increase, and finally it will assume the shape of a pseudo-ellipsoid of three unequal axes, the analogue of the Jacobian ellipsoid for an incompressible fluid. Even after the core has assumed this form, the outer atmosphere, still rotating slowly, would retain its nearly spherical shape were it not for the disturbances transmitted to it as a consequence of the rotation of the core, which, being no longer symmetrical about its axis of rotation, causes an internal upheaval which must be transmitted to the surface of the star, as well as a variation in the gravitational field, to which the surface of the star will adjust itself.

Whatever disturbance is transmitted will travel round the equator in the form of a wave, or system of waves, these passing once round the equator in the time of a complete revolution of the core. Since these waves must travel with a velocity far greater than the velocity of propagation in the outer atmosphere of the star, they may be expected to shew the usual characteristics of such waves, namely a steeply sloping wall-like front and a gradually sloping rear, similar to the bow-wave of a ship, or a tidal bore. Also the position of

---

* *M.N.* LXXXV. (1925), p. 797. The presentation given here varies somewhat from that of the original paper, further mathematical analysis (unpublished) having suggested that some of the details of the original theory need modification.

the wave in longitude will lag behind that of the ends of the major axis of the ellipsoid which are causing the wave.

These two factors necessarily result in an absence of fore-and-aft symmetry both in the radial motion of the star's atmosphere and in the light-curve of the radiation. Moreover, although the details remain to be worked out, the general nature of the asymmetry would seem likely to be of the nature observed in Cepheid variation, a rapid increase both in radial velocity and in light emission being followed by a slow decline. As a very rough approximation, let us suppose that the wave which advances over the surface of the star has an absolutely vertical wall-like front, and that the whole difference in luminosity between maximum and minimum results from high temperature radiation emitted by the front of this wave. Then each time this wave appears from behind the limb of the star, the star's luminosity will increase with great rapidity to its maximum. This maximum is attained as soon as the whole front of the wave is fully visible, after which a foreshortening effect will result in a slow decline to minimum. The spectral lines will shew a maximum velocity of approach exactly when the star is at its maximum, which is precisely what is observed in Cepheid variation.

As a better approximation we can imagine two systems of waves travelling over the surface of the star, one resulting from the variations of the gravitational field at the star's surface, caused by the rotation of the elongated core inside, and the other produced by the transmission of the mechanical upheaval. The succession of two such waves would account for the double maxima observed in Cepheids of the S Sagittae type.

The period of the variation would be half the period of rotation of the ellipsoid. For a Jacobian ellipsoid fairly near to its maximum elongation (which is the configuration in which variability would be most likely of detection), the value of $\omega^2/2\pi\gamma\rho$ is 0·15, so that the period π/ω is equal to $3·2\,(\gamma\rho)^{-\frac{1}{2}}$.

Thus the fission theory leads to the same qualitative relation between P and ρ as the pulsation theory, namely $P \propto \rho^{-\frac{1}{2}}$, but the numerical factor is different. In place of equation (357·1) the fission theory gives

$$P^2\rho = 0·021 \quad\dots\dots\dots\dots\dots\dots\dots\dots(360·1),$$

where P is again measured in days, but ρ is no longer the mean density of the star but the mean density of its core. The values of ρ given by equation (360·1) are about ten times the values of ρ calculated on the pulsation theory, and for the Cepheid variables these were about six times as great as the mean densities estimated by Seares (cf. Table XXXI). Thus for numerical agreement with observed periods, the fission theory requires that the mean densities of the cores shall be about 60 times the mean densities of the stars as estimated by Seares. This factor of 60 is not an unreasonable one; a core which had a mean radius equal to a quarter of that of the star and contained

95 per cent. of its total mass would have 61 times the mean density of the star, while if it had the configuration of the limiting Jacobian ellipsoid, its semi-major axis would be 0·47 times the mean radius of the star. Nevertheless, the general conception is not free from difficulties; for instance, if a Cepheid has a period of a month, equation (360·1) shews that the mean density of its core ought only to be 0·00002, and it is not at present easy to imagine how matter of this low density can shew so little central condensation of mass that a Jacobian ellipsoid can be a possible figure of equilibrium. The difficulty is only one aspect of a wider one which affects the theory of liquid giant stars in general.

362. Since fission commences through a pseudo-ellipsoidal configuration, it can only begin in stars whose central cores are in a liquid or a semi-liquid state. Thus, on the fission theory, Cepheid variation can only occur in stars whose centres are near to the liquid state, and this would restrict it to stars lying on the left-hand edges of the various bands of stability in the temperature-luminosity diagram (cf. fig. 13, p. 161).

If the mean spectral types and absolute magnitudes of normal Cepheids are mapped out on such a diagram* the majority are at once seen to cluster along the extreme left-hand edge of the L-ring area of stability, which is precisely the type of position which the fission theory requires for them. Many of them seem actually to have overstepped the edge so that, if fission is in progress, the final product will be a binary star on the main sequence.

Stars of the β-Cephei type, which are generally regarded as Cepheid variables†, occupy a corresponding position on the main sequence. The more normal cluster-variables conform less well to the anticipations of theory, some lying on the main sequence, some near the left-hand edge of the L-ring branch, and some sprawling over the space between. Bailey‡ and Shapley§ have found that there are three distinct groups of cluster-variables, differentiated by periods, form and range of variation.

363. As the process of fission progresses, the pseudo-ellipsoidal form must become unstable. The core of the star will now undergo the oscillations resulting from secular instability. At this stage the star will have two distinct periodicities, those of its rotation and of its secularly unstable pulsations. Otto Struve has suggested‖ that the rapid changes in the velocity and light curves of stars such as β Cephei, γ Ursae Minoris, 12 Lacertae and others of the β Canis Majoris type, may be explained in terms of the superposition of periodic rotations and pulsations. This would obviously fit in exactly with the fission theory, these stars being interpreted as rotating masses executing the secularly unstable pulsations which immediately precede fission.

* See, for instance, Bruggencate, *Die Naturwissenschaften*, XL. (1926), p. 910.

† Henroteau, *M.N.* LXXXVI. (1926), p. 256, and R. H. Baker, *Ast. Soc. Pac.* XXXVIII. (1926), p. 86.

‡ *Harvard Annals*, XXXVIII. (1902), p. 132. § *Harvard Circular*, CCCXV. (1927).

‖ *M.N.* LXXXV. (1925), p. 75.

Finally the star breaks into two. When this happens, the two periods of rotation and pulsation become equal, both now coinciding with the orbital period of the star. On the fissional theory the sequence of Cepheid variables ought in all respects to join on to the series of spectroscopic binary stars which have just broken up by fission, and this provides further opportunities for the testing of the fissional theory of Cepheid variation.

From Kepler's third law (or equation (215·9)), the orbital velocity K of either component of a binary system must be connected with the period P of the system by a relation of the type

$$K = CP^{-\frac{1}{3}} \quad \dots\dots\dots\dots\dots\dots\dots(363\cdot1),$$

where C is a constant. Otto Struve* has found that this relation is well satisfied in binary systems in which the period is less than about 2·4 days. But for periods of less than this, the law begins to fail. As the period decreases to below this value, C also begins to decrease, and finally K attains a constant value. He has calculated the period at which the law (363·1) must necessarily begin to fail through the two components of the binary system coming into contact, and finds that the period so calculated agrees for each spectral type separately with the observed period at which failure begins. Further the constant value attained by K is found to be equal to the value of K for the Cepheid variables, this being approximately constant for all Cepheids of the same spectral type. The Cepheids are accordingly seen to fit exactly on to the spectroscopic binaries, thus providing a satisfactory confirmation of the fission theory†.

LONG PERIOD VARIABLES.

Regular Variables.

364. The investigation of Adams and Joy to which we have already referred (§ 353) have shewn that the long-period variables, the Cepheids and the cluster-variables form a continuous series in respect of the correlation between period and spectral type, but that the periods fall into three distinct groups centring round periods of approximately 300 days, 10 days and 0·5 days. Otto Struve‡ has carried out a similar count for spectroscopic binaries, and finds that their curve of period-frequencies shews well-defined maxima at 400 days, 3 days and 0·5 days. He interprets this as evidence that a close relationship exists between variable stars and spectroscopic binaries, and considers that the fission theory of variable stars fits in with the various characteristics of the latter remarkably well.

The position of these maxima and minima—or rather *lacunae*—in the frequencies of periods of variable stars makes it clear that they merely

---

* *Astrophys. Journ.* LX. (1924), p. 167 and *M.N.* LXXXVI. (1925), p. 63.

† Zessewitsch (*Ast. Nach.* No. 5534) finds further support for the fission theory in the behaviour of the stars *RW* Draconis and *XZ* Cygni.

‡ *M.N.* LXXXVI. (1925), p. 75.

correspond to alternations of stability and instability in the temperature-luminosity diagram shewn in fig. 12 (p. 159). Fission can only commence when the star has a liquid or semi-liquid centre, and for this reason stars in process of fission are on or near the left-hand edges of the regions of stability, while spectroscopic binaries, which are permanent structures, must of course be actually on the bands of stability.

The foregoing consideration would seem to suggest that the long-period variables may be of the same nature as Cepheids, namely stars in which the process of fission is either about to start or is under way. There is a certain difficulty, however, in supposing that the long periods of these variables can be those of the rotation of Jacobian ellipsoids. Corresponding to a period of 300 days, we have seen that equation (360·1) requires a density of only 0·00002. This hardly suggests matter in a sufficiently liquid state for ellipsoidal configurations, although it must not be overlooked that eclipsing binaries exist, such as W Crucis, in which both components are giants of low density, and these must, so far as we can tell, have broken up by fission. Moreover, there are two distinct types of long-period variables; in one of these, the regular variables, the light-curve repeats itself indefinitely with perfect regularity, as in the Cepheid variables, but in the other types the light-curve is irregular both in period and shape. It seems possible that the fission theory may adequately account for the former. But the irregular variables do not shew the features to be expected in a star breaking up by fission.

Irregular Variables.

365. The irregular long-period variables give various indications which suggest that they are in a state of pulsation. The diameter of Betelgeux has already been found by interferometer observations to vary from 0·047″ to 0·034″. Moreover in most of the irregular long-period variables which have been studied in detail, the spectrum and effective temperature appear to vary exactly in phase with the visual luminosity.

From an exhaustive study of o Ceti, Joy* finds an approximately linear relation between the spectral type (ranging from M 5 to M 9·5) and the logarithm of the visual magnitude (ranging from 0·45 to 1·0). Vogt† has given observational curves for R Scuti in which the maxima and minima of the effective temperature agree exactly with those of the visual magnitude.

If this is accepted as being generally true of long-period variables, then equation (360·2) shews that R, the radius of the star, must be in phase with both the spectral type and the magnitude, which is precisely what is required of a pulsating star. It is true that Joy finds that the displacements of spectral lines of o Ceti, if interpreted as radial velocities, would indicate changes in the star's radius which are out of phase with both these quantities, but

---

* *Astrophys. Journ.* LXIII. (1926), p. 290.
† *Heidelberg (Königsstuhl) Obs.* VIII. (1926) No. 5.

as different spectral lines tell entirely different stories it seems unlikely that their displacements correspond to those of the star's surface.

366. There are cosmogonical reasons for expecting that the long-period variables should be in a state of pulsation. A star first born out of a nebula has a density so low that the gas-laws must be almost exactly obeyed throughout its mass. It must consequently be unstable and will contract until the gas-laws begin to fail.

The analysis of § 108 enables us to trace out the star's motion during this period of instability. To a good enough approximation α and β may each be put equal to zero in the equation of motion (108·1), so that it becomes

$$\frac{d^3}{dt^3}(\delta r) + \frac{(7+n)\,G_0}{C_v T_0}\frac{d^2}{dt^2}(\delta r) + \frac{\gamma M_r}{r_0^3}\left[\frac{\lambda+4}{\lambda+1}\left(\frac{3p_G + 4aT_0^4}{\rho C_v T_0} - 1\right) + \frac{3s\lambda}{\lambda+1}\right]\frac{d}{dt}(\delta r)$$

$$+ \frac{\gamma M_r}{r_0^3}\frac{G}{C_v T_0}\left[-n\frac{\lambda+4}{\lambda+1} + \frac{3s\lambda}{\lambda+1}(7+n)\right](\delta r) = 0 \quad\ldots\ldots\text{(366·1)}.$$

Since $\gamma M_r/r_0^3$ is large in comparison with $G/C_v T_0$, it follows that the time factors for possible oscillations are of the type

$$e^{-Pt \pm (P^2 - Q)^{\frac12}t},\quad e^{-\epsilon t} \quad\ldots\ldots\ldots\ldots\text{(366·2)},$$

where Q is large in comparison with P and ϵ.

Suppose first that the gas-laws are so nearly obeyed that s in equation (366·1) may be neglected. Then the last coefficient in this equation is negative, while the others are positive, so that P and Q are positive while ϵ is negative. Thus there are stable pulsations having a period of the order of $2\pi(\gamma M_r/r_0^3)^{-\frac12}$, which is hundreds of days, and also unstable contractions in which δr is doubled in a period of the order of $C_v T_0/G_0$, which is thousands of years.

With increasing shrinkage the deviations from the gas-laws increase, so that s increases until finally ϵ becomes positive. Both contractional displacements and pulsations are now stable.

The pulsations retain their identity throughout the whole shrinkage of the star, and if they are once excited there is nothing to check them, since the time needed for viscosity and other dissipative forces to produce an appreciable effect is far greater than the whole time of contraction of the star. Thus when the star first reaches a stable configuration, it will be affected by the whole of the pulsations which have been set up during its birth and contraction from nebular density. Although exact calculations are difficult, it seems likely that quite large pulsations must be set up by the stars first falling in from the nebular state. If so, in its first era of stable existence, the star must be executing pulsations of very large amplitude.

367. The duration of these pulsations must of course depend on the magnitude of the dissipative agencies which tend to check them. If viscosity alone were active, it is readily seen, from a consideration of physical dimensions, that the pulsations would be reduced to half amplitude in a time of the order of $\rho r^2/\eta$, which is probably of the order of 10^{11} years, but ordinary conduction of heat, by preventing the pulsations from being strictly adiabatic, must also contribute towards checking the pulsations. Although exact calculations are difficult, it seems very likely that the pulsations are checked mainly by conduction of heat, this reducing the duration of the pulsating stage to a time of the order needed to account for the observed number of irregular long-period variables.

The different possible pulsations of a sphere of gas correspond to its various principal coordinates, and these have periods which are, in general, incommensurable. Thus pulsating stars can shew no clearly defined period, such as is shewn by most of the Cepheid and regular variables, but their light curve must be formed by the superposition of light curves of incommensurable periods. The principal radial pulsation of the star will probably have a greater amplitude than the other vibrations, and this will determine a period in which the light curve ought approximately to repeat itself, although the exact positions of its maxima and minima will always be influenced by the other vibrations of incommensurable periods. This is precisely what is observed in the irregular long-period variables.

CHAPTER XVI

THE SOLAR SYSTEM

368. THE original aim of cosmogony was to discover the origin of the solar system, but the whole history of cosmogony illustrates how nothing fails so surely in science as the direct frontal attack. The plan of action in the present book has been to study the various transformations which astronomical matter must undergo through the action of physical forces, identifying the formations predicted by theory with those observed in the sky when possible. In this way it has proved possible to trace out the origin and evolution of many astronomical objects, including elliptical and spiral nebulae, star clusters of various forms, binary and multiple stars and (conjecturally at least) Cepheid and long-period variables. But nowhere have we come upon anything bearing the least resemblance to the solar system.

If the sun had been unattended by planets, its origin and evolution would have presented no difficulty. It would have been a quite ordinary star, born out of a nebula in the ordinary way, but endowed with insufficient rotation to carry it on to the later stages of fission into a binary or multiple system; it could in fact be supposed to have had precisely the same evolutionary career as half of the stars in the sky. In support of the conjecture that the sun had stopped short of fission on its evolutionary career we should only have had to note the slowness of its present rotation. A simple calculation suggests that the sun has even now only a small fraction of the angular momentum necessary for fission, and in the earlier stages in which its dimensions were greater than now, the fraction must have been still less. Even if we add the angular momenta of all the planets, as we clearly ought if we are supposing that these at one time formed part of the sun, the result is still the same; the sun can never have had more than a fraction of the angular momentum requisite for fission into a binary system.

Angular Momentum of the Solar System.

369. Such a calculation was first made by Babinet in 1861*. Modern investigations have shewn the need for many adjustments to his calculation, but it is difficult to challenge his result. The sun's radiation, as we have seen (Chap. x) is carrying angular momentum away with it continually, so that the sun's angular momentum is not constant. Further, encounters with other stars or systems may change the sun's angular momentum. But the age of the earth is at most some 5000 million years, and in so short a period

---

* *Comptes Rendus*, LII. (1861), p. 481. See also Moulton, *Astrophys. Journ.* (1900). p. 103.

the loss by radiation produces very little change of angular momentum while the chance of disturbance from outside is negligible.

Fouché* estimated the total angular momentum of the solar system to be 28·2 times that of the sun, the latter being supposed to be a homogeneous mass rotating uniformly in the period of its outer layers. The effect of concentrating all this angular momentum in the sun would be to cause it to rotate 28·2 times as fast as now. Such a sun would still be rotating less rapidly than Jupiter. As the densities of the two masses are about the same and the present Jupiter is still very far from breaking up, we may assume the primitive sun also would be. This simple calculation needs some adjustment both because the sun's mass is highly concentrated towards its centre, and because its inner layers rotate faster than its surface. But no reasonable correction can lead to figures which are compatible with the sun having broken up by fission†.

Laplace's Nebular Hypothesis.

370. There remains the possibility of the sun having broken up in the way originally imagined by Laplace, namely by its shrinkage resulting in a shedding of matter in its equatorial plane. There is no limit to the smallness of the total angular momentum with which this can happen; indeed, in Roche's model, which we studied in § 229, it can happen when the angular momentum is zero, since the whole mass is supposed concentrated at the centre. Thus, notwithstanding its present small angular momentum, the sun could have broken up in this way if its mass had been sufficiently concentrated at its centre.

A body with high central condensation of mass begins to break up in the way we are now considering as soon as

$$\frac{\omega^2}{2\pi\gamma\bar{\rho}} = 0\cdot36 \quad\text{..........................(370·1).}$$

If r_0 denotes the mean radius, $\bar{\rho}$ the mean density, k the radius of gyration, M the mass, and $\mathbf{M}$ the total angular momentum of the primitive sun before the supposed break-up, $M = \frac{4}{3}\pi\bar{\rho}r_0^3$, and $\mathbf{M} = Mk^2\omega$, so that

$$\frac{\omega^2}{2\pi\gamma\bar{\rho}} = \frac{2\mathbf{M}^2}{3\gamma M^3}\frac{r_0^3}{k^4} \quad\text{..........................(370·2).}$$

While about 2000 million years would appear to be the most probable age for the earth, various considerations combine to fix its maximum possible age at 5000 million years‡. Thus when the earth was born, the sun must have

---

* *Comptes Rendus*, xcix. (1884), p. 903.

† *Problems of Cosmogony and Stellar Dynamics*, pp. 15 and 269.

‡ A. Holmes, *The Age of the Earth*, Chap. vii.

been pretty much in its present state[*]. Its mass must have been about equal to the total present mass of the solar system, 2×10^{33} grammes, and its angular momentum about equal to the present total angular momentum of the solar system. This can be calculated with fair accuracy, since it is contributed mainly by the orbital momenta of Jupiter and Saturn, and is found to be $3\cdot3 \times 10^{50}$ in c.g.s. units. Comparing equations (370·1) and (370·2) we find that

$$\frac{k^4}{r_0^3} = 3\cdot7 \times 10^8 \dots\dots\dots\dots\dots\dots(370\cdot3).$$

Putting r_0 equal to the sun's present radius, $6\cdot95 \times 10^{10}$ cms., we obtain

$$\frac{k^2}{r_0^2} = 0\cdot072.$$

The value of k^2/r_0^2 for a homogeneous mass having the shape of a Roche's critical lens-shaped figure is found to be $0\cdot523$. If a fraction x of such a mass is concentrated at its centre while the remaining fraction $(1-x)$ is uniformly spread through a critical lens-shaped figure, the value of k^2/r_0^2 is given by

$$\frac{k^2}{r_0^2} = 0\cdot523\,(1-x).$$

This becomes equal to $0\cdot072$ when $x=0\cdot863$. Thus the present total angular momentum of the solar system would have sufficed to break up the sun if $86\cdot3$ per cent. of its mass had been concentrated at its centre while the remaining $13\cdot7$ per cent. had been uniformly distributed throughout its volume.

This shews that if the sun ever broke up under rotation, there must have been great central condensation in the distribution of its mass. This necessity for great central condensation was apparent to Laplace, and has been recognised by most modern cosmogonists[†], although a few, starting from the supposition that the sun must have been of uniform density, have come, naturally enough, to the conclusion that it cannot have broken up by rotation.

Since the sun's density must decrease continuously as we pass from its centre outwards, the numerical result we have just obtained shews that, if ever the sun broke up rotationally, the density at its edge must have been less than 13 per cent. of its mean density.

The theorem of Poincaré given in § 240 has, however, shewn that if a rotating mass sheds a ring of matter this will scatter into space under the disruptive effects of its own rotation, unless its mean density is more than $0\cdot36$ times that of the main mass. Thus the matter ejected by the sun could only form planets if it immediately condensed to about three times its original density.

---

[*] Jeffreys has always maintained this. When I first propounded the Tidal Theory I suggested, on grounds of probability, that the solar system had probably been formed when the sun had a very large diameter. Our knowledge of the ages of the earth and sun now makes this conjecture untenable.

[†] Poincaré, *Leçons sur les Hypothèses Cosmogoniques*, p. 18.

The matter which was first shed by the sun would form a ring of small mass rotating with the sun ; on account of the smallness of its mass, this would have no gravitational cohesion and, so far from increasing in density, would scatter under its own internal gaseous pressure. It seems impossible that such a ring could double its density before the disruptive influence of its rotation could come into play. Thus, on the whole, it seems necessary to admit Babinet's contention that the solar system is not possessed of sufficient angular momentum to have broken up through excessive rotation*. The calculation needs some modification in view of the possibility discussed in Chapter x of the inner layers of the sun rotating faster than the outer, but the modification is found to make rotational break-up still more impossible.

371. The considerations which carry the most obvious condemnation of the theory of rotational break-up are of a somewhat different kind. If the sun once assumed the lenticular shape necessary for the shedding of matter by rotation, it is difficult to see how it could ever abandon it and become as nearly spherical as it now is ; it is also difficult to understand why the planets should be at such widely varying distances from the sun.

These as well as other difficulties again confront the rotational theory when it attempts to explain the origin of the satellites of the planets. Many of these are so small that they can only have escaped scattering into space by liquefying or solidifying immediately after their birth. Thus their birth cannot have been a long drawn out process such as the equatorial shedding of matter by a slowly shrinking mass ; the fact that these satellites survived the birth-process at all proves that they must have been born quickly.

TIDAL THEORIES.

372. As the genesis of the solar system cannot be explained in terms of a single mass rotating by itself in space, the only alternative which remains is to consider whether it can be explained as the result of the interaction of two or more masses. We can fix our attention on two, since encounters of three masses in space must be so rare as to be entirely unimportant.

This brings us naturally and inevitably to tidal theories of the genesis of the solar system. The fundamental conception of these theories is, that at some time in the past a second mass approached so close to our sun as to break it up, through the action of intense tidal forces, into a number of detached masses.

As between rotational and tidal theories, first appearances are wholly in favour of the latter. A rotating mass, and so also a system which has been formed without external interference out of a rotating mass, retains one

* This argument was first given by myself in *M.N.* LXXVII. (1917), p. 186. Jeffreys has re-investigated the question in *M.N.* LXXVIII. (1918), p. 425, slightly modifying the argument, but confirming my original conclusions.

invariable plane which is at right angles to the original axis of rotation. The system must remain symmetrical about the plane, to which the axis of rotation of the central mass remains perpendicular.

Now the invariable plane of the solar system is practically fixed by the orbits of the four outermost planets. Over 99·9 per cent. of the total momentum of the system resides in the orbital motion of these planets, and the orbits of Jupiter, Saturn and Neptune all lie within 45' of this plane. But the sun's axis of rotation is not perpendicular to this plane, but makes an angle of about 6° with the perpendicular. The rotational theory is unable to account for the divergence between the invariable plane and the plane of the sun's rotation; the tidal theory explains it very naturally by identifying the present plane of the sun's rotation with that of the original sun, while supposing the plane of the orbits of the outer planets to mark the plane of passage of the wandering star whose tidal forces tore the planets out of the sun.

THE PLANETESIMAL THEORY.

373. About 1900, Professors Chamberlin and Moulton of Chicago begun to develop a theory of the origin of the solar system, commonly known as the Planetesimal Theory*, which has received a great deal of attention in America.

The authors point out that the sun is even now in a state of constant eruption, fountains of matter occasionally spouting hundreds of thousands of miles above its surface. The matter ejected in this way falls back into the sun, but if a near body were exerting tidal forces on the sun, the fountains might rise much higher than they do. The planetesimal theory supposes that at some past time, while the sun was emitting periodic eruptions, a star came so near that the eruptions formed two long arms of matter extending for enormous distances out from the sun's surface. The sun now formed a sort of spiral nebula; the matter ejected at the eruptions formed its arms, the condensations in the arms representing the ejections of matter at different eruptions. The bits of matter, as they were ejected, solidified and formed the "planetesimals" after which the theory is named. Those ejected at any one eruption aggregated to form a planet; each big eruption had associated with it a lot of little eruptions, and the matter ejected at these little eruptions formed the satellites of the planets.

There is a want of precision about the theory which makes it difficult to submit to the test of precise mathematical calculation. A dynamical investigation which I made in 1916† into the effects of tidal action on a stellar mass indicated, however, that the sequence of actual events would be very different

---

* See T. C. Chamberlin, *Fundamental Problems of Geology* (Year Book No. 3 of the Carnegie Institution); F. R. Moulton, *An Introduction to Astronomy* (New York, 1906), pp. 463 ff; T. C. Chamberlin, *The Origin of the Earth* (Chicago University Press, 1916).

† *Memoirs R.A.S.* LXII. (1916), p. 1.

from those postulated by the planetesimal hypothesis. This investigation further suggested that mere tidal action of itself would suffice to explain the origin of the solar system, without calling in the various complicated and wholly hypothetical mechanisms of intermittent eruptions, of smaller eruptions to form satellites, of planetesimals, and so on, and led me to put forward the simple tidal theory which follows.

THE DYNAMICAL TIDAL THEORY.

374. Just as with rotation, the effects produced in a star by tidal action prove to be very different according as the star is of fairly uniform density or is arranged so as to have high central condensation of mass.

Tidal Effects in an Incompressible Mass.

375. Let us examine first the effects which would be produced in a liquid star of homogeneous incompressible matter.

Following Roche, we examined in § 206 the effect produced on a small mass S of approaching too near to a big mass S' around which S was supposed to revolve in orbital motion. If S and S' were of the same density we found that S could not approach to within 2·45 radii of S' without being broken up. At this distance the difference in the gravitational pull of S' on the nearer and further halves of S became so great that the attraction of the two halves of S for one another failed to keep S together as a single body. The mass S accordingly broke in two, and as the broken halves found themselves again torn to pieces in precisely the same way, the process of disruption continued indefinitely.

A similar situation arises when the two masses only approach one another for a short time and then recede again, as happens in an ordinary gravitational encounter between two stars whose orbits happen to pass fairly close to one another.

When a second star S' approaches the star S whose behaviour we are considering, its first effect is to raise tides of the ordinary kind on S. First suppose that the distance R between S' and S changes very slowly, so that an equilibrium theory is adequate to follow the changes in the tides. In this case the equations of equilibrium are precisely of the type already discussed in § 206; they are in fact equations (206·6) with ω^2 put equal to zero, μ still denoting $\gamma M'/R^3$.

A discussion of the general type already given in § 206[*] shews that the configuration of S remains stable until it has assumed the shape of a prolate spheroid of eccentricity $e = 0·8826$, the value of R at this stage being

$$R = 2·198 \left(\frac{M'}{M} \right)^{\frac{1}{3}} r_0 \quad \dots\dots\dots\dots\dots(375·1),$$

---

[*] For a full account of the whole investigation see the paper already mentioned (p. 391), or *Problems of Cosmogony and Stellar Dynamics*, pp. 43 ff, 118 ff.

where r_0 is the radius of S in its undisturbed state. With a closer approach of S', the mass S becomes unstable and begins to break up.

In passing we may notice that the limit of safety fixed by equation (375·1) is rather less than that obtained by Roche's investigation for the case in which the masses rotate around one another. Clearly this must be so, since in the latter case rotational forces join hands with tidal forces in trying to break up the mass.

I have found it possible to follow out the dynamical motion after instability has set in. No matter whether the mass S' comes closer or not, the mass S rapidly elongates until it has assumed the shape of a spheroid of eccentricity 0·9477. At this stage a new type of instability sets in. Just as when a Jacobian ellipsoid first becomes affected with instability in the rotational problem, so here also a third harmonic displacement supervenes, causing the formation of a waist on the spheroid which would gradually deepen until the mass broke into two. But before this happens other displacements represented by fourth, fifth and higher harmonics become unstable in turn. These form furrows around the spheroidal mass at various points and as these steadily deepen, the mass breaks into a number of detached pieces. The sequence of configurations is shewn in fig. 60 (p. 402), the last configuration being almost entirely conjectural.

376. A somewhat different situation arises when the mass S' moves too rapidly for an equilibrium theory of the tides to give accurate results. In this case we have to add up the impulsive forces communicated to S by the tidal forces at different instants of the passage of S'. If these do not suffice to elongate S to an eccentricity greater than 0·8826, the tides raised in S merely die down after S' has receded. If however the impulses received from S' once elongate S to an eccentricity greater than 0·8826, unstable motion follows, quite independently of the subsequent motion of S', and the sequence of events in S is that described and sketched in fig. 60.

If S' is supposed to describe an approximately rectilinear path past S with a constant relative velocity v, I have found that the closed distance of safe approach R is given by

$$R^2 = \frac{\gamma^{\frac{1}{2}} M'}{0·675\, v \rho^{\frac{1}{2}}} \quad\quad\quad\quad\ldots\ldots\ldots\ldots\ldots\ldots\ldots\ldots(376·1),$$

where M' is the mass of S' and ρ is the density of S, supposed uniform.

If we assign to v the value $v^2 = \gamma (M + M')/R$, which is appropriate to a circular orbit at a distance R, the limiting distance R takes the form

$$R = 2·10 \left(\frac{M'}{M + M'}\right)^{\frac{1}{3}} r_0 \quad\quad\quad\ldots\ldots\ldots\ldots\ldots\ldots\ldots(376·2).$$

This value of R is somewhat less than the value given by equation

(375·1), and necessarily so, since transitory tidal forces are necessarily less potent than permanent ones, but it is not greatly less.

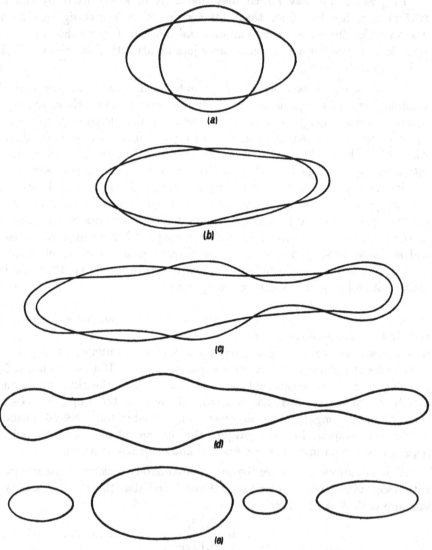

Fig. 60. Configurations of a liquid mass subjected to intense Tidal Forces.

EXPLANATION.

(a) Undistorted sphere, and longest spheroid which is statically stable ($e = 0·8826$).

(b) Longest spheroid which is dynamically stable ($e = 0·9477$), and pear-shaped figure derived by third harmonic displacement.

(c) More elongated pear-shaped figure, and figure derived by fourth harmonic displacement.

(d) The last figure more elongated, and with fifth harmonic displacement superposed.

(e) Conjectural drawing of subsequent configuration.

Actually the value we have assumed for v^2 is impossibly small for two stars in an encounter. The velocity of fall from infinity alone is $\sqrt{2}$ times that appropriate to a circular orbit, so that when two stars meet, the value of v must be greater than that assumed above by a factor of at least $\sqrt{2}$. As a consequence the value of R necessary for break-up is less than 84 per cent. of that given by equation (376·2). But this still does not differ very greatly from the simple equilibrium value given by formula (375·1), at any rate so long as the masses M and M' are comparable.

Equation (375·1) would however suggest that the limiting distance for tidal break-up could be made very large by taking M' very large compared with M. This is true on an equilibrium theory of the tides but equation (376·2) shews it is not so when dynamical factors are taken into account. When M' is very large the two stars shoot past one another with such a high relative velocity that the more intense tidal forces have very little time in which to operate; the shortness of the time of action neutralises the intensity of the forces.

Tidal Effects in a Compressible Mass.

377. When there is great central condensation of mass the problem assumes a different form. Let us consider the extreme case of a body whose whole mass may be treated as concentrated at its centre as in Roche's model (§ 229), so that its gravitational potential is always $\gamma M/r$.

If this mass is under the influence of its own gravitation, and also of a tidal field of force of potential V_T, the total gravitation potential Ω is given by

$$\Omega = \frac{\gamma M}{r} + V_T \dots\dots\dots\dots\dots(377\cdot1).$$

Let the tidal force originate in a second star S' which may be treated as a point of mass M' at a distance R. If r' denotes distance from this point, the whole potential of S' is $\gamma M'/r'$, but only a part of this produces tidal forces. Part goes in producing the acceleration $\gamma M'/R^2$ of S, which may be supposed to originate from a field of force of potential $\gamma x M'/R^2$. Subtracting this, the effective tide generating potential is

$$V_T = \frac{\gamma M'}{r'} - \frac{\gamma x M'}{R^2},$$

and the total potential Ω is

$$\Omega = \gamma \left[\frac{M}{r} + \frac{M'}{r'} - \frac{M'x}{R^2} \right] \dots\dots\dots\dots(377\cdot2).$$

Following an equilibrium theory of the tides, the boundary of the surface of S must be one of the system of equipotentials $\Omega = $ constant. The tidal disruption of S will commence when, if ever, there is no closed equipotential capable of containing the whole volume of S.

Discussing the equipotentials $\Omega =$ constant in the manner adopted in § 229 we find that, whatever the values of M and M' may be, the equipotentials surrounding the point M are at first spheres, but give place to open equipotentials at a certain distance from M. Fig. 61 shews the equipotentials drawn for the special case of $M' = 2M$. The last closed equipotential is drawn thick, and its whole volume is found by quadrature to be equal to that of a sphere of radius 0·348 R.

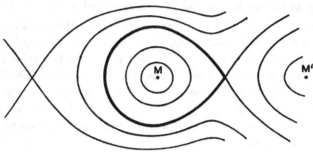

Fig. 61.

Similarly fig. 62 shews the equipotentials drawn for the limiting case of $M'/M = \infty$. The mass M' is now of course at infinity. The outermost curve constitutes the last closed equipotential, and its volume is found by quadrature to be that of a sphere of radius $0·72 \left(\dfrac{M}{2M'}\right)^{\frac{1}{3}} R$.

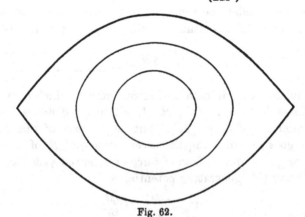

Fig. 62.

The critical equipotential occurs when S' is at a distance R which bears a certain ratio to the undisturbed radius of S. From the figures just given it is found that the critical values of R are

when $M'/M = 2$, $\qquad R = 2·87 \, r_0 = 2·28 \left(\dfrac{M'}{M}\right)^{\frac{1}{3}} r_0$ ....(377·3),

when $M'/M = \infty$, $\qquad R \qquad = 1·75 \left(\dfrac{M'}{M}\right)^{\frac{1}{3}} r_0$ ....(377·4).

PLATE XVI

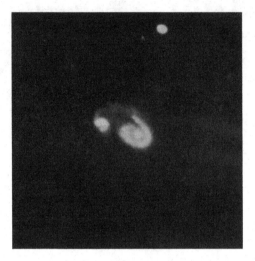

N.G.C. 5278–9

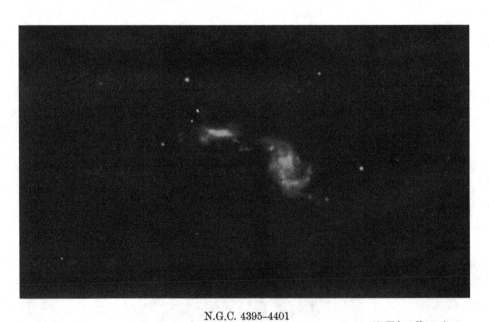

N.G.C. 4395–4401

Mt Wilson Observatory

Nebulae suggestive of Tidal Action

As soon as R becomes less than these critical values, no closed equipotential is capable of containing the whole mass of S. A certain amount begins to spill out through the sharp pointed conical end of the last closed equipotential, and the tidal disruption of S has begun.

We may notice that the critical values of R determined by equations (377·3) and (377·4) do not differ widely from that given by equation (375·1), although the two equations just obtained refer to a body with extreme condensation of mass, while the earlier equation referred to a body having no central condensation of mass at all.

The Birth of Planets.

378. The matter which is ejected through the funnel-shaped end of the last closed equipotential may scatter into space and fall back into the star S unless it is of sufficient amount for its own gravitation to keep it together. If it is of sufficient mass to cohere under its own gravitation, it will form a long filament, pointing approximately towards the tide-generating mass S', or possibly two long filaments starting out from antipodal points of S, the two filaments not generally being symmetrical, and the more massive pointing towards S'. Plate XVI shews two photographs of actual nebulae whose configurations may or may not be due to tidal action but which in any case serve to indicate the type of motion which theory predicts ought to occur under the action of sufficiently intense tidal forces.

The long filament or filaments of matter just described cannot be stable so long as their density remains approximately uniform. They form suitable subjects for the action of gravitational instability of the kind discussed in § 314. As the result of the operation of gravitational instability, condensations will form round which the whole matter of the filaments will ultimately condense into detailed masses. These condensations have an average mass M equal to $\lambda_0{}^3\rho$, where λ is given by formula (316·2), so that

$$M = \left(\frac{\pi\kappa}{3\gamma}\right)^{\frac{3}{2}} c^3\rho^{-\frac{1}{2}} \quad\dots\dots\dots\dots\dots\dots(378\cdot1).$$

Table XXX (p. 350) shews that with reasonable values for the density and molecular velocity of the ejected matter, each condensation would have a mass of the order of magnitude of the actual masses of the planets. It is suggested that the planets originated out of such condensations.

379. In both the extreme cases of an incompressible mass of uniform density and of a mass with its density distributed as in Roche's model, we have found that a sufficiently close encounter with another star will result in the break up of the mass into detached pieces. In the former case, however, the final broken up pieces are each of a mass comparable with that of the parent star; in the latter case, the fragments which are pulled off by tidal action are of comparatively insignificant mass, so that the mass of the parent star is only slightly reduced by the process of disruption. A comparison with the actual facts of the solar system makes it clear that the parent sun must have approximated much more closely to Roche's model than to the incompressible mass.

Thus the sun must have had considerable condensation of mass, and the earth and the other planets must have been formed from its outermost layers in the manner investigated in §§ 377 and 378.

The Birth of Satellites.

380. The principal characteristics of the solar system are reproduced with great fidelity in the smaller systems formed by Jupiter and Saturn with their accompanying families of satellites. Each of these small systems is so exact a replica in miniature of the solar system that no suggested origin for the main system can be accepted unless it can account equally for the smaller planetary systems; any hypothesis which assigned different origins to the main system and the sub-systems would be condemned by its own artificiality.

Immediately after the birth of any planet, say Jupiter, the original situation repeats itself in miniature, Jupiter now playing the part originally assigned to the sun, while the sun or the wandering star, or possibly both together, play the part of the tide-raising disturber. Again we get the emitted filament, again the formation of condensations, and again, as the final result, a chain of detached masses. Since Jupiter, the sun and the disturbing star are all moving originally in the same plane, Jupiter's satellites, when formed, ought also to move in this plane. Not only Jupiter and his satellites but all the other planets and their satellites are observed to move approximately in the same plane, apart from the exceptions occurring on the outer edges of the system which have already been noted in § 2; this plane must then mark the plane of the orbit of the wandering star which was the author of the whole disturbance.

The great disparity in mass between parent and children which prevails in the solar system repeats itself in the planetary systems. The sun's mass is 1047 times that of his greatest satellite, Jupiter, while Jupiter's mass is 12,300 times that of his largest satellite, and the corresponding ratio in the system of Saturn is 4150. The nearest approach to equality of mass is found in the system of the earth and moon with a mass-ratio of 81 to 1. This suggests that in each case there must have been great condensation of mass in the parent body, and that the satellites have been formed by gravitational instability as condensations in filaments of comparatively small mass.

In systems possessing many satellites, namely those of the Sun, Jupiter and Saturn, a general tendency may be detected for the masses to increase up to a maximum as we pass outwards through the system, and subsequently to fall off to a minimum. In the main system, for instance, there is a regular progression of increasing mass through Mercury, Venus, Earth, Mars to the maximum mass of Jupiter, which is broken only by the anomalous mass of Mars; on the descending side the progression through Jupiter, Saturn, Uranus, Neptune fails only through Neptune being some 17 per cent. more massive than Uranus. Clearly when a tidal filament is drawn out of a breaking up star, the matter will be richest near the centre of the filament and will tail off at both ends; this may provide an explanation of the appearance of the more massive planets, Jupiter and Saturn, near the centre of the planetary sequence, and of the similar phenomenon in the planetary systems.

381. So long as the problem is discussed in general terms, it looks as though the process might go on for ever, each satellite of each planet producing a family of sub-satellites to circle around it, and so on *ad infinitum*. Common sense suggests that there must be a limit somewhere, and calculation enables us to fix this limit quite definitely.

The first five satellites of Saturn all have masses of the order of 10^{24} grammes. Formula (378·1) shews that for bodies of this mass to be formed by gravitational instability out of a gaseous filament, the density of the original filament must be many hundreds or thousands of times that of water. Such a density is incompatible with the gaseous state, and the obvious deduction is that these bodies were not born out of a gaseous filament or, at any rate, were not gaseous when born.

We could have reached the same conclusion in another way. The small satellites of Saturn are even now, in their solid states, too small to retain an atmosphere; if they were suddenly transformed into gas they would be still less able to retain their outer layers of gas and would rapidly dissipate into space. This conclusion is entirely independent of any special theories of cosmogony; whatever view we hold as to their origin, it is comparatively certain that most of the asteroids, the majority of the satellites of the planets, and of course the particles of Saturn's rings have been solid or liquid from birth.

This it is that fixes the limit to the birth of endless generations of satellites. Gaseous bodies below a certain limit of mass cannot hold together gravitationally but immediately dissipate into space. A brief reprieve from this law is provided by the possibility of the matter liquefying or solidifying before the process of dissipation is complete, and it is probably owing to the action of this reprieve that most of the satellites of the planets, and possibly the smaller planets themselves owe their present existence. But the very circumstance which saves the lives of these small individuals prevents their giving birth to further generations of astronomical bodies.

382. The more liquid a planet was at birth the less likely it would be to be broken up tidally by the still gaseous sun. But the discussion of § 379 has indicated that if such a tidal break-up occurred, the masses of primary and satellites would be more nearly equal than if the planet had been wholly gaseous. Thus on passing from planets which were wholly gaseous at birth to planets which were wholly liquid either at or immediately after their birth, we ought first to find planets with large numbers of relatively small satellites, and, after passing through the boundary cases of planets with small numbers of relatively large satellites, come to planets with no satellites at all. This is exactly what we find in the solar system. Starting from Jupiter and Saturn, each with nine relatively small satellites, we pass Mars with only two satellites and the earth with its one relatively large satellite, and come to Venus and Mercury which have no satellites at all. Proceeding in the other direction from Jupiter and Saturn, we pass Uranus with four small satellites and come to Neptune with one comparatively big satellite. The earth and Neptune, which have only one satellite each, and those comparatively large ones, form the obvious division between planets which were originally liquid and those which were originally gaseous. Thus we may conjecture that Mercury and Venus must have become liquid or solid immediately after birth, that the Earth and Neptune were partly liquid and partly gaseous, and that Mars, Jupiter, Saturn, and Uranus were born gaseous and remained gaseous during the birth of their families of satellites.

Confirmatory evidence is provided by the circumstance that the masses of Mars and Uranus are abnormally small for their positions in the sequence of planets. If, as we have supposed, the planets were born out of a continuous filament of matter, the mass of Mars at birth ought to have been intermediate between those of the Earth and Jupiter, while the mass of Uranus ought to have been intermediate between those of Neptune and Saturn. If however the two anomalous planets Mars and Uranus were the two smallest planets to be born in the gaseous state, they would be likely to suffer more than the other planets from dissipation of their outer layers. If we suppose that Mars and Uranus are only fragments of planets which were initially far more massive than they now are, then the anomalies disappear and the pieces of the puzzle begin to fit together in a very satisfactory manner.

Effects of a Resisting Medium.

383. The matter which was ejected from the sun by the tidal cataclysm cannot all have immediately condensed into planets; a considerable amount of gas must at first have been scattered throughout the space surrounding the sun, forming a resisting medium through which the new-born planets had to fight their way. The effect of such a resisting medium can be studied by giving a negative sign to G in the analysis of § 261; we see that the presence of a resisting medium must lessen the eccentricity of the orbits of the planets, so that if the medium remains in existence for long enough, all planets and their satellites must acquire approximately circular orbits.

Jeffreys* has studied the rate at which the eccentricities of planetary orbits would be reduced by the action of such a medium, and finds that the orbit of Mercury would be reduced to its present eccentricity in a period of the order of 3000 million years, a time which agrees well with other estimates of the age of the solar system. It is quite likely that this primitive resisting medium has not yet been wholly absorbed by the sun and planets, and the particles which scatter the zodiacal light may well form the last surviving vestiges of it.

The Frequency of Planetary Systems in Space.

384. It appears to be clearly established that, whatever structure we assign to a primitive sun, a planetary system cannot come into being merely as the result of the sun's rotation. If a sun, rotating alone in space, is not able of itself to produce its family of planets and satellites, it becomes necessary to invoke the presence and assistance of some second body. This brings us at once to the Tidal Theory. But our analysis has shewn that the passage of this second body will have no permanent effect on the sun unless the centre of the second body passes with a distance of some 2 or 3 stellar radii of the centre of the sun. The limit is not very much greater than the distance at which a physical collision takes place. Now in § 286 we calculated that, with the present distribution of stars in the neighbourhood of the sun, a given star is only likely to meet with actual collision once in 6×10^{17} years. Its chance of a tidal encounter of sufficient intensity to break it up into a planetary system is somewhat, but not much, greater. We may perhaps suppose that the chance is twice as great, but we must also take into account that not every encounter of the requisite closeness can be expected to form a planetary system.

At a rough estimate we may suppose that a given star's chance of forming a planetary system is one in 5×10^{17} years. Allotting an average age of 5×10^{12} years to the stars, we find that only about one star in 100,000 can have formed a planetary system in the whole of its life, so that only about one star in 100,000 is at present surrounded by planets. Planetary systems must then be of the nature of "freak-formations"; they do not appear in the normal evolutionary course of the normal star.

To a rough approximation we may regard the stars of the galactic system as consisting of about 100 million stars packed approximately as closely as the stars in the neighbourhood of the sun, and a far greater number scattered far more sparsely in space. In the former group of 100 million stars planetary systems must form at the rate of about one per 5×10^{9} years; in the latter the stars are so sparse that the chance of planetary systems coming into being may be almost neglected. We may conclude that in the whole of the galactic system planetary systems only come into being at the rate of about one

* *M.N.* LXXVIII. (1918), p. 424.

per 5000 million years, so that our own system, with an age of the order of 2000 million years, is probably the youngest system in the whole galactic system of stars.

The contrast between the slowness of cosmogonic events as disclosed by the figures just given, and the rapidity with which events move on our earth, leads to some interesting reflections. Let us suppose that civilisation on our earth is 10,000 years old. If each planetary system in the universe contains ten planets, and life and civilisation appear in due course on each, then civilisations appear in the galactic system at an average rate of one per 500 million years. It follows that we should probably have to visit 50,000 galaxies before finding a civilisation as young as our own. And as we have only studied cosmogony for some 200 years, we should have to search through about 25 million galaxies, if they exist, before encountering cosmogonists as primitive as ourselves. We may well be the most ignorant cosmogonists in the whole of space.

CHAPTER XVII

CONCLUSION

385. Now that the detailed discussion of particular problems is ended, we may perhaps attempt to summarise our results and tentative conclusions, sacrificing logical and chronological order in favour of the arrangement which offers the broadest and simplest view of the whole subject.

The easiest part of the problem of cosmogony is the interpretation of the observed shapes of astronomical bodies and formations. Here the effects of rotation have proved to be of primary importance. The earth and many of the planets have the shape of flattened oranges. The degree of flattening is such as would be produced by quite slow rotation about an axis, and there is no room for doubt that this is the actual cause of the observed flattening. It is possible to trace out theoretically the shapes assumed by astronomical bodies having all possible amounts of rotation. Mathematical investigation shews that the flattened-orange shape is assumed by all bodies in slow rotation, no matter what their internal constitution and arrangement may be, but that with more rapid rotation the shape depends on the internal arrangement of the body, being especially affected by the extent to which its mass is concentrated at or near its centre.

Two special and quite extreme types of arrangement have been considered in detail. In the first the body is supposed to consist of matter which cannot be compressed and is of uniform density throughout; to fix our ideas, we may think of a mass of water. As the rotation of such a mass increases, the orange becomes flatter and flatter but retains the shape of an oblate spheroid throughout, until a stage is reached beyond which the flattening cannot go. At this stage the body abandons its circular cross-section; it elongates and concentrates its mass around one of the diameters in its equatorial plane, thus forming an ellipsoid with three unequal axes. This process continues until the mass forms a cigar-shaped figure with a length equal to nearly three times its shorter diameter. At this point the mass begins to concentrate about two distinct points on its longest diameter, a furrow or waist forming near the centre which continually deepens until it cuts the body into two distinct detached masses. These rotate in orbital motion about one another like the earth and moon, except that the two masses are more nearly equal and are closer together.

It is possible for a single rotating mass to assume all these configurations in turn. The concept, first introduced by Laplace, of a mass shrinking and at the same time increasing its speed of rotation, because its rotational momentum must remain constant, remains of the utmost importance to cosmogony to-day. It seems probable that Laplace's concept must be modified in one respect; we no longer contemplate a slow gradual shrinkage, but rather a contraction by spasms, a star remaining of about the same size through a long period of time, after which a fairly rapid contraction occurs, followed by another long epoch of unchanging size, another rapid contraction, and so on.

This jerkiness of contraction is ultimately due to the fact that the atoms out of which the stars are built are not continuous structures. The most important part of an atom, the positively charged nucleus at its centre, is also the smallest. In this, just because of its smallness, the main mass of the atom resides. The atom is a very open structure, being rather more so than the solar system. The nucleus corresponds to the sun, and around this the other constituents of the atom, the negative electrons, revolve like planets. The revolving electrons may form 1, 2, 3 or more rings surrounding the nucleus. At the high temperatures which prevail inside the stars the atoms are much broken up. As the temperature changes, the size of the atom necessarily changes by a whole ring at a time, and these jumps in the sizes of the atoms shew themselves in jumps in the sizes of the stars. The radii of atoms which have $0, 1, 2, 3, \ldots$ rings of electrons revolving round their nuclei are in the proportion $0^2 : 1^2 : 2^2 : 3^2 : \ldots$, and if all the stars of a given mass are graded according to size, we find that they fall into groups in which the radii are in something like these same proportions.

This is of course not the whole story, for the volumes of stars differ substantially from the aggregate volume of the atoms of which they are formed, while theory and observation both shew that the groups of configurations corresponding to the different atomic radii remain distinct only in unusually massive stars. In stars of moderate mass the distinction becomes blurred, so that the various types of configuration merge continuously into one another, but we have found that the diameters of stars of large mass reflect quite clearly the different possible diameters of the atoms of which they are composed. Thus it is possible to say that in one group of stars each atom in the main central mass has two rings of electrons left in orbital motion around it, in another group of smaller size only one ring of electrons survives, while in another group of still smaller size, the "white dwarfs," nearly all the atoms are stripped bare of electrons, and only the positive nuclei are left.

Thus in a sense the secret of the structure of the atom is written across the heavens in the diameters of the stars. For instance, the great disparity in size between the white dwarfs and the group of stars of next larger size, the "main sequence" stars, provides evidence that the positive nucleus has a diameter far smaller than the first ring of electrons surrounding it; we have astronomical proof of the "openness of structure" of the atom.

While a star shrinks, whether continuously or by jerks, its rotation increases, and it can run through the whole sequence of configurations just described, ending as a binary star.

The foregoing discussion has assumed that a star may be treated as a mass of uniform incompressible liquid such as water. Actual stars are not so simple as this, but our detailed mathematical discussion has indicated that they will behave very much as though they were. An actual star cannot rotate at the same speed throughout, for we have found that the continual emission of radiation from its surface must necessarily exercise a braking effect upon its outer layers, which accordingly rotate less rapidly than the inner layers. Moreover, the tenuous gaseous outer layers of a star must have very different physical properties from a uniform incompressible liquid. But their very tenuity makes it permissible to disregard them entirely. It is the massive inner layers that determine the dynamical conduct of the star; the outer layers merely form an obscuring veil drawn round those parts of the star which are dynamically important, conforming to their gravitational field but concealing their motions.

Nevertheless our theoretical investigations have shewn that it will not always be permissible to disregard the effects of the outer layers of a rotating mass. The more massive a star is, the more we have found its mass to be concentrated towards its centre, and the greater is the relative extent of its atmosphere. When a body has a mass far greater than that of a star, its whole mass may be supposed concentrated at one point, namely its centre, while its tenuous atmosphere occupies practically the whole of its volume.

Such masses, when set in rotation, assume shapes very different from those assumed by uniform incompressible masses. With slow rotation they shew the universal flattened-orange formation, but speedier rotation brings about a departure from this shape; they become flatter but do not remain oblate spheroids. When a certain critical speed of rotation is reached, they assume the shape of a double convex lens with a perfectly sharp edge. Still further rotation causes the atmosphere to spill out through the sharp edge of the lens into the equatorial plane, leading to a succession of configurations in which the central mass continually retains the critical lens-shaped configuration, while more and more of its outer layer becomes spilled out and constitutes a thin disc of matter rotating in the equatorial plane.

Fig. 63 shews the contrast between the series (*a*) of configurations assumed by a rotating body of uniform, or nearly uniform, density, and the series (*b*) assumed by a rotating body whose mass is highly condensed towards its centre. In every case the dotted line represents the axis of rotation.

The two chains of configurations are those of two extreme types of mass, one (*b*) having its mass entirely concentrated at its centre, and the other (*a*) having its mass spread uniformly throughout its volume. Actual astronomical masses will lie somewhere between these two extremes, and a mass half-way between might naturally be expected to follow a sequence of configurations half-way between (*a*) and (*b*). Theory shews, however, that it does not. As the degree of central condensation steadily increases, the body continues to follow sequence (*a*), or at any rate a sequence which differs from it only in non-essentials, until, when a certain critical degree of central condensation is reached, it suddenly swings over and follows sequence (*b*), and this same sequence is followed by all masses having a greater degree of concentration than that at which the cross-over occurs. Thus, as regards essential features, there are only the two sequences (*a*) and (*b*) to be considered, and a rotating mass will follow the one or the other, according to the degree to which its mass is concentrated near its centre.

The first sequence is that through which an ordinary star breaks up to form a binary system. The fact that a star follows this sequence rather than sequence (*b*) proves that it cannot have any enormous central condensation of mass, detailed analysis shewing that it cannot be in a purely gaseous state. In its central regions at least the atoms must be jammed together so that the matter approximates to the liquid rather than to the gaseous state; this in turn shews that some at least of its atoms must have rings of electrons left in orbital motion around them.

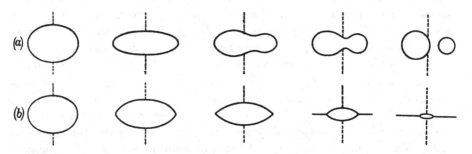

Fig. 63. The sequence of configurations of masses rotating under their own gravitation:
  (*a*) Liquid masses and masses of nearly uniform density.
  (*b*) Gaseous masses and masses whose density is highly concentrated in their central regions.

The second sequence of configurations (*b*) is that exhibited by the spiral and other nebulae whose masses are enormously greater than those of ordinary stars. When Dr Hubble set out to classify the observed nebular forms he tried to disregard all theoretical predictions, but nevertheless found himself compelled to classify the normal nebulae as forming precisely the linear sequence (*b*) predicted by theory (see Plates IX, X and XIII).

We see then that a large proportion of the configurations of astronomical bodies can be explained as the configurations of rotating masses. Binary stars and elliptical and spiral nebulae admit of such an explanation, while we have conjectured that the rotation which just precedes fission may explain the characteristic behaviour of Cepheid variables and of some long-period variables. But rotation has not yet been able to explain the characteristic spiral shape of the arms of spiral nebulae, and definitely fails to explain the distinctive formation observed in the solar system.

386. The extra-galactic nebulae and star clouds are the most massive astronomical formations known, their masses being of the order of a thousand million suns. The masses of the rather enigmatical globular clusters are probably distinctly smaller, but hardly of a different order of magnitude. After these there is a great gap until we come to the stars with masses comparable with the sun. In discussing still smaller masses we are perforce limited to the solar system, since they could not be observed in more distant systems. Here again a great gap appears; after the sun, the next most massive body is Jupiter, whose mass is less than a thousandth part of that of the sun, and then come the planets in general with masses of the order of a ten-thousandth part of the sun's mass. After these there is another great gap, and then a still smaller system of bodies, the satellites, which have masses of the order of only a ten-thousandth part of the masses of their primaries.

We have seen that an explanation of these discontinuities in the sequence of masses is provided by the action of gravitational instability, which also explains how one group of masses is formed out of another. This single concept has proved capable of explaining the births of four successive generations of astronomical bodies, each being born through the action of gravitational instability from the generation of more massive bodies immediately preceding it.

We have found that, as Newton first conjectured, a chaotic mass of gas of approximately uniform density and of very great extent would be dynamically unstable; nuclei would tend to form in it, around which the whole of the matter would ultimately condense. We have obtained a formula which enables us to calculate the average distance apart at which these nuclei would form in a medium of given density, and this determines the average mass which would ultimately condense round each.

If all the matter in those parts of the universe which are accessible to our observation, a sphere of about 140 million light-years radius, were spread out uniformly, it would form a gas of density 10^{-31} or thereabouts. We have calculated that gravitational instability would cause such a medium to break up into detached bodies whose distance apart would be of the same order as the observed distance between the spiral nebulae; the mass of each such body would accordingly be about equal to the mass of the average spiral nebula. We may conjecture, although it is improbable that we shall ever be able to prove, that the spiral nebulae were formed in this way. Any currents in the primaeval chaotic medium would persist as rotations of the nebulae, and, as these would be rotating with different speeds, they might be expected to shew all the various types of configurations of our sequence (b), which is what is actually observed.

Those nebulae whose rotational momentum was sufficient to carry them past the critical lenticular shape in the course of their shrinkage would shed a certain amount of matter in their equatorial plane in the manner indicated in the last two diagrams of sequence (b). Since a uniformly spread-out distribution of matter in this plane would be unstable, gravitational instability would again cause condensations to form and the shed matter would ultimately break up into a series of separate detached bodies. Calculation shews that these would each have a mass of the same order as the masses of the stars. This makes it exceedingly likely that we have here found the birthplaces of the stars. It is significant that stars are observed in abundance in the outer regions of the spiral nebulae, but none in the inner lenticular regions, or in those spherical or elliptical nebulae which, having stopped short of the critical lenticular configurations, have shed no matter out into their equatorial planes. Thus stars are found precisely in those regions in which theory predicts that they should be formed by gravitational instability and nowhere else.

Normally the reign of gravitational instability must end with the birth of stars. The masses of the stars are too small for further shrinkage to carry them again along sequence (*b*), so that if such shrinkage occurs, the stars must follow sequence (*a*) and may finally break up into binary systems.

In a few exceptional cases, however, gravitational instability may come into action again. On rare occasions it may happen that one star passes so near to another that it draws out long arms of matter, these being in effect exaggerated tides caused by the near proximity of the two stars. When this occurs, the matter in these arms is a fit subject for the operation of gravitational instability, and calculation shews that the matter would condense into detached bodies, each of mass about equal to that of the planets. In this way we conjecture that our earth and the other planets were born out of the sun. Such planets as do not liquefy or solidify at once may in their turn be caused to eject long arms of matter which the operation of gravitational instability will break up into detached masses, the satellites of the planets. Our moon forms a rather exceptional case, having a mass far more nearly equal to its primary than is found anywhere else in the solar system. We have found that this indicates that the earth must have been partially liquefied before the moon was born.

The action of gravitational instability must finally end with the birth of satellites. To escape the fate of dissipating away into space these must condense into either liquid or solid form immediately after birth, and when they have done this, gravitational instability can obtain no further hold over them. We have found, however, that gravitational instability accounts for the birth of four generations of astronomical bodies in succession, of nebulae out of chaos, of stars out of nebulae, of planets out of stars and of satellites out of planets. Our conclusion that these successive generations are born by gravitational instability demands no hypotheses beyond the presence of forces which are already known to exist, namely gravitation and gas-pressure, and it survives at every step the test of numerical computation.

387. Another section of cosmogony deals with the time required for these processes to occur and, as a consequence, with the total age of the universe.

Terrestrial evidence indicates that the age of the earth must be between 1500 and 5000 million years, so that this is probably the time which has elapsed since the planets and their satellites were born out of the sun. But stellar evidence demands far greater ages for the stars. We have estimated that the time necessary for the primaeval chaos to condense into nebulae must have been much more than 60,000 million years; the time for the condensations in these nebulae to form stars rotating with some approach to uniformity of angular velocity, such as is shewn by the sun, is of the order of 10^{13} years. The time required for the motions of the stars to acquire the observed approximation to a final steady state of equipartition of energy is again of the order of 10^{13} years. The distribution of the eccentricities and periods in the orbits of binary stars is such as to suggest that these orbits have been moulded by long ages of interaction with other stars, and calculation enables us to fix the length of the necessary time as being of the order of 10^{13} years. Older binary systems shew a closer approach to equality of mass in their two constituents than is shewn by younger systems, and calculation shews that the time necessary to establish the approximation to equality observed in the older system is of the order of 10^{13} years. The galactic system is a confused medley of stars moving nearly at random, but still shewing some faint evidence of ordered motion. A few of the more massive stars move in orderly parallel motion like flights of swans or ducks; the majority move with the random motion of a cloud of starlings. But it seems reasonable to suppose that at one time these latter also shewed the same type of ordered motion as is still shewn by their more massive companions. A dynamical investigation has shewn that the time necessary to break up the formation of the less massive stars would again be of the order of 10^{13} years. Thus all available lines of evidence agree in assigning lives of the order of millions of millions of years to the stars.

388. We observe a star because it is emitting radiation. Modern physical theory teaches that all radiation carries mass with it, so that the emission of radiation necessarily diminishes the mass of a star. The rate of diminution of a star's mass is readily calculated; the rate at which the sun's mass is diminishing is found, for instance, to be 250 million tons a minute.

Neither the sun nor any other star can continue radiating away its mass at such a rate for ever. The mass of a star fixes a limit to the total amount of radiation it can emit, just as surely as the volume of a cistern fixes a limit to the total amount of water it can emit. In the case of the star there is no replenishment, or at least none comparable with the amount lost. The sun loses 250 million tons of mass every minute, and there is no known source of replenishment which can supply new mass to it at even a small fraction of this rate.

Calculation shews that a star whose age is 10^{13} years must in the course of its life have emitted radiation whose total mass is many times greater than the present total mass of the star. Not only, then, are the masses of the stars gradually melting away into radiation, but in most stars the greater part of the mass has already so melted.

The amount of mass which is left to a star provides a measure of the length of time during which it can continue to emit radiation. We find, for instance, that the sun possesses enough mass to continue to radiate at its present rate for 15 million million years. Actually the sun can look forward to a longer life than this, for as a star ages the rate at which it radiates away energy, and so the rate at which it spends its mass, continually diminishes. When allowance is made for this senile tendency to parsimony, we find that stars such as our sun can continue to shine for some hundreds of millions of millions of years.

389. At the beginning of its life a star has a huge store of mass, the greater part of which is destined ultimately to be transformed into radiation. There is only one way known to physics in which such enormous amounts of mass can be stored, namely in the form of electrons and protons which are combined into atoms, although not necessarily atoms of terrestrially known types. And the radiation of the stars must be provided by the annihilation of these atoms. An investigation into the stability of the stars has shewn that for the stars to be stable structures, not liable suddenly to transform their whole mass explosively into radiation, their atoms must liberate energy spontaneously as the radioactive atoms do, the rate of liberation not depending to any great extent on the density or temperature. The relatively cool temperature of the earth's surface proves that terrestrial atoms have no appreciable capacity for liberating energy, so that the atoms which liberate energy in the sun and stars must be of different type from terrestrial atoms.

Thus the future radiation is stored in the form of electrons and protons in the star, and the process of liberation of energy must consist of an annihilation of matter, electrons and protons neutralising one another and setting free radiation of mass equivalent to that of the annihilated matter. Different types of matter must be liable to annihilation at different rates, so that it ought to be possible to estimate the age of a star either from the amount of matter left, or from the proportions in which the different types of matter occur. An important reservation must, however, be made. We have found reasons for thinking that atoms which are completely broken up into their constituent electrons and protons are immune from annihilation. If so we cannot estimate the age of a star unless we know for how long and to what extent its atoms have been preserved from annihilation in this manner.

For instance, if we suppose that the sun has always been a normal star radiating energy at the rate normally appropriate to its mass, calculation shews that it cannot have existed for more than 8 million million years. But this length of life can be extended indefinitely if we suppose part of it to have been spent as a "white dwarf." The matter in these stars is protected from annihilation, with the result that they are so feebly luminous for their mass as to upset all calculations. We have no means of knowing whether the sun, or any other star, has spent part of its life as a white dwarf or not. Theory is not altogether hostile to such a possibility, while observation is necessarily silent on the matter since white dwarfs are so faint that the vast majority of them escape observation entirely.

Similarly, the dense central nuclei of the spiral and other extra-galactic nebulae probably consist of matter which is broken up into its ultimate constituent parts and so is immune from annihilation. These nebulae are in effect vast storehouses of matter which is immune from annihilation and on which age produces no effect. Thus those stars which look the youngest may, although this is hardly likely, actually consist of the oldest atoms, and, in any case, it becomes impossible to divide either stars or atoms up into young and old. It is possible, although again perhaps hardly likely, that all atoms may originally have come into being at the same time.

390. We have seen that the observed radii of stars of large mass fall into distinct detached groups, the different radii being easily identified as corresponding to the various rings of electrons which revolve round the atomic nucleus. The group of stars whose radii are smallest of all, the white dwarfs, consist mainly of atoms stripped bare of electrons right down to their nuclei. The next group, the main sequence stars, have atoms in which only one ring of electrons, the K-ring, is left. The next group, the giant stars, have two rings left, and so on. For the temperatures in the inner regions of the stars to break up the atoms to the extent required for this identification, we have found that the stellar atoms must have atomic numbers in the neighbourhood of 95. If the stars were made of terrestrial atoms, the high interior temperatures of the stars would strip most of the atoms bare down to their nuclei. The completely broken up atoms would now behave almost like a perfect gas, and this would make the star unstable, since we have found that a perfectly gaseous star generating its energy spontaneously cannot be in stable equilibrium. To keep the stars stable two separate conditions are necessary. The first, already mentioned, is that the stellar atoms must liberate their energy spontaneously, as the radioactive elements do; the second requires that they must have atomic numbers somewhere in the neighbourhood of 95 which is just higher than the atomic numbers (84—92) of the radioactive elements.

The atomic number of any atom is the number of electrons which revolve round the nucleus when the atom is complete, and this gives a measure of the complexity of structure of the atom. Our conclusion, then, is that stellar atoms are rather more complex than the most complex of terrestrial atoms, namely, the radioactive atoms. We may regard the stellar atoms, in a sense, as super-radioactive atoms. The sequence from these through the ordinary radioactive atoms to non-radioactive atoms is not only one of decreasing atomic number and decreasing complexity of structure; it is also one of decreasing capacity for liberating energy, the stellar atoms having far more capacity for liberating energy than radium or uranium, whereas of course the ordinary terrestrial atom has much less.

The foregoing statements refer to a sort of average atom of which the star may be supposed to be composed. An actual star must be a mixture of a great number of kinds of atoms, including terrestrial atoms. These latter atoms are probably of little importance in the dynamics and physics of the star as a whole, but they have the special importance that, being the lightest atoms in the star, they float up to its surface and so determine its spectrum.

Terrestrial chemistry, which deals only with these atoms, may properly be described as "surface-chemistry"; it must merge into a wider chemistry on passing inside a star. So also terrestrial physics is a mere "surface-physics." Inside a star we are confronted at once with what appears to be the fundamental physical process of the universe, the wholesale transformation of matter into radiation; and of this terrestrial physics knows nothing. Again, the greater part of the matter of the universe exists in a state of high dissociation of which surface-physics has neither knowledge nor experience. Clearly our physics and chemistry are mere fragments of wider-reaching sciences.

391. In a quite different sense biology is a surface-science, since biology becomes meaningless in the interior of a star. Life implies duration in time, and there can be no life where atoms change their make-up millions of times per second. Life further implies mobility in space, and this restricts it to those small parts of the universe where the physical conditions, temperature in particular, permit of the existence of matter in the liquid state. This does not merely rule out the interior of a star; it rules out the whole star. Every known star pours out such an intense torrent of radiation as to make life quite impossible on its surface. The only possible opportunities for the existence of life would appear to be on a planet some distance from the star's surface, this planet consisting necessarily only of "permanent" matter torn originally from the star's surface and transplanted to a safe, but not too great, distance from its hot surface. Thus biological science may be described as planetary science, since nowhere else in the universe would it seem to have any meaning.

On rare occasions only does one star pass so near to another that planets can be torn out and left to solidify in space, to form the cool ash on which alone life can exist. At a rough computation, about one star in a million may be surrounded by planets, but probably only a small fraction even of these planets are possible abodes of life. Thus life is perforce limited to an amazingly small corner of the universe; whether it even exists where it can we have no means of knowing.

392. The cinematograph film which we set out to construct exhibits the universe as a mass of matter slowly but inexorably dissolving away into intangible radiation. The stars may be compared to icebergs which have broken away from the main ice-packs—the extra-galactic nebulae—and are drifting into warmer seas and extinction. Nothing stands, everything melts away, except for the few permanent atoms which, like the rocks or stones of the iceberg, are destined to survive after all else has dissolved. We have estimated (§ 126) the average duration of the present matter of the universe to be of the order of 10^{12} years only. After some such period the main mass of matter now in existence will have been transformed into radiation. If all matter were completely annihilated, the estimated present mass-density of $1·5 \times 10^{-31}$ grammes per cubic centimetre would produce a mass-density of radiation of $1·35 \times 10^{-10}$ ergs per cubic centimetre and this is the density of radiation at a temperature of 11·5 degrees absolute. But we cannot say how far this radiation may be diluted through invading parts of space in which no matter exists.

While the far end of the film of pictures is fairly clear, the beginning is veiled in obscurity. We have been able to form an estimate of the time during which the stars have existed as stars, but we cannot say for how long their electrons and protons have existed, since, before they formed stars at all, they may have existed in completely dissociated form, immune from annihilation, in the central regions of the great nebulae. Thus the five to ten millions of millions of years which we have estimated as the average age of the stars only provides a sort of lower limit to the ages of the atoms, and so to the age of the universe. We can estimate the future life of the present matter of the universe from the rate at which it is transforming itself into radiation, but we can only fix a lower limit to its past life. Even so, this lower limit is so long as to create a suspicion that we may have appeared on the scene rather late in the history of the universe; possibly the main drama of the universe is over, and our lot is merely to watch the unwanted ends of lighted candles burning themselves out on an empty stage.

An alternative view is that there may be neither beginning nor end, so that we may speak of the ages of the stars but not of the age of the universe. It is difficult, but not impossible, to believe that matter can be continuously in process of creation, or possibly of re-creation out of stray radiation. If, however, this obvious initial difficulty is disregarded, we are free to think of stars and other astronomical bodies as passing in an endless steady stream from creation to extinction, just as human beings pass from birth to the grave, with a new generation always ready to step into the place vacated by the old. Observation cannot finally decide between these two possibilities, but rather frowns upon the view just stated. If the number of objects in the various stages of development had proved to be roughly proportional to the times taken to pass through these stages, this being the characteristic of a steady population, we could have maintained that the present universe shewed signs neither of beginning nor ending. But the galactic system of stars shews too many middle-aged stars, and too few infants or veterans, for such a view to commend itself. There is rather distinct evidence of a special birth of stars at about the time when our sun was born, and this leads naturally to the conjecture that the galactic system was born out of a spiral nebula whose main activity in producing stars centred round that epoch. We are hardly in a position to discuss other stellar systems, or to assign ages to other astronomical objects such as the great nebulae.

The cosmogonist has finished his task when he has described to the best of his ability the inevitable sequence of changes which constitute the history of the material universe. But the picture which he draws opens questions of the widest interest not only to science, but also to humanity. What is the significance of the vast processes it portrays? What is the meaning, if any there be which is intelligible to us, of the vast accumulations of matter which appear, on our present interpretations of space and time, to have been created only in order that they may destroy themselves? What is the relation of life to that universe, of which, if we are right, it can occupy only so small a corner? What, if any, is our relation to the remote nebulae, for surely there must be some more direct contact than that light can travel between them and us in a hundred million years? Do their colossal incomprehending masses come nearer to representing the main ultimate reality of the universe, or do we? Are we merely part of the same picture as they, or is it possible that we are part of the artist? Are they perchance only a dream, while we are brain-cells in the mind of the dreamer? Or is our importance measured solely by the fractions of space and time we occupy—space infinitely less than a speck of dust in a vast city, and time less than one tick of a clock which has endured for ages and will tick on for ages yet to come?

It is not for the cosmogonist to attempt to suggest answers to these wide questions, or even to the more limited questions directly raised by the sequence of events which is his own special study. He will be specially reluctant to attempt either, knowing how dimly most of the sequence of events can be seen, and how much of it cannot be seen at all. His critics may allege that what he sees most clearly is only a creation of his own imagination, and he is only too conscious that it may be so. He can only end with a question; others, more confident or more fortunate, may, if they wish, attempt an answer.

Let us, however, reflect that mankind is at the very beginning of its existence; on the astronomical time-scale it has lived only for a few brief moments, and has only just begun to notice the cosmos outside itself. It is, perhaps, hardly likely to interpret its surroundings aright in the first few moments its eyes are open.

INDEX OF SUBJECTS

The numbers refer to the pages

Taurus cluster, 24, 26, 375
Temperature-luminosity diagram, 60, 61, 161,
183, 184
,, of stellar interiors, 68, 72, 73,
92, 163, 170
,, ,, ,, surfaces, 54, 55, 60
Tidal friction, 231, 293
Tidal theories of origin of solar systems, 398,
400, 409, 416
Trifid nebula, 28, Plate V (p. 28)
Triple star-systems, 23, 311

Uranus, 4, 408
Ursa Major cluster, 24, 375, 380

V Puppis, 58, 109, 129, 288
Variable stars, 23, 382 ff.
Vega (α Lyrae), 8, 31, 41
Velocity-distribution of stellar motions, 302,
361, 368, 369, 371
Venus, 5, 407, 408
Viscosity, 268, 270, 286

W Crucis, 180, 392
White dwarfs, 63, 132, 143, 160, 165, 180, 181,
182, 184, 185, 419

Zodiacal light, 409